Organic Lasers and Organic Photonics (Second Edition)

Online at: https://doi.org/10.1088/978-0-7503-5547-6

IOP Series in Coherent Sources, Quantum Fundamentals, and Applications

About the Editor

F J Duarte is a laser physicist based in Western New York, USA. His career has covered three continents while contributing within the academic, industrial, and defense sectors. Duarte is editor/author of 15 laser optics books and sole author of three books: *Tunable Laser Optics, Quantum Optics for Engineers*, and *Fundamentals of Quantum Entanglement*. Duarte has made original contributions in the fields of coherent imaging, directed energy, high-power tunable lasers, laser metrology, liquid and solid-state organic gain media, narrow-linewidth tunable laser oscillators, organic semiconductor coherent emission, N-slit quantum interferometry, polarization rotation, quantum entanglement, and space-to-space secure interferometric communications. He is also the author of the generalized multiple-prism grating dispersion theory and pioneered the use of Dirac's quantum notation in N-slit interferometry and classical optics. His contributions have found applications in numerous fields, including astronomical instrumentation, dispersive optics, femtosecond laser microscopy, geodesics, gravitational lensing, heat transfer, laser isotope separation, laser medicine, laser pulse compression, laser spectroscopy, mathematical transforms, nonlinear optics, polarization optics, and tunable diode-laser design. Duarte was elected Fellow of the Australian Institute of Physics in 1987 and Fellow of the Optical Society of America in 1993. He has received various recognitions, including the *Paul F Foreman Engineering Excellence Award* and the *David Richardson Medal* from the Optical Society.

Coherent Sources, Quantum Fundamentals, and Applications

Since its discovery the laser has found innumerable applications from astronomy to zoology. Subsequently, we have also become familiar with additional sources of coherent radiation such as the free electron laser, optical parametric oscillators, and coherent interferometric emitters. The aim of this book Series in Coherent Sources, Quantum Fundamentals, and Applications is to explore and explain the physics and technology of widely applied sources of coherent radiation and to match them with utilitarian and cutting-edge scientific applications. Coherent sources of interest are those that offer advantages in particular emission characteristics areas such as broad tunability, high spectral coherence, high energy, or high power. An additional area of inclusion are the coherent sources capable of high performance in the miniaturized realm. Understanding of quantum fundamentals can lead to new and better coherent sources and unimagined scientific and technological applications. Application areas of interest include the industrial, commercial, and medical sectors. Also, particular attention is given to scientific applications with a bright future such as coherent spectroscopy, astronomy, biophotonics, space communications, space interferometry, quantum entanglement, and quantum interference.

A full list of titles published in this series can be found here: https://iopscience.iop.org/bookListInfo/series-in-coherent-sources-and-applications.

Organic Lasers and Organic Photonics (Second Edition)

Edited by
F J Duarte

IOP Publishing, Bristol, UK

ISBN   978-0-7503-5547-6 (ebook)
ISBN   978-0-7503-5545-2 (print)
ISBN   978-0-7503-5548-3 (myPrint)
ISBN   978-0-7503-5546-9 (mobi)

DOI   10.1088/978-0-7503-5547-6

Version: 20240601

IOP ebooks

British Library Cataloguing-in-Publication Data: A catalogue record for this book is available from the British Library.

Published by IOP Publishing, wholly owned by The Institute of Physics, London

IOP Publishing, No.2 The Distillery, Glassfields, Avon Street, Bristol, BS2 0GR, UK

US Office: IOP Publishing, Inc., 190 North Independence Mall West, Suite 601, Philadelphia, PA 19106, USA

In memory of a bright light of optimism: Ruth Virginia Duarte, née Valenzuela *(1920–2018).*

Contents

Preface

Series preface

The laser, the brightest child of the quantum era, awesome emitter of coherent radiation, has given humankind solutions in a vast array of scientific and industrial disciplines. Fields from astronomy to medicine to nanoscience have benefited enormously from the laser. Concurrently, additional sources of coherent radiation such as the free electron laser, optical parametric oscillators, and interferometric emitters, have increased the freedom of choice for scientists and engineers searching for the perfect match between source and application. The aim of this book Series on *Coherent Sources, Quantum Fundamentals, and Applications* is to explore and explain the physics and technology of widely applied sources of coherent radiation and to match them with utilitarian and cutting-edge applications. Coherent sources of interest are those that offer particular advantages in specific emission characteristics areas such as broad tunability, high energy, and high power. Additional areas of interest are miniaturized coherent sources in the micro and nano realms. Understanding of quantum fundamentals can lead to novel coherent sources and unimagined scientific and technological applications. Application areas of interest include the commercial, industrial, and medical sectors. Also, particular attention is given to scientific applications such as astronomy, biophotonics, coherent spectroscopy, quantum entanglement, quantum interference, space communications, and space interferometry.

Book preface

Organic Lasers and Organic Photonics (Second Edition), presents a unique integrated and updated perspective on the field of organic sources of coherent radiation from high-average powers (multi kW) and high-pulse energy (~1 kJ) lasers to miniaturized organic lasers and coherent sources. Although the emphasis is on tunable narrow-linewidth lasers, many other sources, with a variety of organic gain media, are considered. The presentation also focuses on electrically-pumped organic semiconductor coherent sources, their physics and technology. Next, a pragmatic review of organic photonics, including numerous exciting applications, leads to an extensive, authoritative, and notable chapter on organic dyes for optogenetics which provides an up-to-date description of this exciting field. This is followed by a chapter on organic lasers and organic molecules for medicine which also describes light sheet illumination techniques capable of yielding extremely elongated laser beams (up to 3000:1). This second edition concludes with two related quantum optics chapters. First, an updated chapter on quantum organic lasers as sources of ensembles of indistinguishable photons, with orthogonal polarizations, for quantum entanglement communications. This is followed by a brief chapter on quantum coherence, or *Diracian emission*, that is relevant to the preceding chapter and also to the chapter on coherent emission from electrically excited organic semiconductors. In all, thirteen (13) of the sixteen (16) chapters include relevant and original problems.

Acknowledgements

The second-updated edition of *Organic Lasers and Organic Photonics* was only made possible by the timely and expert contributions from our coauthors. The professionalism of Chris Benson and the IoP Productions Team is acknowledged. Support from *Interferometric Optics*, USA, is gratefully recognized.

Editor biography

F J Duarte

Francisco Javier Duarte is a Chilean-born laser physicist, quantum physicist, inventor, and book author. He has been based in the United States since 1983.

Duarte was the first to graduate with First Class Honours in physics from Macquarie University (1978) where he was a student of the quantum physicist J C Ward. His honours thesis was entitled *Excitation Processes in Continuous Wave Rare Gas-Metal Halide Vapour Lasers*. At Macquarie he also completed his PhD research, on optically-pumped molecular lasers (1981) under the guidance of J A Piper. He then became a Postdoctoral Fellow at the University of New South Wales, where he built UV tunable lasers for high-resolution IR–UV double-resonance spectroscopy. Duarte has worked and contributed professionally in the academic, industrial, and defense sectors and has practiced physics in Australia, the Americas, and Europe. Notable in his career is his tenure at the Eastman Kodak Company (1985–2006) where he led the Imaging and Spectral Measurements Laboratory. He is the author of numerous refereed papers and US patents.

Duarte is editor/author of 17 scholarly books including *Dye Laser Principles* (1990), *High-Power Dye Lasers* (1991), *Tunable Lasers Handbook* (1995), *Tunable Laser Applications* (1995, 2009, 2016), and *Organic Lasers and Organic Photonics* (2018, 2024). He is co-author of *Quantum Entanglement Engineering and Applications* (2021) and sole author of *Tunable Laser Optics* (2003, 2015), *Quantum Optics for Engineers* (2014, 2024), and *Fundamentals of Quantum Entanglement* (2019, 2022). Duarte has made original contributions in the fields of coherent imaging, directed energy, extremely-expanded laser beams (up to 3000:1), high-power tunable lasers, laser metrology, liquid and polymer-nanoparticle organic gain media, narrow-linewidth tunable laser oscillators, N-slit quantum interferometry, electrically-pumped organic semiconductor coherent emission, quantum entanglement, and space-to-space inter-ferometric communications. He is also the author of the generalized multiple-prism grating dispersion theory and has pioneered the use of Dirac's quantum notation in interferometry and classical optics.

His contributions have found applications in atomic physics, astronomy, chemistry, cytology, electrophoresis, femtosecond laser microscopy, geodesics, geophysics, gravitational lensing, heat transfer, imaging, laser isotope separation, laser medicine, laser pulse compression, laser spectroscopy, mathematical transforms, nanoengineering, nanophotonics, nonlinear optics, optofluidics, organic semiconductor lasers, phase imaging, polarization rotation, quantum computing, quantum entanglement, quantum fluctuations, quantum philosophy, space exploration, and tunable diode laser design. Current interests include tunable laser physics, interferometry via Dirac's notation, the limits of quantum resolution, and the foundations of quantum entanglement.

Duarte was elected Fellow of the Australian Institute of Physics in 1987 and in 1993 he was elected Fellow of the Optical Society of America. He has received the *Engineering Excellence Award* (1995), for the invention of the N-slit laser interferometer, and the *David Richardson Medal* (2016) 'for seminal contributions to the physics and technology of multiple-prism arrays for narrow-linewidth tunable laser oscillators and laser pulse compression' from the Optical Society (Optica).

Contributing author biographies

Peter Hegemann

Peter Hegemann is Hertie-Professor for Neuroscience at the Humboldt-Universität zu Berlin (since 2016). He studied chemistry at the Ludwig Maximilian Universität Munich (diploma 1980, doctoral degree 1984). Hegemann was then a postdoctoral fellow at the Physics Department, Syracuse University, where he started work on the photobiology of the green alga Chlamydomonas. 1986 to 1992 he was group leader at the Max-Planck-Institute for Biochemistry Martinsried, where he focused on the analysis of photoelectric responses of green algae. He became a Professor for Biochemistry at University of Regensburg in 1993, where he discovered the light-activated ion channel channelrhodopsin. In 2004 Hegemann became Professor of Biophysics at Humboldt-Universität zu Berlin. Currently he works on algal photobiology, molecular characterization of sensory photoreceptors, and optogenetics. He is also focusing on the biophysical characterization of light-regulated enzymes and engineering for optogenetic applications. Hegemann is the author/co-author of some 350 scientific publications and the recipient of various professional awards including the Harvey Prize (2016), the Massry Prize (2016), the Canada Gairdner International Award (2018), the Shaw Prize (2020), the Albert Lasker Award (2021), and the Louisa-Gross-Horwitz Prize (2022).

Suneel Kateriya

Suneel Kateriya is currently working as a professor of Biotechnology at the School of Biotechnology (SBT), Jawaharlal Nehru University (JNU), New Delhi, India, He had also served SBT, JNU as an associate professor for three years (2015–18). Previously (2007–15), he worked as an assistant professor of Biochemistry at the Department of Biochemistry, University of Delhi, India. Kateriya obtained his PhD degree in Biochemistry from the Institute of Biochemistry, University Regensburg, Germany and his postdoctoral training took place at the Department of Pharmacology, Columbia University, NYC, USA. Kateriya participated in the discovery and characterization of the channelrhodopsins from Chlamydomonas (Green Algae) while working as a graduate student with Professor Peter Hegemann. Recently, his team has shown that trafficking of bacterial type rhodopsin (Channelrhodopsin) is modulated by the mammalian-like intraflagellar transport (IFT) machinery in the green algae. He has published more than 70 papers in peer-reviewed journals. He is applying optogenetic tools for studying inflammation pathways, cellular signaling and ciliary signaling. Kateriya has been awarded with the prestigious Young Investigator Biotechnology Award by Department of Biotechnology, Government of India. He is actively pursuing research in the areas of optobiotechnology, optogenetics and ciliopathies.

Alfons Penzkofer

Alfons Penzkofer studied from 1962–65 papermaking at the Oskar von Miller Polytechnikum in Munich (engineering degree: Ing.-grad.). From 1966 to 1971 he studied physics at the Technical University of Munich (degree: Dipl.-Phys.), and from 1971 to 1974 did PhD research under the supervision of Professor Wolfgang Kaiser resulting in the dissertation *Parametric Four-Photon-Interaction in Liquids and Solids with Picosecond Light Pulses* (degree: Dr rer. nat.). His postdoctoral work on nonlinear optics at the Technical University of Munich led to his habilitation in 1977 (treatise: *High Intensity Raman Interactions*, degree: Dr rer. nat. habil). From 1978 to 2008 Penzkofer was a professor of physics at the University of Regensburg. Since 2008 he has performed research at the University of Regensburg on photo-biological spectroscopy towards the characterization of flavin and retinal based photoreceptors applied in optogenetics. During his research career, Penzkofer has published 310 peer-reviewed papers on a wide variety of laser, optics, and spectroscopy subjects including: picosecond and femtosecond laser development; picosecond and femtosecond light continua generation; solid-state organic dye lasers, thin-film electro-luminescent molecule lasers, and thin-film luminescent polymer lasers; linear and nonlinear optical material parameter determination; linear and nonlinear time-integrated and time-resolved optical spectroscopy on liquids, solids, semiconductors, organic dyes, organic polymers, artificial and biological photoreceptors; photocycle dynamics studies on flavoprotein-based photoreceptors (phototropins, BLUF proteins, photoactivated adenyl cyclases, cryptochromes) and rhodopsin-based photoreceptors (histidine kinases, guanylyl cyclases, fluorescent voltage indicators). Penzkofer is a Fellow of Optica (formerly Optical Society of America).

Sergei Popov

Professor Sergei Popov holds an MSc in electrical engineering (laser physics) from Moscow Institute of Physics and Technology (Russia), and DrSc in applied physics from Helsinki University of Technology (Finland). He is a full professor at KTH Royal Institute of Technology (Kungliga Tekniska Högskolan), Sweden, and has diverse research interests in optics: physical optics, optical communication, laser physics, nanophotonics, nanoplasmonics, and novel optical materials. Besides his academic career, Popov has solid experience in industrial research tracking back to Ericsson Telecom AB and Acreo AB, both in Sweden. Popov has an extensive international collaboration network. He was a guest professor at Zhejiang University (ZHJ, China), École Polytechnique Fédérale de Lausanne (EPFL, Switzerland), The University of Tokyo (Japan), and Eidgenössische Technische Hochschule (ETHZ, Switzerland). He has published more than 400 peer-reviewed papers and conference contributions including over 40 invited talks, and he holds 11 patents. Popov has served as Editor-in-Chief for JEOS: RP (*Journal of European Optical Society, Rapid Publications*) and has chaired several international conferences and served as a member of TPC for conferences in optics/photonics community: ECOC, OFC, PIERS, ACP, among others. Popov is a Fellow of Optica (formerly Optical Society of America).

Kathleen M Vaeth

Dr Kathleen M Vaeth earned her BS in Chemical Engineering from Cornell University, where she was a Kodak Fellow, and her Masters and PhD in Chemical Engineering from the Massachusetts Institute of Technology, where she was a Hertz Fellow. Her technical work spans the design, fabrication, and characterization of thin film devices, including OLEDs, piezoelectric vibrational energy harvesters, and MEMS inkjet printeads for the display, automotive, printing, and utilities markets. She is is currently the Senior Engineering R&D Manager for New and Advanced Sensor Products at Qualitrol Corporation. Prior to joining Qualitrol, Vaeth held positions as the Director of User Experience at OLEDWorks, Senior Lecturer in the Robert Frederick Smith School of Chemical and Biomolecular Engineering at Cornell University, Vice President of Engineering at microGen Systems, and Technology Director, Senior Research Scientist, and Project Leader at Eastman Kodak. She has over 20 publications, 30 U.S. patents, and has given numerous invited lectures and talks. Vaeth was elected an Optica Fellow in 2021.

Elena Vasileva

Dr Elena Vasileva has worked for several years for Hübner Photonics, a laser manufacturer, as a Global Product Manager responsible for nanosecond pulsed and diode-pumped solid-state lasers within the company portfolio. Elena has an international educational background. She holds a PhD degree in Applied Laser Physics from the Royal Institute of Technology KTH, Stockholm, Sweden. Her PhD research and the following year of a postdoctoral study at the KTH were centered on light propagation in anisotropically scattering novel optical materials, including dye-based organic composites. Vasileva has obtained two MSc degrees in Electrical Engineering with a focus on Photonics and Nanoelectronics from the Saint Petersburg Electrotechnical University 'LETI', Russia, as well as Technical Mathematics and Technical Physics from Lappeenranta University of Technology, Finland. She has also conducted research on VECSELs in the Optoelectronic Research Center (ORC) at the Tampere University of Technology Finland. Alongside her PhD research, Vasileva was a board member of the Professional Photonics Society—IEEE Photonics Sweden (IPS) and the chair of the Swedish chapter of Women in Engineering (WIE Sweden IEEE), where she provided support and encouragement for young female researchers. Together with the volunteering group and society members, she organized events that aimed to promote optics and photonics for kids and teenagers in Sweden. The events were supported by the IEEE chapters, industrial companies as well as by the Nobel Museum, Sweden. She believes science is about passion.

IOP Publishing

Organic Lasers and Organic Photonics (Second Edition)

F J Duarte

Chapter 1

Introduction

F J Duarte

Basic concepts of organic lasers are introduced with emphasis on tunability and narrow-linewidth oscillation. Major historical developments in organic photonics are highlighted. The second edition of *Organic Lasers and Organic Photonics* is briefly described and contrasted with its first edition.

1.1 Introduction

Organic dye lasers were discovered by Sorokin and Lankard (1966) and Schäfer *et al* (1966). Both these reports refer to pulsed laser-pumped liquid gain media, comprised of diluted organic dye molecules in an organic solvent, situated within a laser resonator comprised of two mirrors.

The next grand entrance into the laser field, via the organic laser, was performed by the *tunable* organic dye laser (Spaeth and Bortfeld 1966, Soffer and McFarland 1967). A few years later, followed the very first *tunable narrow-linewidth* laser introduced by Hänsch (1972). These groundbreaking disclosures set in motion an enormous wave of discoveries with gigantic innovative consequences to analytical chemistry, astronomy, medicine, nuclear industry, physics, spectroscopy, and science in general. A wave of innovation that remains vastly uncelebrated with most of its innovators only tenuously and vaguely recognized.

It is the tunability, throughout the visible spectrum, and the exquisite specificity in wavelength selectivity that made the tunable organic dye laser such a worldwide success in the scientific laboratories of the 1970s-to-1990s epoch. In addition to this tunable coherent ability, the organic dye laser in the liquid state is uniquely apt to remove excess heat from an inherently high-gain molecular active medium that has led to the demonstration of single-pulse energies in the 400–800 J (Baltakov *et al* 1974, Tang *et al* 1987) range and average powers exceeding 25 kW (Bass *et al* 1992). An additional landmark was the introduction of the continuous wave (CW) organic dye laser (Peterson *et al* 1970) that became the developmental workhorse for the femtosecond dye laser (Fork *et al* 1981, 1987, Dietel *et al* 1983) and a background model for ultrafast lasers in general.

The early revolution of tunable organic lasers, and laser dyes, is well covered in various books including *Dye Lasers* (Schäfer 1990), *Laser Dyes* (Maeda 1984), *Dye Laser Principles* (Duarte and Hillman 1990), *High-Power Dye Lasers* (Duarte 1991), and *Selected Papers on Dye Lasers* (Duarte 1992).

The whole gamut of organic lasers now is extended to include solid-state tunable organic dye lasers and semiconductor organic lasers. This latest generation of organic lasers has opened the horizon to the realm of miniaturized devices with its own vast array of applications.

Organic Lasers and Organic Photonics is the very first book to include all these wonderful sources of tunable coherent radiation integrated in a single cohesive, and unique, volume. Furthermore, a pragmatic description of organic photonics, including a plethora of exciting applications of organic lasers, is given in addition to stand-alone chapters on optogenetics, organic laser medicine, and quantum communications.

1.2 Laser linewidth

A laser is differentiated, recognized, and desired, by its ability to emit coherent radiation: that is, spatial coherence and spectral coherence. Spatial coherence is what gives the laser its ability to direct its output energy towards a single confined, minute, area in space equal to the cross section of its emission beam. The first beam divergence from an organic dye laser was quoted by Sorokin and Lankard (1966) and it was 3 mrad (at half angle) at a wavelength of $\lambda \approx 755$ nm.

Spectral coherence is what gives the laser its ability to emit in extremely narrow segments of the electromagnetic spectrum or in extremely pure colors. A measure of how narrow is that segment of the electromagnetic spectrum, or how pure is the color of emission, is the *laser linewidth*. In this section attention is given to what is laser linewidth and what constitutes narrow-linewidth emission in organic lasers. One observation worth mentioning at this stage is that, in regard to the quality of emission, the concept of coherence is more restrictive than the concept of laser. For instance, a device described as a *laser* can have multiple transverse modes and each of those transverse modes can include a multitude of longitudinal modes. Thus a basic laser can be a mediocre source of coherent radiation. On the other hand, a device described as a *coherent source* or *coherent emitter* implies coherence both in the spatial and the spectral domain and thus it can be assumed to emit a single-longitudinal mode in a single-transverse mode.

1.2.1 Laser linewidth in organic dye lasers

As previously mentioned, the first unambiguously measured emission linewidths in organic dye lasers were published by Soffer and McFarland (1967): $\Delta\lambda \approx 6$ nm, at a wavelength of $\lambda \approx 570$ nm for a mirror-mirror cavity, and $\Delta\lambda \approx 0.06$ nm at a wavelength of $\lambda \approx 565$ nm for a mirror-grating cavity. A linewidth of $\Delta\lambda \approx 6$ nm would be classified as broadband laser emission while, on the other hand, a linewidth of $\Delta\lambda \approx 0.06$ nm would begin to *approach the realm of narrow-linewidth* laser emission. A brief literature survey indicates that in high-peak-power pulsed liquid organic dye lasers the laser linewidth, in the absence of intracavity etalons, can be in the $0.00045 \leqslant \Delta\lambda \leqslant 6$ nm range (Soffer and McFarland 1967, Duarte and Piper 1984).

Usually in optimized narrow-linewidth laser oscillators, the narrowness of the laser linewidth $\Delta\lambda$, or $\Delta\nu$ in Hz units, is limited by the pulse duration Δt of the laser emission via a consequence of Heisenberg's uncertainty principle or $\Delta\nu\Delta t \approx 1$.

1.2.2 Narrow-linewidth landmarks in high-power pulsed organic dye lasers

Here, landmark contributions and discoveries pertinent to the development of high-power pulsed tunable narrow-linewidth organic dye lasers are listed. These discoveries albeit originally introduced in the field of organic lasers are found extensively applied throughout laser technology in general.

1967: Intracavity diffraction grating tuning is introduced while also demonstrating dispersion induced linewidth narrowing (Soffer and McFarland 1967).

1971: The distributed feedback (DFB) organic dye laser is discovered (Kogelnick and Shank 1971).

1972: Introduction of the first narrow-linewidth tunable organic dye laser also demonstrating grating dispersion multiplication via intracavity telescopic beam expansion (Hänsch 1972).

1977–78: Discovery of grazing-incidence grating cavities (Shohan *et al* 1977, Littman and Metcalf 1978, Saikan 1978).

1978–80: Introduction of multiple-prism grating oscillators (Kasuya *et al* 1978, Duarte and Piper 1980).

1981–1984: Discovery and refining of multiple-prism grazing-incidence grating cavities (Duarte and Piper 1981, 1984).

1982: Introduction of the generalized multiple-prism grating dispersion theory (Duarte and Piper 1982).

1.2.3 Narrow-linewidth landmarks in CW organic dye lasers

In CW organic dye lasers, where there are no temporal constraints, the laser linewidth varies greatly from system to system and can be approximately in the $0.0000000009 \leqslant \Delta\lambda \leqslant 0.06$ nm range. At one end is the straight-forward linear mirror-mirror cavity, $\Delta\lambda \approx 0.06$ nm, of Peterson *et al* (1970) and at the other extreme the sophisticated stabilized dye laser, $\Delta\lambda \approx 0.0000000009$ nm, of Hough *et al* (1987). Very narrow linewidths from stabilized organic dye lasers can be expressed more readily in the frequency domain where linewidths are minimized to $\Delta\nu \approx 750$ Hz (Hough *et al* 1987) and $\Delta\nu \approx 100$ Hz (Drever *et al* 1983). Parenthetically, the stabilization systems developed by Drever and Hall, for the liquid organic CW dye laser, eventually became crucial to the stabilization of second and third generation lasers applied in the detection of *gravitational waves*. Also, standard CW dye lasers were applied by Aspect *et al* (1981) in the first quantum entanglement experiments in the visible portion of the electromagnetic spectrum.

Important landmark developments in this field include:

1970: The CW organic dye laser is discovered (Peterson *et al* 1970).

1973: Frequency stabilized CW dye lasers are introduced (Berger *et al* 1973).

> 1983: Extremely narrow-linewidth emission in stabilized CW dye lasers is demonstrated (Drever *et al* 1983).

An additional development worth noting in CW organic dye lasers is the demonstration of semiconductor lasers as pump lasers (Scheps 1995).

1.2.4 CW organic dye laser developments for pulse compression

In this subsection, landmark contributions and discoveries pertinent to the development of ultrashort-pulse dye lasers are listed. Ultrashort-pulse lasers, ultrafast lasers, and femtosecond lasers are all homologous lasers developed thanks to the CW organic dye laser. Indeed, it was on the back of the CW organic dye laser that most of the momentous contributions for ultrashort lasers resulted. These discoveries are found applied throughout laser technology, irrespective of gain media, and include:

> 1967: Mode-locking, using intracavity saturable absorbers, is introduced to the organic dye laser (Schmidt and Schäfer 1968).
> 1972: Passively mode-locked CW dye lasers are demonstrated (Ippen *et al* 1972).
> 1976: Introduction of colliding-pulse mode-locking in CW dye lasers (Ruddock and Bradley 1976).
> 1982–87: Generalized multiple-prism dispersion theory, applicable to pulse compression, is developed (Duarte and Piper 1982, Duarte 1987).
> 1983–84: Prismatic negative dispersion for pulse compression is demonstrated (Dietel *et al* 1983, Fork *et al* 1984).
> 1986: Laser pulses as short as 6 fs are reported using multiple-prism and multiple-grating pulse compression (Fork *et al* 1987).
> 1987: Fiber multiple-prism compressors are demonstrated (Kafka and Baer 1987).

Contemporaneous attosecond pulse lasers and few-cycle pulse lasers have, in one way or another, benefited directly from the physics developed to create femtosecond organic dye lasers. For a detailed review on this subject see Diels (1990) and Diels and Rudolph (1996).

1.3 Solid-state organic lasers

Solid-state organic lasers are a class of lasers that include mainly solid-state organic lasers and semiconductor organic lasers. The excitation of these lasers is performed utilizing optical means, although electrical excitation of coherent organic semiconductor sources has been recently demonstrated.

1.3.1 Solid-state organic dye lasers

In this subsection, important contributions and discoveries pertinent to the development of pulsed tunable solid-state organic dye lasers are listed.
> 1967: Laser-pumped solid-state dye lasers are discovered (Soffer and McFarland 1967).

1968: Flashlamp-pumped solid-state dye lasers are introduced (Peterson and Snavely 1968).

1994: First tunable narrow-linewidth solid-state dye laser oscillators are demonstrated (Duarte 1994).

1999: Multiple-prism grating solid-state dye laser oscillator demonstrates a linewidth-temporal performance of $\Delta\nu\Delta t \approx 1.06$ (Duarte 1999).

1999–2000: Distributed feedback (DFB) solid-state dye lasers are introduced (Wadsworth *et al* 1999, Zhu *et al* 2000).

2002: Waveguide solid-state dye lasers are introduced (Oki *et al* 2002).

The tunable narrow-linewidth solid-state dye laser oscillators comprise multiple-prism Littrow grating and hybrid multiple-prism grazing-incidence grating configurations (Duarte 1994).

1.3.2 Further notable developments

Here, additional important contributions and discoveries pertinent to the development of optically pumped organic lasers are listed.

1988: Microcavity organic lasers are discovered (De Martini and Jakobovitz 1988).

1990: Organic fiber lasers are discovered (Knobbe *et al* 1990).

1996: Conjugated polymer lasers are announced (Hide *et al* 1996, Holzer *et al* 1996, Tessler *et al* 1996).

1997: Waveguide organic semiconductor lasers are introduced (Kozlov *et al* 1997).

1997: Hybrid waveguide-DFB organic dye lasers are disclosed (Maeda *et al* 1997).

1998: Organic semiconductor vertical-cavity surface–surface emitting lasers (OVCSELs) are introduced (Bulović *et al* 1998).

2006: Optofluidic DFB lasers are disclosed (Psaltis *et al* 2006).

It should be noted that some of these entries are anticipated by US Patents. Detailed reviews discussing various aspects of these developments are given by Kranzelbinder and Leising (2000), Karnutsch (2007), Samuel and Turnbull (2007), and Grivas and Pollnau (2012).

1.3.3 Coherent emission from electrically-pumped organic semiconductors

Spatially coherent and spectrally coherent emission from an electrically-pumped organic semiconductor was reported for the first time by Duarte *et al* (2005) and Duarte (2007). This was achieved using a high-luminescence OLED device in series, or tandem, using an active medium of coumarin 545 tetramethyl-doped tris(8-hydroxyquinoline) aluminum (Alq_3). This organic semiconductor dye-doped medium was confined within an integrated interferometric configuration. This subject matter is discussed in detail in chapter 11.

1.4 Organic photonics

Organic photonics is an enormous field that utilizes organic materials to generate, transmit, detect, and process light for a large number of applications in various areas of scientific and technological endeavor. In this regard, it is certainly an improvable task to capture the ineffable essence of the field in the confined space of this book.

A pragmatic approach to confront this limitation, however, is to adopt a rather terse reference-based style to describe the various areas of practical interest that have flourished in this extensive field.

1.5 Organic lasers and organic photonics

Organic Lasers and Organic Photonics is composed of fifteen chapters.

The first set of chapters is:

Chapter 2: Organic laser dyes.

Chapter 3: Energetics of organic laser dyes.

Chapter 4: Polymer matrices for lasers.

These chapters include subject matter on organic gain media applicable to organic lasers and organic photonics.

The second set of chapters is:

Chapter 5: Cavity and resonator architectures for high-performance organic laser oscillators.

Chapter 6: Mathematical-physics for tunable narrow-linewidth organic laser oscillators.

These chapters concentrate the attention on the architecture and the physics of high-performance organic lasers.

The next chapter:

Chapter 7: Best performance of organic lasers.

This chapter is a handbook style tabulation summarizing the best performance of optically pumped organic lasers as disclosed in the open literature.

The third set of chapters is:

Chapter 8: Tunable organic lasers for directed energy.

Chapter 9: Polymer-nanoparticle organic lasers.

Chapter 10: Compact and miniaturized organic dye lasers: from glass to bio-based gain media.

Chapter 11: Electrically-pumped organic semiconductor laser emission.

These chapters focus on specific areas of interest within the larger field of organic sources of coherent radiation.

The fourth set of chapters is:

Chapter 12: Organic photonics.

Chapter 13: Organic dyes in optogenetics.
Chapter 14: Tunable organic lasers and organic dyes in medicine.

These chapters focus on use of organic optical media and organic lasers in a variety of utilitarian and exciting applications. In particular, chapters 13 and 14 focus on biomedical applications.

The fifth set of chapters is:

Chapter 15: Organic lasers for N-channel quantum entanglement.
Chapter 16: Intrinsic quantum coherence in electrically-pumped organic interferometric emitters: Diracian emission.

These chapters focus on tunable lasers as a source of entangled ensembles of indistinguishable photons and on the physics of quantum coherence.

1.6 Perspective

Wavelength agility, or tunability, and exquisite high-performance narrow-linewidth emission are two intrinsic characteristics associated with optimized organic lasers. Beyond these characteristics, the field can be divided into large high-energy organic lasers and miniaturized organic lasers. Low capital cost is associated with both these avenues.

More specifically, in regard to optically pumped miniaturized organic lasers: low cost of fabrication and mass production is a major advantage that could have an enormous impact provided these lasers are optimally engineered and optimally coupled to efficient excitation devices such as blue-green diode laser pump sources. Coherent organic integrated interferometric sources directly excited by electrons widen the horizons even further.

Some time ago it was reflected that 'without Maxwell equations and quantum mechanics present day technology including lasers, computers, and the internet, could not have been possible... Quantum mechanics plus electromagnetism provide the bases of the wonderful technologies we enjoy today' (Duarte 2012).

Organic lasers, and organic photonics, are a manifestation of electromagnetism, quantum mechanics, and organic chemistry principles. The emission and transmission of quanta, or photons, play a crucial role in organic lasers and organic photonics. In this regard, the subject matter of this book is not only timely but transcends contemporaneous interests towards the future.

This century has been designated in numerous publications as the 'century of the photon' and that is a most welcome designation. However, the quantum of light, also known as the photon, will certainly dominate a lot more than this century. Photons have been here since the early beginnings ... and will be here for aeons to come. It is certainly an immense pleasure, and privilege, to contribute a small part to this brilliant age of discovery via our account of *organic lasers and organic photonics*.

1.7 Organic Lasers and Organic Photonics (Second Edition)

The second edition of this book has been updated and extended in the following manner:

1. A new chapter (16) entitled Intrinsic quantum coherence in electrically-pumped organic interferometric emitters, has been added.
2. The following chapters have been extended and/or updated: 2, 3, 4, 6, 8, 10, 11, 13, 14, and 15.
3. Problems have been added to the following chapters: 2–9, 11, 12, and 14–16.

Chapters that have not been extended or updated are considered to retain a fair representation of the field at hand. That is the case of chapters 5, 7, and 9. Chapter 13 on organic dyes in optogenetics includes a major extension with approximately 700 new references. This updated and extensive chapter should become a standard reference in this exciting field.

References

Aspect A, Grangier P and Roger G 1981 Experimental tests of realistic local theories via Bell's theorem *Phys. Rev. Lett.* **47** 460–3

Baltakov F N, Barikhin B A and Sukhanov L V 1974 400 J pulsed laser using a solution of rhodamine 6 G in ethyl alcohol *JETP Lett.* **19** 174–5

Barger R L, Sorem M S and Hall J L 1973 Frequency stabilization of a CW dye laser *Appl. Phys. Lett.* **22** 573–5

Bass I L, Bonano R E, Hackel R H and Hammond P R 1992 High-average-power dye laser at Lawrence Livermore National Laboratory *Appl. Opt.* **31** 6993–7006

Bulović V, Kozlov V G, Khalfin V B and Forrest S R 1998 Transform-limited narrow-linewidth lasing action in organic semiconductor microcavities *Science* **279** 553–5

De Martini F and Jakobovitz J R 1988 Anomalous spontaneous-emission-decay phase transition and zero-threshold laser action in a microscopic cavity *Phys. Rev. Lett.* **60** 1711–4

Diels J-C 1990 Femtosecond dye lasers *Dye Laser Principles* ed F J Duarte and L W Hillman (New York: Academic) ch 3

Diels J-C and Rudolph W 1996 *Ultrafast Laser Pulse Phenomena* (New York: Academic)

Dietel W, Fontaine J J and Diels J-C 1983 Intracavity pulse compression with glass: a new method of generating pulses shorter than 60 fs *Opt. Lett.* **8** 4–6

Drever R W P, Hall J L, Kowalski F V, Hough J, Ford G M, Munley A J and Ward H 1983 Laser phase and frequency stabilization using an optical resonator *App. Phys.* B **31** 97–105

Duarte F J 1987 Generalized multiple-prism dispersion theory for pulse compression in ultrafast dye lasers *Opt. Quantum Electron.* **19** 223–9

Duarte F J (ed) 1991 *High-Power Dye lasers* (Berlin: Springer)

Duarte F J 1992 *Selected Papers on Dye Lasers* (Bellingham, WA: SPIE)

Duarte F J 1994 Solid-state multiple-prism grating dye-laser oscillators *Appl. Opt.* **33** 3857–60

Duarte F J 1999 Multiple-prism grating solid-state dye laser oscillator: optimized architecture *Appl. Opt.* **38** 6347–9

Duarte F J 2007 Coherent electrically excited organic semiconductors: visibility of interferograms and emission linewidth *Opt. Lett.* **32** 412–4

Duarte F J 2012 *Laser Physicist* (Rochester, NY: Optics Journal)

Duarte F J and Hillman L W (ed) 1990 *Dye Laser Principles* (New York: Academic)

Duarte F J, Liao L S and Vaeth K M 2005 Coherence characteristics of electrically excited tandem organic light-emitting diodes *Opt. Lett.* **30** 3072–4

Duarte F J and Piper J A 1980 A double-prism beam expander for pulsed dye lasers *Opt. Commun.* **35** 100–4

Duarte F J and Piper J A 1981 Prism preexpanded grazing-incidence grating cavity for pulsed dye lasers *Appl. Opt.* **20** 2113–6

Duarte F J and Piper J A 1982 Dispersion theory of multiple-prism beam expander for pulsed dye lasers *Opt. Commun.* **43** 303–7

Duarte F J and Piper J A 1984 Narrow-linewidth high-prf copper laser-pumped dye laser oscillators *Appl. Opt.* **23** 1391–4

Fork R L, Brito Cruz C H, Becker P C and Shank C V 1987 Compression of optical pulses to six femtoseconds by using cubic phase compensation *Opt. Lett.* **12** 483–5

Fork R L, Greene B I and Shank C V 1981 Generation of optical pulses shorter than 0.1 psec by colliding pulse mode locking *Appl. Phys. Lett.* **38** 671–2

Fork R L, Martinez O M and Gordon J P 1984 Negative dispersion using pairs of prisms *Opt. Lett.* **9** 150–2

Grivas C and Pollnau M 2012 Organic solid-state integrated amplifiers and lasers *Lasers Photon, Rev.* **6** 419–62

Hänsch T W 1972 Repetitively pulsed tunable dye laser for high-resolution spectroscopy *Appl. Opt.* **11** 895–8

Hide F, Schwartz B J, Díaz-García M and Heeger A L 1996 Laser emission from solutions and films containing polymer and titanium dioxide nanocrystals *Chem. Phys. Lett.* **256** 424–30

Holzer W, Penzkofer A, Gong S-H, Bleyer A and Bradley D D C 1996 Laser action in poly (m-phenylenevinylene-co-2, 5-dioctoxy-p-phenylenevinylene) *Adv. Mat.* **8** 974–8

Hough J, Hils D, Rayman M D, Ma L-S, Hollberg L and Hall J L 1987 Dye-laser frequency stabilization using optical resonators *Appl. Phys.* B **33** 179–85

Ippen E P, Shank C V and Dienes A 1972 Passive mode locking of the CW dye Laser *Appl. Phys. Lett.* **21** 348–50

Kafka J D and Baer T 1987 Prism-pair dispersive delay lines in optical pulse compression *Opt. Lett.* **12** 401–3

Karnutsch C 2007 *Low Threshold Organic Thin Film Laser Devices* (Göttingen: Cuvillier)

Kasuya T, Suzuki T and Shimoda K 1978 A prism anamorphic system for Gaussian beam expander *Appl. Phys.* **17** 135–6

Knobbe E T, Dunn B, Fuqua P D and Nishida F 1990 Laser behavior and photostability characteristics of organic dye doped silicate gel materials *Appl. Opt.* **29** 2729–33

Kogelnick H and Shank C V 1971 Stimulated emission in a periodic structure *Appl. Phys. Lett.* **18** 152–4

Kozlov V G, Bulović V, Burrows P E and Forrest S R 1997 Laser action in organic semiconductor waveguide and double-heterostructure devices *Nature* **389** 362–4

Kranzelbinder G and Leising G 2000 Organic solid-state lasers *Rep. Prog. Phys.* **63** 729–62

Littman M G and Metcalf H J 1978 Spectrally narrow pulsed dye laser without beam expander *Appl. Opt.* **17** 2224–7

Maeda M 1984 *Laser Dyes* (New York: Academic)

Maeda M, Oki Y and Imamura K 1997 Utrashort pulse generation from an integrated single-chip dye laser *IEEE J. Quantum Electron.* **33** 2146–9

Oki Y, Aso K, Zuo D, Vasa N J and Maeda M 2002 Wide-wavelength range operation of a distributed-feedback dye laser with a plastic waveguide Japan *J. Appl. Phys.* **41** 6370–4

Peterson O G and Snavely B B 1968 Stimulated emission from a flashlamp-excited organic dyes in polymethyl mehacrylate *Appl. Phys. Lett.* **12** 238–40

Peterson O G, Tuccio S A and Snavely B B 1970 CW operation of an organic dye solution laser *Appl. Phys. Lett.* **17** 245–7

Psaltis D, Quake S R and Yang C 2006 Developing optofluidic technology through the fusion of microfluidics and optics *Nature* **442** 381–6

Ruddock I S and Bradley D J 1976 Bandwidth-limited subpicosecond pulse generation in mode-locked CW dye lasers *Appl. Phys. Lett.* **29** 296–7

Saikan S 1978 Nitrogen-laser-pumped single-mode dye laser *Appl. Phys.* **17** 41–4

Samuel I D W and Turnbull G A 2007 Organic semiconductor lasers *Chem. Rev.* **107** 1272–95

Scheps R 1995 Near-IR dye laser for diode-pumped operation *IEEE J. Quantum Electron.* **31** 126–34

Schäfer F P (ed) 1990 *Dye Lasers* 3rd ed (Berlin: Springer)

Schäfer F P, Schmidt W and Volze J 1966 Organic dye solution laser *Appl. Phys. Lett.* **9** 306–9

Schmidt W and Schäfer F P 1968 Self-mode-locking of dye-laser with saturable absorbers *Phys. Lett.* **26** 558–9

Shohan I, Danon N N and Oppenheim U P 1977 Narrow band operation of pulsed dye laser without beam expansion *J. Appl. Phys.* **48** 4495–7

Soffer B H and McFarland 1967 Continuously tunable narrow-band organic dye lasers *Appl. Phys.* **10** 266–7

Sorokin P P and Lankard J R 1966 Stimulated emission observed from an organic dye, chloro-aluminum phthalocyanine *IBM J. Res. Dev.* **10** 162–3

Spaeth M L and Bortfeld D P 1966 Stimulated emission from polymethine dyes *Appl. Phys.* **9** 179–81

Tang K Y, O'Keefe T, Treacy B, Rottler R and White C 1987 Kilojoule-output XeCl dye laser: optimization and analysis *Proc. Dye Laser/Laser Dye Technical Exchange Meeting 1987* ed J H Bentley (Alabama: U. S. Army Missile Command, Redstone Arsenal) pp 490–502

Tessler N, Denton G J and Friend R H 1996 Lasing from conjugated polymer microcavities *Nature* **382** 695–7

Wadsworth W J, McKinnie I T, Woolhouse A D and Haskell T G 1999 Efficient distributed feedback solid state dye laser with dynamic grating *Appl. Phys.* B **69** 163–5

Zhu X-L, Lam S-K and Lo D 2000 Distributed-feedback dye-doped solgel silica lasers *Appl. Opt.* **39** 3104–7

Chapter 2

Organic laser dyes

F J Duarte

Provides tuning range of known organic laser dye families. Introduces molecular structures and basic physical concepts, such as conversion efficiency, related to laser emission. A comprehensive table including organic laser dyes yielding emission in the visible spectrum is included with molecular formula, molecular weight, and corresponding laser tuning range. Laser spectral performance of various dye-doped solid-state organic matrices is also included and discussed.

2.1 Introduction

Organic dye molecules useful in the emission of coherent radiation in organic lasers comprise only a small fraction of all available organic dyes. Organic dyes that are also *organic laser dyes* require intra-molecular energetic conditions favorable to stimulated emission. For instance, ideally, an organic laser dye should experience little or no self-absorption and no active triplet levels detrimental to laser emission.

As illustrated in figure 2.1, there are at least eight organic laser dye families; these are the oligophenylenes, the oxadiazoles, the stilbenes, the coumarins, the pyrromethenes, the xanthenes, the merocyanines, and the cyanines. Each of these families is composed of a number of laser dyes. One of the most numerous families is the coumarins which includes more than 20 individual laser dyes. In all, these organic laser dye families have been shown to span the emission spectrum from just above 300 nm to nearly 1100 nm. Many individual laser dyes offer an emission spectrum that overlaps the emission spectrum of a laser dye from the same family.

Wavelength agility and the ability to readily emit at hard to reach spectral regions are some of the most salient features, and attractiveness, of tunable organic dye lasers. In addition, liquid gain media is particularly apt to remove heat thus enabling long-pulse high-energy lasing and high-average-power laser emission (Duarte 1991).

Recommended standard books on the subject of organic laser dyes for tunable lasers include *Laser Dyes* (Maeda 1984), *Dye Lasers* (Schäfer 1990), *Dye Laser Principles* (Duarte and Hillman 1990), *Tunable Lasers Handbook* (Duarte 1995), and

doi:10.1088/978-0-7503-5547-6ch2

300	400	500	600	700	800	900	1000	1100

——— Oxadiazoles

——— Oligophenenes

——— Stilbenes

————— Coumarines

——— Pyrromethenes

————— Xanthenes

——— —— Merocyanines

———————————— Cyanines

300	400	500	600	700	800	900	1000	1100

Emission Wavelength (nm)

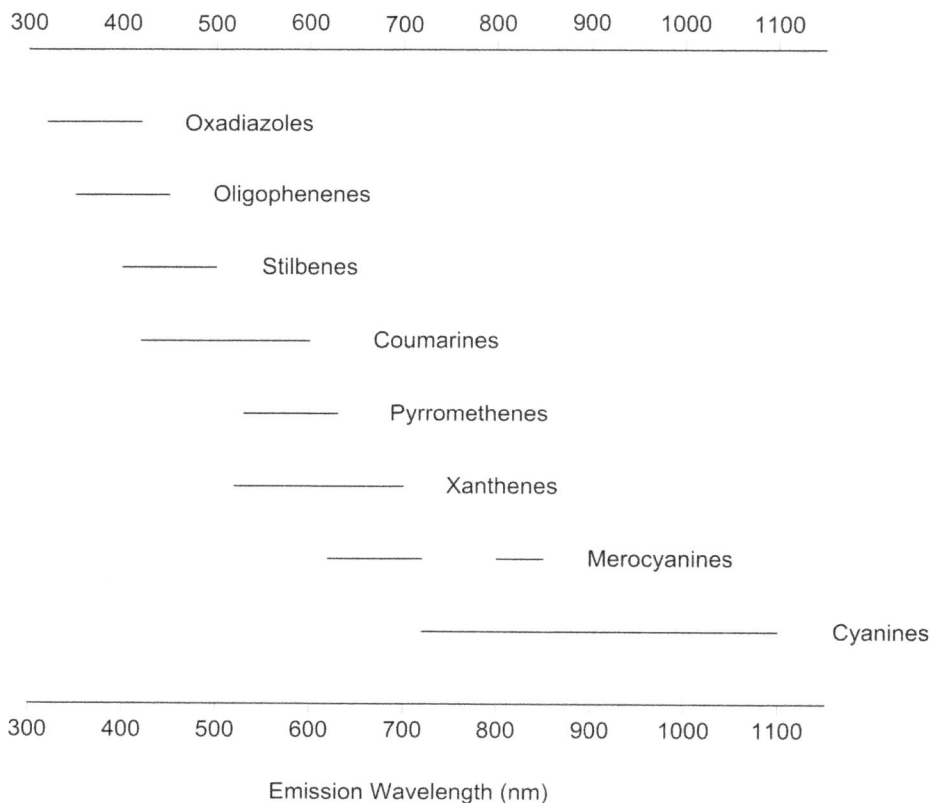

Figure 2.1. Approximate wavelength coverage available from the various known organic laser dye families.

Tunable Laser Optics (Duarte 2003). There are also available several commercials catalogs on organic laser dyes.

2.2 Organic laser dye molecules

Organic laser dye molecules are large molecules with molecular weights in the $174 \leqslant m_u \leqslant 1010$ range. Their organic signature is characterized by carbon–hydrogen ring structures linked with other organic elements such as oxygen and nitrogen. An example of such an organic molecule is coumarin 120, which has a molecular weight of 175.19, has a structure as given in figure 2.2, and lases approximately in the $419 \leqslant \lambda \leqslant 497$ nm range. A molecule more than twice as large is coumarin 314 T, with a molecular weight of 369.35 and a structure depicted in figure 2.3. Coumarin 314 T lases approximately in the $478 \leqslant \lambda \leqslant 525$ nm range. An even larger molecule is coumarin 545 T, which has a molecular weight of 430.56, and a molecular structure illustrated in figure 2.4, and offers a laser tuning range in the $501 \leqslant \lambda \leqslant 574$ nm region.

From an empirical perspective it can be observed that larger laser dye molecules tend to emit toward the red. Indeed, rhodamine 6 G, among the very first laser dyes

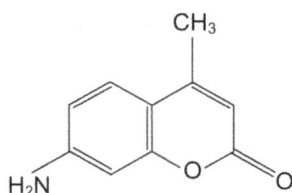

Figure 2.2. The molecular structure for the organic laser dye coumarin 120.

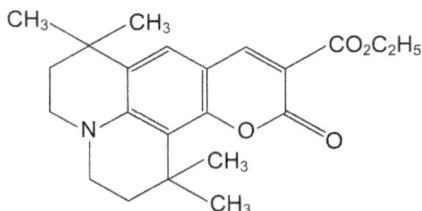

Figure 2.3. The molecular structure for the organic laser dye coumarin 314 T.

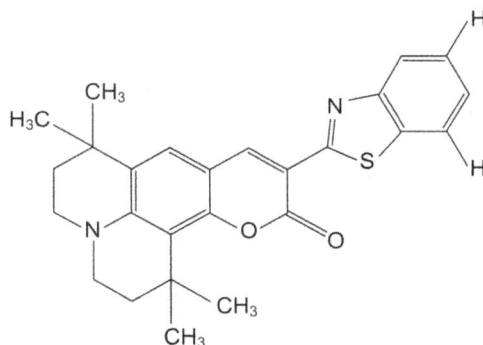

Figure 2.4. The molecular structure for the organic laser dye coumarin 545 T.

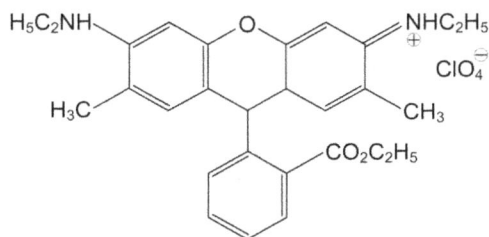

Figure 2.5. The molecular structure from one of the first organic laser dyes rhodamine 6 G.

disclosed in the open literature (Soffer and McFarland 1967; Sorokin and Lankard 1967; Peterson *et al* 1970) has a molecular weight of 479.01, and can emit narrow-linewidth tunable radiation in the $565 \leqslant \lambda \leqslant 605$ nm range (Duarte and Piper 1984). The molecular structure of rhodamine 6 G is depicted in figure 2.5

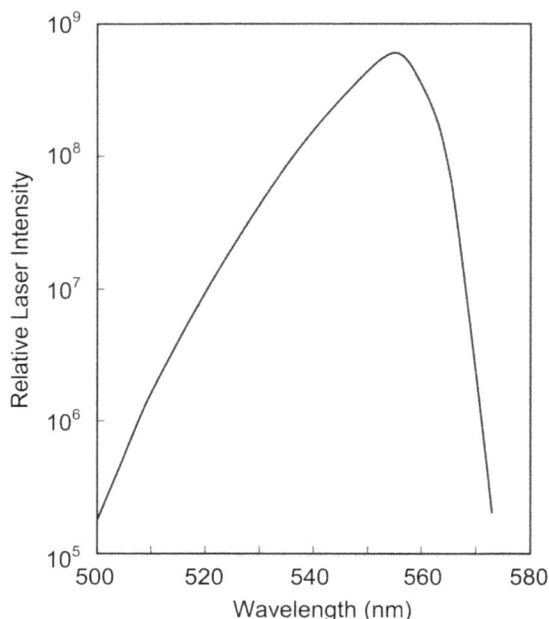

Figure 2.6. Laser tuning range for a mirror-grating cavity using the organic laser dye coumarin 545 T as gain medium (from Duarte *et al* 2006; © IOP Publishing Ltd. All rights reserved.).

Among the various attractive features of organic laser dye molecules as laser gain media is the feature of tunability which is central to tunable dye lasers. Figure 2.6 illustrates the tunability curve measured with a nitrogen laser-pumped coumarin 545 T laser configured by a compact Glan–Thompson polarizer-grating cavity (Duarte *et al* 2006). The laser conversion efficiency was measured to be 14%, and the laser linewidth $\Delta\lambda \approx 3$ nm, at a dye concentration of 3×10^{-3} M in ethanol.

Briefly, optical excitation of organic dye laser gain media leads to $S_0 \rightarrow S_1$ transitions from the ground electronic state S_0 to the first excited electronic state S_1. From the excited electronic state, transitions back to a *range* of higher closely lying vibrational–rotational at the ground electronic state lead to emission at longer wavelengths. The energy difference in this range of vibrational–rotational level can be described, using Planck's quantum energy relation $E = h\nu$, as

$$E_1 - E_2 = h(\nu_2 - \nu_1) \tag{2.1}$$

or

$$E_1 - E_2 = h\delta\nu \tag{2.2}$$

where $h = 6.626\,069\,57 \times 10^{-34}$ Js is Planck's constant, and ν is the frequency of emission in units of Hz. This means that the energy difference ΔE leads to a range of frequencies $\delta\nu$ which using

$$\lambda = \left(\frac{c}{\nu}\right) \tag{2.3}$$

leads to a wavelength range $\delta\lambda$ of emission. It should be noted that there is an inverse relation between frequency and wavelength so that

$$\delta\lambda = \left(\frac{c\delta\nu}{\nu_1\nu_2} \right) \tag{2.4}$$

where $\delta\lambda = (\lambda_1 - \lambda_2)$. For example, if the emission occurs from $\nu_1 = 5.222\ 865\ 12 \times 10^{14}$ to $\nu_2 = 5.983\ 881\ 39 \times 10^{14}$ that means, using equation (2.4), that the tuning range is $\delta\lambda = 73$ nm or in the $501 \leqslant \lambda \leqslant 574$ nm range, since $\lambda_1 = 574$ nm and $\lambda_2 = 501$ nm. This succinctly explains the mechanics behind tunability in organic laser dyes. This topic is considered quantitatively in chapter 3.

From a physical-chemistry perspective, it is said that the excitation occurs from the highest occupied molecular orbital (HOMO), at the lower state, to the lowest unoccupied molecular orbital (LUMO), at the higher state. The energy difference between these levels is referred to as the energy gap. This terminology is used extensively in chapter 11.

2.2.1 Water solubility

For obvious reasons the use of innocuous solvents should be emphasized not just from a biological perspective but also from an engineering perspective. In this regard, except for the testing of a particular laser–dye solvent combination this author has always used ethanol (C_2H_5OH), and methanol (CH_3OH), as the preferred solvents when working on laser cavity development or when using an organic dye laser for a long-term application, with ethanol as the favorite solvent.

Due to the flammable nature of these solvents the use of pure water and water–ethanol combinations is recommended, particularly when working with high-power, or high-energy, dye laser systems. For high-power is meant CW powers of a few Watts or more, or high-pulse repetition-frequency (prf) average powers in the tens of Watts or more. For high-energy is meant single pulse energies of tens of Joules or more. As seen in chapters 1 and 7, these are performance levels easily reached and surpassed by tunable organic dye lasers.

Water solubility in coumarin dyes was first highlighted by Drexhage and co-workers (Tuccio *et al* 1973, Drexhage *et al* 1975). Improved tunable narrow-linewidth performance in the blue-green region of the spectrum was reported by Duarte and colleagues (Chen *et al* 1988, Duarte 1989) using coumarin tetramethyl laser dyes. For a 50:50 water–ethanol mixture the laser output was observed to decline about 30%. Given improvements in the thermo-optical properties of the active medium and the significant minimization in flammability the moderate decline in laser intensity, as compared to the intensity when using pure ethanol, is considered a good tradeoff. Besides coumarin 314 T and C 545 T, other coumarin tetramethyl structures are depicted in figures 2.7–2.10. These include coumarin 102 T, coumarin 338 T, coumarin 334 T, and coumarin 153 T (listed by their appearance in the spectral region, from short to longer wavelengths).

Other organic laser dyes demonstrating emission while dissolved in water include 2-(4-pyridyl)-5-phenyloxazole. This laser dye, developed by Lee and Robb (1980), is

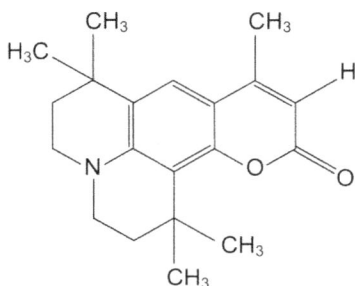

Figure 2.7. The molecular structure for the organic laser dye coumarin 102 T.

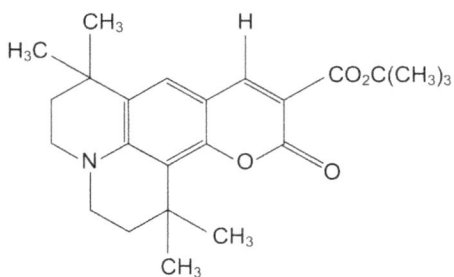

Figure 2.8. The molecular structure for the organic laser dye coumarin 338 T.

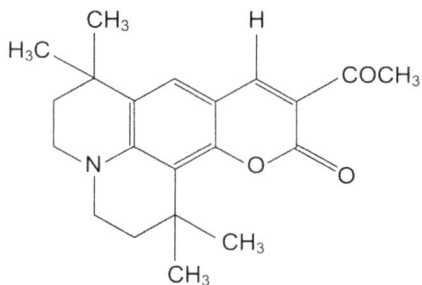

Figure 2.9. The molecular structure for the organic laser dye coumarin 334 T.

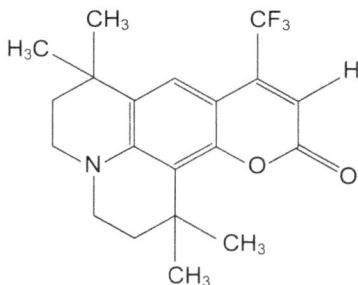

Figure 2.10. The molecular structure for the organic laser dye coumarin 153 T.

also known as 4PyPO and is reported to emit in the $490 \leqslant \lambda \leqslant 530$ nm region (Cotnoir 1981). The development of water soluble or partly water soluble efficient laser dyes for the orange-red portion of the spectrum would be highly desirable.

2.2.2 Criticisms to organic dye as laser gain media

Mainly in the 1990s many articles were published by authors reporting on tunable solid-state gain media that were critical of organic laser dyes. Criticisms centered mainly around the fact that dye lasers use a liquid gain media and the associated dye spills that liberated dangerous toxic chemicals. Two observations are relevant in this regard:
1. The liquid state of the organic laser dye media is crucial, and a highly convenient vehicle, for the efficient removal of excess heat, or cooling, from this media in high-energy lasing or high average-power lasing.
2. So-called dye spills are simply the result of poor, or sloppy, engineering.

2.3 Organic laser dyes in the liquid-state and the solid-state

In this section attention is focused on successful organic lasers dyes emitting mainly in the near ultra-violet-visible-near infrared region. Most of these laser dyes are soluble in solvents such as ethanol, methanol, and in some cases in ethanol/water mixtures. As discussed by Shank (1975), ideally the efficiency of an organic dye laser depends on the ratio of the pump-laser wavelength over the emission-laser wavelength. In other words, the (η_e) is roughly proportional to the ratio of the laser emission frequency (ν_l) over the excitation, or pump, laser frequency (ν_p) so that

$$\eta_e \approx K_\nu \left(\frac{\nu_l}{\nu_p} \right) \tag{2.5}$$

or

$$\eta_e \approx K_\lambda \left(\frac{\lambda_p}{\lambda_l} \right) \tag{2.6}$$

where K_ν and K_λ are the relevant constants of proportionality. Therefore, higher efficiencies are expected for laser systems in which the pump-laser wavelength approaches the laser emission wavelength of the organic dye laser. This is a major factor in facilitating the high efficiency, for instance, of copper-vapor-laser $(\lambda_p \approx 510.554$ nm) pumped dye lasers using the rhodamine 6 G laser dye with a peak emission wavelength of $\lambda_l \approx 580$ nm which report photon emission efficiencies greater than 50% (Bass et al 1992).

Table 2.1 includes a partial list of organic laser dye molecules, in liquid solvents, emitting in the $400 \leqslant \lambda \leqslant 700$ nm region, that is, the visible spectrum. For a complete list of dyes, including near IR dyes, please refer to Duarte and Hillman (1990) and Duarte (2003). Emission wavelength range depends on various parameters, and the dynamic interaction of these parameters, including pump-laser wavelength, solvent refractive index, organic dye concentration, operating temperature, and laser cavity configuration (Duarte 1990a, 1990b). Here it should be remembered that most

Table 2.1. Visible organic laser dyes in liquid media.

Organic dye molecule[a]	Molecular weight (m_u)	Tuning range[b] (nm)	Pump laser solvent[c]	Reference[d]
Carbostyril 124 Carbostyril 7 $C_{10}H_{10}N_2O$	174.20	$400 \leqslant \lambda \leqslant 430$	N_2[e] Methanol	Birge and Duarte (1988)
Coumarin 120 $C_{10}H_9NO_2$	175.19	$418 \leqslant \lambda \leqslant 498$	N_2 Methanol	
Coumarin 2 $C_{13}H_{15}N_2O$	217.27	$430 \leqslant \lambda \leqslant 478$	N_2 Methanol	
Coumarin 339 $C_{13}H_{13}NO_2$	215.25	$437 \leqslant \lambda \leqslant 492$	N_2 Methanol	
Coumarin 1 $C_{14}H_{17}NO_2$	231.29	$438 \leqslant \lambda \leqslant 510$	N_2 Methanol	
Coumarin 138 $C_{14}H_{15}NO_2$	229.27	$441 \leqslant \lambda \leqslant 489$	N_2 Methanol	
Coumarin 102 T $C_{20}H_{25}NO_2$	311.42	$450 \leqslant \lambda \leqslant 510$[f]	XeCl[g] Ethanol Methanol Ethanol/water	Duarte (1989)
Coumarin 102 Coumarin 480 $C_{16}H_{17}NO_2$	255.32	$453 \leqslant \lambda \leqslant 514$	XeCl Ethanol Methanol	Chen et al (1988)
Coumarin 4 $C_{10}H_8O_3$	176.17	$464 \leqslant \lambda \leqslant 534$	N_2 Methanol	Birge and Duarte (1988)
Coumarin 151 $C_{10}H_6NO_2F_3$	229.16	$465 \leqslant \lambda \leqslant 533$	N_2 Methanol	
Coumarin 338 T $C_{24}H_{31}NO_4$	397.51	$477 \leqslant \lambda \leqslant 526$	XeCl Ethanol Methanol Ethanol/water	Chen et al (1988)
Coumarin 30 Coumarin 515 $C_{21}H_{21}N_3O_2$	347.42	$478 \leqslant \lambda \leqslant 518$	N_2 Methanol	Birge and Duarte (1988)
Coumarin 314 T $C_{22}H_{27}NO_4$	369.46	$478 \leqslant \lambda \leqslant 525$	XeCl Ethanol Methanol Ethanol/water	Chen et al (1988)
Coumarin 338 $C_{20}H_{23}NO_4$	341.41	$480 \leqslant \lambda \leqslant 528$	XeCl Ethanol Methanol	
Coumarin 307 $C_{13}H_{12}NO_2F_3$	271.24	$480 \leqslant \lambda \leqslant 552$	N_2 Methanol	Birge and Duarte (1988)
Coumarin 314	313.35	$482 \leqslant \lambda \leqslant 526$	XeCl	Chen et al (1988)

Coumarin 504 $C_{18}H_{19}NO_4$			Ethanol Methanol	
Coumarin 500 $C_{12}H_{10}NO_2F_3$	257.21	$490 \leqslant \lambda \leqslant 540^f$	N_2 Ethanol	Duarte and Piper (1980)
Coumarin 152 $C_{13}H_{10}NO_2F$	257.21	$492 \leqslant \lambda \leqslant 572$	N_2 Methanol	Birge and Duarte (1988)
Coumarin 334 T $C_{21}H_{25}NO_3$	339.43	$500 \leqslant \lambda \leqslant 546$	XeCl Ethanol Methanol Ethanol/water	Chen et al (1988)
Coumarin 334 $C_{17}H_{17}NO_3$	283.33	$501 \leqslant \lambda \leqslant 550$	XeCl Ethanol Methanol	
Coumarin 545 T $C_{26}H_{26}N_2O_2S$	430.56	$501 \leqslant \lambda \leqslant 574$	N_2 Ethanol Methanol Ethanol/water	Duarte et al (2006)
Coumarin 7 $C_{20}H_{19}N_3O_2$	333.39	$507 \leqslant \lambda \leqslant 531$	N_2 Methanol	Birge and Duarte (1988)
Coumarin 153 T $C_{20}H_{22}F_3NO_2$	365.39	$508 \leqslant \lambda \leqslant 588$	XeCl Ethanol Methanol Ethanol/water	Chen et al (1988)
Coumarin 153 $C_{16}H_{14}F_3NO_2$	309.29	$512 \leqslant \lambda \leqslant 585$	XeCl Ethanol Methanol	
Pyrromethene BF$_2$ $C_{22}H_{33}BF_2N_2$	374.32	$528 \leqslant \lambda \leqslant 580$	UV-Flashlamp Methanol	Davenport et al (1990)
Rhodamine 110 Rhodamine 560 $C_{20}H_{15}N_2O_3Cl$	366.80	$539 \leqslant \lambda \leqslant 583$	N_2 Methanol	Birge and Duarte (1988)
Pyrromethene 580 $C_{22}H_{33}BF_2N_2$	374.32	$545 \leqslant \lambda \leqslant 585$	Nd:YAG $(2\nu)^h$ Ethanol	Partridge et al (1994)
Rhodamine 6 G Rhodamine 590 $C_{28}H_{31}N_2O_3Cl$	479.01	$565 \leqslant \lambda \leqslant 605^f$	Cu^i Ethanediol	Duarte and Piper (1984)
Rhodamine B Rhodamine 610 $C_{28}H_{31}ClN_2O_3$	479.01	$595 \leqslant \lambda \leqslant 639$	N_2 Methanol	Birge and Duarte (1988)
Sulforhodamine B Kiton red 620 $C_{27}H_{30}N_2S_2O_7$	558.66	$595 \leqslant \lambda \leqslant 641$	N_2 Methanol	
DCM $C_{19}H_{17}N_3O$	303.36	$600 \leqslant \lambda \leqslant 680$	Nd:YAG (2ν) Ethanol	

(*Continued*)

Table 2.1. (*Continued*)

Organic dye molecule[a]	Molecular weight (m_u)	Tuning range[b] (nm)	Pump laser solvent[c]	Reference[d]
Sulforhodamine 101	606.71		N_2	
Sulforhodamine 640			Methanol	
$C_{31}H_{30}N_2O_7S_2$				
DODC iodide	486.35	$639 \leqslant \lambda \leqslant 659$	Nd:YAG (2ν)	
$C_{23}H_{23}N_2O_2I$			Methanol	

[a] Alternative name, and chemical formula, for the organic laser dye also included.
[b] Order of appearance dictated by shorter wavelength of the tuning range. Given tuning range is approximate unless indicated otherwise.
[c] First solvent is linked to the quoted tuning range. Generally a laser dye soluble in methanol is also soluble in ethanol.
[d] References relate to the literature source for the pump-laser wavelength and emission range.
[e] N_2 laser at $\lambda \approx 337.1$ nm.
[f] Tuning range measured under narrow-linewidth laser emission.
[g] XeCl laser at $\lambda \approx 308$ nm.
[h] Nd:YAG (2ν) laser at $\lambda \approx 532$ nm.
[i] Cu vapor laser at $\lambda \approx 510.554$ nm.

organic laser dyes soluble in ethanol are also soluble in methanol and vice versa. The main difference in the use of these two solvents might be a slight alteration on peak output wavelength and tuning range. The wavelength emission ranges included in table 2.1 are only an approximate range unless specifically indicated otherwise. An alternative name for the laser dye is given with most entries. However, for many dyes there are additional alternative names and it is recommended to verify the molecular weight, or molecular formula, to clearly identify a particular molecule when in doubt.

Organic laser dye concentrations can vary, for example, from 1×10^{-5} M, for narrow-linewidth flashlamp-pumped tunable laser oscillators (Duarte *et al* 1991), to 1×10^{-2} M for high-power laser-pumped narrow-linewidth tunable tunable laser oscillators (Duarte and Piper 1981).

The near IR region, $720 \leqslant \lambda \leqslant 1000$ nm, can be accessed with the following dyes listed in successive peak laser emission wavelength order: DTDC iodide ($C_{23}H_{23}IN_2S_2$, $\lambda_l \approx 698$ nm), DOTC iodide ($C_{25}H_{25}IN_2O_2$, $\lambda_l \approx 762$ nm), HITC iodide ($C_{29}H_{33}IN_2$, $\lambda_l \approx 832$ nm), DTTC iodide ($C_{25}H_{25}IN_2S_2$, $\lambda_l \approx 850$ nm), IR-144 ($C_{56}H_{73}N_5O_8S_2$, $\lambda_l \approx 867$ nm), IR-140 ($C_{39}H_{34}N_3O_4S_2Cl_3$, $\lambda_l \approx 897$ nm), IR-132 ($C_{53}H_{48}N_3O_4S_2ClO_4$, $\lambda_l \approx 914$ nm), and IR-125 ($C_{43}H_{47}N_2O_6S_2Na$, $\lambda_l \approx 915$ nm) (Duarte and Hillman 1990, Duarte 2003). The peak laser wavelength for excitation at $\lambda_p \approx 532$ nm (Birge and Duarte 1988). Organic dyes are also used as saturable absorbers in femtosecond lasers (Diels 1990).

Many organic laser dyes, especially the rhodamines and the pyrromethenes, have also been shown to lase efficiently while contained in polymer solid-state matrices, such as PMMA (Duarte 1994, Duarte *et al* 1998, Costela *et al* 2016), and organic–inorganic matrices such as PMMA-nanoparticle (Duarte and James 2003). Table 2.2 lists some of the organic laser dyes used to dope optical organic polymer matrices and optical polymer organic–inorganic matrices.

Table 2.2. Organic laser dye molecules in solid-state organic and organic–inorganic matrices.

Organic dye molecule	Solid-state matrix	Excitation laser	Tuning range (nm)	Reference
Rhodamine 6 G	MPMMA[a]	Tunable dye $\lambda \approx 520$ nm	$563 \leqslant \lambda \leqslant 610$[b]	Duarte (1994)
Rhodamine 6 G	HEMA:MMA[c]	Tunable dye $\lambda \approx 533$ nm	$564 \leqslant \lambda \leqslant 602$[d]	Duarte et al (1998)
Rhodamine 6 G	PMMA-SiO$_2$ nanoparticles	Tunable dye $\lambda \approx 525$ nm	$567 \leqslant \lambda \leqslant 603$	Duarte and James (2003)
Sulforhodamine B	COP[(HEMA-MMA 7:3)-PETRA 9:1][e]	Nd:YAG (2ν) $\lambda \approx 532$ nm	$575 \leqslant \lambda \leqslant 645$	Costela et al (2016)
Pyrromethene 597	COP[HEMA: TMSPMA][f]	Nd:YAG (2ν) $\lambda \approx 532$ nm	$585 \leqslant \lambda \leqslant 625$	Costela et al (2016)
Rhodamine 640	COP(HEMA-PETA 9:1)[g]	Nd:YAG (2ν) $\lambda \approx 532$ nm	$620 \leqslant \lambda \leqslant 660$	Costela et al (2016)

[a] Modified poly methyl methacrylate.
[b] Narrow-linewidth emission at $\Delta\lambda \approx 1.12$ GHz using a MPL grating oscillator.
[c] 2-Hydroxyethyl methacrylate:methyl methacrylate.
[d] Narrow-linewidth emission at $\Delta\lambda \approx 650$ MHz using a MPL grating oscillator.
[e] Copolymer [2-hydroxyethyl methacrylate-methyl methacrylate)-pentaerythritol tetraacrylate].
[f] Copolymer [2-hydroxyethyl methacrylate: 3-(trimethoxysilyl)propyl methacrylate].
[g] Copolymer [2-hydroxyethyl methacrylate-pentaerythritol tetraacrylate].

2.4 Literature

In the laser dye post 2018 literature attention continues to be focused on pyrromethene dyes with emphasis in eco-friendly solvents (Al-shamiri et al 2023) which is the focus of our section 2.2.1. Work on seminaphthorhodafluor red laser dyes has also been reported (Ming et al 2020).

2.5 Problems

- 2.1 From the list of organic dye molecules given in table 2.1 select, according to your own criterion, the most efficient dye for emission at $\lambda \approx 510$ nm.
- 2.2 From the list of organic dye molecules given in table 2.1 select, according to your own criterion, the most efficient dye for emission at $\lambda \approx 550$ nm.
- 2.3 Assuming that equation (2.6) reduces to

$$\eta_e \approx \left(\frac{\lambda_p}{\lambda_l} \right)$$

estimate the optimum optical efficiency for a system comprised of a copper laser pump $\lambda = 510.554$ nm and a dye laser utilizing rhodamine 6 G lasing at $\lambda_l \approx 590$ nm.

- 2.4 From the various excitation lasers listed in table 2.1 identify the pump laser likely to yield the best optical efficiency for a dye laser utilizing coumarin 545 T as lasant.
- 2.5 Using laser wavelength tuning range, as criterion, identify from table 2.1 a single organic dye molecule that might replace two, or more, competing dyes.

References

Al-shamiri H A S 2023 Experimental and theoretical study of optical properties of pyrromethene (PM-597) laser dye in binary eco-friendly solvent *J. Phys. Org. Chem.* **36** e4445

Bass I L, Bonano R E, Hackel R H and Hammond P R 1992 High-average-power dye laser at Lawrence Livermore National Laboratory *Appl. Opt.* **31** 6993–7006

Birge R R and Duarte F J 1988 *Kodak Laser Dyes* (Rochester, NY: Eastman Kodak)

Chen C H, Fox J L, Duarte F J and Ehrlich J J 1988 Lasing characteristics of new coumarin-analog dyes: broadband and narrow-linewidth performance *Appl. Opt.* **27** 443–5

Costela A, García-Moreno I and Sastre R 2016 Solid-state organic dye lasers *Tunable Laser Applications* ed F J Duarte (New York: CRC Press) 3rd edn ch 3

Cotnoir L J 1981 Tuning curve of water soluble laser dye, 490–530 nm *Appl. Opt.* **20** 2331–2

Davenport W E, Ehrlich J J and Neister S E 1990 Characterization of pyrromethene BF2 complexes as laser dyes *Proc. Int. Conf. on Lasers '89* ed D G Harris and T M Shay (McLean, VA: STS) pp 408–14

Diels J -C 1990 Femtosecond dye lasers *Dye Laser Principles* ed F J Duarte and L W Hillman (New York: Academic) ch 3

Drexhage K H, Erikson G R, Hawks G H and Reynolds G A 1975 Water soluble coumarin dyes for flashlamp-pumped dye lasers *Opt. Commun.* **15** 399–403

Duarte F J 1989 Ray transfer matrix analysis of multiple-prism dye laser oscillators *Opt. Quantum Electron.* **21** 47–54

Duarte F J 1990a Narrow-linewidth pulsed dye laser oscillators *Dye Laser Principles* ed F J Duarte and L W Hillman (New York: Academic) ch 4

Duarte F J 1990b Technology of pulsed dye lasers *Dye Laser Principles* ed F J Duarte and L W Hillman (New York: Academic) ch 6

Duarte F J (ed) 1991 *High-Power Dye Lasers* (Berlin: Springer)

Duarte F J 1994 Solid-state multiple-prism grating dye-laser oscillators *Appl. Opt.* **33** 3857–60

Duarte F J (ed) 1995 *Tunable Lasers Handbook* (New York: Academic)

Duarte F J 2003 *Tunable Laser Optics* (New York: Academic)

Duarte F J, Davenport W E, Ehrlich J J and Taylor T S 1991 Ruggedized narrow-linewidth dispersive dye laser oscillator *Opt. Commun.* **84** 310–6

Duarte F J and Hillman L W (ed) 1990 *Dye Laser Principles* (New York: Academic)

Duarte F J and James R O 2003 Tunable solid-state lasers incorporating dye-doped polymer-nanoparticle gain media *Opt. Lett.* **28** 2088–90

Duarte F J, Liao L S, Vaeth K M and Miller A M 2006 Widely tunable green laser emission using the coumarin 545 tetramethyl dye as the gain *J. Opt. A: Pure Appl. Opt.* **8** 172–4

Duarte F J and Piper J A 1980 A double-prism beam expander for pulsed dye lasers *Opt. Commun.* **35** 100–4

Duarte F J and Piper J A 1981 Prism preexpanded grazing-incidence grating cavity for pulsed dye lasers *Appl. Opt.* **20** 2113–16

Duarte F J and Piper J A 1984 Narrow-linewidth high-prf copper laser-pumped dye laser oscillators *Appl. Opt.* **23** 1391–4

Duarte F J, Taylor T S, Costela A, García-Moreno I and Sastre R 1998 Long-pulse narrow-linewidth dispersive solid-state dye-laser oscillator *App. Opt.* **37** 3987–9

Lee L and Robb R 1980 Water soluble blue-green lasing dyes for flashlamp-pumped dye lasers *IEEE J. Quantum Electron.* **16** 777–84

Maeda M 1984 *Laser Dyes* (New York: Academic)

Ming W, Hu X, Zhang Z, Chang S, Chen R, Tian B and Zhang J 2020 Synthesis and properties of seminaphthorhodafluor red laser dyes *Res. Chem. Intermed.* **46** 1991–2002

Partridge W P, Laurendeau N M, Johnson C C and Steppel R N 1994 Performance of Pyrromethene 580 and 597 in a commercial Nd:YAG-pumped dye-laser system *Opt. Lett.* **19** 1630–2

Peterson O G, Tuccio S A and Snavely B B 1970 CW operation of an organic dye solution laser *Appl. Phys. Lett.* **17** 245–7

Schäfer F P (ed) 1990 *Dye Lasers* 2nd edn (Berlin: Springer)

Shank C V 1975 Physics of dye lasers *Rev. Mod. Phys.* **47** 649–57

Soffer B H and McFarland 1967 Continuously tunable narrow-band organic dye lasers *Appl. Phys.* **10** 266–7

Sorokin P P and Lankard J R 1967 Flashlamp-excitation of organic dye lasers: a short communication *IBM J. Res. Dev* **11** 148

Tuccio S A, Drexhage K H and Reynolds G A 1973 CW laser emission from coumarin dyes in the blue and green *Opt. Commun.* **10** 248–52

IOP Publishing

Organic Lasers and Organic Photonics (Second Edition)

F J Duarte

Chapter 3

Energetics of organic laser dyes

F J Duarte

The physics of excitation and emission of organic dye lasers is described and discussed using detailed semiclassical rate equations. Transitional cross sections are derived quantum mechanically. Thirteen cross sections and six excitation rates are given in tabular form for rhodamine 6 G. Quantum amplified spontaneous emission is described in detail.

3.1 Introduction

In this chapter the excitation dynamics of *optically excited*, or *optically pumped*, organic laser dye molecules is described. The approach considered applies to organic laser dye gain media either in the *liquid* or the *solid state*. In the liquid case the organic laser dye molecules are dissolved in an optically transparent organic solvent such as ethanol, methanol, or ethanol/water mixtures. In the solid-state case the organic laser dye molecules are used to dope optically-transparent organic solid state matrices such as poly methyl methacrylate (PMMA) or modified PMMA, also known as MPMMA (Duarte 1994).

The excitation laser dynamics considered in this chapter is essentially a quantum mechanical phenomenon. However, it is one of those few quantum phenomena that can also be accurately described using classical rate equations that incorporate cross sections that are quantum mechanical quantities. In this regard, this type of approach might be classified as a semi-classical description of laser dynamics.

The optical pumping is performed using broadband flashlamp excitation (Schäfer 1990, Duarte 1991) high-power pulse laser excitation (Duarte and Hillman 1990, Duarte 1991) and continuous wave (CW) laser excitation (Hollberg 1990). Pulse laser excitation can yield long-pulse high-energy tunable laser emission (Duarte 1991) or ultrashort-pulse laser emission in the femtosecond regime (Diels 1990, Diels and Rudolph 1996).

Direct electronic excitation, in the pulse regime, of organic laser dyes is discussed in chapter 7 where organic dye-doped organic semiconductors emitting spatially coherent emission are described.

An additional way to describe transitions in organic laser dyes is to use molecular concepts such as HOMO which stands for highest occupied molecular orbital and LUMO which stands for lowest unoccupied molecular orbital (Jones 1990, Schäfer 1990). Electron transitions occur between these two energy levels via the HOMO–LUMO gap which is equivalent to the S_0–S_1 energy spacing described in the next section. For the coumarin laser dye, Jones (1990) illustrates the promotion of an electron for which the C–N π bond order has increased. The reader is referred to Jones (1990) and Schäfer (1990) for further details. The HOMO–LUMO concept is further utilized in chapter 11.

3.2 Rate equations for generalized multiple-level systems

The energetics literature for optically-pumped dye lasers, via classical rate equations, is summarized in Schäfer (1990), Jensen (1991), Duarte (2003, 2015), and references therein. Haken (1970) provides various theoretical avenues to quantify the dynamics of laser excitation and emission.

A complete energy level diagram applicable to optically-pumped organic laser dye gain media is depicted in figure 3.1. The energy level diagram includes three *electronic states*: the ground state S_0, the first excited state from where laser emission originates S_1, and a higher lying state that might absorb laser emission S_2. In addition, two *triplet*

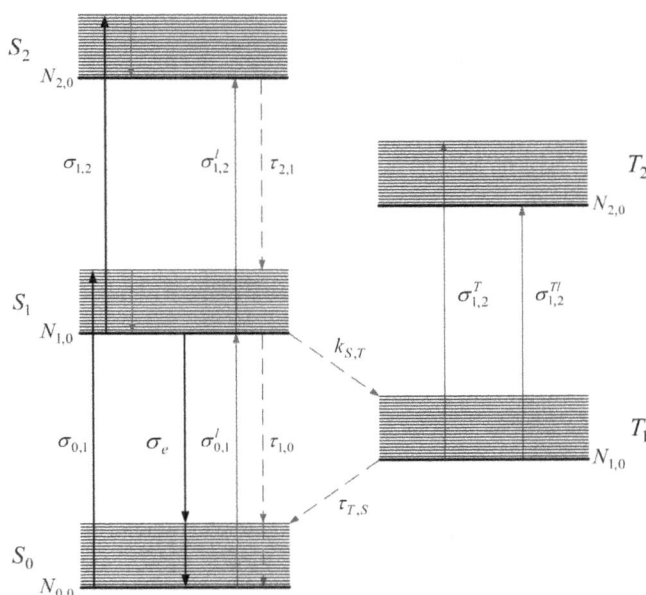

Figure 3.1. Energy level diagram corresponding to an organic laser dye molecule. It includes three electronic levels (S_0, S_1, and S_2), and two triplet levels (T_1 and T_2). Each electronic level contains a large number of vibrational and rotational levels. Laser emission takes place due to $S_1 \rightarrow S_0$ transitions.

states detrimental to laser emission, T_1 and T_2, are also included. Each electronic state contains a large number of overlapping vibrational–rotational levels.

For the sake of pragmatism only vibrational levels are considered in this approach. Assuming transverse laser excitation a detailed and descriptive set of rate equations, as introduced by Duarte (1995a, 1995b), is given as

$$N = \sum_{S=0}^{m} \sum_{v=0}^{m} N_{S, v} + \sum_{T=1}^{m} \sum_{v=0}^{m} N_{T, v} \tag{3.1}$$

$$\begin{aligned}
\frac{\partial N_{1, 0}}{\partial t} &\approx \sum_{v=0}^{m} N_{0, v} \, \sigma_{0, 1_{0}, v} \, I_p(t) + \sum_{v=0}^{m} N_{0, v} \sigma_{0, 1_{0}, v}^{l} I_l(x, t, \lambda_v) + \frac{N_{2, 0}}{\tau_{2, 1}} \\
&\quad - N_{1, 0}\left(\sum_{v=0}^{m} \sigma_{1, 2_{0}, v} \, I_p(t) + \sum_{v=0}^{m} \sigma_{e_{0}, v} \, I_l(x, t, \lambda_v) \right. \tag{3.2} \\
&\quad \left. + \sum_{v=0}^{m} \sigma_{1, 2_{0}, v}^{l} I_l(x, t, \lambda_v) + \left(k_{S, T} + \tau_{1, 0}^{-1} \right) \right)
\end{aligned}$$

$$\frac{\partial N_{T_1, 0}}{\partial t} \approx N_{1, 0} \, k_{S, T} - \frac{N_{T_1, 0}}{\tau_{T, S}} - N_{T_1, 0}\left(\sum_{v=0}^{m} \sigma_{1, 2_{0}, v}^{T} I_P(t) + \sum_{v=0}^{m} \sigma_{1, 2_{0}, v}^{Tl} I_l(x, t, \lambda_v) \right) \tag{3.3}$$

$$c^{-1}\frac{\partial I_P(t)}{\partial t} \approx -\left(N_{0, 0}\sum_{v=0}^{m} \sigma_{0, 1_{0}, v} + N_{1, 0}\sum_{v=0}^{m} \sigma_{1, 2_{0}, v} + N_{T_1, 0}\sum_{v=0}^{m} \sigma_{1, 2_{0}, v}^{T} \right) I_P(t) \tag{3.4}$$

$$\begin{aligned}
c^{-1}\frac{\partial I_l(x, t, \lambda)}{\partial t} + \frac{\partial I_l(x, t, \lambda)}{\partial x} &\approx N_{1, 0}\sum_{v=0}^{m} \sigma_{e_{0}, v} \, I_l(x, t, \lambda_v) \\
&\quad - \sum_{v=0}^{m} N_{0, v} \sigma_{0, 1_{0}, v}^{l} I_l(x, t, \lambda_v) \\
&\quad - N_{1, 0}\sum_{v=0}^{m} \sigma_{1, 2_{0}, v}^{l} I_l(x, t, \lambda_v) \tag{3.5} \\
&\quad - N_{T_1, 0}\sum_{v=0}^{m} \sigma_{1, 2_{0}, v}^{Tl} I_l(x, t, \lambda_v)
\end{aligned}$$

$$I_l(x, t, \lambda) = \sum_{v=0}^{m} I_l(x, t, \lambda_v) \tag{3.6}$$

$$I_l(x, t, \lambda) = I_l^{+}(x, t, \lambda) + I_l^{-}(x, t, \lambda) \tag{3.7}$$

Adopting the style of Duarte (2014) a description of the relevant parameters for this set of rate equations is given in table 3.1, please also refer to figure 3.1.

An intrinsic feature of organic laser dyes is their ability to emit broadband radiation in the absence of a frequency selective laser cavity. It is this very feature that also allows for the emission of tunable narrow-linewidth emission when the

Table 3.1. Description of parameters in rate equations.

Symbol	Description and units
$N_{S,v}$	Population of the S electronic state, at the v vibrational level (cm^{-3})
$N_{T,v}$	Population of the T triplet state, at the v vibrational level (cm^{-3})
$I_p(t)$	Intensity of the pump laser beam (cm^{-2} s^{-1})
$I_l(x, t, \lambda)$	Laser emission intensity from the gain medium (cm^{-2} s^{-1})
$\sigma_{0,\,1_{0,v}}$	Ground state absorption cross section of pump-laser radiation (cm^2)
$\sigma_{1,\,2_{0,v}}^l$	First excited state absorption cross section of laser emission (cm^2)
$\sigma_{1,\,2_{0,v}}^T$	First triplet state absorption cross section of pump-laser radiation (cm^2)
$\sigma_{1,\,2_{0,v}}^{Tl}$	First triplet state absorption cross section of laser emission (cm^2)
$\tau_{1,0}$	Radiationless decay time from the first excited state (s)
$k_{S,T}$	Radiationless decay rate from the excited singlet to the triplet (s^{-1})
$\tau_{T,S}$	Radiationless decay rate of the transition $T_1 \rightarrow S_0$ (s)
$S' \rightarrow S''$	Electronic state transitions
$v' \rightarrow v''$	Vibrational level transitions

organic laser dye gain medium is deployed within a frequency selective laser cavity (Duarte 1991, 2003, 2015).

This broadband-tunable ability is permitted by the fact that emission from the first excited electronic state S_1 transitions on to a broad vibrational manifold at the ground electronic state S_0. This ability to emit in the broadband regime, or to tune the emission whithin a broad frequency range, is mathematically represented by the summation terms of equations (3.5)–(3.7).

Given that organic laser gain media exhibits homogeneous broadening (Hillman, 1990), the introduction of intracavity frequency selective optics (see chapters 5 and 6) enables the molecular gain medium to contribute efficiently towards tunable narrow-linewidth emission.

3.2.1 Rate equations for single-energy levels

A simplified energy level diagram is illustrated in figure 3.2. Here, the vibrational manifolds are replaced by single energy levels and a number of mechanisms are neglected such as spontaneous decay from S_2, and absorption of the pump laser by T_1. These simplifications allow for equations (3.1)–(3.7) to be reduced to (Duarte 1995a, 1995b)

$$N = N_0 + N_1 + N_T \tag{3.8}$$

$$\frac{\partial N_1}{\partial t} \approx N_0\,\sigma_{0,1}\,I_p(t) + \left(N_0\sigma_{0,1}^l - N_1\,\sigma_e - N_1\sigma_{1,2}^l\right)I_l(x, t, \lambda) - N_1\left(k_{S,T} + \tau_{1,0}^{-1}\right) \tag{3.9}$$

$$\frac{\partial N_T}{\partial t} = N_1\,k_{S,T} - N_T\tau_{T,S}^{-1} - N_T\sigma_{1,2}^{Tl}I_l(x, t, \lambda) \tag{3.10}$$

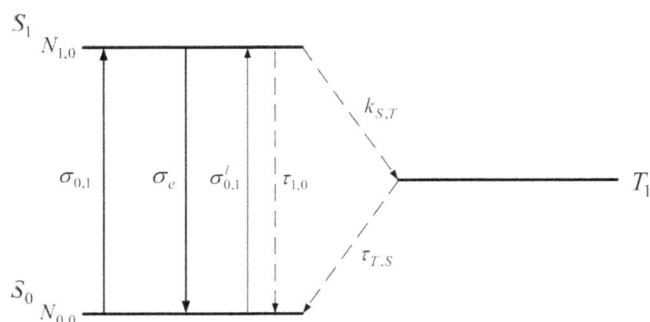

Figure 3.2. Simplified energy level diagram corresponding to an organic laser dye molecule. It includes only two electronic levels (S_0 and S_1) and one triplet levels (T_1). Laser emission takes place due to the $S_1 \to S_0$ transition.

Table 3.2. Measured cross sections for rhodamine 6 G.

Transition parameter symbol	Cross section (cm^2)	λ (nm)	Reference
$\sigma_{0,1}$	1.66×10^{-16}	510	Hargrove and Kan (1980)
$\sigma_{0,1}$	4.50×10^{-16}	530	Everett (1991)
$\sigma_{0,1}$[a]	3.60×10^{-16}	527	Holzer et al (2000)
σ_e	1.86×10^{-16}	572	Hargrove and Kan (1980)
σ_e	1.3×10^{-16}	600	Everett (1991)
σ_e[a]	3.0×10^{-16}	565	Holzer et al (2000)
$\sigma_{0,1}^l$	1.0×10^{-19}	600	Everett (1991)
$\sigma_{1,2}^l$	1.0×10^{-17}	600	Everett (1991)
$\sigma_{1,2}^l$[a]	3.0×10^{-17}		Holzer et al (2000)
$\sigma_{1,2}^T$	1.0×10^{-17}	530	Everett (1991)
$\sigma_{1,2}^{Tl}$	4.0×10^{-17}	600	Everett (1991)

[a]For rhodamine 6 G-doped p(HEMA-MMA) gain media.

$$c^{-1}\frac{\partial I_p(t)}{\partial t} = -(N_0\,\sigma_{0,1} + N_1\,\sigma_{1,2})I_p(t) \tag{3.11}$$

$$c^{-1}\left(\frac{\partial I_l(x,\,t,\,\lambda)}{\partial t}\right) + \left(\frac{\partial I_l(x,\,t,\,\lambda)}{\partial x}\right) = \left(N_1\,\sigma_e - N_0\sigma_{0,1}^l - N_1\sigma_{1,2}^l - N_T\sigma_{1,2}^{Tl}\right)$$
$$\times I_l(x,\,t,\,\lambda) \tag{3.12}$$

This simplified set of equations is similar to the equations considered by Teschke et al (1976). Equations of this form have been solved using numerical tools to simulate the behavior of the laser intensity $I_l(x, t, \lambda)$ as a function of the laser-pump intensity $I_p(t)$ and population parameters. Measured transition cross sections, excitation rates, and decay times are given in tables 3.2 and 3.3.

Table 3.3. Measured decay rates and decay times for rhodamine 6 G.

Transition parameter symbol	Rate (s^{-1})	Time (s)	Reference
$k_{S,T}$	2.0×10^7		Everett (1991)
$\tau_{T,S}$		0.5×10^{-7}	Everett (1991)
$\tau_{1,0}$		3.5×10^{-9}	Everett (1991)
$\tau_{1,0}{}^{a}$		3.3×10^{-9}	Holzer $et\ al$ (2000)
$\tau_{2,1}$		1.0×10^{-12}	Hargrove and Kan (1980)
$\tau_{n,1}{}^{a,b}$		60×10^{-15}	Holzer $et\ al$ (2000)

[a] For rhodamine 6 G-doped p(HEMA-MMA) gain media.
[b] Time decay from higher electronic states.

Further values for these parameters, relevant to excitation with a XeCl excimer laser and emission from a blue laser dye (TBS) are given by Jensen (1991).

3.2.2 Applications of rate equations for single-energy levels

For long pulse, or continuous wave (CW) excitation, the time derivatives approach zero and equations (3.9)–(3.12) reduce to

$$N_0\,\sigma_{0,1}\,I_p + \left(N_0\sigma_{0,1}^l - N_1\,\sigma_e - N_1\sigma_{1,2}^l\right)I_l(x,\,\lambda) = N_1\!\left(k_{S,\,T} + \tau_{1,0}^{-1}\right) \qquad (3.13)$$

$$N_1\,k_{S,\,T} = N_T\tau_{T,\,S}^{-1} + N_T\sigma_{1,2}^{Tl}I_l(x,\,\lambda) \qquad (3.14)$$

$$N_0\,\sigma_{0,1} = -N_1\,\sigma_{1,2} \qquad (3.15)$$

$$\frac{\partial I_l(x,\,\lambda)}{\partial x} = \left(N_1\,\sigma_e - N_0\sigma_{0,1}^l - N_1\sigma_{1,2}^l - N_T\sigma_{1,2}^{Tl}\right)I_l(x,\,\lambda) \qquad (3.16)$$

From these equations some features of long-pulse, or CW, organic dye lasers can be elucidated. For example, from equation (3.13), the population ratio N_1/N_0 just below threshold, that is, when the laser intensity $I_l(x,\,\lambda) \approx 0$, can be approximated to be

$$\frac{N_1}{N_0} \approx I_p\,\sigma_{0,1}\!\left(k_{S,\,T} + \tau_{1,0}^{-1}\right)^{-1} \qquad (3.17)$$

Using the parameters for rhodamine 6 G, given in tables 3.2 and 3.3, it can be estimated that to approach population inversion, pump intensities approaching 10^{24} photons cm^{-2} s^{-1} are necessary.

A problem prevalent in long-pulse and CW dye lasers is intersystem crossing from the first excited level N_1 into the triplet level T_1. Thus, researchers use $triplet\ level$ $quenchers$, such as C_8H_8 and C_7H_8 (Duarte 1990, Jones 1990), to neutralize the triplets. While using an effective triplet quencher, from equation (3.16), the gain factor can be expressed as

$$g = \left(N_1\left(\sigma_e - \sigma_{1,2}^l\right) - N_0\sigma_{0,1}^l\right)L \qquad (3.18)$$

Therefore, in long-pulse and CW organic dye lasers, in the absence of triplet losses, amplification can occur when the ratio of the populations becomes

$$\frac{N_1}{N_0} > \frac{\sigma_{0,1}^l}{\left(\sigma_e - \sigma_{1,2}^l\right)} \qquad (3.19)$$

From the values of the cross sections given by Everett (1991), for rhodamine 6 G in table 3.2, this ratio becomes approximately

$$\frac{N_1}{N_0} > \frac{10^{-19}}{10^{-16}} \approx 0.001 \qquad (3.20)$$

3.3 Quantum approach to transition cross sections

The cross sections utilized in the rate equation dynamics are measured experimentally. Relevant cross sections for the rhodamine 6 G molecule are given in table 3.2. The origin of these cross sections is quantum mechanical and can be outlined using the following Dirac principles (Dirac 1978, Feynman *et al* 1965):

$$\langle \phi | \psi \rangle = \sum_j \langle \phi | j \rangle \langle j | \psi \rangle \qquad (3.21)$$

$$\langle \phi | \psi \rangle = \langle \psi | \phi \rangle^* \qquad (3.22)$$

For $j = 1, 2$

$$\langle \phi | \psi \rangle = \langle \phi | 2 \rangle C_2 + \langle \phi | 1 \rangle C_1 \qquad (3.23)$$

where

$$C_1 = \langle 1 | \psi \rangle \qquad (3.24)$$

$$C_2 = \langle 2 | \psi \rangle \qquad (3.25)$$

These amplitudes are defined by the Hamiltonian

$$i\hbar \frac{dC_j}{dt} = \sum_k^2 H_{jk} C_k \qquad (3.26)$$

In reference to the energy level diagram of figure 3.3

$$\langle N_{0,0} | N_{0,0} \rangle = \langle N_{0,0} | N_{0,n} \rangle \langle N_{0,n} | N_{1,0} \rangle \langle N_{1,0} | N_{1,n} \rangle \langle N_{1,n} | N_{0,0} \rangle \\ + \langle N_{0,0} | N_{0,v} \rangle \langle N_{0,v} | N_{1,1} \rangle \langle N_{1,1} | N_{1,v} \rangle \langle N_{1,v} | N_{0,0} \rangle \qquad (3.27)$$

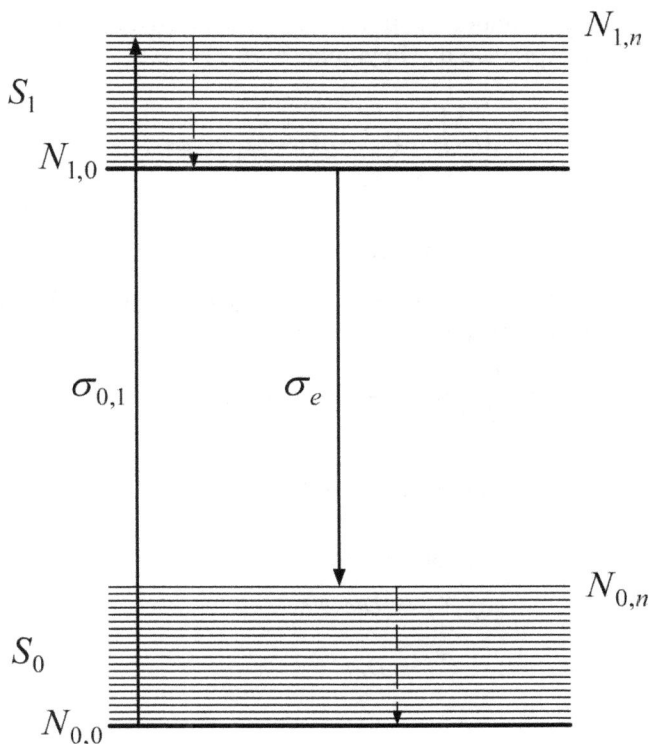

Figure 3.3. Energy level diagram applicable to the quantum transition model.

which can be abstracted into

$$\langle N_{0,0}|N_{0,0}\rangle = \langle N_{0,0}|N_{1,0}\rangle\langle N_{1,0}|N_{0,0}\rangle + \langle N_{0,0}|N_{1,1}\rangle\langle N_{1,1}|N_{0,0}\rangle \tag{3.28}$$

The population probability amplitude described by equation (3.28) has the same form as

$$\langle II|II\rangle = \langle II|1\rangle\langle 1|II\rangle + \langle II|2\rangle\langle 2|II\rangle \tag{3.29}$$

so that

$$\langle II|II\rangle = \langle II|1\rangle\langle II|1\rangle^* + \langle II|2\rangle\langle II|2\rangle^* \tag{3.30}$$

$$\langle II|II\rangle = |C_I|^2 + |C_{II}|^2 \tag{3.31}$$

Since this must equal unity

$$C_{II} = \frac{1}{\sqrt{2}}(C_1 + C_2) \tag{3.32}$$

$$C_I = \frac{1}{\sqrt{2}}(C_1 - C_2) \tag{3.33}$$

For a laser dye molecule under the influence of the electric field $\mathscr{E}$ the matrix of the Hamiltonian can be expressed as (Feynman *et al* 1965)

$$\begin{pmatrix} H_{11} & H_{12} \\ H_{21} & H_{22} \end{pmatrix} = \begin{pmatrix} E_0 + \mu\mathscr{E} & -A \\ -A & E_0 - \mu\mathscr{E} \end{pmatrix} \qquad (3.34)$$

where

$$\mathscr{E} = \mathscr{E}_0(e^{i\omega t} + e^{-i\omega t}) \qquad (3.35)$$

and μ corresponds to the electric dipole moment. Expanding the Hamiltonian given in equation (3.26) then subtracting and adding

$$i\hbar\frac{dC_I}{dt} = (E_0 + A)C_I + \mu\mathscr{E}\, C_{II} \qquad (3.36)$$

$$i\hbar\frac{dC_{II}}{dt} = (E_0 - A)C_{II} + \mu\mathscr{E}\, C_I \qquad (3.37)$$

For a small $\mathscr{E}$, solutions are of the form

$$C_I = D_I e^{-i(E_I/\hbar)t} \qquad (3.38)$$

$$C_{II} = D_{II} e^{-i(E_{II}/\hbar)t} \qquad (3.39)$$

where

$$E_I = E_0 + A \qquad (3.40)$$

and

$$E_{II} = E_0 - A \qquad (3.41)$$

Substitution into the differential equations (3.36) and (3.37), using $(E_I - E_{II}) = 2A = \hbar\omega_0$ and the definition given in equation (3.35), eventually leads to

$$i\hbar\frac{dD_I}{dt} = \mu\mathscr{E}_0(e^{i(\omega+\omega_0)t} + e^{-i(\omega-\omega_0)t})D_{II} \qquad (3.42)$$

$$i\hbar\frac{dD_{II}}{dt} = \mu\mathscr{E}_0(e^{i(\omega-\omega_0)t} + e^{-i(\omega+\omega_0)t})D_I \qquad (3.43)$$

Neglecting the exponential terms associated with the sum of the frequencies ($\omega + \omega_0$), these differential equations reduce to

$$i\hbar\frac{dD_I}{dt} = \mu\mathscr{E}_0\, D_{II} e^{-i(\omega-\omega_0)t} \qquad (3.44)$$

$$i\hbar\frac{dD_{II}}{dt} = \mu\mathscr{E}_0\, D_I e^{i(\omega-\omega_0)t} \qquad (3.45)$$

Assuming that at $t = 0$, $D_I \approx 1$, and integrating equation (3.45) yields (Feynman *et al* 1965)

$$D_{II} = \frac{\mu \mathscr{E}_0}{\hbar} \left(\frac{1 - e^{i(\omega - \omega_0)T}}{(\omega - \omega_0)} \right) \tag{3.46}$$

so that

$$|D_{II}|^2 = \left(\frac{\mu \mathscr{E}_0}{\hbar} \right)^2 \left(\frac{2 - 2\cos(\omega - \omega_0)T}{(\omega - \omega_0)^2} \right) \tag{3.47}$$

which can also be written as (Feynman *et al* 1965)

$$|D_{II}|^2 = \left(\frac{\mu \mathscr{E}_0 T}{\hbar} \right)^2 \frac{\sin^2\left(\frac{1}{2}(\omega - \omega_0)T\right)}{\left(\frac{1}{2}(\omega - \omega_0)T\right)^2} \tag{3.48}$$

which is the probability for the transition for $I \rightarrow II$, or $N_{0,0} \rightarrow N_{1,v}$. It also follows that

$$|D_I|^2 = |D_{II}|^2 \tag{3.49}$$

Using $I = 2\varepsilon_0 c \, \mathscr{E}_0^2$, then $|D_{II}|^2$ can be expressed as

$$|D_{II}|^2 = 2\pi \left(\frac{\mu^2 T^2}{4\pi\varepsilon_0 c \hbar^2} \right) I(\omega_0) \frac{\sin^2\left(\frac{1}{2}(\omega - \omega_0)T\right)}{\left(\frac{1}{2}(\omega - \omega_0)T\right)^2} \tag{3.50}$$

where μ is the dipole moment is in units of Cm, $(1/4\pi\varepsilon_0)$ is in units of Nm^2 C^{-2} $I(\omega)$ is in units of J s^{-1} m^{-2}, or W m^{-2}. Setting $\frac{1}{2}(\omega - \omega_0)T = \pi/2$ (Duarte 2024) then

$$|D_{II}|^2 = \frac{8}{\pi} \left(\frac{\mu^2}{4\pi\varepsilon_0 c \hbar^2} \right) T^2 I(\omega_0) \tag{3.51}$$

$$|D_{II}|^2 = \frac{8}{\pi} \left(\frac{\sigma}{\hbar} \right) T^2 I(\omega_0) \tag{3.52}$$

where the cross section, in units of m^2, is given by

$$\sigma = \left(\frac{\mu^2}{4\pi\varepsilon_0 c \hbar} \right) \tag{3.53}$$

Alternatively, Feynman *et al* (1965) proceeds evaluating the integral of the sine function in equation (3.50), so that

$$|D_{II}|^2 = 2\pi^2 \left(\frac{\sigma}{\hbar} \right) T^2 I(\omega_0) \tag{3.54}$$

Setting

$$\kappa = \frac{\hbar}{\sigma} \qquad (3.55)$$

then, depending on the approach to evaluate equation (3.50) the peak intensity $I(\omega_0)$ can be expressed as

$$I(\omega_0) = \frac{\pi\kappa}{8} T^{-2} |D_{II}|^2 \qquad (3.56)$$

$$I(\omega_0) = \frac{\kappa}{2\pi^2} T^{-2} |D_{II}|^2 \qquad (3.57)$$

More in general,

$$I(\omega_0) = \eta\kappa T^{-2} |D_{II}|^2 \qquad (3.58)$$

where η is a numerical weight that depends on the approach selected to evaluate the sinusoidal function in equation (3.50). Here, the following notes are pertinent:
1. Sargent *et al* (1974) use $3^{-1/2}\mu$ rather than 1μ
2. σ is in units of m^2 rather than the more widely used unit of cm^2, as given in table 3.2
3. the units for the constant κ are J s m^{-2}

Using this cross section an expression for the gain can be stated as (Byer *et al* 1972)

$$g = \sigma NL \qquad (3.59)$$

3.4 Amplified spontaneous emission (ASE)

One of the beauties of organic dye lasers is that they are high-gain sources of coherent tunable radiation. Considering transverse high-power pulse laser excitation, as illustrated in figure 3.4, immediately following the arrival of the pump pulse, *amplified spontaneous emission* (ASE) is emitted in both directions orthogonal to the direction of the pump beam. This ASE has a broad spectral component and it mostly occurs in a well designed tunable narrow-linewidth laser oscillator (as illustrated in figure 3.5), prior

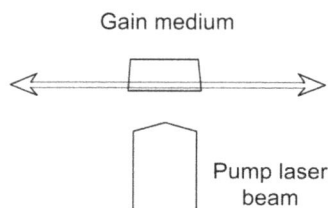

Gain medium

Pump laser
beam

Figure 3.4. Transverse excitation of a laser oscillator.

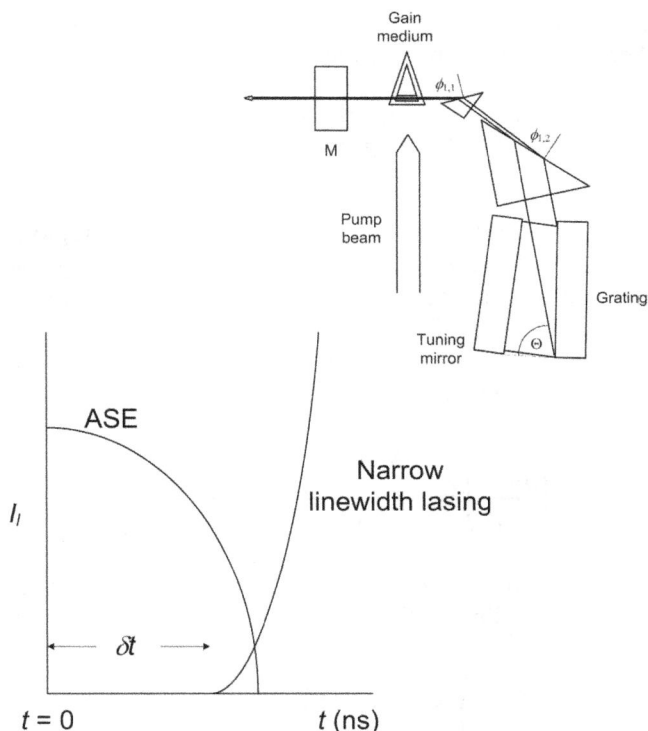

Figure 3.5. Pictorial representation of the temporal emission characteristics of ASE and narrow-linewidth emission in a well-designed narrow-linewidth laser oscillator (upper right inset).

to the onset of laser radiation, even though it might also persist as a low-intensity background emission during the emission of narrow-linewidth lasing. The task of a laser physicist is to keep this ASE, also known as optical noise or background radiation, as low as possible. It is important to minimize the ASE component at the oscillator stage of an oscillator–amplifier system to avoid amplification of the ASE at the amplification stages. In this section the focus in mainly on measurement aspects of ASE and in measures necessary to minimize its presence to enhance the spectral purity of the overall emission. Further detailed technical information on the subject of ASE can be found in Duarte and Piper (1980, 1984) and Duarte (1990).

Previously, the counter propagating laser intensities have been assigned an equation of the form

$$I_l(x, t, \lambda) = I_l^+(x, t, \lambda) + I_l^-(x, t, \lambda)$$

similarly, the pre-laser ASE radiation can be described as

$$I_{\mathrm{ASE}}(x, t, \lambda) = I_{\mathrm{ASE}}^+(x, t, \lambda) + I_{\mathrm{ASE}}^-(x, t, \lambda) \tag{3.60}$$

However, ASE can also occur during the time of narrow-linewidth laser emission. Therefore, it is useful to define the *fraction* of the background ASE (β) as the ratio of:

spectrally integrated energy in broadband ASE/energy in narrow-linewidth laser emission. In other words as defined in Duarte (1990)

$$\beta = \frac{\int_{\Lambda_1}^{\Lambda_2} W(\Lambda)d\Lambda}{\int_{\lambda_1}^{\lambda_2} E(\lambda)d\lambda} \qquad (3.61)$$

here, $W(\Lambda)$ is the broadband ASE energy in the wavelength region from Λ_1 to Λ_2, while $E(\lambda)$ is the narrow-linewidth laser emission in the wavelength segment from λ_1 to λ_2 which is full-width laser linewidth $\Delta\lambda$. The definition in equation (3.61) albeit useful, does not readily give information about the important quality of *spectral brightness*. Hence, it is necessary to introduce the concept of spectral energy density and define a figure of merit for the spectral purity of the laser emission (Duarte 1990)

$$\left(\frac{\rho_{ASE}}{\rho_l}\right) = \frac{(\Delta\Lambda)^{-1}\sum_{n=1}^{r} W(\Lambda)_n \Delta\Lambda_n}{(\Delta\lambda)^{-1}\int_{\lambda_1}^{\lambda_2} E(\lambda)d\lambda} \qquad (3.62)$$

Here, $\Delta\Lambda = |\Lambda_2 - \Lambda_1|$ is the total bandwidth if the broadband ASE radiation and $\Delta\lambda = |\lambda_2 - \lambda_1|$ is the laser linewidth at full width. As $\Delta\Lambda_n$ gets smaller and smaller

$$\left(\frac{\rho_{ASE}}{\rho_l}\right) \approx \frac{(\Delta\Lambda)^{-1}\int_{\Lambda_1}^{\Lambda_2} W(\Lambda)d\Lambda}{(\Delta\lambda)^{-1}\int_{\lambda_1}^{\lambda_2} E(\lambda)d\lambda} \qquad (3.63)$$

For a tunable narrow-linewidth multiple-prism Littrow (MPL) grating oscillator utilizing coumarin 500 as the active medium the ASE characteristics are illustrated in figure 3.6. The left side of figure 3.6 depicts the measured ASE for a laser with an *open cavity configuration*, that is, the laser output is taken from the reflection losses at one of the prisms. The much lower ASE depicted at the right side of figure 3.6 corresponds to a *closed cavity configuration* where the laser output is taken via the output couple mirror (Duarte and Piper 1980).

Measured β and (ρ_{ASE}/ρ_l) coefficients are given in table 3.4 for three high-performance narrow-linewidth oscillators.

Measures to reduce and severely minimize the ASE signature in tunable narrow-linewidth laser oscillators include:

1. The use of trapezoidal geometries at the gain medium.
2. The use of closed cavity configurations.
3. The use of relatively low molecular gain concentrations.
4. Avoiding parasitic broadband reflections back to the gain medium.

Further studies in ASE, relevant to the class of dispersive oscillator designs considered here, are provided by Berik *et al* (1985) and Nair and Dasgupta (1985). Other authors have studied the issue of ASE in distributed feedback lasers (Bor 1981, McIntyre and Dunn 1984).

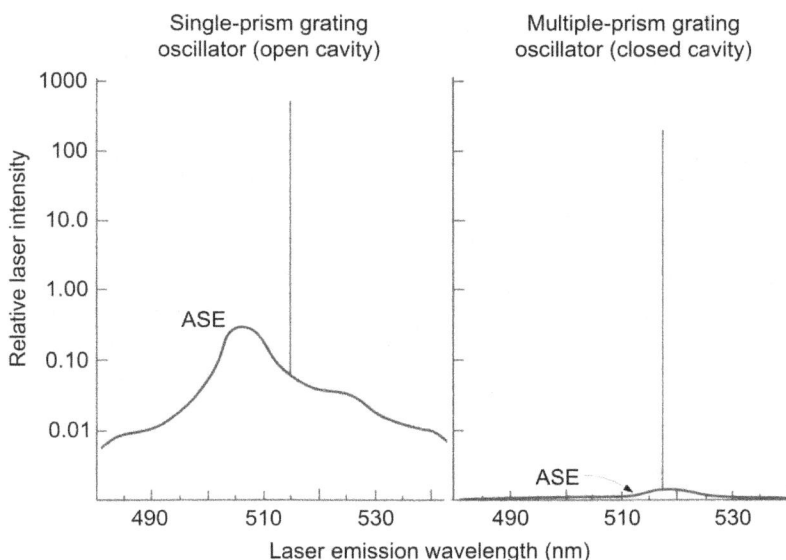

Figure 3.6. ASE emission characteristics of an open cavity laser configuration (left) and a closed-cavity configuration (right) (adapted from Duarte and Piper (1981); courtesy of the Optical Society). Please refer to chapter 5 for a description of open-cavity and closed-cavity designs.

Table 3.4. Amplified spontaneous emission (ASE) characteristics of dispersive laser oscillators.

Oscillator configuration	Laser linewidth	β	(ρ_{ASE}/ρ_l)	Reference
MPL[a]	$\Delta\nu \approx 1610$ MHz $\Delta\lambda \approx 0.0013$ nm @ $\lambda \approx 500$ nm		7.5×10^{-6}	Duarte and Piper (1980)
HMPGI[b]	$\Delta\nu \approx 400$ MHz $\Delta\lambda \approx 0.0004$ nm @ $\lambda \approx 580$ nm	< 0.010	5.0×10^{-7}	Duarte and Piper (1984)
MPL[a]	$\Delta\nu = 714$ MHz $\Delta\lambda \approx 0.0005$ @ $\lambda \approx 475$ nm	~ 0.003	7.6×10^{-8}	Duarte (1987)

[a] Multiple-prism Littow grating oscillator configuration.
[b] Hybrid multiple-prism grazing-incidence grating configuration.

ASE in oscillator–amplifier configurations has been studied by Ganiel *et al* (1975), McKee (1985) and Haag *et al* (1983). An observation of practical importance is that in pulsed lasers most of the ASE occurs in the immediate emission time segment prior to laser threshold. This knowledge can be applied to minimize the overall optical noise and maximize the spectral purity of the laser emission. In studies directly relevant to amplifiers, Duarte (1983) has measured the ASE single-pass broadband output in kW as a function of temperature for dyes such as rhodamine 6 G and coumarin 500. These measurements show that the laser output stays relatively constant until the 28 °C level is reached giving in to a pronounced decline in output power (Duarte 1983).

3.5 Quantum energy

The derivation of quantum transition probabilities (section 3.3), depends on the crucial assumption that $E = h\nu$ as introduced by Planck (1901) without derivation. Here, a heuristic approach to $E = h\nu$ is provided from an experimental perspective.

From empirical observations in quantum optics experiments utilizing ensembles of indistinguishable photons and CCD detectors, it has been observed that the measured intensity is proportional to the emission probability (Duarte 2022)

$$I \propto |\langle d|s \rangle|^2 \tag{3.64}$$

where the intensity I is in units of J s^{-1} m^{-2}. Working per meter square $I \rightarrow P$, so that

$$P \propto |\langle d|s \rangle|^2 \tag{3.65}$$

where the power P is in units of J s^{-1} or W. This relation can be expressed as

$$P = \zeta |\langle d|s \rangle|^2 \tag{3.66}$$

At $|\langle d|s \rangle|^2 = 1$, $P = \zeta$ so that the units of the parameter ζ are J s^{-1}. For a straight forward dimensional procedure we set $\zeta \rightarrow \xi\nu$ which allow us write

$$E = \frac{\xi\nu}{\nu} = \left(\frac{\xi}{\nu} \right) \cdot \nu \tag{3.67}$$

where (ξ/ν) becomes a new parameter η in units of J s as already suggested by Duarte (2024) so that

$$E = \eta \cdot \nu \tag{3.68}$$

What is this parameter η in units of J s? Some hints can be derived from observing that we are dealing with the propagation of quanta and the generation of electronic charge at the detector. Thus, parameters such as e, ε_0, and c must be involved. Drawing a parallel with the equations describing transition probabilities, see equation (3.51) for example, we can write

$$E = \gamma \left(\frac{e^2}{\varepsilon_0 c} \right) \cdot \nu \tag{3.69}$$

where $(e^2/\varepsilon_0 c)$ has units of J s and γ is just a number. Insertion of numerical values, limited to six decimal places, for e, ε_0, and c, yield

$$E \approx \gamma (9.670554 \times 10^{-36}) \cdot \nu \tag{3.70}$$

where it can be observed that $9.6705540 \times 10^{-36}$ J s is approximately 69 times smaller than the numerical value of Planck's constant h. Under closer scrutiny, using h as a reference, it is found that

$$\gamma \approx 68.517999 \approx \frac{\alpha^{-1}}{2} \tag{3.71}$$

where $\alpha^{-1} \approx 137.035999$ is the reciprocal of the dimensionless fine structure constant α. Using this approximation into equation (3.68) and (3.67) enables us to infer that

$$E \approx \left(\frac{e^2}{2\alpha\varepsilon_0 c}\right) \cdot \nu \tag{3.72}$$

and

$$E \approx h \cdot \nu \tag{3.73}$$

where

$$h \approx \frac{\alpha^{-1}}{2}(9.670554 \times 10^{-36}) \tag{3.74}$$

or

$$h \approx \left(\frac{e^2}{2\alpha\varepsilon_0 c}\right) \tag{3.75}$$

It should be noted that this approximation for h can also be deduced via Dirac's temporal ratios approach (Dirac 1938) as already indicated by Duarte (2022). Although this exercise provides a glimpse into the dependent entanglement of h and α, it does not explain the origin of these fundamental constants of Nature.

3.6 Problems

- 3.1 The simplified energetic system described in equations (3.8)–(3.12) is a more deterministic version of the original system described in equations (3.1)–(3.7). Since quantum cross sections are only available from manifold to manifold, as given in table 3.1, the simplification is justified. Can you think of a situation where the detailed mathematical description of the manifolds given in equations (3.1)–(3.7) is necessary?
  Note: you might leave this problem for the end.
- 3.2 Ignoring the triplet level, show that from equation (3.16) the gain can be expressed as equation (3.18).
- 3.3 Verify that the ratio $(N_1/N_0) \approx 0.001$ using the appropriate cross-sections given in table 3.2.
- 3.4 Verify that equation (3.27) can be abstracted into equation (3.28).
- 3.5 Verify that integration of equation (3.45) yields equation (3.46).
- 3.6 Using equation (3.46) verify that $|D_{II}|^2$ is given by equation (3.47).

- 3.7 Show that equation (3.48) follows from equation (3.47).
- 3.8 Show that integration of equation (3.50) yields $|D_{II}|^2$, as given in equation (3.54).

References

Berik E, Davidenko B, Mihkelsoo V, Apanasevish P, Grabchikov A and Orlovich V 1985 Stimulated Raman scattering of dye laser radiation in hydrogen: improvement of spectral purity *Opt. Commun.* **56** 283–7

Bor Z 1981 Amplified spontaneous emission from N_2-laser pumped dye laser *Opt. Commun.* **39** 383–9

Byer R L, Herbst R L, Kildal H and Levenson M D 1972 Optically pumped molecular iodine vapor-phase laser *Appl. Phys. Lett.* **20** 463–6

Diels J-C 1990 Femtosecond dye lasers *Dye Laser Principles* ed F J Duarte and L W Hillman (New York: Academic) ch 3

Diels J-C and Rudolph W 1996 *Ultrafast Laser Pulse Phenomena* (New York: Academic)

Dirac P A M 1978 *The Principles of Quantum Mechanics* 4th edn (London: Oxford University Press)

Dirac P A M 1938 A new basis for cosmology *Proc. R. Soc.* A **165** 199–208

Duarte F J 1983 Thermal effects in double-prism dye laser cavities *IEEE J. Quantum Electron.* **QE-19** 1345–7

Duarte F J 1987 Critical assessment of ASE in laser-pumped pulsed dye lasers *Proc. Int. Conf. on Lasers '86* ed R W McMillan (McLean, Va: STS Press) pp 416–19

Duarte F J 1990 Narrow-linewidth pulsed dye laser oscillators *Dye Laser Principles* ed F J Duarte and L W Hillman (New York: Academic) ch 4

Duarte F J (ed) 1991 *High-Power Dye Lasers* (Berlin: Springer)

Duarte F J 1994 Solid-state multiple-prism grating dye-laser oscillators *Appl. Opt.* **33** 3857–60

Duarte F J 1995a Dye lasers *Tunable Lasers Handbook* ed F J Duarte (New York: Academic) ch 5

Duarte F J 1995b Solid-state dispersive dye laser oscillator: very compact cavity *Opt. Commun.* **117** 480–4

Duarte F J 2003 *Tunable Laser Optics* (New York: Elsevier-Academic)

Duarte F J 2014 *Quantum Optics for Engineers* (New York: CRC)

Duarte F J 2015 *Tunable Laser Optics* (New York: CRC) 2nd edn

Duarte F J 2022 *Fundamentals of Quantum Entanglement* 2nd edn (Bristol: IOP Publishing)

Duarte F J 2024 *Quantum Optics for Engineers: Quantum Entanglement* 2nd edn (New York: CRC)

Duarte F J and Hillman L W (ed) 1990 *Dye Laser Principles* (New York: Academic)

Duarte F J and Piper J A 1980 A double-prism beam expander for pulsed dye lasers *Opt. Commun.* **35** 100–4

Duarte F J and Piper J A 1981 Prism preexpanded grazing-incidence grating cavity for pulsed dye lasers *Appl. Opt.* **20** 2113–6

Duarte F J and Piper J A 1984 Narrow-linewidth high-prf copper laser-pumped dye laser oscillators *Appl. Opt.* **23** 1391–94

Everett P N 1991 Flashlamp-excited dye lasers *High Power Dye Lasers* ed F J Duarte (Berlin: Springer) ch 5

Feynman R P, Leighton R B and Sands M 1965 *The Feynman Lectures on Physics* **vol III** (Reading, MA: Addison-Wesley)

Ganiel U, Hardy A, Neumann G and Treves D 1975 Amplified spontaneous emission and signal amplification in dye-laser systems *IEEE J. Quantum Electron.* **QE-11** 881–91

Haag G, Munz M and Marowski G 1983 Amplified spontaneous emission (ASE) in laser oscillators and amplifiers *IEEE J. Quantum Electron.* **QE-19** 1149–60

Haken H 1970 *Light and Matter* (Berlin: Springer)

Hargrove R S and Kan T K 1980 High power efficient dye amplifier pumped by copper vapor lasers *IEEE J. Quantum Electron.* **QE-16** 1108–13

Hillman L W 1990 Laser dynamics *Dye Laser Principles* ed F J Duarte and L W Hillman (New York: Academic) ch 2

Hollberg L 1990 CW dye lasers *Dye Laser Principles* ed F J Duarte and L W Hillman (New York: Academic) ch 5

Holzer W, Gratz H, Schmitt T, Penzkofer A, Costela A, Garca-Moreno I, Sastre R and Duarte F J 2000 Photo-physical characterization of rhodamine 6 G in a 2-hydroxyelthyl-methacrylate methyl-methacrylate copolymer *Chem. Phys.* **256** 125–36

Jensen C 1991 Pulsed dye laser gain analysis and amplifier design *High Power Dye Lasers* ed F J Duarte (Berlin: Springer) ch 3

Jones G II 1990 Photochemistry of laser dyes *Dye Laser Principles* ed F J Duarte and L W Hillman (New York: Academic) ch 7

McIntyre I A and Dunn M H 1984 Amplified spontaneous emission in distributed feedback dye lasers *Opt. Commun.* **50** 169–72

McKee T J 1985 Spectral-narrowing techniques for excimer laser oscillators *Can. J. Phys.* **63** 214–9

Nair L G and Dasgupta K 1985 Amplified spontaneous emission in narrow-band pulsed dye laser oscillators—theory and experiment *IEEE J. Quantum Electron.* **QE-21** 1782–94

Planck M 1901 Ueber das gesetz der energieverteilung im normalspectrum *Ann. Phys.* **4** 553–63

Sargent M, Scully M O and Lamb W E 1974 *Laser Physics* (Reading, MA: Addison Wesley)

Schäfer F P (ed) 1990 *Dye Lasers* 2nd edn (Berlin: Springer)

Teschke O, Dienes A and Whinnery J R 1976 Theory and operation of high-power CW and long-pulse dye lasers *IEEE J. Quantum Electron.* **QE-12** 383–95

IOP Publishing

Organic Lasers and Organic Photonics (Second Edition)

F J Duarte

Chapter 4

Polymer matrices for lasers

F J Duarte and K M Vaeth

Properties of polymer matrices for solid-state organic lasers are discussed in terms of molecular structure relevant physical parameters and $\partial n/\partial T$. cross sections and rates are given for rhodamine 6G-doped HEMA-MMA.

4.1 Introduction

In this chapter a brief and pragmatic introduction to the main polymers utilized in laser and photonic applications is provided. The emphasis is primarily from a physical-engineering perspective. For further details the reader should refer to the comprehensive review of Costela *et al* (2016).

Poly(methyl methacrylate), also known as PMMA, whose chemical nomenclature is $(C_5O_2H_8)_n$, was registered in the 1930s under the commercial name of Perspex. It is also known as Lucite, Acryle, and Pexiglass.

PMMA, whose molecular structure is depicted in figure 4.1, has the distinction of being the first polymer to be used in photonics applications. In this regard, PMMA was doped with the laser dye rhodamine 6 G to create the very first organic gain medium for laser applications: optically pumped solid-state organic dye lasers were introduced by Soffer and McFarland (1967) and Peterson and Snavely (1968). It was the early days of lasers.

Monomers utilized as matrix constituents in solid-state dye lasers include methyl methacrylate (MMA) and 2-hydroxyethyl methacrylate (HEMA). The combination of HEMA-MMA is another matrix that becomes a gain medium when doped with an organic laser dye such as rhodamine 6 G (Costela *et al* 2016). Table 4.1 lists chemical formulae and molecular weights for the three organic laser matrices already mentioned.

The molecular structures of these monomers are shown in figure 4.2.

doi:10.1088/978-0-7503-5547-6ch4

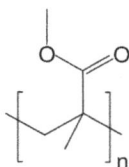

Figure 4.1. Molecular structure of PMMA.

Table 4.1. Monomers and polymers for organic laser matrices.

Monomer or polymer	Chemical formula	Molecular weight (amu)
HEMA	$C_6O_3H_{10}$	130.14
MMA	$C_5O_2H_8$	100.12
PMMA	$(C_5O_2H_8)_n$	

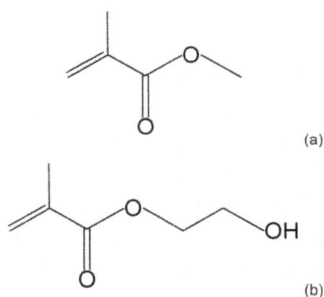

(a)

(b)

Figure 4.2. Molecular structure of (a) MMA and (b) HEMA.

4.2 Physical parameters of PMMA

Optical–thermal property parameters of PMMA relevant to organic laser perform-ance are listed in table 4.2.

4.3 Polymer matrices for organic lasers

Polymer matrices utilized in solid state laser applications are given by Costela *et al* (2016) and include:

1. 2-Hydroxyethyl methacrylate methyl methacrylate (HEMA-MMA);
2. 2-Hydroxyethyl methacrylate pentaerythritol tetraacrylate (HEMA-PETRA);
3. 2-Hydroxyethyl methacrylate pentaerythritol triacrylate (HEMA-PETA);
4. 2-Hydroxyethyl methacrylate trimethylolpropane (HEMA-TMPTMA);
5. Modified poly(methyl methacrylate) (MPMMA).

Table 4.2. Physical–optical parameters for PMMA.

Parameter	Value	Reference
n^a	1.492	Wunderlich (1989)
$\partial n/\partial T$	$\sim 1.1 \times 10^{-4}$ K^{-1}	Wunderlich (1989)
Heat capacity[b]	1.42 kJ kg^{-1} K^{-1}	Brandrup et al (1975)
Coeff. thermal exp. (linear)[c]	$\sim 7 \times 10^{-5}$ K^{-1}	Brandrup et al (1975)
Thermal conductivity[c]	0.193 W m^{-1} K^{-1}	Brandrup et al (1975)
Coefficient of friction	0.30–0.40	Nuño et al (2006)
Density[b]	1.188×10^3 kg m^{-3}	Brandrup et al (1975)
Poisson's ratio	0.34–0.40	Bushan and Burton (2005)
Shear modulus[d]	1.7 GPa	Brandrup et al (1975)
Tensile strength	$\sim$40 MPa	Vallittu (2009)
Young's modulus	5.0 GPa	Bushan and Burton (2005)

[a] At $\lambda = 589$ nm.
[b] At 298 K.
[c] 273–323 K.
[d] At 298 K; dynamic measurement.

In the case of HEMA it should be noted that the designations of 2-hydroxyethyl methacrylate and (hydroxyethyl) methacrylate are equivalent. Besides the type of organic laser matrices listed above, biological matrices have also been demonstrated as a laser host material, as described in chapter 10 (Vasileva et al 2017).

A particularly successful form of PMMA, for laser applications, is modified PMMA or MPMMA which was used in the first single-transverse-mode single-longitudinal-mode tunable organic lasers (Duarte 1994, 1995). This polymer was made under a proprietary method (Maslyukov et al 1995) and its optical-laser characteristics are summarized as follows:

1. Surface quality better than $\lambda/4$ (Duarte 1994);
2. Surface damage threshold better than 13 J cm^{-2} (Gromov et al 1985, Dyumaev et al 1992);
3. Laser dye bleaching threshold $\sim$1 J cm^{-2} (Duarte 1994).

One extraordinary quality of laser organic dye-doped MPMMA, which might also manifest itself in homologous gain media, is the fact that when dye bleaching takes place at a particular site, then over time, the dye from the matrix diffuses back into the vacant site thus enacting a process of 'self healing' (Duarte 1997, Duarte and James 2016).

Optical–thermal parameters important to organic laser dye-doped polymer matrices are listed in table 4.3.

As seen from table 4.3 the $\partial n/\partial T$ coefficient for laser dye-doped polymer (DDP) matrices is in the $|1.3 \times 10^{-4}| \leqslant \partial n/\partial T \leqslant |1.4 \times 10^{-4}|$ range which leads to mildly higher than desired laser beam divergences and the presence of thermal lensing effects at the gain medium. An alternative avenue to atone for this divergent effect is

Table 4.3. Refractive index and $\partial n/\partial T$ for rhodamine 6 G-doped polymers.

Dye-doped polymer matrix	n^{a}	$\partial n/\partial T^{\mathrm{b}}$ (K^{-1})	Reference
MPMMA	1.4943	$(-1.4 \pm 0.2) \times 10^{-4}$	Duarte *et al* (2000)
HEMA-MMA	1.5039	$(-1.3 \pm 0.2) \times 10^{-4}$	Duarte *et al* (2000)
Bz-MA HEMA-MMA		$(-1.4 \pm 0.3) \times 10^{-4}$	Duarte *et al* (2000)

[a] Measured at $\lambda = 593.93$ nm and $T = 297$ K.
[b] Measured in the $297 \leqslant T \leqslant 337$ K.

Table 4.4. Excitation parameters for rhodamine 6 G-doped HEMA-MMA matrix.[a]

Parameter	Cross section (cm^2)	Measurement wavelength (nm)	Time (s)
$\sigma_{0,1}$	3.60×10^{-16}	527	
σ^e	3.0×10^{-16}	565	
$\sigma^l_{1,2}$	3.0×10^{-17}		
$\tau^{1,0}$			3.3×10^{-9}
$\tau_{n,1}{}^{\mathrm{b}}$			60×10^{-15}

[a] Measurements by Holzer *et al* (2000)
[b] Time decay from higher electronic states, see chapter 3.

to employ dye-doped organic-inorganic gain media, in particular laser dye-doped polymer-nanoparticle (PN) gain media. This approach was taken by Duarte and James (2003) who found that the new laser dye-doped PN gain media had a marked influence in minimizing beam divergence to

$$\Delta\theta \approx 1.3\left(\frac{\lambda}{\pi w}\right) \tag{4.1}$$

that is ~1.3 times the diffraction limit. This was accomplished by a reduction in the $\partial n/\partial T$ coefficient characterized by (Duarte and James 2003)

$$\left|\frac{\partial n}{\partial T}\right|_{\mathrm{DDP}} \approx 1.83 \left|\frac{\partial n}{\partial T}\right|_{\mathrm{PN}} \tag{4.2}$$

In other words, the $|\partial n/\partial T|$ for the PN gain media is almost half the corresponding coefficient for the pure polymer gain media. Further details on this subject are given in chapter 9.

4.3.1 Excitation parameters for dye-doped polymer laser matrices

Table 4.4 lists excitation parameters measured in rhodamine 6 G in a 2-hydrox-yethyl-methacrylate methyl-methacrylate copolymer gain media by Holzer *et al* (2000) at a dye concentration of 0.1 mM. These results are directly relevant to the long-pulse tunable narrow-linewidth solid-state organic dye laser described by Duarte *et al* (1998).

4.4 Longevity of polymer matrices for organic lasers

Modified poly(methyl mathacrylate) matrices doped with rhodamine 6 G, in trapezoid geometry, as reported by Duarte (1994, 1999) have survived nearly three decades with no signs of degradation whatsoever despite having been used in numerous experiments. One of these matrices is depicted in figure 4.3 as photographed in 2023. A similar triumphant history of preservation can be documented in regard to rhodamine 6 G doped HEMA: MMA matrices utilized by Duarte et al (1998) in narrow-linewidth long-pulse tunable oscillators. The matrix used in those experiments is shown in figures 4.4 and 4.5. Figure 4.4 shows one of the exit optical surfaces while figure 4.5 depicts the excitation optical surface.

4.5 Problems

- 4.1 Given that the measured surface damage energy threshold for MPMMA is reported to be ~ 13 J cm^{-2}, find a transparent inorganic optical material with comparable or higher surface damage energy threshold.
- 4.2 Given that the measured surface damage energy threshold for MPMMA is reported to be ~ 13 J cm^{-2}, find an inorganic crystal laser material with comparable or higher surface damage energy threshold.
- 4.3 Given that the measured bleaching energy threshold for dye-doped MPMMA is reported to be ~ 1 J cm^{-2}, find an inorganic crystal laser material with comparable or higher bleaching energy threshold. This crystal laser material should yield tunable laser emission around $\lambda \approx 590$ nm.

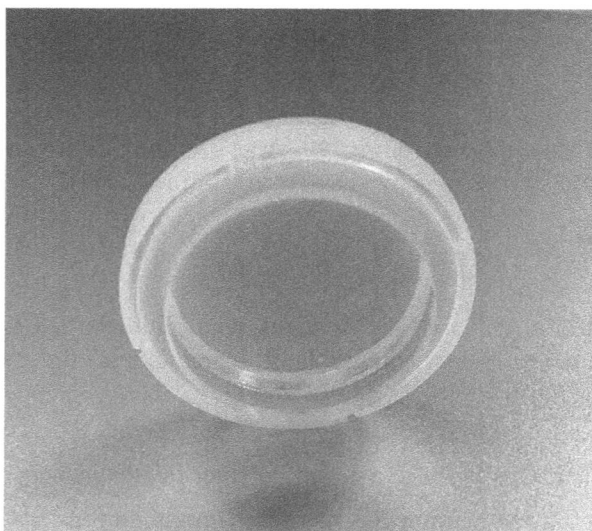

Figure 4.3. Rhodamine 6 G doped MPMMA laser matrix, in trapezoidal geometry, as utilized by Duarte (1994, 1999). In nearly three decades this matrix had been used in numerous laser experiments and configurations and continues to survive unblemished. This photograph was captured for the updated edition of this book in 2023.

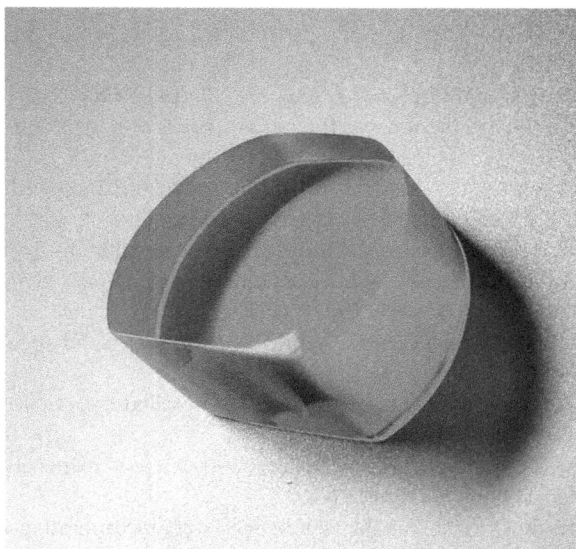

Figure 4.4. Rhodamine 6 G doped HEMA:MMA laser matrix as utilized by Duarte *et al* (1998) in narrow-linewidth long-pulse laser oscillator experiments. This image shows one of the laser exit optical surfaces. This photograph was captured for the updated edition of this book in 2023.

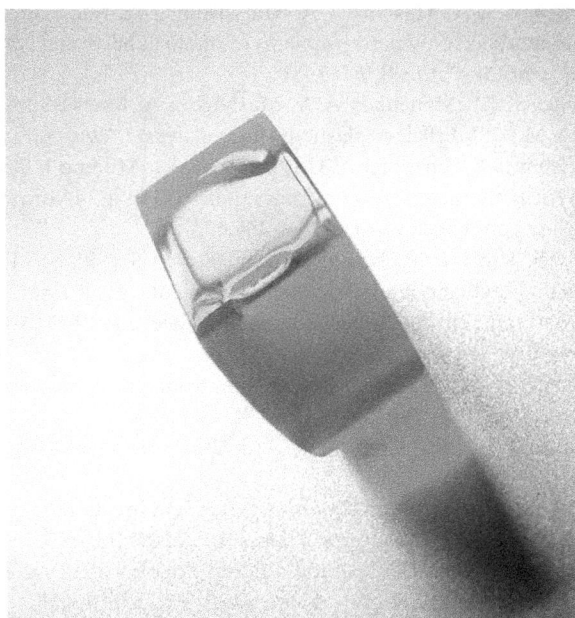

Figure 4.5. Rhodamine 6 G doped HEMA:MMA laser matrix as utilized by Duarte *et al* (1998) in narrow-linewidth long-pulse laser oscillator experiments. This image shows the laser excitation surface. At the lower part of the excitation surface and at the upperright vertex thermal distortions are visible due to the use of excess pump-energy in initial experiments. This photograph was captured for the updated edition of this book in 2023.

References

Brandrup J, Immergut E H and Grulke E A (ed) 1975 *Polymer Handbook* (New York: Wiley)

Bushan B and Burton Z 2005 Adhesion and friction properties of polymers in microfluidic devices *Nanotech.* **16** 467–78

Costela A, García-Moreno I and Sastre R 2016 Solid-state organic dye lasers *Tunable Laser Applications* ed F J Duarte (New York: CRC) 3rd edn ch 3

Duarte F J 1994 Solid-state multiple-prism grating dye-laser oscillators *Appl. Opt.* **33** 3857–60

Duarte F J 1995 Solid-state dispersive dye laser oscillator: very compact cavity *Opt. Commun.* **117** 480–4

Duarte F J 1997 Multiple-prism near-grazing-incidence grating solid-state dye-laser oscillator *Opt. Laser Tech.* **29** 513–6

Duarte F 1999 Multiple-prism grating solid-state dye laser oscillator: optimized architecture *Appl. Opt.* **38** 6347–9

Duarte F J, Costela A, García-Moreno I and Sastre R 2000 Measurements of $\partial n/\partial T$ in solid-state dye-laser gain media *App. Opt.* **39** 6522–3

Duarte F J and James R O 2003 Tunable solid-state lasers incorporating dye-doped polymer-nanoparticle gain media *Opt. Lett.* **28** 2088–90

Duarte F J and James R O 2016 Organic dye-doped polymer-nanoparticle tunable lasers *Tunable Laser Applications* ed F J Duarte (New York: CRC) 3rd edn ch 4

Duarte F J, Taylor T S, Costela A, García-Moreno I and Sastre R 1998 Long-pulse narrow-linewidth dispersive solid-state dye-laser oscillator *App. Opt.* **37** 3987–9

Dyumaev K M, Manenkov A A, Maslyukov A, Matyushin G A, Nechilaito V S and Prokhorov A M 1992 Dyes in modified polymers: problems of photostability and conversion efficiency at high intensities *J. Opt. Soc. Am.* B **9** 143–51

Gromov D A, Dyumaev K M, Manenkov A A, Maslyukov A, Matyushin G A, Nechilaito V S and Prokhorov A M 1985 Efficient plastic-host dye lasers *J. Opt. Soc. Am.* B **2** 1028–31

Holzer W, Gratz H, Schmitt T, Penzkofer A, Costela A, Garca-Moreno I, Sastre R and Duarte F J 2000 Photo-physical characterization of rhodamine 6 G in a 2-hydroxyelthyl-methacrylate methyl-methacrylate copolymer *Chem. Phys.* **256** 125–36

Maslyukov A, Sokolv S, Kailova M, Nyholm K and Popov S 1995 Solid-state dye laser with modified poly(methyl methacrylate)-doped active elements *Appl. Opt.* **34** 1516–8

Nuño N 2006 Static coefficient of friction between stainless steel and PMMA used in cemented hip and knee implants *Clin. Biomech.* **21** 956–62

Peterson O G and Snavely B B 1968 Stimulated emission from a flashlamp-excited organic dyes in polymethyl mehacrylate *Appl. Phys. Lett.* **12** 238–40

Soffer B H and McFarland B B 1967 Continuously tunable narrow-band organic dye lasers *Appl. Phys.* **10** 266–7

Vallittu P K 2009 Some aspects of tensile strength of unidirectional glass fibre-polymethyl methacrylate composite used in dentures *J. Oral Rehab.* **25** 100–5

Vasileva E, Li Y, Sychugov I, Mensi M, Berglund L and Popov S 2017 Lasing from organic dye molecules embedded in transparent wood *Adv. Opt. Mat.* **5** 1700057

Wunderlich W 1989 Physical constants of poly(methyl methacrylate) *Polymer Handbook* (New York: Wiley) pp 77–80

IOP Publishing

Organic Lasers and Organic Photonics (Second Edition)

F J Duarte

Chapter 5

Cavity and resonator architectures for high-performance organic laser oscillators

F J Duarte

Laser cavity and resonator architectures for high-performance narrow-linewidth lase oscillators are described and discussed in detail. Emphasis is given to laser oscillators yielding single-transverse-mode and single-longitudinal-mode oscillation. Multiple-prism grating cavities oscillating near the limit allowed by Heisenberg's uncertainty principle $\Delta \nu \Delta t \approx 1.05$ are described in detail. Unstable resonators for amplifier stages are also considered.

5.1 Introduction

In this chapter, various cavity configurations are reviewed. The emphasis is pragmatic rather than historical, or chronological, and only the cavities that have led directly, or serve as a transitional stage, to high-performance laser configurations are considered.

By high-performance is meant a laser output characterized by:

1. Single-transverse-mode (TEM_{00}) emission.
2. Near diffraction-limited beam divergence.
3. Tunable narrow-linewidth emission, preferable single-longitudinal-mode (SLM).
4. Low amplified spontaneous emission (ASE).

In this regard, sections 5.2 and 5.3 belong mainly in the transitional category and only the cavity architectures described in section 5.4 are intrinsically high-performance.

At this stage it is appropriate to mention that the terms laser cavity and laser resonators are nearly synonymous. At a fundamental level a laser cavity can be thought of as the reflective elements, for instance the two mirrors, necessary to accomplish laser emission while a laser resonator can be envisioned as the two mirrors plus the gain medium in between. However, most laser physicists tend to use these two terms interchangeably.

A laser oscillator is a laser resonator that has the ability to emit low divergence tunable narrow-linewidth emission. By narrow-linewidth emission is meant pulsed emission for whose the product of its pulse temporal duration (Δt) by its linewidth in frequency units ($\Delta \nu$) is nearly unity, that is

$$\Delta \nu \Delta t \approx 1 \qquad\qquad (5.1)$$

that is an alternative for Heisenberg's uncertainty principle. In other words, for a given pulse duration there is a limit on how narrow the laser linewidth can be. For example, for a tunable laser oscillator with a pulse duration $\Delta t \approx 3$ ns, at approximately full-width half-maximum (FWHM), its laser linewidth is measured to be $\Delta \nu \approx 350$ MHz (FWHM) thus yielding $\Delta \nu \Delta t \approx 1.05$ (Duarte 1999). In other words, the output of this oscillator is said to be nearly limited by Heisenberg's uncertainty principle. The beauty is that the peak power made available by this emission is ~ 33 kW thus providing an enormous spectral brightness.

The various laser cavities and resonators described in this chapter are applicable to any gain medium in the gas, liquid, or solid state. They are also applicable to lasers excited by optical or electrical means. In the case of optically-pumped lasers they are apt for either longitudinal or transverse excitation. Albeit the word organic is used in the title, the cavity configurations described here apply equally well to crystalline, inorganic, or organic–inorganic gain media.

In section 5.6 a class of unstable resonator used in tunable organic dye laser design is described. This is particularly relevant since these resonators can be used in the amplification stage of oscillator–amplifier configurations (Duarte and Conrad 1987). Here, it should also be indicated that multiple-prism grating oscillators could also, under certain conditions, be classified as unstable resonators (Duarte *et al* 1997, Duarte 2001).

In chapter 6, the mathematical physics applicable to the laser cavities and resonators described here is described from an engineering non-derivational perspective. Chapter 7 summarizes the reported best performance of various tunable organic dye lasers and organic lasers in terms of key parameters such as efficiency, beam divergence, and laser linewidth.

Finally, only intracavity prismatic beam expansion is considered since two-dimensional telescopic beam expansion adds to the cavity length and introduces thermal variability (Duarte and Piper 1980, Duarte 1990a). In addition, tunable narrow-linewidth oscillators including an intracavity interferometer, or etalon, in addition to an intracavity beam expander and diffraction grating are not included in this discussion since the intracavity etalon adds an extra degree of complexity and tuning limitations. Besides, the all important goal of reaching single-longitudinal-mode oscillation can be accomplished without having to resort to including an intracavity interferometer (Duarte 1999). Detailed descriptions on the theory, design, and use of intracavity etalons in tunable lasers can be found in Duarte (2003, 2015).

5.2 Mirror–mirror cavities

The most basic and most pervasive laser cavity, or laser resonator, is the mirror–mirror cavity. In this cavity generally one mirror is a total reflector while the other mirror is a partial reflector or *output coupler*. This cavity usually also incorporates an intracavity aperture to control the transverse mode structure. In well designed laser optically-pumped laser resonators the aperture is configured by the tight dimensions of the focused pump beam. As quantified exactly via the N-slit interferometric equation introduced in chapter 6 (section 6.2) the ratio of the aperture radius (w) to cavity length (L) determines the number of transverse modes and the necessary dimensions to achieve single-transverse-mo (TEM$_{00}$). In this regard, a very useful expression that can be used as an initial step is the Fresnel Number (Siegman 1986)

$$N_F = \left(\frac{w^2}{L\lambda} \right) \qquad (5.2)$$

where λ is the wavelength of emission. For TEM$_{00}$ this number should be at most $N_F \approx 1$ and ideally $N_F \leqslant 1$.

As observed in figure 5.1 the laser resonator is composed of the two mirrors and the gain medium. An intracavity aperture might be used if the focusing of the excitation beam does not confine the emission to a single-transverse mode. In the case of optically-pumped organic lasers, the gain medium can be excited either transversely (figure 5.1) or longitudinally (figure 5.2). At this stage it is important to mention that for best results the windows of the gain medium should be at an angle, in a trapezoidal geometry (Duarte and Piper 1980), relative to the propagation axis to avoid uncontrolled broadband laser emission. This detail becomes very important in suppressing unwanted ASE, as emphasized in chapter 3.

Although an intracavity aperture can always be deployed, it is much more preferable to confine the dimensions of the emission beam by the focusing of the excitation or pump-laser. The dimensions of the diameter of the output beam for this class of lasers is in the $100 \leqslant 2w \leqslant 300$ μm range, where w is the beam's radius of the beam at the gain medium (see, for example, Duarte and Piper 1984).

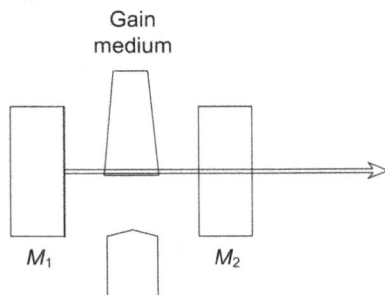

Figure 5.1. Basic optically-pumped organic laser resonator comprised of a mirror–mirror cavity. The excitation is performed transversely.

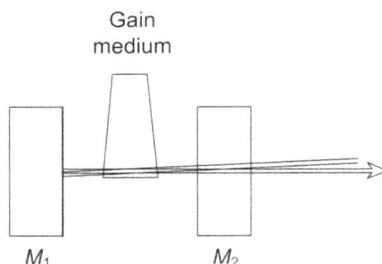

Figure 5.2. Basic optically-pumped organic laser resonator comprised of a mirror–mirror cavity. The excitation is performed longitudinally.

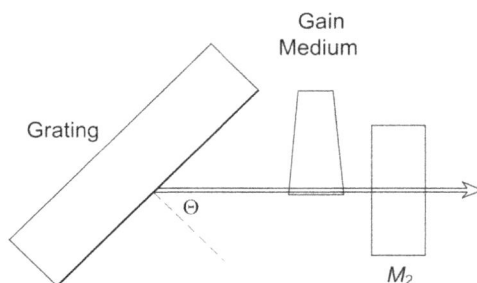

Figure 5.3. Mirror–grating cavity with the grating deployed in a Littrow configuration.

5.3 Mirror-grating cavities

The mirror–mirror cavities considered in the previous section do not offer optical means to tune the laser emission. Under those circumstances wavelength tuning requires major physical changes such as changing the pump laser, varying the temperature of the active medium, or varying the concentration of the active species. Even then, tuning is limited.

The full-potential of wavelength tunability can be attained via the use of intracavity tuning elements such as etalons, prisms, and diffraction gratings, or a combination of these (see Duarte 2015 for a detailed technical discussion). Given the pragmatic style of this chapter it is mentioned from the onset that diffraction gratings offer the largest dispersion power as compared to the previously mentioned optics and therefore are the element of choice for wavelength tuning in various classes of tunable lasers.

5.3.1 Littrow grating cavities

A quintessential mirror–grating cavity with its grating deployed in a Littrow configuration is depicted in figure 5.3. In a grating mounted in Littrow configuration the angle of incidence (Θ) equals the angle of refraction (Φ) so that $\Theta = \Phi$. Tunability of the laser radiation is accomplished via rotation of the grating, about its center, on the axis perpendicular to the cavity propagation axis. The emission wavelength is thus geometrically controlled according to the Littrow grating equation (see chapter 6)

$$m\lambda = 2d \sin \Theta \qquad (5.3)$$

where m is the order of diffraction and d is the groove density of the grating. Usually gratings for the visible spectrum have 600–4000 lines/mm and are 5–15 cm in their diffractive width. For example, for a grating with 1800 lines/mm, deployed in first order ($m = 1$) at a wavelength of $\lambda = 590$ nm the angle of incidence becomes $\Theta \approx 32.07°$.

As the grating tunes continuously from one wavelength region to another the *free spectral range* (*FSR*), or the longitudinal mode spacing, of the cavity changes according to

$$\text{FSR} = \left(\frac{\lambda^2}{2L}\right) \qquad (5.4)$$

in wavelength units and according to

$$\text{FSR} = \left(\frac{c}{2L}\right) \qquad (5.5)$$

in frequency units. This means that in order to maintain the intracavity longitudinal mode spacing constant, and thus avoid *mode hopping*, the cavity length must be adjusted in synchronicity with the rotation of the diffraction grating. This is known as *synchronous tuning*. For further details in various wavelength tuning techniques the reader is referred to Duarte (2015).

For a Littrow cavity in the absence of intracavity beam expansion the single-pass laser linewidth is given by (Duarte 1992)

$$\Delta\lambda \approx \Delta\theta \left(\frac{\partial\Theta}{\partial\lambda}\right)^{-1} \qquad (5.6)$$

where $\Delta\theta$ is the beam divergence and $(\partial\Theta/\partial\lambda)$ is the grating's dispersion

$$\left(\frac{\partial\Theta}{\partial\lambda}\right) = \left(\frac{2\tan\Theta}{\lambda}\right) \qquad (5.7)$$

In these equations the diameter of illumination on the diffraction grating is determined by the radius of the intracavity beam, or beam waist w, introduced via the beam divergence equation that in its limiting, and ideal, form is given by

$$\Delta\theta \approx \left(\frac{\lambda}{\pi w}\right) \qquad (5.8)$$

Further details on these equations are given in chapter 6. Useful references in interferometry are those of Steel (1967) and Born and Wolf (1999). A good reference on diffraction gratings is Meaburn (1976) while a comprehensive review of

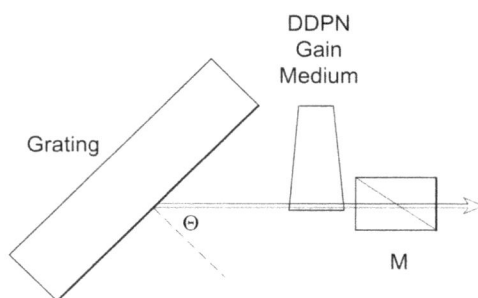

Figure 5.4. Output-coupler Glan–Thompson polarizer-grating cavity with the grating deployed in a Littrow configuration.

interferometric and diffraction principles applied to laser cavities is given by Duarte (2015).

A variant of the mirror–grating grating cavity derived from high-performance tunable laser oscillators (Hänsch 1972, Duarte *et al* 1991, Duarte 1999) is the replacement of the output-coupler mirror by a specially adapted Glan–Thompson polarizer, as illustrated in figure 5.4. The Glan–Thompson polarizer has its inner surface antireflection coated while the outer surface, that serves as the traditional output-couple mirror is coated with partial reflectivity, usually in the 10–30% range (Duarte and James 2003, Duarte *et al* 2006). The mission of the output-couple polarizer is to minimize ASE and to reinforce the polarization of the laser emission parallel to the plane of incidence.

5.3.2 Grazing-incidence grating cavities

One limitation of laser resonators incorporating gratings in Littrow configuration, in the absence of intracavity beam expansion, is their limited ability to yield narrow-linewidth lasing and their vulnerability to optical damage due to the incidence, on their diffracvive surface, of high-intracavity power densities. These limitations, in cavities not including intracavity beam expansion, are overcome by adding an extra component, a tuning mirror, and deploying the grating at a very high-angle of incidence (Shohan *et al* 1977, Littman and Metcalf 1978) as depicted in figure 5.5. In these cavities the laser linewidth is controlled via the diffraction grating equation

$$d \left(\sin \Theta \pm \sin \Phi \right) = m\lambda \tag{5.9}$$

while the grating dispersion is given by

$$\left(\frac{\partial \Theta}{\partial \lambda} \right) = \left(\frac{\sin \Theta \pm \sin \Phi}{\lambda \cos \Theta} \right) \tag{5.10}$$

Since the laser linewidth is given by

$$\Delta \lambda \approx \Delta \theta \left(\frac{\partial \Theta}{\partial \lambda} \right)^{-1}$$

Grating

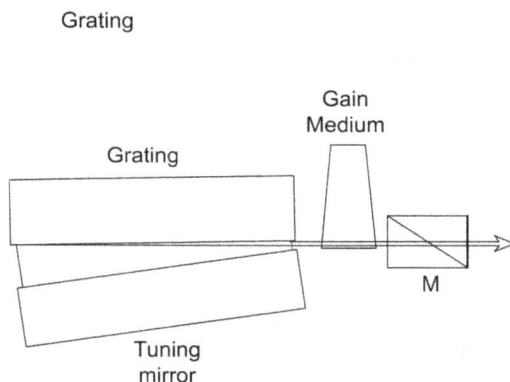

Figure 5.5. Mirror–grating cavity with the grating deployed at grazing-incidence. The mirror is an output-coupler Glan–Thompson polarizer.

it can be easily seen that as the angle of incidence gets very large, that is $\Theta \to 90°$, the grating dispersion $(\partial\Theta/\partial\lambda)$ gets very large and the laser linewidth $\Delta\lambda$ gets very small.

Grazing-incidence cavities have become widely used, especially in the semiconductor laser area. In this regard, the designer should be aware of the following:

1. The efficiency of a grating to diffract, at high angles of incidence, decreases considerably (Duarte and Piper 1981) so that the efficiency of narrow-linewidth tunable lasers incorporating diffraction gratings at grazing-incidence is rather low.

2. Deployment of diffraction gratings at grazing-incidence can lead to relatively large amounts of uncontrollable broadband ASE, which can be particularly detrimental to efforts to efficiently inject amplifier stages in oscillator–amplifier systems.

3. Although the laser linewidth performance of a laser oscillator configured with a grazing-incidence grating cavity is far superior to that of a Littrow grating cavity, in the absence of intracavity beam expansion, the grazing-incidence configuration requires an extra mirror. This aspect is considered further in the next section.

5.3.3 Open cavity versus closed cavity configurations

An *open cavity* is defined as a laser cavity in which the laser output is coupled via the reflection losses of an intracavity element such as a prism or diffraction grating. On the other hand, a closed cavity is defined as a laser cavity in which the laser emission is coupled along the laser cavity axis through an output couple mirror. All the laser cavities, or laser resonators considered in this chapter, belong to the closed cavity category. The advantages of closed cavities over open cavities have been discussed in detail in the published literature (Duarte and Piper 1980, 1981, Duarte 1990a) and include:

1. Significantly lower ASE.
2. Minimization of optical coupling risks with external optics via back reflections.
3. A higher degree of polarization as defined by the intracavity optics.

The ASE observations made here not only apply to both, Littrow-grating cavities and grazing-incidence grating cavities, but to purely prismatic cavities and laser cavities and resonators in general.

Please refer to chapter 3 for further information on ASE that includes spectral measurements that beautifully illustrate the massive advantage of closed-cavity designs to yield high spectral purity narrow-linewidth emission. It should be emphasized that this ultra-important aspect of laser design, that is, the need to generate low-noise high-spectral purity laser emission, is ignored and not even mentioned in most of the contemporaneous literature of laser emission. A glimpse of this situation is discretely revealed in chapter 7.

5.4 Output-coupler polarizer multiple-prism grating oscillators

In this section high-performance tunable narrow-linewidth laser oscillators are described. In general, these oscillators are of the multiple-prism grating category and they have been demonstrated with a plethora of laser gain media in the gas, the liquid, and the solid state. Although the emphasis of this book is tunable organic lasers, it should be mentioned that a recent review describing the application of these laser oscillator architectures to high-power gas lasers, on the one hand, and semiconductors lasers, at the other extreme, is given by Duarte (2015).

In terms of narrow linewidth performance, tunability, and low ASE content, the organic lasers described in this section remain as *best in class*.

Finally, it is appropriate to remind the reader that in this chapter the focus is on laser oscillator architecture. The mathematical-physics aspects are presented in chapter 6 while a broadly based documented organic laser performance review, mostly in tabular form, is given in chapter 7.

5.4.1 Hybrid multiple-prism grazing-incidence (HMPGI) grating configurations

Inspired by the dilemma of either low ASE and low efficiency or higher efficiencies and high ASE, prismatic beam expansion was introduced into the realm of grazing-incidence cavity configurations (Duarte and Piper 1981). A hybrid multiple-prism grazing-incidence (HMPGI) grating cavity is a cavity in which intracavity prismatic pre-expansion is introduced to reduce the incidence angle on the diffraction grating and thus bring the grating to an operational angle where the diffraction efficiency is much higher (Duarte 1990a). For example, a pure grazing-incidence grating cavity, configured in a closed configuration, incorporating a diffraction grating with 1800 lines/mm deployed at an angle of incidence of $\Theta \approx 89.5°$ yields a conversion efficiency of ~1.22% for a laser linewidth of $\Delta\lambda \approx 0.009$ nm at $\lambda = 510$ nm. Introducing a small intracavity prismatic beam expansion ($M \approx 20$) brings the grating to a higher efficiency operating angle $\Theta \approx 85°$ resulting in a laser conversion

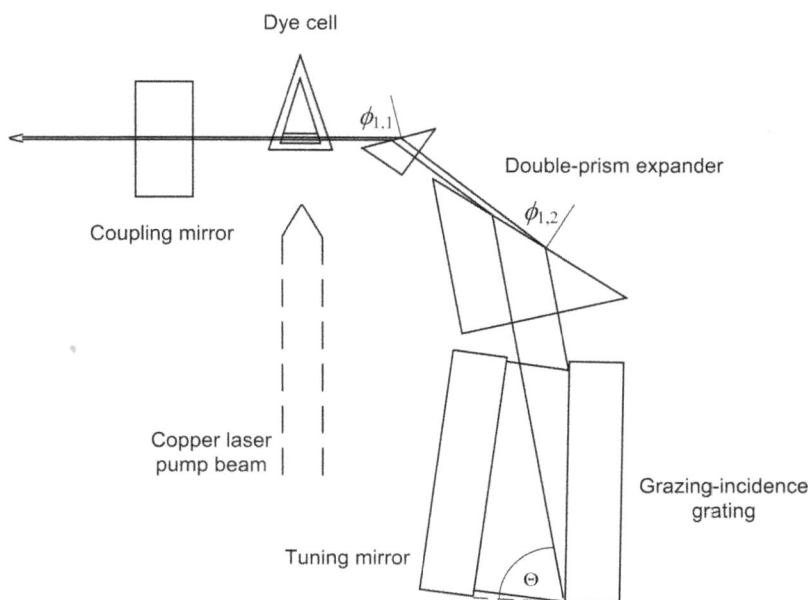

Figure 5.6. Copper-vapor laser-pumped hybrid multiple-prism grazing-incidence (HMPGI) grating oscillator (adapted from Duarte and Piper (1984); courtesy of the Optical Society).

efficiency of $\sim$7% for a laser linewidth of $\Delta\lambda \approx 0.0010$ nm at $\lambda = 510$ nm (Duarte and Piper 1981).

A high-performance liquid organic HMPGI grating tunable narrow-linewidth laser oscillator pumped by a copper vapor laser, operating at a prf of $\sim$8 kHz, is depicted in figure 5.6. For an intracavity beam expansion of $M \approx 26$, and a grating with 2400 lines/mm, this oscillator delivers laser conversion efficiencies in the 4–5% range for a laser linewidth of $\Delta\nu \approx 400$ MHz at $\lambda = 575$ nm. The laser tuning range is reported to be $565 \leqslant \lambda \leqslant 603$ nm (Duarte and Piper 1984) and the ASE level has been estimated, using the ρ_{ASE}/ρ_l definition, to be approximately 5×10^{-7} (Duarte 1990b).

High-performance organic solid-state multiple-prism grating oscillators were first introduced using as gain medium dye-doped MPMMA (Duarte 1994).

A high-performance solid-state organic MPGI, or HMPGI, grating tunable narrow-linewidth laser oscillator pumped by liquid dye laser, of the class belonging to that depicted in figure 5.7, was demonstrated by Duarte (1997). For an intracavity beam expansion of $M \approx 30$, and a grating with 2400 lines/mm, this class of oscillator delivers laser conversion efficiencies in the 3–4% range for a laser linewidth of $\Delta\nu \approx 375$ MHz at $\lambda = 580$ nm. The laser tuning range is reported to be $565 \leqslant \lambda \leqslant 610$ nm and the quoted ASE level is $\rho_{ASE}/\rho_l \approx 10^{-7}$ (Duarte 1997).

Common to both these HMPGI grating tunable narrow-linewidth laser oscillators is their strongly polarized, $\sim$100% parallel to the plane of incidence, laser emission. This polarization preference has its origin in the multiple-prism grating polarization preference (Duarte 1990b, 2015) and in the case of the solid-state oscillator, it is reinforced by the Glan–Thompson output-coupler polarizer.

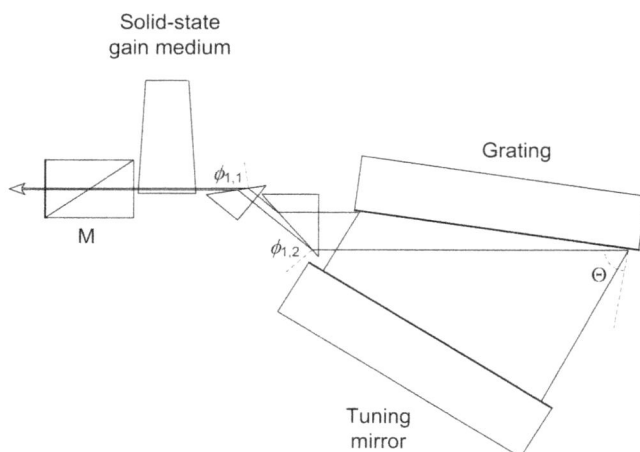

Figure 5.7. Organic solid-state hybrid multiple-prism grazing-incidence (HMPGI) grating oscillator. The laser emission is in a single-transverse-mode nearly diffraction limited beam with a linewidth of $\Delta\nu \approx 375$ MHz.

Although these oscillators are being referred to here as hybrid multiple-prism grazing-incidence (HMPGI) grating configurations, in the literature they are also known as hybrid multiple-prism pre-expanded grazing-incidence grating configurations. The pre-expanded adjective refers to the relatively smaller beam magnification M utilized given the high angle of incidence at the grating.

5.4.2 Multiple-prism Littrow (MPL) grating configurations

Although this class of cavity was demonstrated earlier than 1984 (Duarte and Piper 1980), the schematics of the copper-laser-pumped tunable organic dye laser displayed in figure 5.8 is highly instructional. This is a quintessential high-performance multiple-prism Littrow (MPL) grating oscillator utilizing a liquid organic dye gain medium. Here, the beam at the gain medium has a waist of $w \approx$ 100 μm and it is magnified by the multiple-prism beam expander by a factor of $M \approx$ 95, so that the width illuminating the Littrow grating is $2wM \approx 19$ mm (Duarte and Piper 1984). This means that for the 632 lines/mm grating deployed in its fifth order at a central wavelength of $\lambda \approx 575$ nm, about 91% of the available diffractive surfaced is utilized.

This MPL grating tunable narrow-linewidth laser oscillator was excited transversely by a copper-vapor laser, operating at a prf of ~8 kHz. This oscillator delivers a laser conversion efficiency ~5% for a laser linewidth of $\Delta\nu \approx 1.4$ GHz at $\lambda = 575$ nm. The laser tuning range is reported to be $565 \leqslant \lambda \leqslant 605$ nm (Duarte and Piper 1984) and the ASE level has been estimated to be $\rho_{ASE}/\rho_l \approx 10^{-6}$. The use of a trapezoidal dye cell and the disposition of the intracavity optics yield a laser beam that is >90% polarized parallel to the incidence angle.

Flashlamp-pumped liquid organic tunable narrow-linewidth lasers are uniquely interesting given their intrinsic ability to emit laser radiation in long pulses. In this type

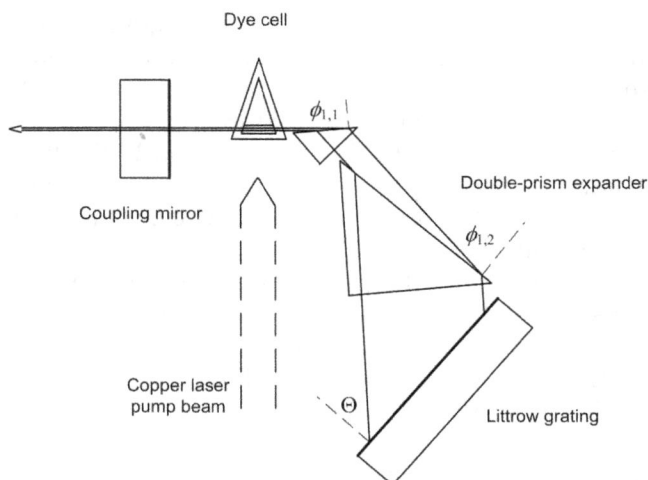

Figure 5.8. Copper-vapor laser-pumped multiple-prism Littrow (MPL) grating oscillator (adapted from Duarte and Piper (1984); courtesy of the Optical Society).

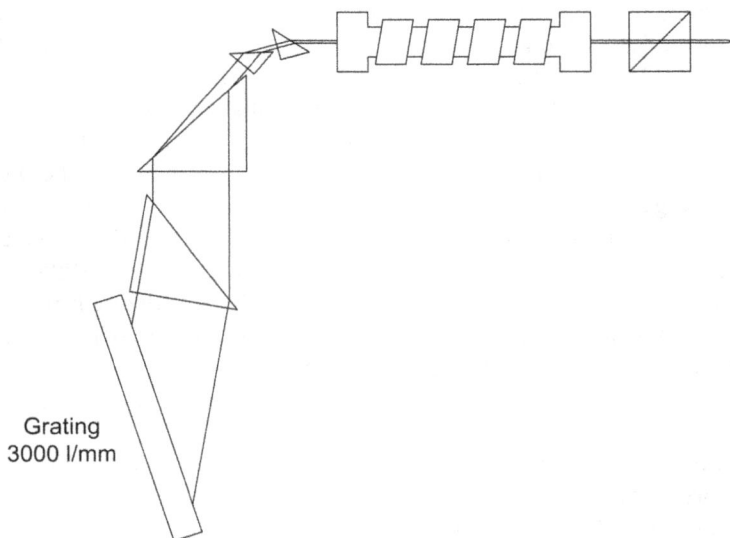

Figure 5.9. Flashlamp-pumped multiple-prism Littrow (MPL) grating oscillator configuration apt to be used as a ruggedized high-performance tunable laser oscillator.

of lasers, long pulse is synonymous with high energy once the oscillator is used to inject an amplifier stage. Following a series of developments and innovations, in flashlamp-pumped laser oscillators, at the US Army, a high-performance ruggedized multiple-prism grating oscillator was developed by Duarte *et al* (1991) whose performance remains top in its class. A schematic applicable to one of these exquisitely engineered multiple-prism grating laser oscillators is depicted in figure 5.9.

For an intracavity beam expansion of $M \approx 34$, and a grating with 3000 lines/mm, 140 mm wide, this oscillator delivers a TEM_{00} beam with a measured divergence of $\Delta\theta \approx 0.35$ mrad and laser linewidth of $\Delta\nu \approx 300$ MHz at $\lambda \approx 590$ nm. At pulse durations of $\Delta t \approx 150$ ns the output energy is at least $E \approx 3$ mJ and the wavelength stability of the emission $(\delta\lambda/\lambda) \approx 4.63 \times 10^{-7}$ (Duarte et al 1991). The use of very low dye concentrations, 1×10^{-5} M, facilitates the attainment of very low ASE figures down to $\rho_{ASE}/\rho_l \approx 10^{-8}$ levels and the laser emission is ~100% polarized parallel to the plane of incidence.

Two features are worth highlighting about the ruggedized MPL grating oscillator described here. First, the whole cavity structure was engineered on solid super invar, about 11 mm thick. Secondly, the whole liquid dye flow system was engineered using high-quality stainless steel and Teflon. Even though the laser dye solution was pumped by hard UV from the coaxial flashlamp, the organic dye gain medium retained its original performance throughout the duration of the experiments and beyond. This is mentioned to remind the reader that properly engineered liquid organic lasers do not have to suffer the tribulations associated with them in some sections of the literature. More details are given in chapter 8.

A unique high-performance long-pulse solid-state organic MPL grating tunable narrow-linewidth laser oscillator excited by a flashlamp-pumped liquid dye laser is depicted in figure 5.10. The gain medium was a HEMA:MMA polymer doped with the rhodamine 6 G laser dye at 0.5×10^{-3} M. For an intracavity beam expansion of $M \approx 92$, and a grating with 3000 lines/mm, this oscillator delivers a laser conversion efficiency of ~2%, and an output energy of $E \approx 0.4$ mJ, for a laser linewidth of $\Delta\nu \approx$ 650 MHz at $\lambda \approx 590$ nm. The pulse duration is $\Delta t \approx 105$ ns and the TEM_{00} laser beam has a divergence of $\Delta\theta \approx 3.5$ mrad. The laser tuning range is reported to be 564 $\leqslant \lambda \leqslant 602$ nm and the quoted ASE level is $\rho_{ASE}/\rho_l \approx 10^{-4}$. The laser emission is ~100% polarized parallel to the plane of incidence (Duarte et al 1998).

The larger $\Delta\theta$ value as compared to the MPL oscillator utilizing liquid gain media is attributed to thermal lensing at the organic polymer gain matrix: $\partial n/\partial T$ measurements of this class of gain media are reported by Duarte et al (2000). This effect is augmented given the long-pulse, and therefore higher energy, characteristics of the emission.

The epitome of high-performance solid-state organic tunable narrow-linewidth laser oscillators is the MPL grating oscillator incorporating a rhodamine 6 G-doped polymer gain matrix of the MPMMA type. This oscillator utilized semi-longitudinal laser excitation provided by liquid dye laser and is depicted in figure 5.11. For an intracavity beam expansion of $M \approx 44$, and a grating with 3300 lines/mm, this oscillator delivers conversion efficiencies of 4% for single-longitudinal-mode laser emission with a measured linewidth of $\Delta\nu \approx 350$ MHz at $\lambda = 590$ nm. The measured beam divergence of the TEM_{00} laser beam is $\Delta\theta \approx 2.2$ mrad that corresponds to $\times$ 1.46 its diffraction limit $(\lambda/\pi w)$. The laser tuning range is reported to be 550 $\leqslant \lambda \leqslant 603$ nm and the quoted ASE level is $\rho_{ASE}/\rho_l \approx 10^{-6}$ (Duarte 1999).

The near-Gaussian temporal pulse with $\Delta t \approx 3$ ns (FWHM) is shown in figure 5.12 and the photographically recorded interferogram showing a laser linewidth of $\Delta\nu \approx$ 350 MHz, $\Delta\lambda \approx 0.0004$ nm at $\lambda = 590$ nm, is depicted in figure 5.13. Therefore, this

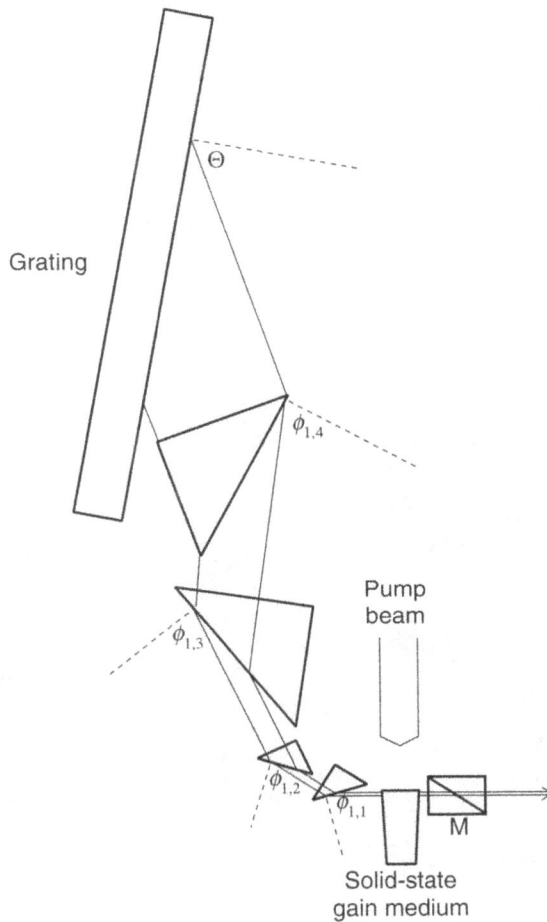

Figure 5.10. Long-pulse organic solid-state multiple-prism Littrow (MPL) grating oscillator (Duarte *et al* 1998; courtesy of the Optical Society).

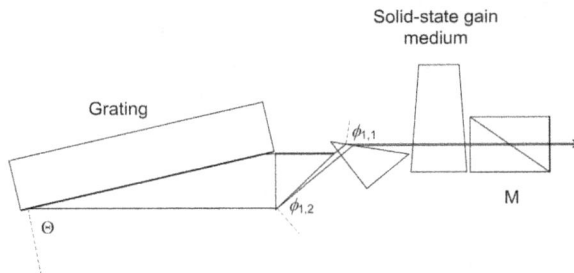

Figure 5.11. Optimized organic solid-state multiple-prism Littrow (MPL) grating oscillator. The laser emission is in a single-transverse-mode nearly diffraction limited beam with a linewidth of $\Delta\nu \approx 350$ MHz. The single-longitudinal-mode linewidth is only limited by Heisenberg's uncertainty principle (from Duarte (1999); courtesy of the Optical Society).

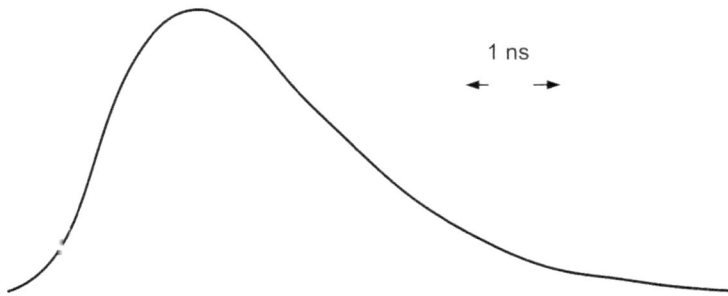

Figure 5.12. Near-Gaussian temporal pulse from the optimized organic MPL grating oscillator. Temporal scale is 1 ns per division (from Duarte (1999); courtesy of the Optical Society).

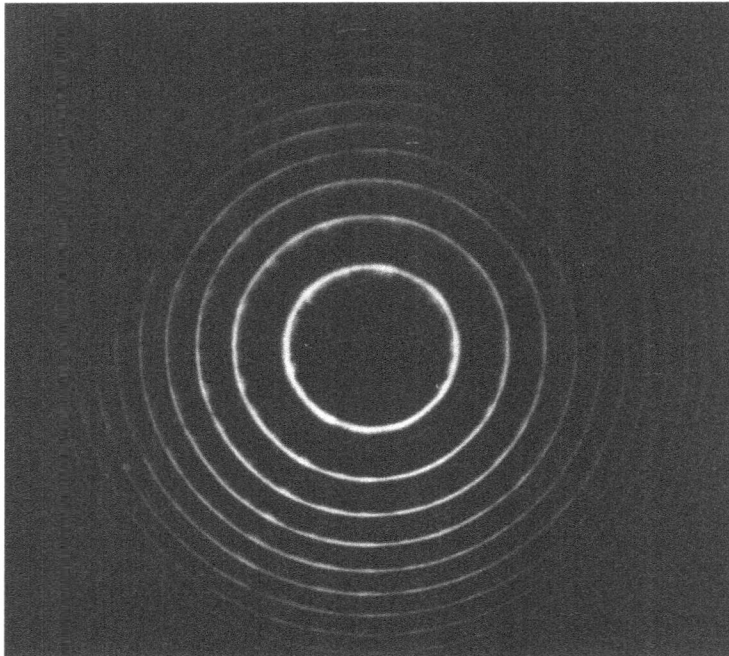

Figure 5.13. Frabry–Perot interferogram of the single-longitudinal-mode emission from the optimized organic MPL grating oscillator (from Duarte (1999); courtesy of the Optical Society).

organic laser oscillator performs at the limit allowed by Heisenberg's uncertainty principle since $\Delta t \Delta \nu \approx 1.05$.

The peak power delivered by this oscillator is $P_l \approx 33.33$ kW, which at a laser linewidth of $\Delta \lambda \approx 0.0004$ nm at $\lambda = 590$ nm, means that the spectral power density ρ_{SP} of the emission is

$$\rho_{SP} \approx \left(\frac{P_l}{\Delta \lambda} \right) \approx 8.33 \times 10^7 \text{ W nm}^{-1} \tag{5.11}$$

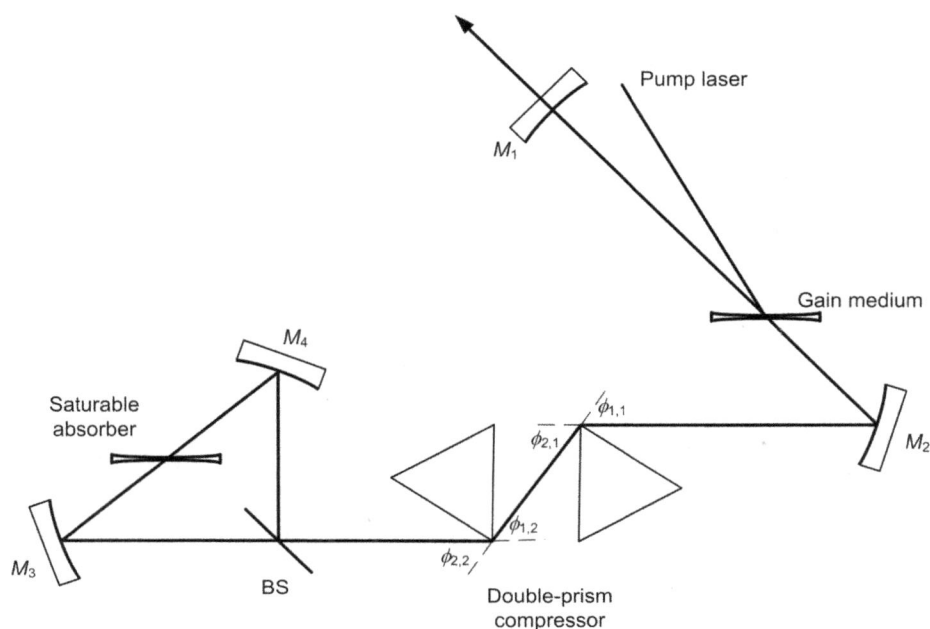

Figure 5.14. Generic linear CW laser cavity including a double-prism pulse compressor. The double-prism pulse compressor adds negative dispersion to the overall intracavity dispersion thus minimizing the overall value of $\nabla_\lambda \theta$ (see chapter 6).

In the absence of intracavity etalons, the performance of this MPL grating oscillator is as good as, or even better, than the performance reported using the same gain medium in a HMPGI grating oscillator. Better still, it does so with one less optical component (the tuning mirror shown in figure 5.6). This is allowed by the use of a high density diffraction grating which is deployed at $\Theta \approx 77°$. In other words, in terms of laser linewidth performance, MPL grating tunable laser oscillators are equivalent to grazing-incidence grating configurations, a fact not yet recognized in the specialized literature. In terms of laser conversion efficiencies, MPL grating oscillator architectures have the edge.

5.5 Linear and ring laser cavities

A linear laser cavity engineered for a CW organic dye laser is depicted in figure 5.14. This cavity includes a double-prism pulse compressor. A ring laser cavity configuration applicable to a CW organic laser, and other broad spectral gain laser gain media, is illustrated in figure 5.15. This ring cavity configuration includes a four-prism pulse compressor.

5.6 Unstable resonators as laser amplifiers

Unstable resonator configurations are of interest in the field of organic lasers since the unstable resonator geometry is ideally suited to design and construct co-linear

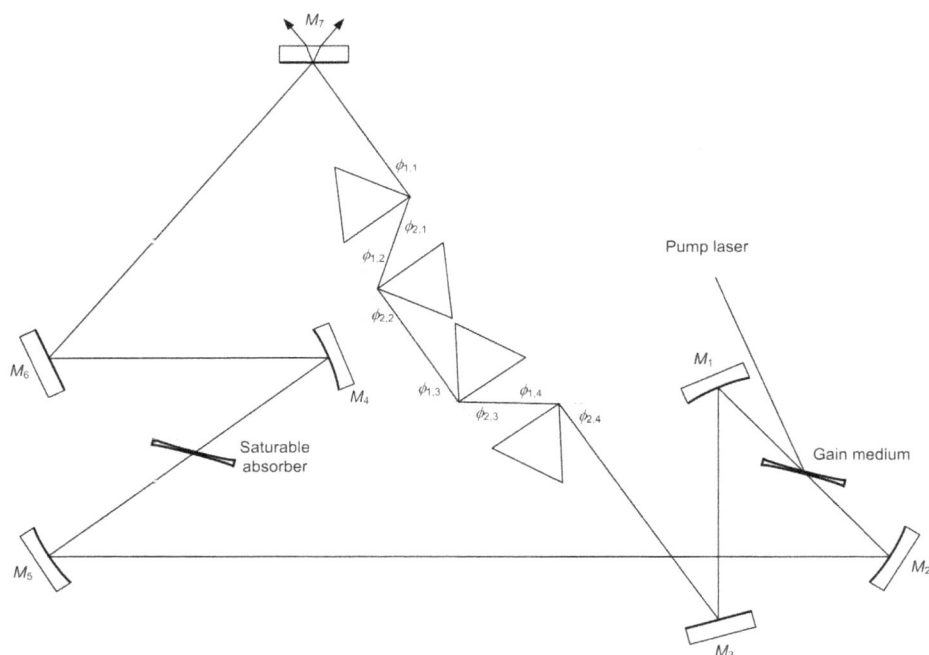

Figure 5.15. Generic ring CW laser cavity including a four-prism pulse compressor. The 2 × double-prism, or quad-prism, pulse compressor adds negative dispersion to the overall intracavity dispersion thus minimizing the overall value of $\nabla_\lambda \theta$ (see chapter 6). Notice the symmetry of the quad-prism compressor.

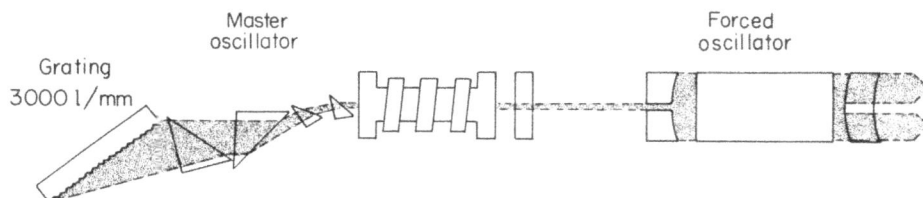

Figure 5.16. Oscillator–amplifier co-linear configuration comprised of an MPL grating master oscillator and an unstable-resonator power amplifier (from Duarte and Conrad (1987); courtesy of the Optical Society).

amplification stages to be injected with narrow-linewidth laser emission from a high-performance oscillator.

A co-linear liquid organic oscillator–amplifier configuration comprised of an MPL grating oscillator and an unstable-resonator amplifier stage is illustrated in figure 5.16. The rhodamine 6 G dye concentration used at the oscillator stage was 5×10^{-5} M while at the amplifier stage it was 2.5×10^{-5} M in methanol–water. The amplifier stage, that has a 10 mm diameter active region, when using a flat–flat mirror configuration with a 20% reflectivity output coupler, delivered an output energy of $E \approx 2$ J. This output energy decreased to 130 mJ when a 5 mm intracavity aperture was introduced.

The MPL grating oscillator delivers an output energy of $E \approx 11$ mJ in a TEM_{00} beam profile with a measured divergence of $\Delta\theta \approx 0.5$ mrad. The pulse duration of the oscillator emission is $\Delta t \approx 200$ ns and its single-longitudinal-mode laser linewidth $\Delta\nu \approx 375$ MHz (Duarte and Conrad 1987).

The cavity of the amplifier stage was a positive branch confocal unstable resonator. The planoconcave total reflector has a radius of curvature of $R_2 = 400$ cm, and a 5 mm diameter perforation at its center. The output-coupler has a radius of curvature of $R_1 = 200$ cm with a 5 mm diameter high-reflectivity center. The magnification provided by this class of *Cassegrainian telescope* unstable resonator is $M = 2$. The unstable resonator amplifier stage provided high-fidelity reproduction of the oscillator injected beam, with a linewidth of $\Delta\nu \approx 375$ MHz, at an output energy of $E \approx 600$ mJ thus demonstrating an amplification factor of ~55.

In a 2×2 propagation $ABCD$ matrix the two terms of interest are the A and D terms since a resonator is considered to be unstable if (Siegman 1986)

$$\left| \frac{A + D}{2} \right| > 1 \tag{5.12}$$

The ray matrix applicable to the Cassegrainian telescope used here is given by (Siegman 1986)

$$\begin{pmatrix} A & B \\ C & D \end{pmatrix} = \begin{pmatrix} M & (M + 1)L/M \\ 0 & M^{-1} \end{pmatrix} \tag{5.13}$$

where

$$M = -\left(\frac{R_2}{R_1} \right) \tag{5.14}$$

$$L = \left(\frac{R_1 + R_2}{2} \right) \tag{5.15}$$

It can be readily be verified that $(|A + D|/2) > 1$ thus certifying the condition of instability for the Cassegrainian telescope used in the amplifying stage.

Note that it is not necessary to have a reflective type telescope to achieve the condition of unstable resonator. Duarte *et al* (1997) have indicated that a multiple-prism grating oscillator can achieve the condition of instability when incorporating a gain medium that exhibits the condition of thermal lensing. This topic is discussed further in chapter 6.

A comprehensive review of unstable resonator principles is given by Siegman (1986). Propagation ray matrices applicable to unstable resonators are discussed in detail by Duarte (2003, 2015) and relevant mathematical aspects are discussed in chapter 6.

5.7 Laser-pumped amplifier stages

Laser-pumped organic oscillator–amplifier laser systems are conceptually and pragmatically simpler than their flashlamp-pumped oscillator–amplifier counter parts. Laser amplifiers can be excited longitudinally, transversely, or doubly transversely, as illustrated in figure 5.17. The main prerequisites associated with the

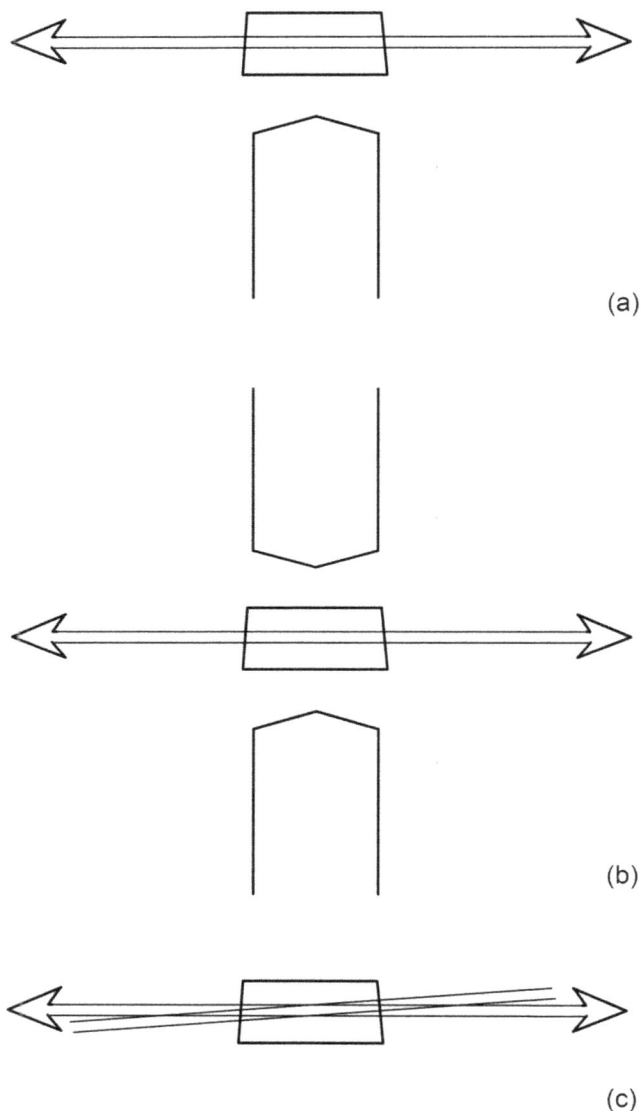

Figure 5.17. Various pumping geometries applicable to organic laser amplification stages: (a) transverse excitation, (b) double-transverse excitation, and (c) longitudinal excitation.

design of an efficient organic narrow-linewidth oscillator amplifier system are the following:

1. The availability of a tunable narrow-linewidth, preferably SLM, extremely low ASE ($\rho_{\mathrm{ASE}}/\rho_l \leqslant 10^{-5}$), oscillator stage.
2. Given that most excitation lasers for this type of task emit pulses in the $10 \leqslant \Delta t \leqslant 100$ ns range, it is important to time correctly, via the use of proper optical delay lines, the arrival of the excitation pulse with the arrival of the oscillator injection signal.
3. Dividing the available excitation laser pulse energy appropriately between the oscillator stage and successive amplification stages is also important.

Dupre (1987) describes an MPGI grating oscillator amplifier system, with two amplification stages, delivering a laser linewidth of $\Delta\nu \approx 650$ MHz, at $\lambda \approx 440$ nm, with a total gain of ~700. This was a low prf system excited by a XeCl laser and the overall conversion efficiency is reported as 9%.

Bass *et al* (1992) describe an MPL grating oscillator amplifier system, with three to four amplification stages in four amplification chains. The reported linewidth performance is in the $500 \leqslant \Delta\nu \leqslant 5000$ MHz range, at $\lambda \approx 590$ nm. This was a high prf system, operating at 13.2 kHz, excited by a copper-vapor laser system. The overall output average power was 26 kW at a conversion efficiency in the 50–60% range.

These systems are also referred in the literature as master-oscillator power-amplifier (MOPA) systems. Occasionally the amplifier stage is comprised of a CW laser oscillator (Farkas and Eden 1993).

5.8 Distributed feedback configurations

Distributed feedback (DFB) tunable organic laser configurations are quite different to the resonators and oscillators considered up to this point. In first generation DFB lasers the pump laser beam is divided into two beams that interfere at a common point at the gain medium and the emission, along the gain medium optical axis, exits in two directly opposite directions (see figure 5.18). Variation of the angle θ enables wavelength tuning via mechanical means, albeit tuning via thermal means is also utilized.

In second generation organic DFB lasers the gain medium itself is configured as a diffraction grating, as illustrated in figure 5.19, this reduces significantly the complexity of the optical set up.

Macroscopic DBF solid-state organic lasers have demonstrated laser linewidths of $\Delta\lambda \approx 0.06$ nm at $\lambda \approx 482$ nm (Zhu *et al* 2000) while an optofluidic DFB organic dye laser chip is reported to yield $\Delta\lambda \approx 0.1$ nm at $\lambda \approx 567$ nm (Li and Psaltis 2008).

5.9 Vertical cavity surface emitting lasers (VCSELs)

Vertical cavity surface emitting laser (VCSEL) configurations were introduced to the world of organic lasers via the optically-pumped organic semiconductor laser and

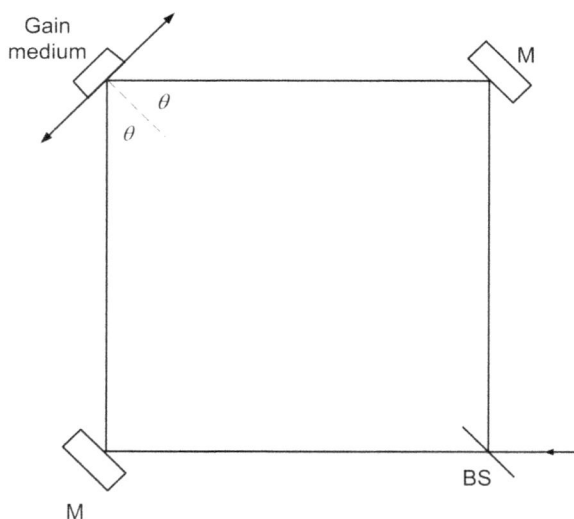

Figure 5.18. First generation DFB laser configuration. It should be noted that in the early DFB lasers the interferometric arrangement was necessary to create a grating undulation in the liquid gain medium.

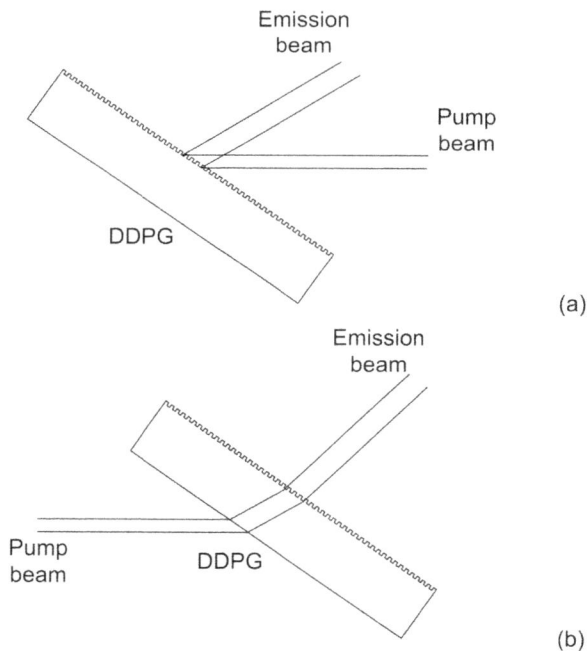

Figure 5.19. Second generation DFB lasers obviate the interferometric arrangement simply by imposing the periodic structure of a diffraction grating on the gain medium. The arrangement in (a) depicts a dye-doped polymer grating (DDPG) gain medium being excited in a reflection configuration. The arrangement in (b) depicts a DDPG gain medium being excited in a transmission configuration.

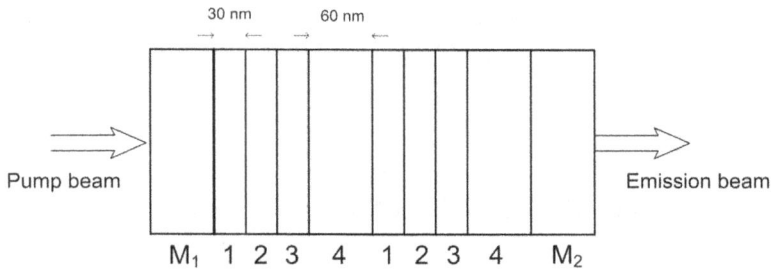

Figure 5.20. OVCSEL structure configured for optical excitation and emission in the horizontal direction. Regions 2 are light emitting regions such as C 545 T dye-doped Alq$_3$. M$_1$ and M$_2$ are dichroic mirrors that allow transmission of the pump laser beam while reflecting the emission from the organic semiconductor. Often M$_1$ is a total reflector, for the emission, while M$_2$ is a partial reflectivity output-coupler. Alternatively, M$_1$ and M$_2$ can be distributed Bragg reflectors (DBRs). The material composition of regions 1–4 are described in chapter 11.

were designated as organic VCSELs or OVCSELs (Bulović *et al* 1998). Essentially, these structures are comprised of a multilayered sandwich of semiconducting films and one or more of those layers is comprised of a light-emitting compound such as an organic dye-doped Alq$_3$ region. Specific examples of such light-emitting regions are DCM dye-doped Alq$_3$ (Bulović *et al* 1998) and C 545 T dye-doped Alq$_3$ (Duarte *et al* 2005).

For optically-excited OVCSEL devices the reflectors at each end of the multi-layered structure are partially reflecting. These reflectors can be straight-forward mirrors or distributed Bragg reflectors (DBRs) (see, for example, Yang *et al* 2015).

In figure 5.20 multilayered VCSEL configurations are illustrated for emission in the vertical direction as well as in the horizontal direction.

VCSEL structures are also applied beyond the semiconductor laser arena into the organic dye-doped polymer realm (Yang *et al* 2015).

Electrically-pumped vertical cavity surface emitting coherent interferometric emitters are discussed in chapter 11.

5.10 Perspective

It becomes clear now (2024) that the majority of transcendental innovations in laser cavity designs and laser oscillator architectures took place from the late 1970s and throughout the 1980s and 1990s. This observation is also relevant in regard to the physics and theory of narrow-linewidth laser oscillators.

5.11 Problems

- 5.1 For a cavity length of $L = 10$ cm and a beam radius of $w = 0.5$ mm calculate the Fresnel number N_F at $\lambda = 590$ nm.
- 5.2 For a cavity length of $L = 10$ cm and a beam radius of $w = 100$ μm calculate the Fresnel number N_F at $\lambda = 590$ nm. Comment.

- 5.3 Verify that the equation for the FSR given in equation (5.5) is equivalent to equation (5.4).
- 5.4 Using the linewidth equation (5.6) write explicit equations for the linewidth of a laser in Littrow configuration and a laser in grazing-incident configuration. Compare and comment.
- 5.5 Verify that for a laser oscillator yielding a peak power of $P_l \approx 33.33$ kW, and a laser linewidth of $\Delta\lambda = 0.0004$ nm, the spectral power density ρ_{SP} is as given in equation (5.11).

References

Bass I L, Bonanc R E, Hackel R H and Hammond P R 1992 High-average-power dye laser at Lawrence Livermore National Laboratory *Appl. Opt.* **31** 6993–7006

Born M and Wolf E 1999 *Principles of Optics* 7th edn (Cambridge: Cambridge University)

Bulović V, Kozlov V G, Khalfin V B and Forrest S R 1998 Transform-limited narrow-linewidth lasing action in organic semiconductor microcavities *Science* **279** 553–5

Duarte F J 1990a Narrow-linewidth pulsed dye laser oscillators *Dye Laser Principles* ed F J Duarte and L W Hillman (New York: Academic) ch 4

Duarte F J 1990b Technology of pulsed dye lasers *Dye Laser Principles* ed F J Duarte and L W Hillman (New York: Academic) ch 6

Duarte F J 1992 Cavity dispersion equation $\Delta\lambda \approx \Delta\theta(\partial\theta/\partial\lambda)^{-1}$: a note on its origin *Appl. Opt.* **31** 6979–82

Duarte F J 1994 Solid-state multiple-prism grating dye-laser oscillators *Appl. Opt.* **33** 3857–60

Duarte F J 1997 Multiple-prism near-grazing-incidence grating solid-state dye-laser oscillator *Opt. Laser Tech.* **29** 513–6

Duarte F J 1999 Multiple-prism grating solid-state dye laser oscillator: optimized architecture *Appl. Opt.* **38** 6347–9

Duarte F J 2001 Multiple-return-pass beam divergence and the linewidth equation *Appl. Opt.* **40** 3038–41

Duarte F J 2003 *Tunable Laser Optics* (New York: Academic)

Duarte F J 2015 *Tunable Laser Optics* 2nd edn (New York: CRC)

Duarte F J and Conrad R W 1987 Diffraction limited single-longitudinal-mode multiple-prism flashlamp-pumped dye laser oscillator: linewidth analysis and injection of amplifier system *Appl. Opt.* **26** 2567–71

Duarte F J, Costela A, García-Moreno I, Sastre R, Ehrlich J J and Taylor T S 1997 Dispersive solid-state dye laser oscillators *Opt. Quantum Electron.* **29** 461–72

Duarte F J, Costela A, García-Moreno I and Sastre R 2000 Measurements of $\partial n/\partial T$ in solid-state dye-laser gain media *Appl. Opt.* **39** 6522–3

Duarte F J, Davenport W E, Ehrlich J J and Taylor T S 1991 Ruggedized narrow-linewidth dispersive dye laser oscillator *Opt. Commun.* **84** 310–6

Duarte F J and James R O 2003 Tunable solid-state lasers incorporating dye-doped polymer-nanoparticle gain media *Opt. Lett.* **28** 2088–90

Duarte F J, Liao L S and Vaeth K M 2005 Coherence characteristics of electrically excited tandem organic light-emitting diodes *Opt. Lett.* **30** 3072–4

Duarte F J, Liao L S, Vaeth K M and Miller A M 2006 Widely tunable green laser emission using the coumarin 545 tetramethyl dye as the gain *J. Opt. A: Pure Appl. Opt.* **8** 172–4

Duarte F J and Piper J A 1980 A double prism beam expander for pulsed dye lasers *Opt. Commun.* **35** 100–4

Duarte F J and Piper J A 1981 Prism preexpanded grazing-incidence grating cavity for pulsed dye lasers *Appl. Opt.* **20** 2113–6

Duarte F J and Piper J A 1984 Narrow-linewidth high-prf copper laser-pumped dye laser oscillators *Appl. Opt.* **23** 1391–4

Duarte F J, Taylor T S, Costela A, García-Moreno I and Sastre R 1998 Long-pulse narrow-linewidth dispersive solid-state dye-laser oscillator *App. Opt.* **37** 3987–9

Dupre P 1987 Quasiunimodal tunable pulsed dye laser at 440 nm: theoretical development for using a quad prism beam expander and one or two gratings in a pulsed dye laser oscillator cavity *Appl. Opt.* **26** 860–71

Farkas A M and Eden J G 1993 Pulsed dye laser amplification and frequency doubling of single-longitudinal-mode semiconductor lasers *IEEE J. Quantum Electron.* **29** 2917–23

Hänsch T W 1972 Repetitively pulsed tunable dye laser for high-resolution spectroscopy *Appl. Opt.* **11** 895–8

Li Z and Psaltis D 2008 Optofluidic dye lasers *Microfluid. Nanofluid.* **4** 145–58

Littman M G and Metcalf H J 1978 Spectrally narrow pulsed dye laser without beam expander *Appl. Opt.* **17** 2224–7

Meaburn J 1976 *Detection and interferometry of Faint Light* (Boston, MA: Reidel)

Steel W H 1967 *Interferometry* (Cambridge: Cambridge University)

Shohan I, Danon N N and Oppenheim U P 1977 Narrow band operation of pulsed dye laser without beam expansion *J. Appl. Phys.* **48** 4495–7

Siegman A E 1986 *Lasers* (Mill Valley, CA: University Science Books)

Yang Y, Zhou Y, Liao Z, Yu J, Cui Y, García-Moreno I, Wang Z, Costela A and Qian G 2015 Mechanically tunable organic vertical-cavity surface emitting lasers (VCSELs) for highly sensitive stress probing in dual modes *Opt. Express* **23** 4385–96

Zhu X-L, Lam S-K and Lo D 2000 Distributed-feedback dye-doped solgel silica lasers *Appl. Opt.* **39** 3104–7

IOP Publishing

Organic Lasers and Organic Photonics (Second Edition)

F J Duarte

Chapter 6

Mathematical-physics for tunable narrow-linewidth organic laser oscillators

F J Duarte

The mathematical physics needed to design tunable narrow-linewidth laser oscillators is described and discussed in detail. The treatment begins with a discussion of the Dirac-Feynman interferometric principle that leads to the generalized quantum interferometric probability equation that describes the spatial characteristics of laser emission and that be used to derive the cavity linewidth equation $\Delta\lambda \approx \Delta\theta(\nabla_\lambda\theta)^{-1}$ and the generalized grating diffraction and the generalized refraction equation. Next, the dispersion equations applicable to linewidth narrowing and laser pulse compression are derived. The rth-derivative generalized multiple-prism dispersion equation $\nabla_n^r\phi_{2,n}$ for laser pulse compression is also derived.

6.1 Introduction

In this chapter the mathematical-physics of high-performance narrow-linewidth organic oscillators is reviewed from a pragmatic perspective rather than with an emphasis on derivation. Background reviews to this physics can be found in Duarte (1990a, 2014, 2015).

6.2 The generalized interferometric equations

The probabilistic Dirac principle (Dirac 1978, Feynman *et al* 1965)

$$\langle d|s \rangle = \sum_{j=1}^{N} \langle x|j \rangle \langle j|s \rangle \tag{6.1}$$

is the probability amplitude that allows the analysis and quantification of the propagation of a single quanta from a source (s), via an array of N-slits (j), toward an interferometric plane x. This probability amplitude also applies to an ensemble of indistinguishable photons as available from narrow-linewidth lasers (Duarte 1991, 1993). Following Dirac (1978) and substituting the individual probability amplitudes $\langle j|s\rangle$ and $\langle x|j\rangle$ with wave functions

$$\langle j|s\rangle = \Psi(r_{j,s})e^{-i\theta_j} \tag{6.2}$$

$$\langle d|j\rangle = \Psi(r_{x,j})e^{-i\phi_j} \tag{6.3}$$

lead to

$$\langle d|s\rangle = \sum_{j=1}^{N}\Psi(r_j)e^{-i\Omega_j} \tag{6.4}$$

where $\Psi(r_j) = \Psi(r_{x,j})\Psi(r_{j,s})$ and

$$\Omega_j = (\theta_j + \phi_j) \tag{6.5}$$

Multiplying the probability amplitude given in equation (6.4) by its complex conjugate leads to

$$|\langle d|s\rangle|^2 = \sum_{j=1}^{N}\Psi(r_j)\sum_{m=1}^{N}\Psi(r_m)e^{i(\Omega_m-\Omega_j)} \tag{6.6}$$

and eventually to the measurable probability (Duarte 1991, 1993)

$$|\langle d|s\rangle|^2 = \sum_{j=1}^{N}\Psi(r_j)^2 + 2\sum_{j=1}^{N}\Psi(r_j)\left(\sum_{m=j+1}^{N}\Psi(r_m)\cos(\Omega_m-\Omega_j)\right) \tag{6.7}$$

For the two-dimensional case, for propagation in the x-axis, the Dirac probability amplitude can be expressed as

$$\langle d|s\rangle = \sum_{j_z=1}^{N}\sum_{j_y=1}^{N}\langle x|j_{zy}\rangle\langle j_{zy}|s\rangle \tag{6.8}$$

Abstracting the j from j_{zy}, equation (6.8) can be written as

$$\langle d|s\rangle = \sum_{z=1}^{N}\sum_{y=1}^{N}\Psi(r_{zy})e^{-i\Omega_{zy}} \tag{6.9}$$

and the measurable probability is given by (Duarte 1995)

$$|\langle d|s\rangle|^2 = \sum_{z=1}^{N}\sum_{y=1}^{N}\Psi(r_{zy})\sum_{q=1}^{N}\sum_{p=1}^{N}\Psi(r_{pq})e^{i(\Omega_{qp}-\Omega_{zy})} \tag{6.10}$$

For a three-dimensional N-slit array the measurable probability is given by (Duarte 1995)

$$|\langle d|s\rangle|^2 = \sum_{z=1}^{N}\sum_{y=1}^{N}\sum_{x=1}^{N}\Psi(r_{zyx})\sum_{q=1}^{N}\sum_{p=1}^{N}\sum_{r=1}^{N}\Psi(r_{qpr})e^{i(\Omega_{qpr}-\Omega_{zyx})} \tag{6.11}$$

As outlined by Lamb and colleagues (Sargent *et al* 1974) and already explained in chapter 3, measurable probabilities are proportional to the intensity of propagating interferograms (Duarte 2014) which is what is ultimately measured by the experimenter. The one-dimensional N-slit interferometric probability given in equation (6.6), also known as the N-slit interferometric equation, can be used to:

1. Calculate transverse mode structure from a laser resonator, laser cavity, or laser oscillator (Duarte 2003). This can be used to determine the necessary intracavity aperture dimension, relative to a cavity length, to obtain single-transverse-mode (TEM_{00}) emission which is a must to ultimately achieve narrow-linewidth laser emission and single-longitudinal-mode (SLM) emission. In this regard, the calculation proceeds by dividing the intracavity aperture, often in the $100 \leqslant 2w \leqslant 1000$ µm range (where w is the radius of the aperture), by hundreds of equally-spaced imaginary slits j (Duarte 1993).
2. Calculate the beam divergence of single-transverse-mode (TM_{00}) laser emission.
3. Calculate the beam divergence of multiple-transverse-mode laser emission.

6.2.1 The uncertainty principle in optics

From the *phase term* of the generalized N-slit interferometric equation the diffraction identity

$$\Delta\lambda \approx \left(\frac{\lambda^2}{\Delta x}\right) \tag{6.12}$$

can be derived. From this identity, while using the geometry of a diffraction grating, the approximation

$$\Delta p\Delta x \approx h \tag{6.13}$$

can be arrived at (Duarte 2003) which is known as *Heisenberg's uncertainty principle* (Dirac 1978, Feynman *et al* 1965).

Alternative forms of the uncertainty are (Duarte 2003)

$$\Delta\lambda\Delta x \approx \lambda^2 \tag{6.14}$$

$$\Delta\nu\Delta x \approx c \tag{6.15}$$

$$\Delta E\Delta t \approx h \tag{6.16}$$

and

$$\Delta\nu\Delta t \approx 1 \tag{6.17}$$

6.2.2 Beam divergence

From the uncertainty principle relation $\Delta p \Delta x \approx h$ it can be shown that the angular spread of a laser beam is given by

$$\Delta\theta \approx \left(\frac{\lambda}{\Delta x}\right) \tag{6.18}$$

Equation (6.18) should be compared with the classical equation for beam divergence (Duarte 1990a)

$$\Delta\theta = \frac{\lambda}{\pi w}\left(1 + \left(\frac{L_{\mathscr{R}}}{B}\right)^2 + \left(\frac{AL_{\mathscr{R}}}{B}\right)^2\right)^{1/2} \tag{6.19}$$

here w is the beam waist, and

$$L_{\mathscr{R}} = \left(\frac{\pi w^2}{\lambda}\right) \tag{6.20}$$

is known as the Rayleigh length, and A and B are spatial propagation parameters from propagation the $ABCD$ matrices as given in section 6.5 (Duarte 1990a, 2003). For optimum conditions

$$\Delta\theta \approx \frac{\lambda}{\pi w} \tag{6.21}$$

which is known as the *diffraction limit* of the beam divergence.

6.3 The cavity linewidth equation

From the generalized N-slit interferometric equation (6.7) the generalized diffraction equation can be derived (Duarte 2006)

$$d_m(\pm n_1 \sin\Theta_m \pm n_2 \sin\Phi_m) = m\lambda \tag{6.22}$$

For positive diffraction, $n_1 = n_2$, and for $\Theta_m \approx \Phi_m(=\Theta)$, the well-known diffraction grating equation for a grating deployed in Littrow configuration follows

$$2d \sin\Theta = m\lambda \tag{6.23}$$

Following Duarte (1992) and considering two slightly different wavelengths,

$$\Delta\lambda = \frac{2d}{m}(\sin\Theta_1 - \sin\Theta_2) \tag{6.24}$$

for $\Theta_1 \approx \Theta_2(=\Theta)$ equation (6.24) can be rewritten as

$$\Delta\lambda \approx \frac{2d}{m}\Delta\theta\left(1 - \frac{3\Theta^2}{3!} + \frac{5\Theta^4}{5!} - \frac{7\Theta^6}{7!} + \cdots\right) \tag{6.25}$$

Differentiation of the Littrow diffraction grating equation, that is equation (6.23), leads to

$$\frac{\partial \theta}{\partial \lambda} \cos \Theta = \frac{m}{2d} \tag{6.26}$$

and substitution into equation (6.25) yields

$$\Delta \lambda \approx \Delta \theta \left(\frac{\partial \theta}{\partial \lambda}\right)^{-1} \left(1 - \frac{\Theta^2}{2!} + \frac{\Theta^4}{4!} - \frac{\Theta^6}{6!} \cdots\right) (\cos \Theta)^{-1} \tag{6.27}$$

This equation reduces to the single-pass *cavity linewidth equation* (Duarte 1992)

$$\Delta \lambda \approx \Delta \theta \left(\frac{\partial \theta}{\partial \lambda}\right)^{-1} \tag{6.28}$$

or

$$\Delta \lambda \approx \Delta \theta (\nabla_\lambda \theta)^{-1} \tag{6.29}$$

where $\nabla_\lambda \theta = (\partial \theta / \partial \lambda)$ is the overall intracavity dispersion. In summary: the cavity linewidth equation originates from the generalized N-slit interference equation and incorporates $\Delta \theta$ whose value can be determined via Heisenberg's uncertainty principle. An alternative, more heuristic, derivation of the cavity linewidth equation using quantum principles is given by Duarte (1992). The multiple-return-pass version of this equation is given in section 6.5.

6.4 The diffraction equations

As mentioned previously from the phase term of the generalized N-slit interferometric equation, that is equation (6.7), the generalized diffraction equation can be stated, in a slightly reduced form, as (Duarte 2006)

$$d_m(\pm n_1 \sin \Theta_m \pm n_2 \sin \Phi_m) = m\lambda$$

For $n_1 = n_2$ this equation further reduces to

$$d_m(\pm \sin \Theta_m \pm \sin \Phi_m) = m\lambda \tag{6.30}$$

where $m = 0, 1, 2, \ldots$ and d_m is the sum of the dimensions of the individual slit plus its corresponding slit-to-slit separation. For incidence above the normal, this equation reduces further to

$$d_m(\sin \Theta_m \pm \sin \Phi_m) = m\lambda \tag{6.31}$$

where Θ_m is the angle of incidence and Φ_m is the angle of diffraction. This is the quasi-generalized equation of diffraction used to quantify diffraction grating deployed at *grazing incidence*. For $\Theta_m \approx \Phi_m (= \Theta)$ equation (6.31) reduces to the grating equation for Littrow configuration given in equation (6.23).

By differentiating equation (6.31) the dispersion for a grating deployed in the grazing-incidence configuration is given by

$$\left(\frac{\partial \Theta_m}{\partial \lambda}\right) = \left(\frac{\sin \Theta_m \pm \sin \Phi_m}{\lambda \cos \Theta_m}\right) \tag{6.32}$$

or

$$\left(\frac{\partial \Theta_m}{\partial \lambda}\right) = \left(\frac{m}{d_m \cos \Theta_m}\right) \tag{6.33}$$

For deployment in Littrow configuration $\Theta_m \approx \Phi_m (= \Theta)$ the diffraction grating dispersion becomes

$$\left(\frac{\partial \Theta}{\partial \lambda}\right) = \left(\frac{2 \tan \Theta}{\lambda}\right) \tag{6.34}$$

Equations (6.32)–(6.34) are the intracavity dispersion equations to be used in conjunction with the cavity linewidth equation (6.28) to estimate the single-pass linewidth in cavities incorporating only a grating as the intracavity dispersion element.

6.5 The generalized multiple-prism equations

The generalized cumulative single-pass multiple-prism equation is given by (Duarte and Piper 1982, Duarte 2006)

$$\nabla_\lambda \phi_{2,m} = \pm \mathcal{H}_{2,m} \nabla_\lambda n_m \pm (k_{1,m} k_{2,m})^{-1} \left(\mathcal{H}_{1,m} \nabla_\lambda n_m (\pm) \nabla_\lambda \phi_{2,(m-1)}\right) \tag{6.35}$$

where $\nabla_\lambda = \partial/\partial\lambda$ and

$$k_{1,m} = \frac{\cos \psi_{1,m}}{\cos \phi_{1,m}} \tag{6.36}$$

$$k_{2,m} = \frac{\cos \phi_{2,m}}{\cos \psi_{2,m}} \tag{6.37}$$

$$\mathcal{H}_{1,m} = \frac{\tan \phi_{1,m}}{n_m} \tag{6.38}$$

$$\mathcal{H}_{2,m} = \frac{\tan \phi_{2,m}}{n_m} \tag{6.39}$$

In this notation m refers to the mth prism and $(m - 1)$ refers to the previous prism. $\phi_{1,m}$ is the angle of incidence at the mth prism and $\psi_{1,m}$ the corresponding angle of refraction. Similarly $\phi_{2,m}$ is the angle of exit, or emergence, at the mth prism and $\psi_{2,m}$ the corresponding angle of refraction. In this notation $k_{1,m}$ and $k_{2,m}$ refer to the physical beam expansion of the incidence and the emergence beams, respectively. The signs in parentheses (±) refer to deployment in either a positive (+) or compensating (−) configuration, while the simple ± indicates either positive or negative refraction (Duarte 2006).

The generalized cumulative single-pass dispersion equation for positive refraction, that is,

$$\nabla_\lambda \phi_{2,r} = \mathcal{H}_{2,m}\nabla_\lambda n_m \pm (k_{1,m}k_{2,m})^{-1}\left(\mathcal{H}_{1,m}\nabla_\lambda n_m(\pm)\nabla_\lambda\phi_{2,(m-1)}\right) \tag{6.40}$$

can be written in a more engineering-style notation (Duarte 1985, 1989a, 1990a)

$$\nabla_\lambda \phi_{2,r} = \sum_{m=1}^{r}(\pm 1)\mathcal{H}_{1,m}\left(\prod_{j=m}^{r}k_{1,j}\prod_{j=m}^{r}k_{2,j}\right)^{-1}\nabla_\lambda n_m$$
$$+ (M_1 M_2)^{-1}\sum_{m=1}^{r}(\pm 1)\mathcal{H}_{2,m}\left(\prod_{j=1}^{m}k_{1,j}\prod_{j=1}^{m}k_{2,j}\right)\nabla_\lambda n_m \tag{6.41}$$

where

$$M_1 = \prod_{j=1}^{r}k_{1,j} \tag{6.42}$$

$$M_2 = \prod_{j=1}^{r}k_{2,j} \tag{6.43}$$

In this notation r refers to the total number or prism in the array while M_1 denotes the total beam expansion induced by illumination at the incidence surfaces and M_2 denotes the total beam expansion beam expansion induced on the beam at the exit surfaces. It should be noted that most multiple-prism beam expanders are engineered so that the angle of emergence $\phi_{2,m} \approx 0$ so that $M_2 \approx 1$.

6.5.1 Return-pass multiple-prism intracavity dispersion

The double-pass dispersion of multiple-prism beam expanders uses the mirror-image of the original passage of the laser beam to quantify the dispersion for the first return-pass. In this notation the dispersion for the first return-pass is identified as

$$\frac{\partial \phi'_{1,m}}{\partial \lambda} = \nabla_\lambda \phi'_{1,m} \tag{6.44}$$

and the generalized cumulative multiple-prism return-pass dispersion is given by (Duarte and Piper 1982, 1984)

$$\nabla_\lambda \phi'_{1,r} = \mathcal{H}'_{1,m}\nabla_\lambda n_m + (k'_{1,m}k'_{2,m})\left(\mathcal{H}'_{2,m}\nabla_\lambda n_m \pm \nabla_\lambda \phi'_{1,(m+1)}\right) \tag{6.45}$$

where

$$k'_{1,m} = \frac{\cos\psi'_{1,m}}{\cos\phi'_{1,m}} \tag{6.46}$$

$$k'_{2,m} = \frac{\cos \phi'_{2,m}}{\cos \psi'_{2,m}} \tag{6.47}$$

$$\mathcal{H}'_{1,m} = \frac{\tan \phi'_{1,m}}{n_m} \tag{6.48}$$

$$\mathcal{H}'_{2,m} = \frac{\tan \phi'_{2,m}}{n_m} \tag{6.49}$$

In this notation, $\nabla_\lambda \phi'_{1,(m+1)}$ provides the cumulative single-pass multiple-prism dispersion in addition to the dispersion from the diffractive element returning the beam back into its original path

$$\nabla_\lambda \phi'_{1,(m+1)} = \left(\nabla_\lambda \Theta_G \pm \nabla_\lambda \phi_{2,r} \right) \tag{6.50}$$

In general, this diffractive element is a reflective diffraction grating either in grazing-incidence or Littrow configuration with a dispersion of $\nabla_\lambda \Theta_G$. Depending on the particular configuration this dispersion is given either by equation (6.32) or (6.34). In the special case of replacement of the diffraction grating by a dispersionless mirror then only the prismatic dispersion counts and

$$\nabla_\lambda \phi'_{1,(m+1)} = \nabla_\lambda \phi_{2,r} \tag{6.51}$$

Defining $\nabla_\lambda \phi'_{1,m} = \nabla \Phi_P$, where the capital ϕ stands for return pass and P for a plurality of prisms, the explicit-engineering version of the generalized cumulative double-pass, or single return-pass, dispersion for a multiple-prism mirror system is given by (Duarte 1985, 1989a)

$$\begin{aligned}
\nabla_\lambda \Phi_P = 2M_1 M_2 \sum_{m=1}^{r} (\pm 1)\mathcal{H}_{1,m} \left(\prod_{j=m}^{r} k_{1,j} \prod_{j=m}^{r} k_{2,j} \right)^{-1} \nabla_\lambda n_m \\
+ 2 \sum_{m=1}^{r} (\pm 1)\mathcal{H}_{2,m} \left(\prod_{j=1}^{m} k_{1,j} \prod_{j=1}^{m} k_{2,j} \right) \nabla_\lambda n_m
\end{aligned} \tag{6.52}$$

6.5.2 Multiple-return-pass multiple-prism intracavity dispersion

The multiple-return-pass intracavity dispersion of a multiple-prism diffraction grating ensemble is given by (Duarte and Piper 1984)

$$(\nabla_\lambda \theta)_R = (RM \nabla_\lambda \Theta_G + R\nabla_\lambda \Phi_P) \tag{6.53}$$

where R is the number of return passes. The number of return passes refers to the number of intracavity passes the photons can achieve between the leading edge of the excitation pulse to the onset of laser emission at the excited oscillator, resonator, or cavity. Equation (6.53) indicates that the grating dispersion is multiplied by the factor RM, where M is the overall beam magnification of the multiple-prism beam expander thus enormously augmenting the intrinsic intracavity dispersion provided by the diffraction grating $\nabla_\lambda \Theta_G$. If the grating is replaced by a dispersionless mirror, that is, $\nabla_\lambda \Theta_G = 0$, the overall intracavity dispersion reduces to

$$(\nabla_\lambda \theta)_R = R \nabla_\lambda \Phi_P \tag{6.54}$$

6.5.3 Multiple-return-pass cavity linewidth

The single-pass cavity linewidth equation

$$\Delta\lambda \approx \Delta\theta (\nabla_\lambda \theta)^{-1}$$

can be re stated as (Duarte 1990a, 2001, 2003)

$$\Delta\lambda = \Delta\theta_R (RM \nabla_\lambda \Theta_G + R \nabla_\lambda \Phi_P)^{-1} \tag{6.55}$$

where

$$\Delta\theta_R = \frac{\lambda}{\pi w}\left(1 + \left(\frac{L_\mathscr{R}}{B_R}\right)^2 + \left(\frac{A_R L_\mathscr{R}}{B_R}\right)^2\right)^{1/2} \tag{6.56}$$

here, $L_\mathscr{R}$ is the Rayleigh length defined in section 6.2 and the propagation matrix elements A_R and B_R are given in the next subsection. For a well designed cavity the beam divergence approaches its diffraction limit as the number of intracavity passes, R, increase

$$\Delta\theta_R \rightarrow \left(\frac{\lambda}{\pi w}\right) \tag{6.57}$$

6.5.4 Multiple-return-pass propagation matrix

Exact calculation of the multiple return-pass $ABCD$ matrix elements requires a detailed mathematical representation of the tunable laser oscillator of interest. Then, the cavity is unfolded at the exit surface of the output-coupler mirror as illustrated in figure 6.1. Then the matrix elements representing each space and each optical element of the cavity are written down. For the single return-pass of the optimized multiple-prism grating cavity laser oscillator (figure 6.1) incorporating a polymeric organic solid-state gain medium (Duarte 2001) represented by a matrix $\alpha\beta\chi\delta$ that means the multiplication of 14 matrices.

In reference to figure 6.1, defining L_1 the distance between the inner surface of the output coupler-polarizer to the gain medium, L_2 the distance between the gain medium and the entrance of the multiple-prism expander, and L_3 the distance between the exit surface of the multiple-prism beam expander and the diffraction

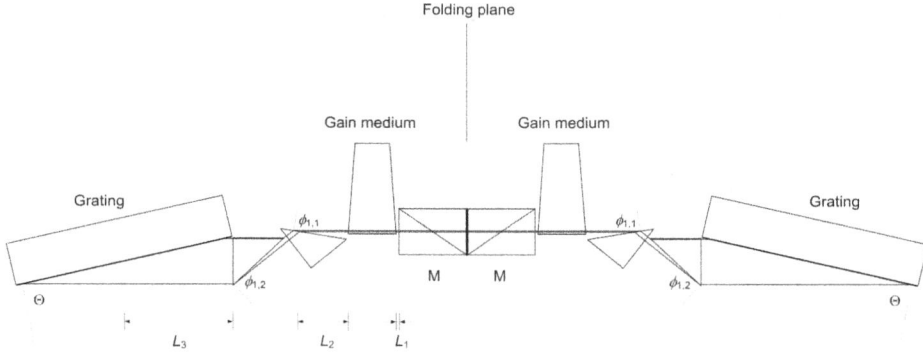

Figure 6.1. Unfolded high-performance narrow-linewidth tunable narrow-linewidth laser oscillator utilized for the multiple-pass propagation matrix analysis.

grating deployed in Littrow configuration, then a single-return pass leads to the following quantities

$$\Lambda = \frac{L_0}{n_0} + L_1 \tag{6.58}$$

$$\Xi = 2L_2 + \frac{2B}{M} + \frac{2L_3}{M^2} \tag{6.59}$$

where L_0 and n_0 are the length and refractive index of the output-coupler polarizer while M is the overall beam magnification provided by the multiple-prism beam expander and B is the corresponding term in its single-pass propagation matrix (Duarte 1989b, 2003)

$$\begin{pmatrix} A & B \\ C & D \end{pmatrix} = \begin{pmatrix} M_1 M_2 & B \\ 0 & (M_1 M_2)^{-1} \end{pmatrix} \tag{6.60}$$

where

$$M_1 = \prod_{m=1}^{r} k_{1,m} \tag{6.61}$$

$$M_2 = \prod_{m=1}^{r} k_{2,m} \tag{6.62}$$

$$B = M_1 M_2 \sum_{m=1}^{r-1} L_m \left(\prod_{j=1}^{m} k_{1,j} \prod_{j=1}^{m} k_{2,j} \right)^{-2} + \frac{M_1}{M_2} \sum_{m=1}^{r} \frac{l_m}{n_m} \left(\prod_{j=1}^{m} k_{1,j} \right)^{-2} \left(\prod_{j=m}^{r} k_{2,j} \right)^{2} \tag{6.63}$$

and the overall beam expansion is given by

$$M = M_1 M_2 \tag{6.64}$$

The multiple-return pass analysis of the optimized high-performance oscillator described by Duarte (1999) leads to (Duarte 2001)

$$A_R = (\alpha A_{R-1} + \chi \mathcal{L}_{R-1})(\alpha + \chi(\Xi - L_2)) + \chi \mathcal{L}_{R-1}(\chi L_2 + \delta) \\ + \chi A_{R-1}(\alpha L_2 + \beta)$$ (6.65)

and

$$B_R = A_R \Lambda + (\alpha A_{R-1} + \chi \mathcal{L}_{R-1})(\beta + \delta(\Xi - L_2)) \\ + \delta \mathcal{L}_{R-1}(\chi L_2 + \delta) + \delta A_{R-1}(\alpha L_2 + \beta)$$ (6.66)

where

$$\mathcal{L}_{R-1} = \Lambda A_{R-1} + B_{R-1}$$ (6.67)

In summary, the matrix elements A_R and B_R are used in the equation for the multiple-return pass beam divergence $\Delta\theta_R$, that is equation (6.56), and the matrix elements $\alpha\beta\chi\delta$ represent the ABCD terms of the gain medium that ideally can be assumed to be $\alpha \approx \delta \approx 1$, $\chi = 0$, and $\beta = \beta$.

6.5.5 Multiple-prism mathematical series

As discussed by Duarte (2014) the generalized first order multiple-prism dispersion equation for positive refraction, that is equation (6.40), leads to mathematical power series that have yet to find applications. Specifically,

$$\nabla_\lambda \phi_{2,m} = \mathcal{H}_{2,m} \nabla_\lambda n_m + (k_{1,m} k_{2,m})^{-1} \left(\mathcal{H}_{1,m} \nabla_\lambda n_m \pm \nabla_\lambda \phi_{2,(m-1)} \right)$$

for the *special case* of r identical prisms deployed at the same angle of incidence ($\phi_{1,1} = \phi_{1,2} = \cdots = \phi_{1,m}$) and orthogonal beam exit ($\phi_{2,1} = \phi_{2,2} = \cdots = \phi_{2,m} = 0$), reduces to the power series (Duarte and Piper 1982, Duarte 1990)

$$\nabla_\lambda \phi_{2,r} = \tan \psi_{1,1} \nabla n_1 \left(1 \pm k_{1,1}^{-1} \pm k_{1,1}^{-2} \pm k_{1,1}^{-3} \pm \cdots \pm k_{1,1}^{-(r-1)} \right)$$ (6.68)

Furthermore, for orthogonal beam exit equation (6.40) reduces to (Duarte 1985, 2014)

$$\nabla_\lambda \phi_{2,r} = \sum_{m=1}^{r} (\pm 1) \mathcal{H}_{1,m} \left(\prod_{j=m}^{r} k_{1,j} \right)^{-1} \nabla_\lambda n_m$$ (6.69)

which can be expressed in its longhand version (Duarte 2012)

$$\nabla_\lambda \phi_{2,r} = \pm \mathcal{H}_{1,1}(k_{1,1}k_{1,2} \ldots k_{1,r})^{-1} \nabla_\lambda n_1 \pm \mathcal{H}_{1,2}(k_{1,2} \ldots k_{1,r})^{-1} \\ \nabla_\lambda n_2 \pm \cdots \pm \mathcal{H}_{1,r}(k_{1,r})^{-1} \nabla_\lambda n_r$$ (6.70)

6.6 The generalized prismatic equations for laser pulse compression

From the cavity linewidth equation

$$\Delta\lambda \approx \Delta\theta(\nabla_\lambda\theta)^{-1}$$

it is clear that a very large intracavity dispersion $\nabla_\lambda\theta$ is necessary to induce very narrow-linewidth emission. For pulsed lasers emitting in the ultrashort pulse regime, such as femtosecond lasers, broadband emission in a single pulse is necessary. Therefore, the control of intracavity dispersion becomes of paramount importance. In this regard, prisms and prism sequences are used to add *negative dispersion* so that the overall dispersion in the cavity is minimized (Dietel *et al* 1983, Fork *et al* 1984). For a review on this subject the reader is referred to Diels (1990) and Diels and Rudolph (1996).

In cavities incorporating prismatic and multiple-prism pulse compressors, control of the intracavity dispersion requires control of the first order dispersion $\nabla_n\phi_{2,m}$ and higher order derivatives. Using the identity

$$\nabla_\lambda\phi_{2,m} = \nabla_n\phi_{2,m}(\nabla_\lambda n_m)^{-1} \tag{6.71}$$

the generalized cumulative single-pass dispersion given in equation (6.40) (Duarte and Piper 1982) can be restated as (Duarte 1987, 2009)

$$\nabla_n\phi_{2,m} = \mathcal{H}_{2,m} + (\mathcal{M})^{-1}\left(\mathcal{H}_{1,m} \pm \nabla_n\phi_{2,(m-1)}\right) \tag{6.72}$$

where

$$(\mathcal{M})^{-1} = k_{1,m}^{-1}k_{2,m}^{-1} \tag{6.73}$$

The second derivative of the refraction angle $\phi_{2,m}$, or first derivative of the dispersion $\nabla_n\phi_{2,m}$, is given by (Duarte 1987)

$$\begin{aligned}
\nabla_n^2\phi_{2,m} = {} & \nabla_n\mathcal{H}_{2,m} \\
& + (\nabla_n\mathcal{M}^{-1})\left(\mathcal{H}_{1,m} \pm \nabla_n\phi_{2,(m-1)}\right) \\
& + (\mathcal{M}^{-1})\left(\nabla_n\mathcal{H}_{1,m} \pm \nabla_n^2\phi_{2,(m-1)}\right)
\end{aligned} \tag{6.74}$$

The third derivative of the refraction angle $\phi_{2,m}$, or second derivative of the dispersion $\nabla_n\phi_{2,m}$, is given by (Duarte 2009)

$$\begin{aligned}
\nabla_n^3\phi_{2,m} = {} & \nabla_n^2\mathcal{H}_{2,m} \\
& + (\nabla_n^2\mathcal{M}^{-1})\left(\mathcal{H}_{1,m} \pm \nabla_n\phi_{2,(m-1)}\right) \\
& + 2(\nabla_n\mathcal{M}^{-1})\left(\nabla_n\mathcal{H}_{1,m} \pm \nabla_n^2\phi_{2,(m-1)}\right) \\
& + (\mathcal{M}^{-1})\left(\nabla_n^2\mathcal{H}_{1,m} \pm \nabla_n^3\phi_{2,(m-1)}\right)
\end{aligned} \tag{6.75}$$

The fourth derivative of the refraction angle $\phi_{2,m}$, or third derivative of the dispersion $\nabla_n\phi_{2,m}$, is given by (Duarte 2009)

$$
\begin{aligned}
\nabla_n^4\phi_{2,m} = \nabla_n^3\mathcal{H}_{2,m} \\
+ (\nabla_n^3\mathcal{M}^{-1})\big(\mathcal{H}_{1,m} \pm \nabla_n\phi_{2,(m-1)}\big) \\
+ 3(\nabla_n^2\mathcal{M}^{-1})\big(\nabla_n\mathcal{H}_{1,m} \pm \nabla_n^2\phi_{2,(m-1)}\big) \\
+ 3(\nabla_n\mathcal{M}^{-1})\big(\nabla_n^2\mathcal{H}_{1,m} \pm \nabla_n^3\phi_{2,(m-1)}\big) \\
+ (\mathcal{M}^{-1})\big(\nabla_n^3\mathcal{H}_{1,m} \pm \nabla_n^4\phi_{2,(m-1)}\big)
\end{aligned}
\tag{6.76}
$$

The fifth derivative of the refraction angle $\phi_{2,m}$, or fourth derivative of the dispersion $\nabla_n\phi_{2,m}$, is given by (Duarte 2009)

$$
\begin{aligned}
\nabla_n^5\phi_{2,m} = \nabla_n^4\mathcal{H}_{2,m} \\
+ (\nabla_n^4\mathcal{M}^{-1})\big(\mathcal{H}_{1,m} \pm \nabla_n\phi_{2,(m-1)}\big) \\
+ 4(\nabla_n^3\mathcal{M}^{-1})\big(\nabla_n\mathcal{H}_{1,m} \pm \nabla_n^2\phi_{2,(m-1)}\big) \\
+ 6(\nabla_n^2\mathcal{M}^{-1})\big(\nabla_n^2\mathcal{H}_{1,m} \pm \nabla_n^3\phi_{2,(m-1)}\big) \\
+ 4(\nabla_n\mathcal{M}^{-1})\big(\nabla_n^3\mathcal{H}_{1,m} \pm \nabla_n^4\phi_{2,(m-1)}\big) \\
+ (\mathcal{M}^{-1})\big(\nabla_n^4\mathcal{H}_{1,m} \pm \nabla_n^5\phi_{2,(m-1)}\big)
\end{aligned}
\tag{6.77}
$$

The sixth derivative of the refraction angle $\phi_{2,m}$, or fifth derivative of the dispersion $\nabla_n\phi_{2,m}$, is given by (Duarte 2009)

$$
\begin{aligned}
\nabla_n^6\phi_{2,m} = \nabla_n^5\mathcal{H}_{2,m} \\
+ (\nabla_n^5 M^{-1})\big(\mathcal{H}_{1,m} \pm \nabla_n\phi_{2,(m-1)}\big) \\
+ 5(\nabla_n^4 M^{-1})\big(\nabla_n\mathcal{H}_{1,m} \pm \nabla_n^2\phi_{2,(m-1)}\big) \\
+ 10(\nabla_n^3 M^{-1})\big(\nabla_n^2\mathcal{H}_{1,m} \pm \nabla_n^3\phi_{2,(m-1)}\big) \\
+ 10(\nabla_n^2 M^{-1})\big(\nabla_n^3\mathcal{H}_{1,m} \pm \nabla_n^4\phi_{2,(m-1)}\big) \\
+ 5(\nabla_n M^{-1})\big(\nabla_n^4\mathcal{H}_{1,m} \pm \nabla_n^5\phi_{2,(m-1)}\big) \\
+ (\mathcal{M}^{-1})\big(\nabla_n^5\mathcal{H}_{1,m} \pm \nabla_n^6\phi_{2,(m-1)}\big)
\end{aligned}
\tag{6.78}
$$

Observing equations (6.74)–(6.78) it can be seen that the numerical factors in the equations, beyond the first term, follow the pattern of Pascal's triangle

$$
\begin{array}{ccccccc}
& & & 1 & & & \\
& & 1 & & 1 & & \\
& 1 & & 2 & & 1 & \\
1 & & 3 & & 3 & & 1 \\
\end{array}
$$

1 4 6 4 1
1 5 10 10 5 1

for N where $N + 1$ is the order of the derivative. Following the equation sequence disclosed here, a generalized format for these equations can be expressed following Duarte (2013):

$$\nabla_n^r \phi_{2,m} = \nabla_n^{r-1} \mathcal{H}_{2,m} + (\mathcal{M})^{-1}(\nabla_n + \zeta)^{r-1} \tag{6.79}$$

where

$$\zeta^s = \nabla_n^s \mathcal{H}_{1,m} \pm \nabla_n^{s+1} \phi_{2,(m-1)} \tag{6.80}$$

$$\zeta^0 = 1 = \mathcal{H}_{1,m} \pm \nabla_n \phi_{2,(m-1)} \tag{6.81}$$

In equation (6.79), when writing the expansion in r, the term $\zeta^0 = 1$ must be included as defined in equation (6.81). The maximum value of the s exponent is $s = (r - 1)$.

Although for femtosecond laser applications the use of $\nabla_n^1 \phi_{2,m}, \nabla_n^2 \phi_{2,m}, \nabla_n^3 \phi_{2,m}$, is necessary, higher order derivatives are thought to be applicable in nonlinear optics. Detailed numerical calculations, for lower order derivatives, have been performed by Duarte (1990b) and comparison with experimentally determined dispersion values directly applicable to double-prism compressors in femtosecond lasers have been performed, by Osvay et al (2004, 2005), with good agreement between theory and experiment.

6.6.1 Fundamentals of pulse compression

As mentioned previously (section 6.2.1) the phase term in the generalized interferometric equation leads directly to the interferometric identity, equation (6.12) (Duarte 2003, 2015)

$$\Delta\lambda = \frac{\lambda^2}{\Delta x}$$

which in turn leads to Heisenberg's uncertainty principle $\Delta p \Delta x \approx h$. Using the interferometric identity in conjunction with $E = mc^2$ and $E = h\nu$, it can be shown that (Duarte 2003, 2015)

$$\Delta\nu\Delta t \approx 1$$

which succinctly and beautifully teaches that in order to have a very short pulse Δt it is imperative to have a very broad $\Delta\nu$. Note that this relation was previously introduced as equation (6.17).

The linewidth equation

$$\Delta\lambda \approx \Delta\theta(\nabla_\lambda\theta)^{-1}$$

can be expressed in the frequency domain as

$$\Delta\nu \approx \Delta\theta(\nabla_\lambda\theta)^{-1}\left(\frac{c}{\lambda^2}\right) \tag{6.82}$$

which indicates that for a very broad $\Delta\nu$ it is necessary that the intracavity dispersion should be minimized to an extreme, or

$$(\nabla_\lambda\theta) \to 0 \tag{6.83}$$

And that is precisely the role of the intracavity multiple-prism sequences mathematically described in the previous section. For a diagram of multiple-prism pulse compressors please refer to chapter 5.

One final point: ultrashort pulse lasers, ultrafast pulse lasers, femtosecond lasers, or attosecond lasers are homologous. A new terminology surfacing in the literature lately is few-cycle pulse lasers. This represents a departure from simply stating the duration of the pulse directly in femtosecods (10^{-15}) or attoseconds (10^{-18}). This few-cycle terminology appears to be based on the limiting concept of $\Delta\nu\Delta t \approx 1$: for example, a wavelength emission of $\lambda \approx 600$ nm corresponds to a frequency of $v \approx 4.99654097 \times 10^{14}$ Hz or $v \approx 5 \times 10^{14}$. This means that the half cycle has a temporal length of $\delta t \approx 2$ fs. For a measured pulse duration of $\Delta t \approx 6$ fs this would correspond to a pulsed laser yielding over one-cycle and less than two-cycles. In this regard, stating the measurement of just $\Delta t \approx 6$ fs is far more informative and accurate.

6.6.2 Grating pair pulse compression

The use of grating pairs in laser pulse compression was introduced and described by Treacy (1969). In the Treacy grating pair configuration the incidence laser beam is diffracted from the first grating on to the second grating and the doubly diffracted beam is then reflected back from a mirror for a second pass through the grating pair. This technique was applied by Strickland and Mourou (1985) in what became known as chirped pulse amplification (CPA). A review of grating pulse compression, in the larger context of intracavity laser pulse compression, is given by Diels and Rudolph (1996).

6.7 Distributed feedback

Distributed feedback (DFB) laser configurations play a significant role in the field of tunable organic lasers. Therefore, some of the basic physics related to DFB lasers is briefly reviewed here. DFB laser architectures do not belong to any of the types of resonators previously described, since they do not incorporate mirrors in the optical axis of the emission (Kogelnick and Shank 1971). Furthermore, first generation DFB lasers were optically pumped by two counter-propagating waves. Even though these devices have been replaced by simple gain-regions configured in a diffraction grating format, they are considered here since they provide an instructive perspective on the physics involved.

Kogelnick and Shank (1971) consider the scalar wave equation for the electric field

$$\frac{\partial^2 E}{\partial z^2} + \kappa^2 E = 0 \tag{6.84}$$

The electric field is then described as the sum of two counter-propagating waves

$$E(z) = R(z)e^{-i\kappa z/2} + S(z)e^{i\kappa z/2} \tag{6.85}$$

where z is in the direction of propagation. These authors also assume a periodic spatial variation in the refractive index

$$n(z) = n + n_1 \cos \kappa z \tag{6.86}$$

where κ is related to the fringe spacing Λ by

$$\kappa = \frac{2\pi}{\Lambda} \tag{6.87}$$

while the laser wavelength is defined by the *Bragg condition*

$$\lambda_l = 2n\Lambda \tag{6.88}$$

The passive linewidth of a DFB laser is given by Bor (1979)

$$\Delta \lambda_l = \left(\frac{\lambda_l}{\lambda_p}\right)\Delta \lambda_p \tag{6.89}$$

where λ_l is the wavelength of the DBF laser, λ_p is the wavelength of the excitation laser, and $\Delta \lambda_p$ its linewidth.

Tuning DFB lasers can be accomplished by either varying the refractive index of the gain medium or by changing Λ. For a DFB laser configuration allowing rotation of the excitation mirrors, so that the angle of incidence on the gain medium (θ) can be varied, the fringe separation is given by

$$\Lambda = \frac{\lambda_p}{2 \sin \theta} \tag{6.90}$$

and Shank *et al* (1971)

$$\lambda_l = \frac{n\lambda_p}{\sin \theta} \tag{6.91}$$

which allow the laser to be tuned by geometrical means.

Notice that equation (6.90) is strikingly similar to the diffraction grating equation in Littrow configuration, that is, equation (6.23) $2d \sin \Theta = m\lambda$, for $m = 1$. Indeed, second generation DFB lasers obviate the interferometric arrangement simply by configurationally imposing the periodic structure of a diffraction grating on the gain medium. In other words an organic dye-doped MPMMA diffraction grating becomes a DFB laser via optical excitation at an appropriate angle.

Please refer to chapter 5 for diagrams on DFB configurations. Also, the interferometric physics of diffraction gratings, both in the transmission and reflection domain, is elucidated in revealing detail by Duarte (2006) and reviewed in Duarte (2015).

6.8 Longitudinal tuning in laser microcavities

Wavelength tuning in dispersive tunable lasers utilizes grating rotation techniques that are based on the diffraction grating equation and synchronous tuning that allows for a laser cavity change whist the diffraction grating is rotated. These techniques are well covered and explained in previous quantitative reviews (Duarte 2003, 2015). Here, a compact set of equations designed to describe wavelength tuning in microcavity lasers, using the change in the longitude of the cavity, is described. This set of equations, and methodology, was introduced and refined by Duarte (2003, 2014, 2015).

Longitudinal tuning is based on the interferometric identity

$$\delta\lambda = \frac{\lambda^2}{\delta x} \tag{6.92}$$

that can be derived from Heisenberg's uncertainty principle (Duarte 2003). For two relatively close but different wavelengths, λ_1 and λ_2, corresponding to cavity lengths L and $L \pm \Delta L$ (Duarte 2014, 2015)

$$\delta\lambda_1 = \frac{\lambda_1^2}{2L} \tag{6.93}$$

and

$$\delta\lambda_2 = \frac{\lambda_2^2}{2(L \pm \Delta L)} \tag{6.94}$$

where $\delta\lambda_1$ and $\delta\lambda_2$ are the corresponding longitudinal mode spacing. Also, the number of longitudinal modes for each case is given by

$$N_1 = \frac{\Delta\lambda_1}{\delta\lambda_1} \tag{6.95}$$

$$N_2 = \frac{\Delta\lambda_2}{\delta\lambda_2} \tag{6.96}$$

Assuming that the laser linewidth remains approximately the same at these two wavelengths, that is $\Delta\lambda_1 \approx \Delta\lambda_2$, then

$$\lambda_2 \approx \lambda_1 \left(\frac{\delta\lambda_2}{\delta\lambda_1}\right)^{1/2} \left(1 \pm \frac{\Delta L}{L}\right)^{1/2} \tag{6.97}$$

or

$$\lambda_2 \approx \lambda_1 \left(\frac{N_1}{N_2}\right)^{1/2}\left(1 \pm \frac{\Delta L}{L}\right)^{1/2} \tag{6.98}$$

For single-longitudinal-mode oscillation $N_1 = N_2 = 1$ and equation (6.98) reduces to

$$\lambda_2 \approx \lambda_1\left(1 \pm \frac{\Delta L}{L}\right)^{1/2} \tag{6.99}$$

Thus, changing the cavity length by the ratio $(\Delta L/L)$ leads to direct wavelength tuning. For the case of wavelength tuning when there is multi-longitudinal-mode lasing, and the inclusion of numerical examples, see Duarte (2015).

6.9 Laser linewidth of microcavity emission

Microcavities, or nanocavities, provide very special confines, with unique properties, for laser radiation since the length of the cavity can be in the $(\lambda/n) \leqslant L \leqslant \lambda$ range, where n is a number greater than 1. The exposition given here refines previously published basic concepts by Duarte (2008, 2010, 2016).

Assuming that the transverse-mode structure in the emission from a microcavity can be confined, using geometrical means, to a single-transverse-mode then the task is entirely reduced to the characterization of the longitudinal-mode emission. And it is in this characterization of the longitudinal-mode emission where the unique physics of the dimensions of the microcavity, or nanocavity, come into play.

The interferometric identity introduced in the previous section can be re-expressed as

$$\delta\lambda_e = \frac{\lambda_e^2}{2L} \tag{6.100}$$

which sets the conditions for the intracavity longitudinal mode spacing $\delta\lambda_e$. Here, λ_e is the emission wavelength, and L is the length of the microcavity. Although excessively pedagogical, the best way to proceed is by means of an example. Let us assume that emission wavelength is $\lambda_e \approx 600$ nm and the cavity length $L \approx 300$ nm. Under these conditions the intracavity mode spacing becomes $\delta\lambda_e \approx 600$ nm, which is an extraordinarily enormous intracavity mode spacing! The implications are that any emission linewidth in the $\Delta\lambda_e \leqslant 300$ nm range, let us say, corresponds to *single-longitudinal-mode emission*.

6.10 Linewidth equivalence

The interferometric identity equation (6.14)

$$\Delta\lambda = \frac{\lambda^2}{\Delta x}$$

is given in meter units (m). The equivalent identity in the frequency domain is equation (6.15)

$$\Delta \nu = \frac{c}{\Delta x}$$

given in units cf Hz, where the speed of light is $c = 2.997\ 924\ 58$ m s^{-1}. These two expressions allow the conversion of linewidths from the spatial to the frequency domain and vice versa. In this regards, experimentalists often convert $\Delta \nu$ in MHz to $\Delta \lambda$ in nm. For example: $\Delta \nu = 300$ MHz becomes $\Delta \lambda \approx 0.000\ 36$ nm at $\lambda = 600$ nm.

On the other hand, the spectroscopists use the reciprocal cm, or cm^{-1}, that is defined from

$$\frac{\Delta \nu}{c} = \frac{1}{\Delta x} \tag{6.101}$$

where the unit of Δx is meters (m). Hence, reaching down to the cm^{-1} requires the use of (Duarte 2015)

$$\frac{1}{\Delta x'} = \frac{1}{100}\left(\frac{1}{\Delta x}\right) \tag{6.102}$$

Using the previous conversion example: $\Delta \nu = 300$ MHz becomes $(1/\Delta x') \approx 0.01$ cm^{-1}.

6.11 Problems

- 6.1 Show that equation (6.6) and equation (6.7) are equivalent, or that equation (6.7) follows from equation (6.6).
- 6.2 Show that equation (6.40) and equation (6.41) are equivalent, or that equation (6.41) follows from equation (6.40).
- 6.3 Show that the factor B in the matrix equation (6.60) can be expressed as equation (6.63).
- 6.4 Show that equation (6.35) can be expressed as the power series given in equation (6.68).
- 6.5 Show that equation (6.69) can be expressed as the power series given in equation (6.70).
- 6.6 Show that the second derivative of $\phi_{2,m}$ is given by equation (6.74).
- 6.7 Show that the third derivative of $\phi_{2,m}$ is given by equation (6.75).
- 6.8 Show that the generalized rth derivative of $\phi_{2,m}$, that is $\nabla_n^r \phi_{2,m}$, is given by equation (6.79).
- 6.9 Show that equation (6.97) and equation (6.98) are equivalent.

References

Bor Z S 1979 A novel pumping arrangement for tunable single picosecond pulse generation with a N$_2$ laser pumped distributed feedback dye laser *Opt. Commun.* **29** 103–8

Diels J-C 1990 Femtosecond dye lasers *Dye Laser Principles* ed F J Duarte and L W Hillman (New York: Academic) ch 3

Diels J-C and Rudolph W 1996 *Ultrafast Laser Pulse Phenomena* (New York: Academic)

Dietel W, Fontaine J J and Diels J-C 1983 Intracavity pulse compression with glass: a new method of generating pulses shorter than 60 fs *Opt. Lett.* **8** 4–6

Dirac P A M 1978 *The Principles of Quantum Mechanics* 4th edn (London: Oxford)

Duarte F J 1985 Note on achromatic multiple-prism beam expanders *Opt. Commun.* **53** 259–62

Duarte F J 1987 Generalized multiple-prism dispersion theory for pulse compression in ultrafast dye lasers *Opt. Quantum Electron.* **19** 223–9

Duarte F J 1989a Transmission efficiency in achromatic nonorthogonal multiple-prism laser beam expanders *Opt. Commun.* **71** 1–5

Duarte F J 1989b Ray transfer matrix analysis of multiple-prism dye laser oscillators *Opt. Quantum Electron.* **21** 47–54

Duarte F J 1990a Narrow-linewidth pulsed dye laser oscillators *Dye Laser Principles* ed F J Duarte and L W Hillman (New York: Academic) ch 4

Duarte F J 1990b Prismatic pulse compression: beam deviations and geometrical perturbations *Opt. Quantum Electron.* **22** 467–71

Duarte F J 1991 Dispersive dye lasers *High Power Dye Lasers* ed F J Duarte (Berlin: Springer) ch 2

Duarte F J 1992 Cavity dispersion equation $\Delta\lambda \approx \Delta\theta(\partial\theta/\partial\lambda)^{-1}$: a note on its origin *Appl. Opt.* **31** 6979–82

Duarte F J 1993 On a generalized interference equation and interferometric measurements *Opt. Commun.* **103** 8–14

Duarte F J 1994 Solid-state multiple-prism grating dye-laser oscillators *Appl. Opt.* **33** 3857–60

Duarte F J 1995 Interferometric imaging *Tunable Laser Applications* ed F J Duarte (New York: Marcel-Dekker) 1st edn ch 5

Duarte F J 1999 Multiple-prism grating solid-state dye laser oscillator: optimized architecture *Appl. Opt.* **38** 6347–49

Duarte F J 2001 Multiple-return-pass beam divergence and the linewidth equation *Appl. Opt.* **40** 3038–41

Duarte F J 2003 *Tunable Laser Optics* (New York: Elsevier-Academic)

Duarte F J 2006 Multiple-prism dispersion equations for positive and negative refraction *Appl. Phys.* B **82** 35–8

Duarte F J 2008 Coherent electrically exited organic semiconductors: coherent or laser emission? *Appl. Phys.* B **90** 101–8

Duarte F J 2009 Generalized multiple-prism dispersion theory for laser pulse compression: higher order phase derivatives *Appl. Phys.* B **96** 809–14

Duarte F J 2010 Electrically-pumped organic-semiconductor coherent emission: a review *Coherence and Ultrashort Pulse Laser Emission* ed F J Duarte (Rijeka: InTech)

Duarte F J 2012 Tunable organic dye lasers: physics and technology of high-performance liquid and solid-state narrow-linewidth oscillators *Prog. Quantum Electron.* **36** 29–50

Duarte F J 2013 Tunable laser optics: applications to optics and quantum optics *Prog. Quantum Electron.* **37** 326–47

Duarte F J 2014 *Quantum Optics for Engineers* (New York: CRC)

Duarte F J 2015 *Tunable Laser Optics* 2nd edn (New York: CRC)

Duarte F J 2016 Coherent electrically-excited organic semiconductors *Tunable Laser Applications* ed F J Duarte (New York: CRC) 3rd edn ch 12

Duarte F J and Piper J A 1982 Dispersion theory of multiple-prism beam expander for pulsed dye lasers *Opt. Commun.* **43** 303–7

Duarte F J and Piper J A 1984 Multi-pass dispersion theory of prismatic pulsed dye lasers *Optica Acta* **31** 331–5

Feynman R P, Leighton R B and Sands M 1965 *The Feynman Lectures on Physics* **vol III** (Reading, MA: Addison-Wesley)

Fork R L, Martinez O M and Gordon J P 1984 Negative dispersion using pairs of prisms *Opt. Lett.* **9** 150–2

Kogelnick H and Shank C V 1971 Stimulated emission in a periodic structure *App. Phys. Lett.* **18** 152–4

Osvay K, Kovács A P, Heiner Z, Kurdi G, Klebniczki J and Csatári M 2004 Angular dispersion and temporal change of femtosecond pulses from misaligned pulse compressors *IEEE J. Selec. Top. Quantum Electron* **10** 213–20

Osvay K, Kovács A P, Kurdi G, Heiner Z, Divall M, Klebniczki J and Ferincz I E 2005 Measurement of non-compensated angular dispersion and the subsequent temporal lengthening of femtosecond pulses in a CPA laser *Opt. Commun.* **248** 201–9

Sargent M, Scully M O and Lamb W E 1974 *Laser Physics* (Reading, MA: Addison-Wesley)

Shank C V, Bjorkholm J E and Kogelnik H 1971 Tunable distributed-feedback dye laser *Appl. Phys. Lett.* **18** 395–6

Strickland D and Mourou G 1985 Compression of amplified chirped optical pulses *Opt. Commun.* **56** 219–21

Treacy B E 1969 Optical pulse compression with diffraction grating *IEEE J. Quantum Electron.* **QE-5** 454–60

Chapter 7

Best performance of organic lasers

F J Duarte

Best performance for liquid and solid-state organic tunable lasers is provided in tabular form. Output parameters emphasized include: laser linewidth, beam divergence, amplified spontaneous emission levels, tuning range, and conversion efficiency.

7.1 Introduction

In this chapter the best performance of the organic laser class is presented in tabular form. Best performance is defined from a perspective of measured narrow-linewidth lasing correlated with single-longitudinal-mode (SLM) emission or with double-longitudinal-mode (DLM) emission at the least. This class of highly-coherent emission, of course, implies lasing in a single-transverse-mode beam. Other attributes of best in class include highly-polarized emission, a wide tunablity range, and low ASE levels.

Although a great deal has been published on organic lasers, see reviews by Kranzelbinder and Leising (2000), Samuel and Turnbull (2007), and Karnutsch (2007), relatively few works report emission with the salient characteristics outlined above.

A caveat: although some classes of lasers included in this chapter do not measure up in performance to the standards described above, they are included nevertheless for the sake of completeness and given their potential for future high-performance tunable narrow-linewidth development.

In chapter 5 the architecture of high-performance oscillators was presented. In chapter 6 the theory necessary to design these narrow-linewidth oscillators, and quantify their emission performance, was outlined. In this chapter the performance of narrow-linewidth oscillators and narrow-linewidth tunable organic laser systems is tabulated.

doi:10.1088/978-0-7503-5547-6ch7

7.2 Laser-pumped liquid organic dye lasers

In this section the performance of tunable narrow-linewidth laser-pumped liquid organic dye lasers is tabulated. Although the emphasis here is on exquisite narrow-linewidth performance, it should be remembered that laser-pumped liquid organic dye lasers have an enormous capacity for the generation of raw high-energy per pulse as documented by the 800 J obtained in an excimer-laser-pumped coumarin 480 dye laser (Tang *et al* 1987).

7.2.1 Tunable narrow-linewidth laser oscillators

The first high-performance narrow-linewidth tunable organic laser was that reported by Hänsch in 1972. This laser configuration used a two-dimensional intracavity telescope as a beam expander to illuminate the diffraction grating. Compactness and significantly shorter cavities were facilitated by the adoption of multiple-prism beam expanders (see Duarte (1990a) for a review).

In table 7.1 the emission characteristics of various high-performance tunable organic lasers is tabulated. Particular attention is given to wavelength tuning range, laser linewidth $\Delta\nu$, ASE level in terms of spectral density ratios ρ_{ASE}/ρ_l, and laser conversion efficiency $\eta(\%)$.

Two important footnotes refer to the entries related to the telescopic cavity including an intracavity etalon, yielding $\Delta\nu \approx 300$ MHz, and the HMPGI cavity configuration yielding $\Delta\nu \approx 400$ MHz: these two laser linewidths correspond to SLM emission.

An additional important fact relevant to the copper-vapor-laser pumped MPL and HMPGI oscillators (Duarte and Piper 1984) is that their performance was

Table 7.1. High-performance laser-pumped organic dye tunable laser oscillators.

Cavity configuration	Tuning range (nm)	$\Delta\nu$	ρ_{ASE}/ρ_l	$\eta(\%)$	Reference
Telescopic		2.5 GHz		20	Hänsch (1972)
Telescopic		300 MHz[a]		2–4	Hänsch (1972)
MPL[b]	$490 \leqslant \lambda \leqslant 530$	1.61 GHz	7.5×10^{-6}	14	Duarte and Piper (1980)
MPL		60 MHz[a]		5	Benhardt and Rassmusen (1981)
HMPGI[c]	$490 \leqslant \lambda \leqslant 530$	1.15 GHz		4	Duarte and Piper (1981)
MPL[d]	$565 \leqslant \lambda \leqslant 605$	1.4 GHz		5	Duarte and Piper (1984)
HMPGI[d]	$565 \leqslant \lambda \leqslant 603$	400 MHz[e]	5.0×10^{-7}	4	Duarte and Piper (1984)

[a] Linewidth obtained via the introduction of an intracavity etalon.
[b] Multiple-prism Littrow (MPL) grating configuration.
[c] Hybrid multiple-prism grazing-incidence (HMPGI) grating configuration.
[d] Lasing performance recorded at a prf of 8.2 KHz.
[e] SLM emission at a beam divergence of $\Delta\theta \approx 2$ mrad.

recorded at a prf of 8 kHz and thus it includes frequency jitter, which for an optimized laminar flow, at ~5 m s^{-1}, should be fairly low, nevertheless.

An important feature of these oscillators is their ~100% polarization parallel to the plane of incidence. This is accomplished by a careful matching of the copper-laser polarization to intrinsic polarization preference of the rhodamine 6 G molecules (Duarte 1990b) plus the reinforcement from multiple-prism grating ensemble (Duarte 1990a).

At this stage it should be mentioned that this class of high-performance high-prf copper-vapor laser pumped tunable organic laser oscillator is of central interest to the important application of atomic vapor laser isotope separation (AVLIS) (Duarte and Piper 1984, Broyer *et al* 1984, Webb 1991, Bass *et al* 1992, Singh *et al* 1994, Dasgupta *et al* 1995, Maruyama *et al* 1995, Sugiyama *et al* 1996, Takehisa 1997). Further published works on this subject are given by Bokhan (2006), Singh (2006), Ray (2007), and Chaube (2008). A contemporaneous review on the laser aspects of laser isotope separation is given by Duarte (2016a).

7.2.2 Master-oscillator power-amplifiers

Laser-pumped master-oscillator power-amplifier (MOPA) laser systems incorporate a high-performance master oscillator and one or several amplification stages often pumped transversely by the same laser (Bos 1981, Dupre 1987), or chains of highly synchronized excitation lasers (Bass *et al* 1992). Perhaps the most iconic laser system in this category is the former copper-vapor-laser pumped organic dye laser system at Lawrence Livermore (Bass *et al* 1992) which used to run 24 h per day at a prf of ~13 kHz and delivering narrow-linewidth emission at a combined total average power of 2.5 kW. It should be mentioned that this laser, despite being a significant technical success, was decommissioned once sold to private interests.

In table 7.2 the emission characteristics of three high-performance tunable organic MOPA laser systems is tabulated. Particular attention is given to output

Table 7.2. High-performance organic dye master-oscillator (MO) power-amplifiers (PA).

Oscillator	$\Delta\nu$	Amplification stages	Gain η^{b}	Energy[c] or power	Reference
Telescopic[a]	320 MHz @ 590 nm	3	229 55%	165 mJ	Bos (1981)
HMPGI	650 MHz @ 440 nm	2	700 9%	3.5 mJ	Dupre (1987)
MPL	0.5 – 5.0 GHz @ 590 nm	3 – 4[d]	50 – 60%	2400 W[e]	Bass *et al* (1992)

[a] Linewidth obtained with the introduction of an intracavity etalon.
[b] Optical conversion efficiency.
[c] Energy per pulse.
[d] At each of four amplification chains.
[e] Average power at a prf of 13.2 kHz.

laser linewidth $\Delta\nu$, number of amplification stages, gain from the oscillator to the amplified output, conversion efficiency $\eta(\%)$, and output energy per pulse, or total average power.

It should be noticed that, for the Lawrence Livermore laser, although the laser linewidth range is quoted as $500 \leqslant \Delta\nu \leqslant 5000$ MHz, most likely the preferred linewidth range was $500 \leqslant \Delta\nu \leqslant 1000$ MHz given their dedicated isotope separation application. A further feature well worth mentioning is the fairly high conversion efficiency which is reported in the 50–60% range.

7.2.3 CW lasers

Detailed reviews on the physics and technology of CW organic dye lasers are given by Hollberg (1990) and Johnston and Duarte (2002). Ultra narrow linewidth oscillation in this class of lasers is reported to be as narrow as $\Delta\nu \approx 750$ Hz (Hough *et al* 1987) and $\Delta\nu \approx 100$ Hz (Drever *et al* 1983) whist using stabilization techniques.

In table 7.3 the performance of two CW organic dye lasers is summarized. This table includes cavity configuration, laser linewidth ($\Delta\nu$), tuning range with a single dye output laser power in Watts, and laser efficiency (η). Both lasers use rhodamine 6 G as the gain medium and Ar^+ laser excitation. For a description of linear and ring CW laser cavities please refer to chapter 5.

7.2.4 Femtosecond pulse lasers

The linear and ring cavity configurations of CW organic dye lasers used in conjunction with a saturable absorber were the workhorses on which femtosecond pulse lasers were developed (see, for example, Diels (1990), Diels and Rudolph (1996)). As illustrated in the schematics described in chapter 5, both linear and ring CW dye laser cavity configurations, including a saturable absorber, were also made to incorporate prismatic pulse compressors (Dietel *et al* 1983). The performance of these lasers is summarized in table 7.4 which includes cavity configuration, emission wavelength, type of pulse compressor, and pulse duration Δt. It should be noted that according to the new terminology in the contemporaneous literature, the pulse length of $\Delta t \approx 6$ fs, at $\lambda \approx 620$ nm would correspond to an emission just over the one-cycle.

Table 7.3. High-performance CW organic dye lasers.

Cavity	$\Delta\nu$ (MHz)	Tuning range (nm)	Output power (W)	$\eta(\%)$	Reference
Linear		$560 \leqslant \lambda \leqslant 650$	33	17	Baving *et al* (1982)
Ring	3–4	$575 \leqslant \lambda \leqslant 640$	5.6	23	Johnston and Duarte (2002)

Table 7.4. High-performance femtosecond organic dye lasers.

Cavity configuration	λ (nm)	Pulse compressor	Δt (fs)	Reference
Ring	~600	Single-prism	60	Dietel *et al* (1983)
Cavity-dumped	570–630	Double-prism plus fiber	50	Kafka and Baer (1992)
Linear	615	2 × (double-prism)	29	Kubota *et al* (1988)
	620	2 × (double-prism) plus grating pairs	6	Fork *et al* (1987)

The multiple-prism pulse compressors developed for CW organic dye lasers are used pervasively throughout the laser field including in Ti:sapphire (Osvay *et al* 2005) and semiconductor lasers (Pang *et al* 1992).

7.3 Flashlamp-pumped organic dye lasers

In this section the performance of tunable narrow-linewidth flashlamp-pumped liquid organic dye lasers is tabulated. Again, although the emphasis here is on specifically narrow-linewidth performance, it should be emphasized that flashlamp-pumped liquid organic dye lasers have an intrinsic capacity for the generation of extraordinary high-energy per pulse as documented by the 400 J obtained using a coaxial device and rhodamine 6 G dye (Baltakov *et al* 1974). Another flashlamp-pumped dye laser in the same energy category is reported to yield 140 J per pulse using transverse excitation and rhodamine 6 G dye (Klimek *et al* 1992). A further feature of this work is that operating at a prf of 10 Hz the output average power was 1.4 kW from this single device.

7.3.1 Tunable narrow-linewidth laser oscillators

Various oscillator configurations, including intracavity etalons, yielding relatively narrow-linewidth emission in the $346 \leqslant \Delta\nu \leqslant 9000$ MHz range are discussed in the review literature (Duarte 1990b). Here, the focus is specifically on dispersive cavities that do not need the added feature of one or more intracavity etalons to reach narrow-linewidth performance status.

In table 7.5 the emission characteristics of two high-performance tunable organic laser oscillators is tabulated. Attention is given to output laser linewidth $\Delta\nu$, laser beam divergence $\Delta\theta$, ASE level as determined by the ratio ρ_{ASE}/ρ_l, and output energy per pulse. The performance of the two multiple-prism grating organic laser oscillators included in table 7.5 remain at the top of the scale for this class of tunable laser oscillators.

Features not included in table 7.5 and worth mentioning are that these oscillators provide a laser output ~100% polarized parallel to the plane of incidence. Both these oscillators emitted in the long-pulse domain. The first oscillator, under linear flashlamp excitation, emitted pulses as long as $\Delta t \approx 1000$ ns while the second oscillator, under coaxial flashlamp excitation, emitted relatively shorter pulses at Δt

Table 7.5. High-performance flashlamp-pumped organic dye tunable laser oscillators.

Cavity configuration	$\Delta\nu$ (MHz)	$\Delta\theta$ (mrad)	ρ_{ASE}/ρ_l	Energy (mJ)	Reference
MPL	138[a]	0.57	3×10^{-11}	3.6	Duarte *et al* (1990)
MPL	300[b]	0.35	1×10^{-8}	3.4	Duarte *et al* (1991)

[a] SLM oscillation at $\Delta t \approx 1000$ ns and $(\delta\lambda/\lambda) \approx 1.2 \times 10^{-6}$.
[b] Near-SLM oscillation at $\Delta t \approx 150$ ns and $(\delta\lambda/\lambda) \approx 4.63 \times 10^{-7}$.

≈ 150 ns. The wavelength stability of the first oscillator was measured at $(\delta\lambda/\lambda) \approx 1.2 \times 10^{-6}$ while this figure is $(\delta\lambda/\lambda) \approx 4.63 \times 10^{-7}$ for the ruggedized oscillator. The linewidth reported for the first oscillator, $\Delta\nu \approx 138$ MHz, is SLM oscillation while the linewidth quoted for the ruggedized oscillator $\Delta\nu \approx 300$ MHz is an upper limit corresponding to the double-longitudinal-mode spacing of the multiple-prism grating cavity. For both lasers the dye was rhodamine 6 G.

7.3.2 Flashlamp-pumped master-oscillator forced-oscillators

Various master-oscillator forced-oscillator (MOFO) configurations are discussed in the review literature (Duarte 1990b). Here, the focus is specifically on MOFO systems that deliver both narrow-linewidth emission and laser outputs in the range of hundreds of mJ.

In table 7.6 the emission characteristics of two high-performance MOFOs is tabulated. Attention is given to output laser linewidth $\Delta\nu$, gain, and output energy per pulse. Besides the information included in tabular form the following characteristics of the MOFO system injected by the MPL grating oscillator are useful to highlight: the beam divergence was $\Delta\theta \approx 0.5$ mrad and the pulse length $\Delta t \approx 250$ ns. The quoted linewidth of $\Delta\nu \approx 375$ MHz was the average observed figure although it could be as low as $\Delta\nu \approx 333$ MHz with smooth temporal pulses consistent with SLM oscillation (Duarte and Conrad 1987).

7.4 Solid-state tunable organic dye lasers

In this section attention is focused on tunable solid-state narrow-linewidth organic lasers. This is a relatively new field initiated with the demonstration of near single-longitudinal-mode emission, at $\Delta\nu \approx 1.12$ GHz and a conversion efficiency better than 9%, using rhodamine 6 G dye-doped modified poly methyl methacrylate (MPMMA) gain media at a concentration of 5×10^{-4} M, within multiple-prism oscillator architectures by Duarte (1994).

The high optical quality polymeric gain media for these experiments was specifically manufactured using trapezoidal geometries and a set of predetermined dye concentrations, as designed by Duarte (1994) and by Maslyukov *et al* (1995). It should be added that these dye-doped polymeric matrices are still today (2018) beautifully functional and that they tend to 'heal themselves' by diffusion if dye bleaching occurs when pumped at energy densities at or above 1 J cm^{-2} (Duarte 1997, Duarte and James 2016).

Table 7.6. High-performance flashlamp-pumped organic dye MOFOs.

MO Reference	configuration	FO configuration	$\Delta\nu$	Gain	Energy[a]
Etalons	Flat mirror cavity	346 MHz @ 589 nm		300 mJ	Flamant *et al* (1984)
MPL	Unstable resonator	375 MHz[b] @ 590 nm	51	600 mJ	Duarte and Conrad (1987)

[a] Energy per pulse.
[b] Narrowest linewidth measured in these experiments was $\Delta\nu \approx 333$ MHz consistent with SLM oscillation.

7.4.1 Narrow-linewidth tunable solid-state laser oscillators

As previously mentioned, there is a very large body of literature associated with the subject of organic lasers as can be established by perusing reviews on the subject (Kranzelbinder and Leising 2000, Samuel and Turnbull 2007, Karnutsch 2007). However, it appears that the bulk of this literature is concerned with organic gain media, methods of excitation, and micro devices. In this regard, the literature dedicated to tunable organic lasers with a high degree of spatial and spectral coherence, is scarce. In this section the focus is precisely on high-performance tunable narrow-linewidth organic lasers. In particular, the focus is on longitudinally-excited HMPGI and MPL grating oscillator architectures and on a vertical cavity-type configuration. The gain medium in the case of the dispersive oscillators is dye-doped MPMMA while the vertical cavity lasers uses dye-doped thin-film PMMA.

First, the performance of the dispersive laser oscillators is tabulated in table 7.7 which refers to spatial and spectral coherence of the emission, and table 7.8 which summarizes the energetic properties of the emission.

In table 7.7 attention is given to output laser linewidth $\Delta\nu$, laser beam divergence $\Delta\theta$, and tuning range. Also included in this table is the $\Delta\nu\Delta t$ product that determines how close the emission from the laser is to the limit established by Heisenberg's uncertainty principle $\Delta\nu\Delta t \approx 1$ (see chapter 6). Table 7.8 includes laser peak power P_l, spectral power density ρ_{SP}, ASE level as determined by the ratio ρ_{ASE}/ρ_l, and conversion efficiency η.

Additional information necessary to assess the performance of these oscillators is the time duration of the laser pulse which, for the MPL grating configuration, is $\Delta t \approx 3$ ns in a smooth near-Gaussian profile (Duarte 1999). From table 7.7 it is seen that, for the optimized MPL grating configuration in terms of linewidth performance is near the limit allowed by Heisenberg's uncertainty principle, $\Delta t\,\Delta\nu \approx 1.05$ while offering enormous spectral brightness as illustrated by the descriptive spectral power density parameter $\rho_{SP} \approx 8.33 \times 10^7$ W nm^{-1} (see table 7.8).

For the thin-film vertical external-cavity surface emitting organic laser (VECSOL) configuration (see tables 7.9 and 7.10), $\Delta t \approx 20$ ns in a non-symmetrical oscillatory pulse (Mhibik *et al* 2016) and $\Delta t\,\Delta\nu \approx 4$. As far as spectral brightness is concerned a spectral power density coefficient of $\rho_{SP} \approx 4.78 \times 10^5$ W nm^{-1} can be derived from the data provided.

Table 7.7. High-performance laser-pumped solid-state organic laser oscillators.

Cavity configuration	$\Delta\nu$(MHz)	$\times (\Delta t\Delta\nu)$	$\Delta\theta$ (mrad)	Tuning range (nm)	Reference
MPL	420	1.26	2.3	$562 \leqslant \lambda \leqslant 613$	Duarte (1995)
HMPGI	375[a]	1.13	2.3	$565 \leqslant \lambda \leqslant 610$	Duarte (1997)
MPL	350[b]	1.05	2.2	$550 \leqslant \lambda \leqslant 603$	Duarte (1999)

[a] Narrowest linewidth measured in these experiments was $\Delta\nu \approx 200$ MHz.
[b] Narrowest linewidth measured in these experiments was $\Delta\nu \approx 290$ MHz.

Table 7.8. Energetics of high-performance laser-pumped solid-state organic laser oscillators.

Cavity configuration	P_l (kW)[a]	ρ_{SP} (W nm^{-1})	ρ_{ASE}/ρ_l	η (%)	Reference
MPL	~33.00	~6.76×10^7	7×10^{-7}	5	Duarte (1995)
HMPGI	~27.00	~6.41×10^7	1×10^{-7}	3 – 4	Duarte (1997)
MPL	~33.33	~8.33×10^7	1×10^{-6}	4	Duarte (1999)

[a] Laser emission pulse power.

Table 7.9. Performance of laser-pumped thin-film organic laser oscillator.

Cavity configuration	$\Delta\nu$ (MHz)	$\times (\Delta t\Delta\nu)$	$\Delta\theta$ (mrad)[a]	Tuning range[a] (nm)	Reference
VBG-VECSOL[b]	200[c,d]	~4			Mhibik et al (2016)

[a] Empty space means no data available.
[b] Volume Bragg grating vertical external-cavity surface-emitting organic laser.
[c] Linewidth measured indirectly.
[d] Structure of non-Gaussian temporal pulse indicative of double-longitudinal-mode lasing.

Table 7.10. Energetics of high-performance thin-film organic laser oscillator.

Cavity configuration	P_l[b] (kW)	ρ_{SP} (W nm^{-1})	ρ_{ASE}/ρ_l[c]	η (%)	Reference
VBG-VECSOL[a]	0.128	~ 4.78×10^5		~1.8	Mhibik et al (2016)

[a] Volume Bragg grating vertical external-cavity surface-emitting organic laser.
[b] Laser pulse power.
[c] Empty space means no data available.

The temporal pulse exhibited is modulated and typical of double-longitudinal-mode emission. The authors do not provide information on beam quality or transverse-mode structure while the cavity length is quoted at ~3 mm (Mhibik et al 2016).

In summary, in terms of emission linewidth: organic solid-state HMPGI and MPL grating oscillators provide single-longitudinal-mode laser linewidths in the $200 \leqslant \Delta\nu \leqslant 350$MHz range near the $\Delta t \Delta\nu \approx 1$ limit. The organic solid-state

VBG-VECSOL configuration has been reported as providing a double-longitudinal-mode linewidth of $\Delta\nu \approx 200$ MHz at $\Delta t\, \Delta\nu \approx 4$.

7.4.2 Long-pulse solid-state tunable laser oscillators

In order to generate narrow-linewidth tunable laser emission at the Joule level using organic solid-state gain media, it is necessary to have long pulse narrow-linewidth master oscillators to inject either power amplifiers (PA) or forced oscillators (FO).

In this section the performance of such an oscillator is presented in two tables. Table 7.11 includes output laser linewidth $\Delta\nu$, pulse length Δt, laser beam divergence $\Delta\theta$, and tuning range. Table 7.12 includes laser pulse energy, spectral power density ρ_{SP}, ASE level as determined by the ratio ρ_{ASE}/ρ_l, and conversion efficiency η. The laser beam is $\sim$100% polarized parallel to the plane of incidence, or plane of propagation.

The gain medium in these experiments was rhodamine 6 G doped 2-hydroxyethyl methacrylate:methyl methacrylate (HEMA:MMA) at a concentration of 5×10^{-4} M (Duarte *et al* 1998).

7.5 Additional solid-state organic lasers

In this section a survey on the performance of a variety of organic laser architectures, and organic laser gain media, is given. Albeit the output characteristics of these lasers is not quite in the same class as the performance of the MPL- and HMPGI-grating laser oscillators surveyed in the previous sections, they offer the potential to reach that class of high-spectral and high-spatial coherence performance, albeit in the low-power domain. In a sense, the lasers included in this section might be considered as *second generation* organic lasers.

One peculiarity of this section is that, given the variety of laser configurations applied in organic lasers, it is difficult to maintain the order of presentation adopted throughout this book that mainly adopts the style of presenting a gain medium followed by various laser configurations using that particular gain medium. Here,

Table 7.11. High-performance long-pulse solid-state organic master-oscillator.

Cavity configuration	$\Delta\nu$ (MHz)	Δt (ns)	$\Delta\theta$ (mrad)	Tuning range (nm)	Reference
MPL	650[a]	105	3.5	$564 \leqslant \lambda \leqslant 602$	Duarte *et al* (1998)

[a] Narrowest linewidth measured in these experiments was $\Delta\nu \approx 600$ MHz.

Table 7.12. Energetics of high-performance long-pulse solid-state organic laser oscillator.

Cavity configuration	E_l (mJ)[a]	ρ_{SP} (W nm^{-1})	ρ_{ASE}/ρ_l	η (%)	Reference
MPL	0.4	$\sim 5.4 \times 10^6$	1×10^{-4}	2	Duarte *et al* (1998)

[a] Laser pulse energy.

Table 7.13. Laser linewidth and tuning range of various additional organic lasers.

Gain medium	Laser configuration	Tuning range or emission λ(nm)	$\Delta\lambda$ (nm)	Reference
Rhodamine 6 G-doped fiber	Mirror-mirror cavity	568	7	Gvishi et al (1996)
Perylene red-doped PMMA[a]	DFB	$604 \leqslant \lambda \leqslant 649$	0.01	Wadsworth et al (1999)
LPPP[b]	DFB	491	0.25	Riechel et al (2000)
T3[c]	DBF	$425 \leqslant \lambda \leqslant 442$	0.25	Herrnsdorf et al (2010)
Rhodamine 6 G	DFB optofluidic waveguide	$565 \leqslant \lambda \leqslant 595$	0.1	Li et al (2006b)
EHCz:C153[d]	Fabry-Perot optofluidic	548, 564, 580[e]	3–4	Choi et al (2013)
Pyrromethene 567-doped PMMA[a]	DFB waveguide	$541 \leqslant \lambda \leqslant 570$	0.14	Yang et al (2010)
DCM-doped Alq$_3$[f]	VCSEL		0.02	Bulović et al (1998)
Dye-doped PDMS[g]	VCSEL	$597 \leqslant \lambda \leqslant 605$	0.35	Zhou et al (2015)

[a] Poly(methyl methacrylate).
[b] Poly(p-phenylene).
[c] Tris(trifluorene)truxenes.
[d] 9-(2-Ethylhexyl)carbazole:coumarin 153.
[e] Discrete wavelengths.
[f] Tris(8-hydroxyquinolinato)aluminium.
[g] Polydimethilsiloxane.

that order is abandoned to highlight the importance of various physical laser configurations common to several gain media.

The performance of salient laser configurations in this section is highlighted in table 7.13. Entries include gain medium, laser configuration, tuning range, and laser linewidth.

7.5.1 Microcavity and optofluidic organic lasers

Liquid organic dye lasers operating within the boundaries of a micro cavity are known to offer some unusual characteristics including near 'zero-laser-threshold' emission (De Martini and Jakobovitz 1988). It should be noticed that the concept of microcavities and even nanocavities is revisited in more detail in chapter 11.

The term 'fluidic' in relation to miniaturized liquid organic dye lasers was adopted after 2000 (Helbo et al 2003). The interest in this class of lasers is that they are compatible with 'lab-on-chip' or 'lab-on-a-chip' architectures. These optofluidic lasers have been demonstrated to deliver a single-mode laser linewidth of $\Delta\lambda \approx 0.21$ nm, at $\lambda \approx 567$ nm, utilizing a DFB waveguide configuration (Li et al 2006a). Using rhodamine 6 G as the gain medium a tuning range of $565 \leqslant \lambda \leqslant 595$ nm has been reported (Li et al 2006b) (see table 7.13).

A pulsed tunable microfluidic waveguide organic dye laser is described by Vezenov *et al* (2005). These authors report an approximate tuning range of $610 \leqslant \lambda \leqslant 640$ nm and a linewidth of $\Delta\lambda \approx 4$ nm, at $\lambda \approx 652$ nm. A Fabry–Perot optofluidic microresonator, tunable approximately in the $685 \leqslant \lambda \leqslant 705$ nm range, is reported by Lahos *et al* (2016). Solid-state organic waveguide lasers are considered in section 7.5.4.

A ring optofluidic dye laser has been reported to yield linewidths below 1 nm at $\lambda \approx 573$ nm (Shopova *et al* 2007). A review on optofluidic DBF and ring configurations is given by Li and Psaltis (2008) and the physics of microfluidics is reviewed by Song *et al* (2018).

7.5.2 Organic fiber lasers

Organic dye-doped organic fiber lasers made their appearance around the mid to late 1990s. Gvishi *et al* (1996) used rhodamine 6 G to dope a sol–gel fiber while utilizing the excitation of a frequency-doubled Nd:YAG laser at $\lambda \approx 532$ nm. They report laser emission at $\lambda \approx 568$ nm with a linewidth of $\Delta\lambda \approx 7$ nm (see table 7.13). Peng *et al* (1996), report on a rhodamine B-doped PMMA fiber laser.

Using a rhodamine 6 G doped poly(methyl methacrylate-co-2-hydroxyethyl methacrylate fiber and the excitation of a frequency doubled Nd:YAG laser (Kuruki *et al* 2000a) report a broadband output energy of 640 μJ for a pump energy of 1.5 mJ. Using the same type of polymer fiber, Kuruki *et al* (2000b) report lasing from rhodamine B, rhodamine 6 G, perylene orange, and pyrromethene 567. For transverse excitation with a frequency-doubled Nd:YAG laser they report on a 24% efficiency whilst using rhodamine 6 G.

As of 2018 there appear to be no published reports on the characterization of tunable organic dye-doped organic fiber lasers. The same appears to be the case for narrow-linewidth organic dye-doped organic fiber lasers and for narrow-linewidth organic fiber lasers at large. A review on narrow-linewidth tunable fiber lasers is given by Shay and Duarte (2016). The application of the intracavity techniques described by Shay and Duarte to demonstrate tunable narrow-linewidth *organic* fiber lasers is long overdue.

7.5.3 Solid-state distributed feedback organic lasers

Early liquid distributed feedback (DFB) lasers with linewidth characteristics of $\Delta\nu \approx 45$ GHz (Bor 1979), became well-known sources of pulses in the picosecond regime. More recently, DFB laser configurations have also been introduced with solid-state dye laser gain media (Wadsworth *et al* 1999, Zhu *et al* 2000).

The work of Wadsworth *et al* (1999) demonstrated a solid-state organic dye laser, using perylene red doped PMMA as gain medium, with a linewidth of $\Delta\lambda \approx 0.01$ nm and tunable in the $604 \leqslant \lambda \leqslant 649$ nm region (see table 7.13). Zhu *et al* (2000) using a coumarin 460 doped sol–gel silica gain medium report a linewidth of $\Delta\lambda \approx 0.06$ nm at $\lambda \approx 482$ nm. Gindre *et al* (2006), using a rhodamine 6 G-doped polymer thin-film gain medium report picosecond emission at a linewidth of $\Delta\lambda \approx 0.5$ nm at $\lambda \approx 562$ nm.

A linewidth of $\Delta\lambda \approx 0.2$ nm at $\lambda \approx 587.4$ nm is reported using a gain medium of rhodamine 6 G-doped cellulose acetate waveguide (Tsutsumi *et al* 2016).

Other relevant work on solid-state organic dye DFB lasers includes the reports of Oki *et al* (2002a) and Watanabe *et al* (2010). In this regard, Oki *et al* (2002a) have demonstrated a 'palm top size' microchip laser-pumped DBF-waveguide laser system tunable, with various laser dyes, in the 575–945 nm range.

An additional member of the organic laser class is the conjugated polymer laser (Hide *et al* 1996, Holzer *et al* 1996a, 1996b, Tessler *et al* 1996, Frolov *et al* 1997). Spectroscopic studies of conjugated polymers and neat films have been performed by Holzer *et al* (1999).

Using a conjugated ladder-type poly(*p*-phenilene) (LPPP) as the gain medium and a frequency-doubled Ti:sapphire emitting at $\lambda \approx 400$ nm, Riechel *et al* (2000) report laser emission at $\lambda \approx 491$ nm exhibiting a linewidth of $\Delta\lambda \approx 0.25$ nm and a beam divergence of $\Delta\theta \approx 2.4$ mrad (see table 7.13). An interesting feature of this work is that the excitation was performed using a pulse duration of 150 fs. Also, the laser configuration is a two-dimensional DFB arrangement made possible by the 2-D structure of the LPPP film. The authors indicate that the 2-D LPPP DFB laser offers lower thresholds and a higher efficiency than the equivalent 1-D arrangement (Riechel *et al* 2000).

Yet another type of organic semiconductor laser uses tris(trifluorene)truxenes (T3) as gain medium in a DFB configuration (Herrnsdorf *et al* 2010). These authors report on an emission linewidth of $\Delta\lambda \approx 0.25$ nm at $\lambda \approx 427$ nm while using the third harmonic from a Nd:YAG laser at $\lambda \approx 355$ nm as excitation. The tuning range is $425 \leqslant \lambda \leqslant 442$ nm using a 3600 lines mm^{-1} grating structure (see table 7.13).

7.5.4 Solid-state waveguide organic lasers

Hybrid waveguide-DFB organic dye lasers are discussed by Oki *et al* (2002a, 2002b, 2002c), and Watanabe *et al* (2005, 2010).

Oki *et al* (2002c) employed a gain medium comprised of rhodamine 6 G-doped PMMA spin coated on a PMMA substrate waveguide. They report a laser linewidth of $\Delta\lambda \approx 0.06$ nm at $\lambda \approx 590$ nm and an output energy in the 1–7 μJ range. More recently, Yang *et al* (2010) report on highly-stable laser emission using pyrromethene dye-doped polymer gain media also in a DFB waveguide configuration. These researchers report a laser linewidth of $\Delta\lambda \approx 0.14$ nm tunable in the $541 \leqslant \lambda \leqslant 570$ nm range and an output energy per pulse in the 10–80 nJ range at a prf of 500 Hz (see table 7.13).

Laser emission from waveguide configurations incorporating dye-doped polymer-nanoparticle gain media, in the absence of a resonator, has been reported by Cerdán *et al* (2010).

7.5.5 Solid-state microcavity organic lasers

The concept of microcavity to the organic laser field appears to have been introduced by De Martini and Jakobovits (1988) who also noticed that in a dye laser microcavity the laser thresholds were near 'zero.' As will be explained in chapter 11, an additional feature of microcavity organic lasers is that, given the ultrashort cavity length, once a transverse mode has been geometrically confined, then the emission within that TEM_{00} is also single-longitudinal-mode even though its linewidth is not necessarily narrow (Duarte 2016). In this regard, mode-suppression in organic laser microcavities has also been studied by Popov et al (2007).

Wang et al (2017) discuss coherent molecule coupling to a scanning Fabry–Perot microcavity and demonstrate single-molecule stimulated emission.

Wavelength tuning, via cavity length changes or variations, in microcavity lasers is quantified by Duarte (2003, 2015) and relevant equations are also included in chapter 6. For tuning using thermal means in microcavity solid-state dye lasers the reader can refer to Ricciardi et al (2007).

7.5.6 Organic vertical cavity surface emitting lasers (VCSELs)

An additional class of gain media for tunable organic lasers is that of optically-excited organic semiconductors. Comprehensive reviews on optically-pumped organic semiconductor lasers are given by Kranzelbinder and Leising (2000), Samuel and Turnbull (2007), Karnutsch (2007), and Grivas and Pollnau (2012).

In this area, using what is known as an organic vertical-cavity surface–surface emitting laser (OVCSEL) linewidths in the $\Delta\lambda \approx 0.02 \pm 0.01$ nm range were reported using DCM dye-doped Alq_3 thin film at a peak power of 3 W (Bulović et al 1998) (see table 7.13). An organic semiconductor laser, in a DFB configuration for biosending applications, is reported to yield a linewidth of $\Delta\lambda \approx 0.13$ nm at $\lambda \approx$ 430 nm while exhibiting a threshold of 2.5 nJ (Haughey et al 2016).

Using MEMS driven tuning and a DCM dye-doped Alq_3 gain medium in a VCSEL configuration, Chang et al (2015) report on lasing in the $628 \leqslant \lambda \leqslant 637$ nm range. No information on laser linewidth is given although, from the lineshape presented, it could be surmise to be below the nm range.

Using pyrromethene dye-doped polydimethylsiloxane (PDMS) films in a vertical cavity surface emitting laser (VCSEL) configurations, Zhou et al (2015) report tunable emission in the $597.3 \leqslant \lambda \leqslant 605.8$ range using the dye PM597-8C9. This wavelength coverage was attained using mechanical stress on the microcavity and the laser linewidth is reported at $\Delta\lambda \approx 0.35$ nm for output energies up to ~300 nJ.

An organic laser based on a longitudinally-excited VECSEL configuration and using a gain medium composed of a single molecular layer of dye-doped PMMA, an 'organic monolayer,' is reported by Palatnik et al (2017).

In addition to their use in optically-pumped organic semiconductor lasers, VCSELs configurations are used in various other solid-state organic laser gain media.

Using a pyrromethene 597-doped polydimethilsiloxane (PDMS) gain medium in an optically-pumped VCSEL configuration Yang et al (2015) reports a mechanically driven approximate tunability range of $625 \leqslant \lambda \leqslant 630$ nm at double-mode emission with the modes separated by $7.11 \leqslant \delta\lambda \leqslant 8.33$ nm. The reflectors in this VCSEL configuration are distributed Bragg reflectors (DBR).

As this book goes to press, there has been a scarcity of published material on tunable narrow-linewidth optically-pumped organic semiconductor lasers.

7.6 Problems

- 7.1 What is the crucial physical characteristic of liquid dye lasers that enables the generation of average powers in the kW regime?
- 7.2 What emission parameter benefits directly from this crucial physical characteristic?
- 7.3 What is the main physical advantage of high-performance narrow-linewidth solidstate dye lasers oscillators?
- 7.4 Besides the MPL solid-state dye laser oscillator: can you identify another class of pulsed laser oscillator, organic or inorganic, capable of yielding $\rho_{SP} \approx 8.33 \times 10^7$ W nm^{-1} tunable in the $560 \leq \lambda \leq 600$ nm nm range?
- 7.5 A laser with a linewidth of $\Delta\nu \approx 350$ MHz, at $\lambda \approx 590$ nm, is directly applicable to perform high resolution excitation spectroscopy of molecules such as I_2. Name at least two other molecules that would benefit from this class of emission.

References

Baltakov F N, Barikhin B A and Sukhanov L V 1974 400 J pulsed laser using a solution of rhodamine 6 G in ethyl alcohol *JETP Lett.* **19** 174–5

Bass I L, Bonano R E, Hackel R H and Hammond P R 1992 High-average-power dye laser at Lawrence Livermore National Laboratory *Appl. Opt.* **31** 6993–7006

Baving H J, Muuss H and Skolaut W 1982 CW dye laser operation at 200W pump power *Appl. Phys.* B **29** 19–21

Berhardt A F and Rasmussen P 1981 Design criteria and operating characteristics of a single-mode pulsed dye laser *App. Phys.* B **26** 141–6

Bokhan P A 2006 *Laser Isotope Separation in Atomic Vapor* (Weinheim: Wiley)

Bos F 1981 Versatile high-power single-longitudinal-mode pulsed dye laser *Appl. Opt.* **20** 1886–90

Broyer M, Chevaleyre J, Delacretaz G and Woste L 1984 CVL-pumped dye laser for spectroscopic applications *Appl. Phys.* B **35** 31–6

Bulović V, Kozlov V G, Khalfin V B and Forrest S R 1998 Transform-limited narrow-linewidth lasing action in organic semiconductor microcavities *Science* **279** 553–5

Cerdán L, Costela A, Garca-Moreno I, Garca O and Sastre R 2010 Laser emission from mirrorless waveguides based on photosenthesyzed polymers incorporating POSS *Opt. Ex.* **18** 10247–56

Chang W, Wang A, Murarka A, Akselrod G B, Packard C, Lang J H and Bulović V 2015 Electrically tunable organic vertical-cavity surface-emitting laser *Appl. Phys. Lett.* **105** 073303

Chaube R 2008 Design criteria and numerical analysis of a stable dye laser with a curved flow cell *Opt. Eng.* **47** 014301

Choi E Y, Mager L, Cham T T, Dorkenoo K D, Fort A, Wu J W, Barsella A and Ribierre J-C 2013 Solvent-free fluid organic dye lasers *Opt. Ex.* **21** 11368–75

Dasgupta K, Kundu S and Nair L G 1995 Extraction efficiency of saturated-gain high-average-power dye laser amplifiers: effect of nonlinear signal absorption *Appl. Opt.* **34** 1982–8

De Martini F and Jakobovitz J R 1988 Anomalous spontaneous-emission-decay phase transition and zero-threshold laser action in a microscopic cavity *Phys. Rev. Lett.* **60** 1711–4

Diels J-C 1990 Femtosecond dye lasers *Dye Laser Principles* ed F J Duarte and L W Hillman (New York: Academic) ch 3

Diels J-C and Rudolph W 1996 *Ultrashort Laser Pulse Phenomena* (New York: Academic)

Dietel W, Fontaine J J and Diels J-C 1983 Intracavity pulse compression with glass: a new method of generating pulses shorter than 60 fs *Opt. Lett.* **8** 4–6

Drever R W P, Hall J L, Kowalski F V, Hough J, Ford G M, Munley A J and Ward H 1983 Laser phase and frequency stabilization using an optical resonator *App. Phys.* B **31** 97–105

Duarte F J 1990a Narrow-linewidth pulsed dye laser oscillators *Dye Laser Principles* ed F J Duarte and L W Hillman (New York: Academic) ch 4

Duarte F J 1990b Technology of pulsed dye lasers *Dye Laser Principles* ed F J Duarte and L W Hillman (New York: Academic) ch 6

Duarte F J 1994 Solid-state multiple-prism grating dye-laser oscillators *Appl. Opt.* **33** 3857–60

Duarte F J 1995 Solid-state dispersive dye laser oscillator: very compact cavity *Opt. Commun.* **117** 480–4

Duarte F J 1997 Multiple-prism near-grazing-incidence grating solid-state dye-laser oscillator *Opt. Laser Tech.* **29** 513–16

Duarte F J 1999 Multiple-prism grating solid-state dye laser oscillator: optimized architecture *Appl. Opt.* **38** 6347–9

Duarte F J 2003 *Tunable Laser Optics* (New York: Academic)

Duarte F J 2015 *Tunable Laser Optics* 2nd edn (New York: CRC)

Duarte F J 2016a Tunable laser atomic vapor laser isotope separation *Tunable Laser Applications* ed F J Duarte (New York: CRC) 3rd edn ch 11

Duarte F J 2016b Coherent electrically-excited organic semiconductors *Tunable Laser Applications* ed F J Duarte (New York: CRC) 3rd edn ch 12

Duarte F J and Conrad R W 1987 Diffraction limited single-longitudinal-mode multiple-prism flashlamp-pumped dye laser oscillator: linewidth analysis and injection of amplifier system *Appl. Opt.* **26** 2567–71

Duarte F J, Davenport W E, Ehrlich J J and Taylor T S 1991 Ruggedized narrow-linewidth dispersive dye laser oscillator *Opt. Commun.* **84** 310–6

Duarte F J, Ehrlich J J, Davenport W E and Taylor T S 1990 Flashlamp pumped narrow-linewidth dispersive dye laser oscillators: very low amplified spontaneous emission levels and reduction of linewidth instabilities *Appl. Opt.* **29** 3176–9

Duarte F J and James R O 2016 Organic dye-doped polymer-nanoparticle tunable lasers *Tunable Laser Applications* ed F J Duarte (New York: CRC) 3rd edn ch 4

Duarte F J and Piper J A 1980 A double prism beam expander for pulsed dye lasers *Opt. Commun.* **35** 100–4

Duarte F J and Piper J A 1981 Prism preexpanded grazing-incidence grating cavity for pulsed dye lasers *Appl. Opt.* **20** 2113–6

Duarte F J and Piper J A 1984 Narrow-linewidth high-prf copper laser-pumped dye laser oscillators *Appl. Opt.* **23** 1391–4

Duarte F J, Taylor T S, Costela A, García-Moreno I and Sastre R 1998 Long-pulse narrow-linewidth dispersive solid-state dye-laser oscillator *App. Opt.* **37** 3987–9

Dupre P 1987 Quasiunimodal tunable pulsed dye laser at 440 nm: theoretical development for using a quad prism beam expander and one or two gratings in a pulsed dye laser oscillator cavity *Appl. Opt.* **26** 860–71

Flamant P H, Josse D and Maillard D J M 1984 Transient injection frequency-locking of a microsecond-pulsed dye laser for atmospheric measurements *Opt. Quantum Electron.* **16** 179–82

Fork R L, Brito Cruz C H, Becker P C and Shank C V 1987 Compression of optical pulses to six femtoseconds by using cubic phase compression *Opt. Lett.* **12** 483–5

Frolov S V, Gellerman W, Ozaki M, Yoshino K and Vardeny Z V 1997 Cooperative emission in π–conjugated polymer thin films *Phys. Rev. Lett.* **78** 729–31

Gindre D, Vesperini A, Nunzi J-M, Leblond H and Dorkenoo K D 2006 Refractive-index saturation-mediated multiple line emission in polymer thin-film distributed-feedback lasers *Opt. Lett.* **31** 1657–9

Grivas C and Pollnau M 2012 Organic solid-state integrated amplifiers and lasers *Lasers Photon. Rev.* **6** 419–62

Gvishi R, Ruland G and Prasad P N 1996 New laser medium: dye-doped sol–gel fiber *Opt. Commun.* **126** 66–72

Hänsch T W 1972 Repetitively pulsed tunable dye laser for high-resolution spectroscopy *Appl. Opt.* **11** 895–8

Haughey A-M, McConnell G, Guilhabert B, Burley A B, Dawson M D and Laurand N 2016 Organic semiconductor laser biosensor: design and performance discussion *IEEE Select. Top. Quantum Electron.* **22** 1300109

Helbo B, Kristensen A and Menon A 2003 A microcavity fluidic dye laser *J. Micromech. Microeng.* **13** 696–701

Herrnsdorf J, Guilhabert B, Chen Y, Kanibolotski A L, Mackintosh A R, Pethrick R A, Skabara P J, Gu E, Laurand N and Dawson M D 2010 Flexible blue-emitting encapsulated organic semiconductor DFB laser *Opt. Express* **18** 25535–45

Hide F, Schwartz B J, Díaz-García M and Heeger A L 1996 Laser emission from solutions and films containing polymer and titanium dioxide nanocrystals *Chem. Phys. Lett.* **256** 424–30

Hollberg L 1990 CW dye lasers *Dye Laser Principles* ed F J Duarte and L W Hillman (New York: Academic) ch 5

Holzer W, Penzkofer A, Gong S-H, Bleyer A and Bradley D D C 1996a Laser action in poly (m-phenylenevinylene-co-2, 5-dioctoxy-p-phenylenevinylene) *Adv. Mat.* **8** 974–8

Holzer W, Pichlmaier M, Penzkofer A, Bradley D D C and Blau W J 1999 Fluorescence spectroscopic behavior of neat and blended conjugated polymer thin Flims *Chem. Phys.* **246** 445–62

Holzer W, Pichlmaier M, Penzkofer A, Bradley D D C and Blau W J 1996b Fluorescence spectroscopy behavior of neat and blended conjugated polymer thin films *Chem. Phys.* **248** 445–62

Hough J, Hils D, Rayman M D, Ma L-S, Hollberg L and Hall J L 1987 Dye-laser frequency stabilization using optical resonators *Appl. Phys.* B **33** 179–85

Johnston T F and Duarte F J 2002 Lasers, dye *Encyclopedia of Physical Science and Technology* ed R A Meyers (New York: Academic) 3rd ednvol 8 pp 315–59

Kafka J D and Baer T 1992 Prism-pair delay lines in optical pulse compression *Opt. Lett.* **12** 401–3

Karnutsch C 2007 *Low Threshold Organic Thin Film Laser Devices* (Göttingen: Cuvillier)

Klimek D E, Aldag H R and Russell J 1992 High-energy high-power flashlamp-pumped dye laser *Conference on Lasers and Electro Optics* (Washington DC: Optical Society of America) p 332

Kranzelbinder G and Leising G 2000 Organic solid-state lasers *Rep. Prog. Phys.* **63** 729–62

Kubota H, Kurokawa K and Nakazawa M 1988 29-fs pulse generation from a linear-cavity synchronously pumped dye laser *Opt. Lett.* **13** 749–51

Kuruki K, Kobayashi T, Imai N, Tamura T, Nishihara S, Nishizawa Y, Tagaya A and Koike Y 2000a High-efficiency organic dye-doped polymer optical fiber laser *Appl. Phys. Lett.* **77** 331

Kuruki K, Kobayashi T, Imai N, Tamura T, Koike Y and Okamoto Y 2000b Organic dye-doped polymer optical fiber laser *Poly. Adv. Tech.* **11** 612–6

Lahos F, Martn I R, Gil-Rostra J, Oliva-Ramirez M, Yubro F and Gonzalez-Elipe A R 2016 Portable IR laser optofluidic microresonator as a temperature and chemical sensor *Opt. Express* **24** 14382–83

Li Z, Zhang Z, Emery T, Scherer A and Psaltis D 2006a Single mode optofluidic distributed feedback dye laser *Opt. Express* **14** 696–701

Li Z, Zhang Z, Scherer A and Psaltis D 2006b Mechanically tunable optofluidic distributed feedback dye laser *Opt. Express* **14** 10494–99

Li Z and Psaltis D 2008 Optofluidic dye lasers *Microfluid. Nanofluid.* **4** 145–58

Maruyama Y, Kato M, Ohba M and Arisawa T 1995 A narrow-linewidth dye laser pumped by a high-repetition-rate long pulse Nd:YAG laser *Japan. J. Appl. Phys.* **34** L1045–7

Maruyama Y, Kato M and Arizawa T 1996 Effects of excited-state absorption and amplified spontaneous emission in a high-average-power dye laser amplifier pumped by copper vapor lasers *Opt. Eng.* **35** 1084–7

Maslyukov A, Sokolov S, Kaivola M, Nyholm K and Popov S 1995 Solid-state dye laser with modified poly (methyl methacrylate)-doped active elements *Appl. Opt.* **34** 1516–8

Mhibik O, Forget S, Ott D, Venus G, Divliansky I, Glebov L and Chénais S 2016 An ultra-narrow linewidth solution-processed organic laser *Light Sci. Appl.* **5** e16026

Oki Y, Aso K, Zuo D, Vasa N J and Maeda M 2002a Wide-wavelength range operation of a distributed-feedback dye laser with a plastic waveguide *Japan. J. Appl. Phys.* **41** 6370–4

Oki Y, Miyamoto S, Tanaka M, Zuo D and Maeda M 2002b Long lifetime and high repetition rate operation from distributed feedback plastic waveguided dye lasers *Opt. Commun.* **214** 277–83

Oki Y, Yoshiura T, Chisaki Y and Maeda M 2002c Fabrication of distributed feedback dye laser with a grating structure in its plastic waveguided *Appl. Opt.* **41** 5030–5

Osvay K, Kovács A P, Kurdi G, Heiner Z, Divall M, Klebniczki J and Ferincz I E 2005 Measurements of non-compensated angular dispersion and the subsequent temporal lengthening of femtosecond pulses in a CPA laser *Opt. Commun.* **248** 201–9

Palatnik A, Aviv H and Tischler Y R 2017 Microcavity laser based on a single molecule thick high gain layer *Am. Chem Soc. Nano* **11** 4514–20

Pang L Y, Fujimoto J G and Kintzer E S 1992 Ultrashort-pulse generation from high-power diode arrays by using intracavity optical nonlinearities *Opt. Lett.* **17** 1599–601

Peng D G, Chu P K, Xiong Z, Whitbread T W and Chaplin R P 1996 Dye-doped step-index polymer optical fiber for broadband optical amplification *J. Light. Tech.* **14** 2215–23

Popov S, Ricciardi S, Friberg A T and Sergeyev S 2007 Mode suppression in a microcavity solid-state dye laser *J. Euro. Opt. Soc.* **2** 07023

Ray A K 2007 A binary solvent of water and propanol for use in high-average power dye lasers *Appl. Phys.* B **87** 489–95

Ricciardi S, Popov S, Friberg A T and Sergeyev S 2007 Thermally induced wavelength tunability of microcavity solid-state dye lasers *Opt. Exp.* **15** 12971–8

Riechel S, Kallinger C, Lemmer U, Feldmann J, Gombert A, Wittwer V and Scherf U 2000 A nearly diffraction limited surface emitting conjugated polymer laser utilizing a two-dimensional photonic band structure *Appl. Phys. Lett.* **77** 2310–2

Samuel I D W and Turnbull G A 2007 Organic semiconductor lasers *Chem. Rev.* **107** 1272–95

Shay T M and Duarte F J 2016 Tunable fiber lasers *Tunable Laser Applications* ed F J Duarte (New York: CRC) 3rd edn ch 6

Shopova S I, Zhou H, Fan X and Zhang P 2007 Optofluidic ring resonator based dye laser *Appl. Phys. Lett.* **90** 221101

Singh N 2006 Influence of optical inhomogeneity in the gain medium on the bandwidth of a high-repetition-rate dye laser pumped by copper vapor laser *Opt. Eng.* **45** 104204

Singh S, Dasgupta K, Kumar S, Manohar K G, Nair L G and Chatterjee U K 1994 High-power high-repetition-rate copper vapor-pumped dye laser *Opt. Eng.* **33** 1894–904

Song Y, Cheng D and Zhao L (ed) 2018 Microfluidics: Fundamentals *Devices, and Applications* (Weinheim: Wiley)

Sugiyama A, Nakayama T, Kato M, Maruyama Y and Arisawa T 1996 Characteristics of a pressure-tuned single-mode dye laser oscillator pumped by a copper vapor oscillator *Opt. Eng.* **35** 1093–7

Takehisa K 1997 Scaling up of a high average power dye laser amplifier and its new pumping designs *Appl. Opt.* **36** 584–92

Tang K Y, O'Keefe T, Treacy B, Rottler R and White C 1987 Kilojoule-output XeCl dye laser: optimization and analysis *Proceedings: Dye Laser/Laser Dye Technical Exchange Meeting 1987* ed J H Bentley (Alabama: U. S. Army Missile Command, Redstone Arsenal) pp 490–502

Tessler N, Denton G J and Friend R H 1996 Lasing from conjugated polymer microcavities *Nature* **382** 695–7

Tsutsumi N, Nagi S, Kinashi K and Sakai W 2016 Re-evaluation of all-plastic organic dye laser with DFB structure fabricated using photoresists *Sci. Rep.* **6** 34741

Vezenov D V, Mayers B T, Conroy R S, Whitesides G M, Snee P T, Chan Y, Nocera D G and Bawendi M G 2005 A low-threshold, high-efficiency, microfluidic waveguide laser *J. Am. Chem. Soc.* **127** 8952–3

Wadsworth W J, McKinnie I T, Woolhouse A D and Haskell T G 1999 Efficient distributed feedback solid state dye laser with dynamic grating *Appl. Phys.* B **69** 163–5

Wang D, Kelkar H, Martin-Cano D, Utikal T, Gtzinger S and Sandoghdar V 2017 Coherent coupling of a single molecule to a scanning Fabry–Perot microcavity *Phys. Rev. X* **7** 021014

Watanabe H, Oki Y and Maeda M 2005 Waveguide dye laser including a SiO_2 nanoparticle-dispersed random scattering active media *Appl. Phys. Lett.* **86** 151123

Watanabe H, So H, Oki Y, Akine S and Omatsu T 2010 Picosecond-pulse-pumped distributed-feedback thick-film waveguide blue laser using fluorescent brightener 135 *Japan. J. Appl. Phys.* **48** 072105

Webb C E 1991 High-power dye lasers pumped by copper vapor lasers *High Power Dye Lasers* ed F J Duarte (Berlin: Springer) ch 5

Yang Y, Goto R, Omi S, Yamashita K, Watanabe H, Miyazaki M and Oki Y 2010 Highly photo-stable solid-state distributed feedback (DFB) channeled waveguide lasers by a pen-drawing technique *Opt. Express* **18** 22081–9

Yang Y, Zhou Y, Liao Z, Yu J, Cui Y, García-Moreno I, Wang Z, Costela A and Qian G 2015 Mechanically tunable organic vertical-cavity surface emitting lasers (VCSELs) for highly sensitive stress probing in dual modes *Opt. Express* **23** 4385–96

Zhu X-L, Lam S-K and Lo D 2000 Distributed-feedback dye-doped solgel silica lasers *Appl. Opt.* **39** 3104–7

Zhou Y, Zhang J, Hu Q, Liao Z, Cui Y, Yang Y and Qian G 2015 Stable and mechanically tunable vertical-cavity surface-emitting lasers (VECSLs) based on dye-doped elastic thin films *Dyes Pigm.* **116** 114–8

IOP Publishing

Organic Lasers and Organic Photonics (Second Edition)

F J Duarte

Chapter 8

Tunable organic lasers for directed energy

F J Duarte

Emission characteristics of tunable laser systems compatible with directed energy applications are quoted and discussed. Emphasis is given to high-performance narrow-linewidth oscillators and unstable resonator architectures for amplifier stages. Relevant issues such as heat removal and average power versus continuous wave (CW) power are discussed.

8.1 Introduction

At present the lasers applied in the *directed energy* (DE) sphere belong almost exclusively to the crystalline solid-state class and the fiber laser class. However, this was not always the case. Initially, large gas lasers primarily of CO_2 lasers and excimer lasers were developed for this application. Additional lasers of interest were chemical lasers and organic dye lasers. Although the trend now favors fiber laser systems, technology trends are unpredictable. This unpredictability justifies consideration of alternative laser systems such as organic lasers.

The concept of directed energy utilizes spectral and spatial coherence principles to focus energy of a specific wavelength on a particular and predetermined point, area, or target, in space.

In order to maximize effectiveness the lasers used in DE applications need to:

1. Possess wavelength agility. In other words, be tunable. This feature is necessary to increase significantly the difficulty of countermeasures and to exploit spectral transmission windows.
2. Have a high degree of spatial coherence. In other words, be capable of single-transverse-mode emission and demonstrate low beam divergence.
3. Have a high degree of spectral coherence. In other words, be capable of narrow-linewidth emission, and preferably single-longitudinal-mode emission. Narrow $\Delta \nu$ emission, in the sub GHz range, is desirable to successfully exploit narrow spectral transmission windows and to effectively increase the spectral power density ρ_{SP}.
4. Be capable of delivering high pulsed energies on target.

8-1

The organic lasers mentioned in this chapter can be used to successfully meet these emission demands.

8.1.1 Atmospheric propagation

The use of lasers as DE devices in outer space is relatively straight forward from a propagation perspective. In an Earthly environment, however, this is not the case. In this regard Earth's atmosphere, via clouds and turbulence, is by far the most significant neutralizer of DE coherent emitters. The phenomenon of the effect of atmospheric turbulence on laser propagation in the atmosphere has attracted the interest of researchers for some time (Priest 1980, Taylor and Gregory 2002).

N-slit laser interferometry has been shown to be highly sensitive to the presence of clear air turbulence and therefore offer an effective pathway for the detection and characterization of atmospheric turbulence affecting the propagation of coherent radiation through Earth's atmosphere (Duarte *et al* 2010).

8.2 Organic laser oscillators for directed energy

Flashlamp-pumped organic dye lasers are suitable for DE applications given that they deliver long-pulse high-energy radiation. Therefore, operation at an even moderate pulse repetition frequency (prf) can deliver relatively high average powers. In other words, well engineered flashlamp-pumped tunable organic dye lasers meet the energetic requirements for some important DE applications, especially in the areas of detector and sensor countermeasures.

Here, relevant emission characteristics for DE applications of tunable narrow-linewidth organic master oscillators are tabulated. Table 8.1 includes output laser linewidth $\Delta\nu$, laser pulse length Δt, laser beam divergence $\Delta\theta$, and output laser energy per pulse E_l. Common to all these master oscillators is that they all emit in a single-transverse-mode (TEM$_{00}$), they are all tunable, and all yield very low $\rho_{\mathrm{ASE}}/\rho_l$ ratios. In other words, their spectral emission is very pure.

Although all these master oscillators were designed to inject forced oscillators or power amplifiers, some of them are delivering energies in the tens of mJ range, which might be useful to some limited DE applications without the need for an amplification stage. It is also worth noticing that the ruggedized master oscillator,

Table 8.1. Flashlamp-pumped organic dye tunable master oscillators.

MO[a]	$\Delta\nu$(MHz)	Δt(ns)	$\Delta\theta$(mrad)	E_l (mJ)	References
HMPGI	260[b]	200		60	Duarte and Conrad (1986b)
MPL	375[c]	250	0.25	11.7	Duarte and Conrad (1987)
MPL	138[d]	1000	0.57	3.6	Duarte *et al* (1990)
MPL	300[e]	150	0.35	3.0	Duarte *et al* (1991)

[a] These master oscillators deliver tunable laser radiation in a TEM$_{00}$ beam.
[b] SLM oscillation.
[c] SLM oscillation observed at $\Delta\nu \approx 333$ MHz.
[d] SLM oscillation and $(\delta\lambda/\lambda) \approx 1.2 \times 10^{-6}$.
[e] Near-SLM oscillation and $(\delta\lambda/\lambda) \approx 4.63 \times 10^{-7}$.

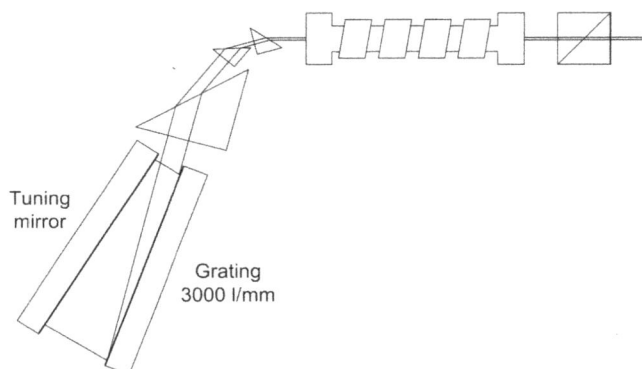

Figure 8.1. Cavity configuration of a *ruggedized* tunable narrow-linewidth hybrid multiple-prism grazing-incidence (HMPGI) grating organic dye laser oscillator.

delivering a linewidth of $\Delta\nu \approx 300$ MHz, and a wavelength jitter of $(\delta\lambda/\lambda) \approx 4.63 \times 10^{-7}$, performed splendidly in outside terrain testing (Duarte *et al* 1991). A diagram of a multiple-prism grating cavity configuration, within the ruggedized oscillator class, is depicted in figure 8.1. It should be noted that the ruggedized narrow-linewidth tunable laser oscillator reported by Duarte *et al* (1991) was the first visible tunable laser oscillator of its kind. For further details visit chapter 5.

8.3 Organic master-oscillator forced-oscillator (MOFO) for directed energy

The MOFO system described here is comprised of a master oscillator of the multiple-prism Littrow (MPL) grating class and a forced oscillator configured by an unstable resonator (see chapter 5 for schematics). The output laser performance of this MOFO system is described in table 8.2 and includes information on the laser linewidth of the master oscillator $\Delta\nu_{MO}$, the energy output per pulse of the master oscillator E_{MO}, the energy output per pulse of the MOFO, E_{MOFO}, the half-angle beam divergence $\Delta\theta_{MOFO}$, and the overall gain of the system.

The linewidth of the MOFO system $\Delta\nu_{MOFO} \approx 700$ MHz was measured interferometrically and it was consistent with SLM lasing. That is, the interferograms recorded from the MOFO system are identical to the interferograms from the master oscillator which is known to emit SLM lasing at $\Delta\nu_{MO} \leqslant 375$ MHz. The

Table 8.2. High-performance flashlamp-pumped organic dye MOFO.

$\Delta\nu_{MO}$(MHz)	E_{MO}(mJ)[a]	E_{MOFO}(mJ)[a]	$\Delta\theta_{MOFO}$(mrad)	Gain	References
375[b]	11	600	0.7	51	Duarte and Conrad (1987)

[a] Energy per pulse.
[b] Narrowest linewidth measured in these experiments was $\Delta\nu \approx 333$ MHz consistent with SLM oscillation.

observation of a laser linewidth at $\Delta\nu_{MOFO} \approx 700$ MHz is attributed to the insufficient finesse of the Fabry–Perot interferometer utilized in the measurements.

Although the overall energy measured in these experiments, at a rhodamine 6G dye concentration of 2.5×10^{-5} M, was $E_{MOFO} = 600$ mJ, the pulse energy provided by the master oscillator ($E_{MO} = 11$ mJ) should be sufficient to effectively inject a forced oscillator of the multi Joule class (Duarte and Conrad 1987). An interesting phenomenon observed in these experiments was the slight dynamic oscillation in the laser linewidth which is characterized and explained in detail by Duarte *et al* (1988).

The demonstration of a high-performance ruggedized master oscillator (Duarte *et al* 1991) indicates that the integration of MOFO systems delivering tunable narrow-linewidth multi Joule radiation for DE applications is simply reduced to an engineering endeavor.

8.4 High-energy amplification stages

Liquid organic dye lasers possess an undisputed inherent capacity to generate unparalleled tunable high-energy pulses directly in the visible spectrum. In the introduction of *High-Power Dye Lasers* energies per pulse approaching the 1 kJ level are mentioned while using ultraviolet excimer laser excitation (Duarte 1991). The interest in this section, however, is focused on flashlamp-pumped organic dye lasers given their long-pulse characteristics. Of several potential entries, three such lasers are included in table 8.3.

From table 8.3 it can be deduced that liquid flashlamp-pumped organic dye lasers have a demonstrated capability to deliver tens and even hundreds of Joules per pulse thus easily reaching average powers beyond the kW level.

Deploying this class of raw broadband energy gain region within unstable resonator systems can lead to formidable MOFO configurations for DE applications. A relevant MOFO configuration is depicted in figure 8.2. To a physicist-engineer skilled in the field, it is not difficult to recognize that a kW class tunable MOFO laser system, as outlined in figure 8.2, is suitable for deployment on a dedicated vehicle. This is all that can be disclosed in this area of research within the confines of the open literature.

It should be noted that all types of high-energy and high-average-power lasers, regardless of gain media, need some kind of cooling system. The excess heat has to be removed. Some crystalline high-power solid state lasers use water cooling while others use cryogenic cooling (Brown 1997). In regard to high-power fiber lasers it is said that 'heat generation is currently one of the main factors limiting output power' (Richardson *et al* 2010). High-power and high-energy liquid dye lasers are capable of inherent cooling simply by recirculating the excited gain volume to the gain media reservoir.

Table 8.3. High-energy flashlamp-pumped organic dye lasers.

Organic dye	Pulse energy (J)	Average power (W)	prf (Hz)	References
Rhodamine 6G	400		Low	Baltakov *et al* (1974)
Rhodamine 6G	1.4	1200	850	Morton and Draggoo (1981)
Rhodamine 6G	140	1400	10	Klimek *et al* (1992)

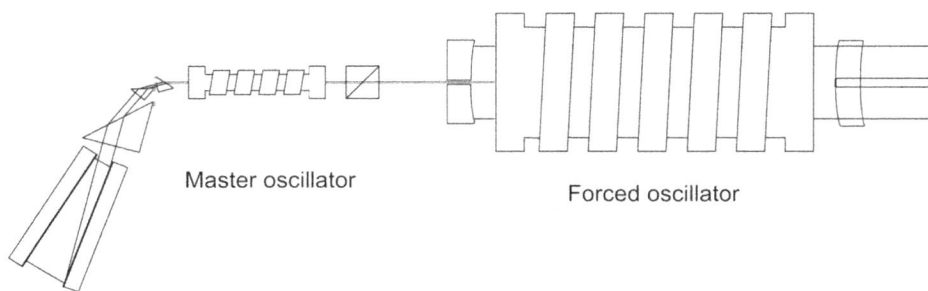

Figure 8.2. High-energy MOFO organic laser configuration. The large volume FO, or amplifying, stage can be excited coaxially or transversely.

8.5 Further development for solid-state tunable organic lasers

Following the introduction of high-performance compact narrow-linewidth solid-state organic tunable laser oscillators (see chapters 5 and 7) emitting at laser pulse durations the $3 \leqslant \Delta t \leqslant 10$ ns regime (Duarte 1994, 1995, 1997, 1999), and at the $\Delta t \approx 105$ ns pulse regime (Duarte *et al* 1998), it is possible to envision further developments in the area of solid-state tunable organic lasers for certain specific DE applications requiring wavelength agility in the visible spectrum and low-to-moderate average powers.

The applications considered here require narrow-linewidth, low beam divergence, radiation tunable throughout the visible spectrum at pulse energy levels at the 1 J level. In this regard, it is possible to apply available high-performance oscillator configurations, with improved laser excitation technologies, to design and build compact solid-state tunable MOPA systems delivering the desired emission characteristics.

Here, it should be noted that the peak power P_p, directly in the visible spectrum, of the narrow-linewidth single-longitudinal-mode *oscillator*, delivering a laser linewidth of $\Delta \nu \approx 350$ MHz (Duarte 1999), is $P_p \approx 34$ kW whilst the peak power of the long-pulse oscillator, delivering a laser linewidth of $\Delta \nu \approx 650$ MHz (Duarte *et al* 1998), is $P_p \approx 3.8$ MW. These figures translate into spectral power densities $\rho_{SP} \approx 8.33 \times 10^7$ W nm^{-1} and $\rho_{SP} \approx 5 \times 10^9$ W nm^{-1}, respectively.

The availability of blue–green semiconductor lasers bring them within reach the development and engineering of compact blue–green diode excitation of solid-state organic gain media emitting in the yellow–orange-red region of the electromagnetic spectrum.

More specifically in MOPA systems, for the pre-amplification stage, and further amplification stages, synchronous diode-laser arrays should be configured to excite organic dye-doped polymer, or polymer- nanoparticle, gain media within unstable resonators. The emission information being provided by the pulsed high-performance narrow-linewidth single-longitudinal-mode oscillators described here.

A highly-reflective metal grazing-incidence waveguide geometry for a diode-array excitation of dye-doped polymer, or polymer-nanoparticle, gain media is described in the literature by Duarte (2005, 2016).

8.6 Pulse average power versus CW power

Tunable organic lasers for DE applications, either in the liquid state or the solid state, are inherently pulsed lasers. Their average power P_A is dependent on their pulse repetition frequency (prf). This means that for a given average power the peak pulse power P_p can be fairly high depending on their prf. This is quite different to the situation arising from a fixed CW power P_{CW}. For instance, consider the 1.2 kW average power from the laser entry in table 8.3. At the quoted prf of 850 Hz that corresponds to an energy per pulse of 1.4 J. For a quoted pulse duration of 1500 ns at FWHM (Morton and Draggoo 1981) that translates into a peak power per pulse of $P_p \approx 933$ kW. On the other hand an equivalent CW laser offers just the quoted average power at 1.2 kW.

It has been known for quite a while that heat needed for vaporization is 10–25 × larger that those needed for fusion for common metals (Fox 1975). This means that removal of molten material is facilitated by high-peak power pulses that, among other effects, can cause *petaling* on the reverse side of the irradiation (Fox 1975).

A given average power P_A can be attained either via high prf train pulses with low peak power pulses P_{LP} or via lower prf and high-peak power pulses P_{HP}. Equivalent P_{CW} and P_A can heat and melt an irradiated surface. However, the same average power exhibiting P_{HP} characteristics can be expected to heat, melt, and perforate.

8.7 Thermal dissipation

In section 8.4 the issue of removal of excess heat was mentioned. This is an inescapable basic physics problem facing every laser and all high-power lasers. For instance, in high-power fiber lasers 'heat generation is currently one of the main factors limiting output power' (Richardson *et al* 2010). This problem not only affects the lasant, or laser active medium, such as the fiber, but it is also relevant in regard to the diode-laser arrays being increasingly used as excitation sources. And this introduces another basic physics problem since it is well known that *heat means death to a semiconductor*. This is a problem for all lasers utilizing diode-laser technology as excitation sources. Advances in heat transfer technology include the development heat sinks, for diode-laser arrays, that employ liquid metal, or $Ga_{68}In_{20}Sn_{12}$ (Li *et al* 2020).

In regard to heat removal in liquid versus solid-state gain media: crystalline high-power solid state lasers use liquid cooling while others use cryogenic cooling (Brown 1997). High-power and high-energy liquid dye lasers are capable of inherent cooling simply by recirculation of the excited gain volume back to the gain media reservoir. It remains to be determined whether removal of heat via radial methods is more, or less, efficient than recirculation of the gain volume into the gain media reservoir.

8.8 Problems

- 8.1 For a MOFA solid-state organic dye laser yielding an energy of 1 J per pulse, work out its average power if operating at a prf of (a) 10 Hz and (b) 50 Hz.

- 8.2 If the pulse duration of the tunable MOFA organic laser in the previous problem is $\Delta t \approx 100$ ns at FWHM, work out its peak power.
- 8.3 What configurational-engineering characteristic of pulsed liquid organic dye lasers makes them useful for directed energy applications?
- 8.4 Identify one emission characteristic that enable flashlamp-pumped pulsed liquid organic dye lasers to reach kW average powers at relatively low pulse repetition frequencies.

Acknowledgments

All the material disclosed in this article is based on papers published in the open literature. Most of the published material, by the author, disclosed in this chapter, was made possible via the support of the US Army Missile Command and the US Army Space and Missile Defense Command. Some of this research also focused on the development of a high-energy tunable narrow-linewidth multiple-prism grating 500 J e-beam CO_2 laser (Duarte and Conrad 1986a).

References

Baltakov F N, Barikhin B A and Sukhanov L V 1974 400 J pulsed laser using a solution of rhodamine 6 G in ethyl alcohol *JETP Lett.* **19** 174–5

Brown D C 1997 Ultra-high power diode-pumped Nd:YAG and Yb:YAG lasers *IEEE J. Quantum Electron.* **33** 861–73

Duarte F J 1991 *High-Power Dye Lasers* (Springer Series in Optical Sciences vol 65) (Cham: Springer)

Duarte F J 1994 Solid-state multiple-prism grating dye-laser oscillators *Appl. Opt.* **33** 3857–60

Duarte F J 1995 Solid-state dispersive dye laser oscillator: very compact cavity *Opt. Commun.* **117** 480–4

Duarte F J 1997 Multiple-prism near-grazing-incidence grating solid-state dye-laser oscillator *Opt. Laser Tech.* **29** 513–6

Duarte F J, Taylor T S, Costela A, García-Moreno I and Sastre R 1998 Long-pulse narrow-linewidth dispersive solid-state dye-laser oscillator *Appl. Opt* **37** 3987–9

Duarte F J 1999 Multiple-prism grating solid-state dye laser oscillator: optimized architecture *Appl. Opt.* **38** 663–5

Duarte F J 2005 Light emitting diode-pumped laser and method of excitation *US Patent Application* 2005/00833986 A1

Duarte F J 2016 Organic dye-doped polymer-nanoparticle tunable lasers *Tunable Laser Applications* ed F J Duarte (New York: CRC Press) 3rd edn ch 4

Duarte F J and Conrad R W 1986a Evaluation of multiple-prism techniques for linewidth narrowing in large-scale CO_2 lasers *Proc. Int. Conf. on Lasers'85* ed C P Wang (McLean, VA: STS Press) pp 145–52

Duarte F J and Conrad R W 1986b Single-mode flashlamp-pumped dye laser oscillators *Appl. Opt.* **25** 2567–71

Duarte F J and Conrad R W 1987 Diffraction limited single-longitudinal-mode multiple-prism flashlamp-pumped dye laser oscillator: linewidth analysis and injection of amplifier system *Appl. Opt.* **26** 2567–71

Duarte F J, Ehrlich J J, Patterson S P, Russell S D and Adams J E 1988 Linewidth instabilities in narrow-linewidth flashlamp-pumped dye laser oscillators *Appl. Opt.* **27** 843–6

Duarte F J, Ehrlich J J, Davenport W E and Taylor T S 1990 Flashlamp pumped narrow-linewidth dispersive dye laser oscillators: very low amplified spontaneous emission levels and reduction of linewidth instabilities *Appl. Opt.* **29** 3176–9

Duarte F J, Davenport W E, Ehrlich J J and Taylor T S 1991 Ruggedized narrow-linewidth dispersive dye laser oscillator *Opt. Commun.* **84** 310–6

Duarte F J, Taylor T S, Clarck A B and Davenport W E 2010 The *N*-slit interferometer: an extended configuration *J. Opt.* **12** 015705

Fox J A 1975 A method for improving continuous wave laser penetration of metal targets *Appl. Phys. Lett.* **26** 682–4

Klimek D E, Aldag H R and Russell J 1992 High-energy high-power flashlamp-pumped dye laser *Conf. on Lasers and Electro Optics* (Washington, DC: Optical Society of America) p 332

Li X-P *et al* 2020 Wavelength-stable 1.1-kW diode laser array cooled by liquid metal *IEEE Photon. Tech. Lett.* **32** 434–7

Morton R G and Draggoo V G 1981 Reliable high average power high pulse energy dye laser *IEEE J. Quantum Electron.* **QE17** 222

Priest T H 1980 *Atmospheric Sensitivities of High Energy Lasers* (Scott AFB, IL: Air Weather Service)

Richardson D J, Nilsson J and Clarkson W A 2010 High power fiber lasers:current status and future perspectives *J. Opt. Soc. Am.* **27** B63–92

Taylor T S and Gregory D A 2002 Laboratory simulation of atmospheric turbulence-induced optical wavefront distortion *Opt. Laser Tech.* **34** 665–9

IOP Publishing

Organic Lasers and Organic Photonics (Second Edition)

F J Duarte

Chapter 9

Polymer–nanoparticle organic lasers

F J Duarte

Organic-dye-doped polymer-nanoparticle lasers are described and their emission characteristics discussed. Emphasis is given to the homogeneous gain media providing favorable $\partial n/\partial T$ factors leading to homogeneous low divergence laser beams.

9.1 Introduction

The advent of highly homogeneous high-surface quality organic laser dye-doped polymeric gain matrices enabled a major advance in the field of organic lasers. Indeed, it was access to this class of laser-grade material that enabled the first demonstration of TEM_{00} beam quality near single-longitudinal-mode emission from multiple-prism grating oscillators (Duarte 1994). Following a configuration iteration of oscillator architectures this gain media was instrumental in achieving low divergence TEM_{00} laser beams confining single-longitudinal-mode emission at $\Delta \nu \approx 350$ MHz ($\Delta \lambda \approx 0.00039$ nm at $\lambda \approx 580$ nm) for a $550 \leqslant \lambda \leqslant 603$ nm tuning range (Duarte 1999). This remains the most powerful organic narrow-linewidth oscillator in the literature with a $\Delta \nu \Delta t \approx 1.05$ near the edge allowed by Heisenberg's uncertainty principle and a spectral power density of $\rho_{SP} \approx 8.33 \times 10^7$ W nm^{-1}.

Besides the remarkable optical homogeneity and optical surface quality (better than $\lambda/4$) these modified poly (methyl methacrylate) (MPMMA) exhibited a dye bleaching threshold of ~ 0.7 J cm^{-2} (Duarte 1994). Besides this high dye bleaching threshold, this dye-doped polymer (DDP) exhibits a 'self-healing' mechanism by which laser dye molecules slowly diffuse in the vacant space created by the excess laser radiation. As this review is written today (early 2018) the DDP gain media used in numerous experiments in the 1994–2000 period looks new and in experiment-ready conditions. One caveat associated with these superb tunable laser gain media, however, is their relatively high $\partial n/\partial T$ value which for laser-grade rhodamine 6 G-doped MPMMA is $\partial n/\partial T = (-1.4 \pm 0.2) \times 10^{-4}$ K^{-1} (Duarte *et al* 2000). This $\partial n/\partial T$ value results in limitations in pulse repetition frequency (prf) and in the inducement of higher beam divergences when lasing in the long-pulse regime at or above $\Delta t \approx 105$ ns is desired (Duarte *et al* 1998).

A solution to this problem is to use laser dye-doped organic–inorganic gain media such as ORMOSILS and nanocomposites. However, as these solid-state gain media were investigated for their usefulness in narrow-linewidth laser oscillators (Duarte *et al* 1993) some of these organic–inorganic materials were found to lack the inner nanoscopic homogeneity necessary to support homogeneous TEM_{00} lasing and henceforth narrow-linewidth emission (Duarte and Pope 1995). In other words, there were nanoscopic refractive index inhomogeneties to overcome.

An almost coincidental, and yet inspiring, solution to this problem was demonstrated by incorporating SiO_2 nanoparticles into the dye-doped polymer matrix (Duarte and James 2003, 2004). This discovery enabled the attainment of internally highly homogeneous gain media that enabled the demonstration of homogeneous TEM_{00} laser beams (Duarte and James 2003).

The utilization of nanoparticles in laser gain media has expanded beyond SiO_2 nanoparticles to include other elements from the periodic table notably gold nanoparticles in liquid (Dong *et al*, 2012) and solid-state matrices (Hoa *et al* 2016).

Additional work on organic–inorganic nanocomposite laser gain media include: polymeric matrices containing silicon atoms (Costela *et al* 2004), development of hybrid nanocomposites (García-Moreno *et al* 2005), the introduction of polyhedral oligomeric silsesquioxane (POSS)-modified polymeric matrices (Sastre *et al* 2009). A detailed review on this subject is given by Costela *et al* (2016). Besides their use as laser gain media, nanoparticle-polymer materials have found applications in non-linear optics (Dolgaleva and Boyd 2012).

9.2 Laser dye-doped polymer–nanoparticle gain media

Besides the original disclosures on this subject (Duarte and James 2003, 2004) working details on the synthesis and fabrication for laser dye-doped polymer–nanoparticle gain media are given in a US Patent and an exhaustive review article Duarte and James (2005, 2016).

Here, the focus of attention is on relevant physical characteristics of laser dye-doped polymer–nanoparticle gain media. The focus is on the thermal properties of the gain matrices and on their internal homogeneity. These are critical parameters that have a direct effect on laser beam homogeneity (that is, multiple-transverse-mode emission versus TEM_{00}), laser beam divergence ($\Delta\theta$), and prf capabilities. Table 9.1 compares the derivative $\partial n/\partial T$ for the pure polymer and the polymer–nanoparticle gain media. The polymer is laser-grade PMMA.

The data in table 9.1 clearly indicates that there is a significant difference between the $|\partial n/\partial T|$ for the pure polymer versus the $|\partial n/\partial T|$ for the polymer–nanoparticle matrices, more specifically

$$\left| \frac{\partial n}{\partial T} \right|_{DDP} \approx 1.83 \left| \frac{\partial n}{\partial T} \right|_{PN} \tag{9.1}$$

which means that for a given change in temperature there is a much larger change in the refractive index of the pure polymer than the refractive index of the polymer–

Table 9.1. Measured $\partial n/\partial T$ for DDP and PN solid matrices[a]

Matrix	λ (nm)[b]	$\partial n/\partial T$	Reference
DDP[c]	593.93	$-1.4 \pm 0.2 \times 10^{-4}$	Duarte et al (2000)
PN[d,e] 30% SiO$_2$	632.82	-0.8840×10^{-4}	Duarte and James (2003)
PN[d,e] 50% SiO$_2$	632.82	-0.6484×10^{-4}	Duarte and James (2003)

[a] Adapted from Duarte and James (2003).
[b] Measurement wavelength.
[c] Dye-doped polymer.
[d] Polymer–nanoparticle.
[e] The SiO$_2$ nanoparticles used in these experiment had a diameter, on average, of ~10 nm.

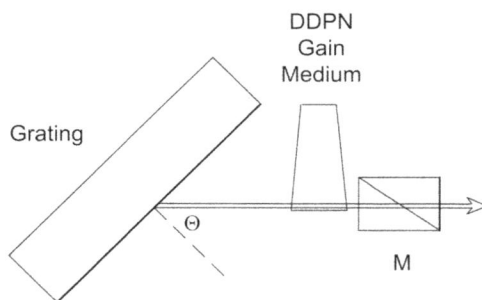

Figure 9.1. Schematics of the Glan–Thompson output-coupler-polarizer Littrow grating oscillator used to compare the performance of DDP and DDPN gain media.

nanopartice composite. Ideally, the $|\partial n/\partial T|$ needs to be as small as possible to minimize the changes in n as the temperature is changed.

In order to make a meaningful and valid comparison between the laser performance of the polymer versus the polymer–nanoparticle materials, a very straightforward and simple tunable laser oscillator, as illustrated in figure 9.1, was utilized. This oscillator was excited transversely by a tunable coumarin 152 dye laser that delivered ~2 mJ per pulse with a pulse duration of $\Delta t \approx 3.5$ ns (FWHM) tunable in the $520 \leqslant \lambda \leqslant 555$ nm range.

The oscillator described in figure 9.1 was configured by a 2400 lines/mm diffraction grating and a Glan–Thompson output-coupler polarizer with a 20% reflectivity at its outer window. The cavity length of this oscillator is ~75 mm and it allowed easy interchange of different gain media in trapezoidal geometries. The gain length was ~10 mm (Duarte and James 2003).

9.3 Organic dye-doped polymer–nanoparticle tunable lasers

Table 9.2 compares the performance of laser emission from three different laser matrices, using the identical oscillator illustrated in figure 9.1. The first gain medium used is a representative pure dye-doped polymer (DDP) gain matrix. The following two gain media are dye-doped polymer–nanoparticle (DDPN) matrices. This table includes dye concentration C, excitation wavelength λ_p, wavelength tuning range, measured beam divergence $\Delta\theta$, and laser conversion efficiency η.

Table 9.2. Laser performance for DDP and DDPN solid matrices[a]

Gain Matrix	C (M)[b]	λ_p (nm)[c]	Tuning range (nm)	$\Delta\theta$ (mrad)	η(%)
DDP[d]	5.0×10^{-4}	~525	$563 \leqslant \lambda \leqslant 610$	2.3	49
DDPN[e] 30% SiO$_2$	3.1×10^{-4}	~525	$567 \leqslant \lambda \leqslant 603$	1.9	63
DDPN[e] 50% SiO$_2$	3.1×10^{-4}	~550	$575 \leqslant \lambda \leqslant 600$	1.6	9

[a] Adapted from Duarte and James (2003).
[b] Initial concentration of rhodamine 6 G dye.
[c] Laser pump wavelength.
[d] Dye-doped polymer.
[e] Dye-doped polymer–nanoparticle. SiO$_2$% quoted is weight per weight (w/w).

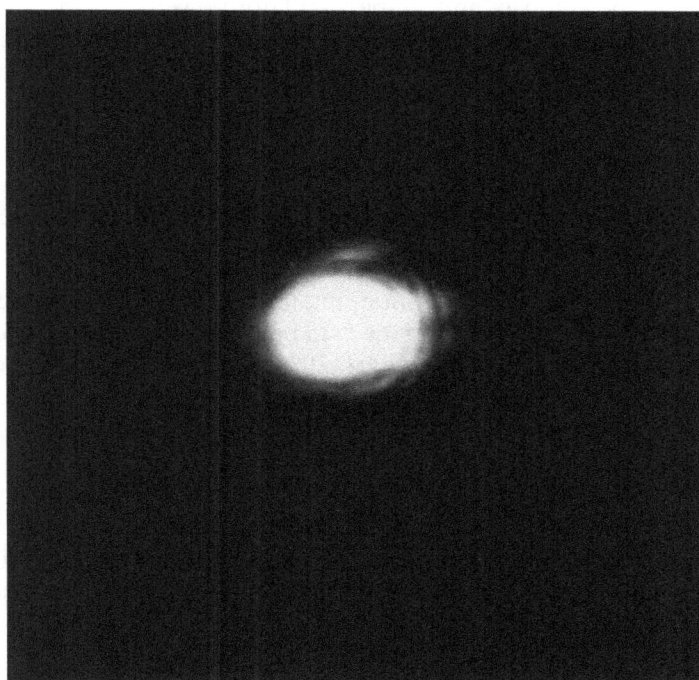

Figure 9.2. Homogeneous near-TEM$_{00}$ laser emission beam profile obtained at $\lambda \approx 580$ nm for a DDPN gain medium at 30% w/w SiO$_2$. The beam divergence is $\Delta\theta \approx 1.9$ mrad, the tuning range $567 \leq \lambda \leq 603$ nm, and the conversion efficiency $\eta \approx 63\%$. It should be noted that inclusion of an intracavity multiple-prism assembly should readily yield narrow-linewidth emission at conversion efficiencies in the 5–10% range (from Duarte and James (2003); courtesy of the Optical Society).

The oscillator using the DDP gain medium delivered a TEM$_{00}$ beam profile with a divergence of $\Delta\theta \approx 2.3$ mrad. This divergence is ~1.6 times the divergence limit. Using the dye-doped polymer–nanoparticle (DDPN) gain medium with a concentration of 30% SiO$_2$ nanoparticles yielded a laser beam with a clean near-TEM$_{00}$ beam profile, as shown in figure 9.2, with a beam divergence of $\Delta\theta \approx 1.9$ mrad. This divergence is ~1.3 times the divergence limit. For a DDPN matrix with a

50% SiO_2 nanoparticles yields a TEM_{00} beam profile with a beam divergence of $\Delta\theta$ ≈ 1.6 mrad. This divergence is ~ 1.1 times the divergence limit. These results clearly show, as indicated in table 9.2, that there is a clear correlation between the presence of SiO_2 nanoparticles in the gain medium and improved beam divergence characteristics.

In terms of energetics, the increase in laser efficiency to $\eta \approx 63\%$ whist using the DDPN matrix at a 30% SiO_2 appears to be related to the increase in dye concentration per unit volume experience in these materials following the initial synthesis (Duarte and James 2003). However, the gain in efficiency ceases to be observed as the number of nanoparticles is increased and it does so in a nonlinear fashion as only a $\eta \approx 9$ % is observed for a SiO_2 nanoparticle presence.

Going back to the compact and simple oscillator powered by the DDPN matrix at 30% SiO_2: its overall performance is summarized by a laser linewidth of $\Delta\lambda \approx 3$ nm, a homogeneous laser beam with a divergence of $\Delta\theta \approx 1.9$ mrad which is $\times 1.3(\lambda/\pi w)$, a tuning range of $567 \leqslant \lambda \leqslant 603$nm, and a remarkable conversion efficiency of $\eta \approx 63\%$.

9.4 Interferometric interpretation on polymer–nanoparticle matrix homogeneity

This section on the interferometric interpretation of the internal structure of the DDPN gain media is largely based on the data presented in the paper and the review on this subject by Duarte and James (2004, 2008, 2016). These authors even extend the discussion into nanoparticle invisibility and core–shell particles. Here, the discussion is focused on laser beam homogeneity and its interferometric interpretation. Some new insights on the physics are included.

Duarte and Pope (1995) suggested that the origin of laser beam inhomogeneities, illustrated in figure 9.3, whilst using laser dye-doped organic–inorganic gain media was *interference* derived from the interaction of coherent radiation and a randomized diffraction grating structure internal to the gain medium. It was also suggested that these three-dimensional diffraction grating structures originated in microscopic refractive index diffrentials.

The origin of this interferometric effect in polymer–silicate matrices can be initially quantified by comparing the refractive index for rhodamine 6 G dye-doped MPMMA, $n(\lambda) = 1.4953$, to that of fused silica, $n(\lambda) = 1.4582$, at $\lambda = 593.93$ nm and $T = 297$ K (Duarte *et al* 2000). Thus in an organic–inorganic gain matrix refractive index gradients of the order of $\Delta n \approx 0.0371$ can be present. This mismatch in refractive indices is the origin of the randomized microscopic grating structures that induce the beam inhomogeneities. In other words the 'beam breakup' is a randomized interference pattern.

Paradoxically, the existence of a diffractive index differential, that is $\Delta n \approx 0.0371$, also leads to an ultimate improvement in organic–inorganic gain media via the incorporation of SiO_2 nanoparticles into the internal structure of the polymeric gain matrices. Indeed, in the previous section two major effects resulting

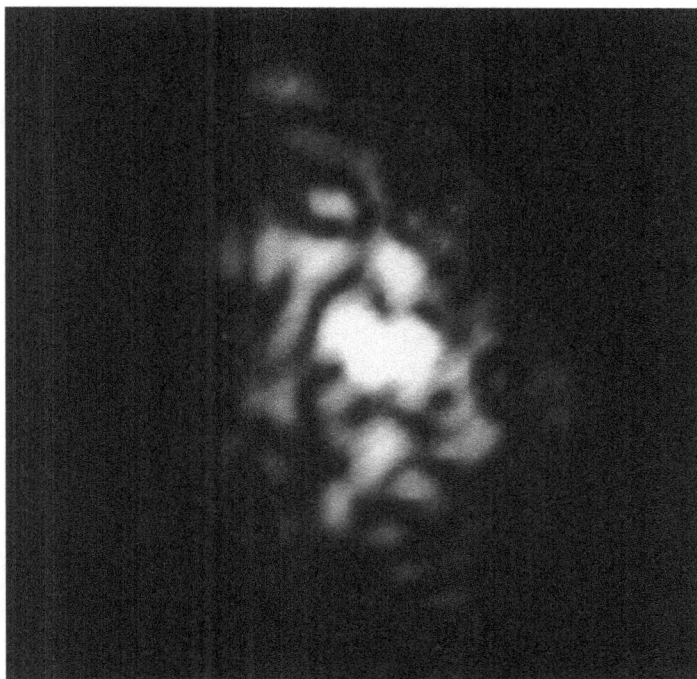

Figure 9.3. Beam profile following propagation of a TEM$_{00}$ beam through a non-nanoparticle laser dye-doped silica–polymer composite.

from the introduction of SiO$_2$ nanoparticles into the synthesis of the organic–inorganic gain matrix are described:

1. Improved $\partial n/\partial T$ characteristics leading to decreased laser beam divergences to $\Delta\theta$ values in the $(1.1 \leqslant \Delta\theta \leqslant 1.3) \times (\lambda/\pi w)$ range.
2. The generation of homogeneous TEM$_{oo}$ laser beams in the absence of internal interference effects. The preservation of laser beam homogeneity is illustrated in figure 9.4.

How was the microscopic interference neutralized? This most interesting effect can be nicely explained by observing the nanoscopic–microscopic structure of the DDPN gain media. This was done, via an electron microscope, for coumarin 500 DDPN and rhodamine 6 G-DDPN. By observing the nanographs it was established that the nanoparticles were dispersed into the space of the polymer in homogeneous distributions. Moreover, the inner-space *with* and *without* nanoparticles of the average gap dimension, that is, *slit* dimension, and the average non-gap dimension, that is, *interslit* dimension, can be quantified. This slit non-slit morphology can be easily recognized as a non-deterministic volumetric transmission diffraction grating as originally anticipated by (Duarte and Pope 1995). Results are given in table 9.3.

This organic–inorganic gain media interferometric morphology can be understood via three-dimensional, two dimensional, or one-dimensional generalized N-slit

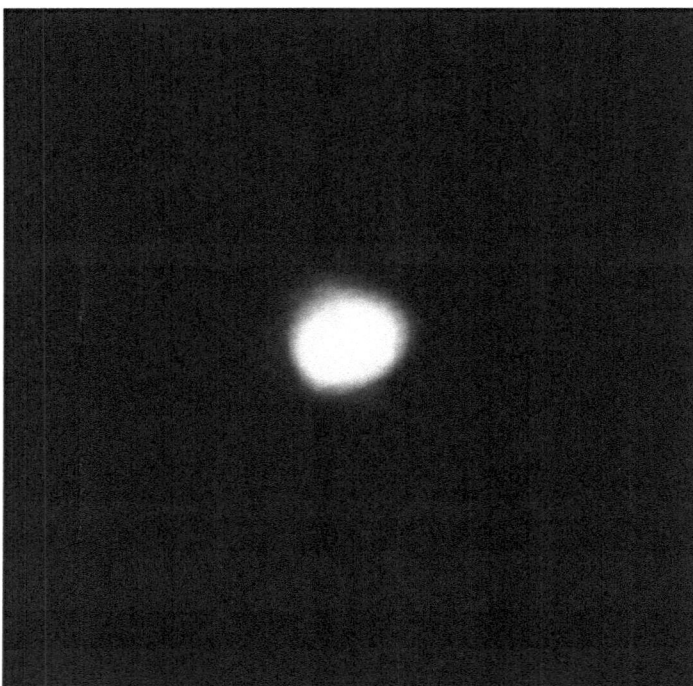

Figure 9.4. Beam profile following propagation of a TEM$_{00}$ beam through a DDNP homogeneous gain medium capable of yielding single-longitudinal-mode oscillation (from Duarte and James (2003); courtesy of the Optical Society).

Table 9.3. Average slit dimensions in DDPN solid matrices[a]

Laser matrix	w/w SiO$_2$ nanoparticle	Slit dimensions[b] (nm)	Interslit dimensions[b] (nm)
Rhodamine 6 G DDPN	0.3	49 ± 29	42 ± 31
Coumarin 500 DDPN	0.3	55 ± 32	57 ± 41

[a] Adapted from Duarte and James (2004).
[b] Measured with a JEOL electron microscope (JEM 100CX II).

interference equations (Duarte 1991, 1993). In this regard, the question of interest becomes: given the laser wavelength range of interest ($500 \leqslant \lambda \leqslant 700$ nm, let's say) what is the range of slit and interslit dimensions under which internal interference will cease to be observed?

To answer this question, the generalized one-dimensional interferometric equation for indistinguishable quanta (Duarte 1991, 1993)

$$|\langle d|s\rangle|^2 = \sum_{j=1}^{N}\Psi(r_j)^2 + 2\sum_{j=1}^{N}\Psi(r_j)\left(\sum_{m=j+1}^{N}\Psi(r_m)\cos(\Omega_m - \Omega_j)\right) \qquad (9.2)$$

should be considered. From this equation the phase term gives origin to the generalized diffraction expresion (Duarte 1993, 2015)

$$d_j(\sin \Theta_j \pm \sin \Phi_j) = m\lambda \qquad (9.3)$$

where λ is the laser wavelength, m the diffraction order, d_j the sum of the dimensions of the slits plus the interslit dimensions, Θ_j the angle of incidence, and Φ_j the angle of diffraction.

For the rhodamine 6 G DDPN matrix the internal nano-structure is illustrated in figure 9.5.

From table 9.3 it can be seen that $d_j \approx 91$ nm which, for $\lambda \approx 580$ nm, and $m = 1$, leads to (Duarte and James 2004)

$$\frac{m\lambda}{d_j} \approx 6.37 \qquad (9.4)$$

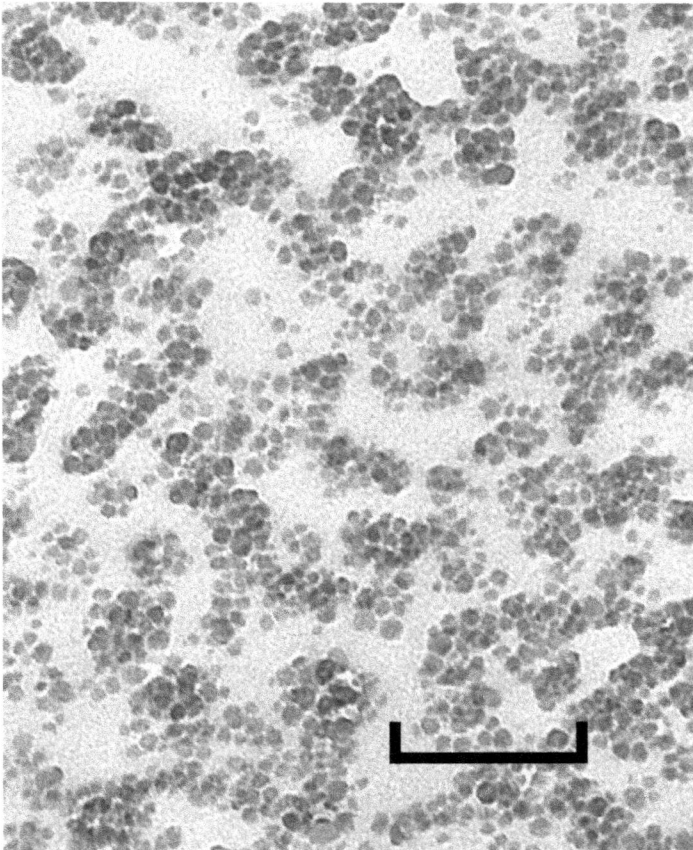

Figure 9.5. Nanograph of the rhodamine 6 G DDNP gain medium. The scale shown corresponds to 200 nm (from Duarte and James (2004); courtesy of the Optical Society).

For the coumarin 500 DDPN matrix the internal nano-structure is illustrated in figure 9.6.

Here, $d_j \approx 112$ nm, $\lambda \approx 510$ nm, and $m = 1$, so that (Duarte and James 2016)

$$\frac{m\lambda}{d_j} \approx 4.55 \tag{9.5}$$

From the generalized diffraction expression, the condition for diffraction is satisfied only if (Duarte 1997)

$$\frac{m\lambda}{d_j} < 2 \tag{9.6}$$

From a different perspective, this is an extremely important condition that states that for a given wavelength λ the diffraction grating will cease to diffract below the condition $d_j \leqslant \lambda$. Under those circumstances the grating will just behave as a refraction medium (Duarte 1997, 2015).

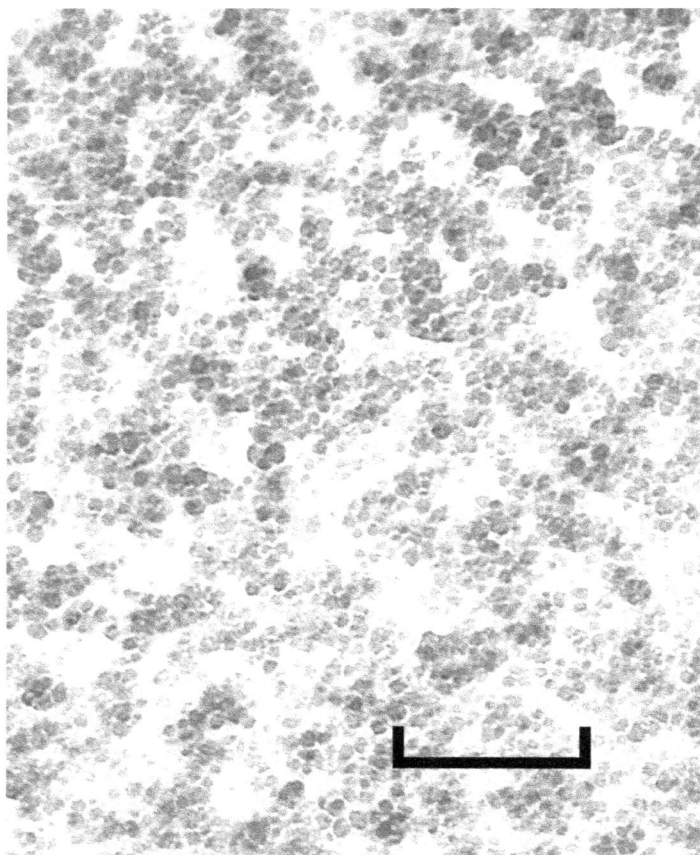

Figure 9.6. Nanograph of the coumarin 500 DDNP gain medium. The scale shown corresponds to 200 nm (from Duarte and James (2004); courtesy of the Optical Society).

Therefore, neither the rhodamine 6 G-DPN gain medium, at $\lambda \approx 580$ nm, nor the coumarin 500-DPN matrix, $\lambda \approx 510$ nm, meet the requirement for intra-gain medium diffraction.

In conclusion, it can be confidently stated that the space dimensions of the nanoparticle aggregates and the space dimensions of the voids (that is, pure DDP space) yield an overall space distribution that does not meet the conditions for internal diffraction. This absence of internal interference leads to homogeneous laser beams from DDPN gain media. In an extension of this research, Duarte and James (2008, 2016) discuss the use of core–shell nanoparticles to augment invisibility of nanoparticle distributions.

9.5 Problems

- 9.1 Use the $\partial n / \partial T$ data given in table 9.1 to verify that equation (9.1) holds.
- 9.2 Use the data given in table 9.3 to verify that for rhodamine 6 G DDPN $(m\lambda / d_j) \approx 6.37$.
- 9.3 Use the data given in table 9.3 to verify that for coumarin 500 DDPN $(m\lambda / d_j) \approx 4.55$.
- 9.4 Use the generalized diffraction given in equation (9.3) to show that the condition for diffraction is given by $(m\lambda / d_j) < \lambda$.
- 9.5 Show that a diffraction grating will cease to diffract below $d_j \leq \lambda$.

References

Costela A, García-Moreno I, del Agua D, García O and Sastre R 2004 Silicon-containing organic matrices as hosts for highly-photostabe solid-state dye lasers *Appl. Phys. Lett.* **85** 2160

Costela A, García-Moreno I and Sastre R 2016 Solid-state organic dye lasers *Tunable Laser Applications* 3rd edn ed F J Duarte (New York: CRC Press) ch 3

Dolgaleva K and Boyd R W 2012 Local field effects in nanostructured photonics materials *Adv. Opt. Photon.* **4** 1–77

Dong F Y, Chughtai A, Popov S, Friberg A T and Mamoun M 2012 Photostability of lasing process from water solution of rhodamine 6 G with gold nanoparticles *Opt. Lett.* **37** 34–6

Duarte F J 1991 Dispersive dye lasers *High Power Dye Lasers* ed F J Duarte (Berlin: Springer) ch 2

Duarte F J 1993 On a generalized interference equation and interferometric measurements *Opt. Commun.* **103** 8–14

Duarte F J 1994 Solid-state multiple-prism grating dye-laser oscillators *Appl. Opt.* **33** 3857–60

Duarte F J 1997 Interference, diffraction, and refraction, via Dirac's notation *Am. J. Phys.* **65** 637–40

Duarte F J 1999 Multiple-prism grating solid-state dye laser oscillator: optimized architecture *Appl. Opt.* **38** 6347–9

Duarte F J 2015 *Tunable Laser Optics* 2nd edn (New York: CRC)

Duarte F J, Costela A, García-Moreno I and Sastre R 2000 Measurements of $\partial n / \partial T$ in solid-state dye-laser gain media *App. Opt.* **39** 6522–3

Duarte F J, Ehrlich J J, Davenport W E, Taylor T S and McDonald J C 1993 A new tunable dye laser oscillator: preliminary report *Proc. of the Int. Conf. on Lasers '92.* ed C P Wang (McLean, VA: STS) 293–6

Duarte F J and James R O 2003 Tunable solid-state lasers incorporating dye- doped polymer-nanoparticle gain media *Opt. Lett.* **28** 2088–90

Duarte F J and James R O 2004 Spatial structure of dye-doped polymer nanoparticle laser media *Appl. Opt.* **43** 4088–90

Duarte F J and James R O 2005 Dye-doped nanoparticle gain medium *US Patent* 6888862

Duarte F J and James R O 2008 Tunable lasers based on dye-doped gain media incorporating homogeneous distributions of functional nanoparticles *Tunable Laser Applications* 2nd edn ed F J Duarte (New York: CRC) ch 4

Duarte F J and James R O 2016 Organic dye-doped polymer-nanoparticle tunable lasers *Tunable Laser Applications* 3rd edn ed F J Duarte (New York: CRC) ch 4

Duarte F J and Pope E J A 1995 Optical inhomogeneities in sol–gel derived ORMOSILS and nanocomposites *Ceram. Trans.* **55** 267–73

Duarte F J, Taylor T S, Costela A, García-Moreno I and Sastre R 1998 Long-pulse narrow-linewidth dispersive solid-state dye-laser oscillator *App. Opt.* **37** 3987–9

García-Moreno I, Costela A, Cuesta A, García O, del Agua D and Sastre R 2005 Synthesis, structure, and physical properties of hybrid nanocomposites for solid-state dye lasers *J. Phys. Chem.* B **109** 21618–26

Hoa D Q, Lien N T H, Duong V T T, Duong V and An N T M 2016 Optical features of spherical gold nanoparticle-doped solid-state dye laser medium *J. Electron. Mater.* **45** 2484–9

Sastre R, Martín V, Garrido L, Chiarra J L, Trastoy B, García O, Costela A and García-Moreno I 2009 Dye-doped polyhedral olygomeric silsesquioxane (POSS)-modified polymeric matrices of highly efficient and photostable solid-state lasers *Adv. Funct. Mat.* **19** 3307–16

IOP Publishing

Organic Lasers and Organic Photonics (Second Edition)

F J Duarte

Chapter 10

Compact and miniaturized organic dye lasers: from glass to bio-based gain media

S Popov and E Vasileva

The technology of compact and miniaturized organic lasers is reviewed with emphasis on VECSOL, VECSEL, DFB, and organic fiber laser configurations. Organic gain media is also described and discussed. The discussion then focuses on laser dye-doped transparent wood gain media and emission characteristics of tunable lasers powered by such organic media.

10.1 Introduction: from liquid to solid gain media

Since the time of their initial study and rapid development in the 1960s, organic dyes were in the focus of research as materials with strong optical gain suitable for efficient lasing (Brock *et al* 1961, Rautian and Sobel'Man 1961). Very soon, comprehensive investigations led to experimental demonstration of dye lasers with gain materials in different physical states: liquid (Sorokin and Lankard 1966), solid, pumped with either another laser (Soffer and McFarland 1967) or flash lamp (Peterson and Snavely 1968) and gas phase (Borisevich *et al* 1973). However, the latter did not get high popularity and wide implementation due to the complexity of design related to the use of gas vapor-phase (Steyer and Schäfer 1974), and, thus, dyes as the optical gain media mainly in liquid and solid states made a significant contribution in science and technology.

After first demonstrations of organic dyes as substances with optical gain, liquid dye lasers became very popular systems in the research world. Their preference over other types of dye lasers was proved by a relatively simple preparation process of lasing media of high quality and homogeneity, as well as the simplicity of the system design and cooling solutions to prevent photo- and thermal-degradation of the gain material (a challenging problem for dye-containing materials). Over the decades, liquid dye solutions were investigated extensively by using a variety of available pump sources: lasers (McFarland 1967, Sorokin *et al* 1967, Schäfer *et al* 1967), flash lamps (Sorokin and Lankard 1966, Peterson and Snavely 1968, McFarland 1967, Sorokin et al 1967, Bass et al 1968, Schäfer et al 1966, Spaeth and Bortfeld 1966, Bass and Steinfeld 1968, Murakawa et al 1968), pulse LEDs (Lupton 2008, Yang et al 2008), laser diodes (Samuel and Turnbull 2007). Different types and concentration of dyes (Sorokin and Lankard 1966, Schäfer et al 1966, Spaeth and Bortfeld 1966, Lempicki and Samelson 1966, Sorokin et al 1966, McFarland 1967, Sorokin et al 1967, Schäfer et al 1967, Snavely et al 1967, Deutsch et al 1967, Lempicki and Samelson 1967, Stepanov et al 1967, Bass et al 1968) were tested for implementation in either CW (Schäfer et al 1966, Peterson et al 1970), or pulsed operation regimes (Bass et al 1968). However, for practical application, liquid solutions were also far from an ideal option and had significant drawbacks related to employing large volumes of organic solvents which could be toxic, hazardous and flammable, experienced flow fluctuations and solvent evaporation, and were expensive to use and dispose of (Schäfer 2013). Moreover, a large operating volume of dyes made the system very bulky and rather expensive for commercial usage outside research laboratories. Due to these reasons, interest in solid-state dye lasers was renewed in the 1990s when organic dyes incorporated in polymer host materials demonstrated reasonable efficiency. They became alternative solutions to overcome the disadvantages of liquid dye lasers. Solid-state dye lasers have obvious technological and implementation advantages such as compactness, lack of toxicity and flammability, and easy handling/maintenance. The exploration of new dyes (Duarte et al 1993, Duarte 1994, Maslyukov et al 1995) and improvement of host media with higher laser damage threshold (Pacheco et al 1988, Dunn et al 1990) revealed great potential for the development of many different types of dye solid-state gain media. Moreover, the variety of dyes and their broad emission spectra made it possible to create tunable solid-state dye lasers covering the wavelength range from UV to mid infrared (300–1800 nm (Lo et al 1992, Mhibik et al 2013)).

10.2 Tunable solid-state dye lasers

The main features distinguishing dyes from other lasing media are high quantum yield (Sorokin et al 1968), as well as broad emission and absorption spectra owing to numerous vibrational and rotational energy states. The combination of such properties provides favorable conditions for realization of tunable narrow-linewidth lasers. One needs to notice that there is no essential difference in the nature of light generation between solid-state and liquid-state dye lasers (O'Connell and Saito 1983). However, energy diagrams and life times of energy levels for each particular dye and, as a result, the properties of lasers based on their use, depend on dye concentration (Bass et al 1968, Sorokin et al 1968) and host materials. Thus, there has been much research effort dedicated to development of host materials to establish the most favorable combinations of dyes and hosting media (Costela et al 2003). The basic requirements for a host matrix are high optical quality with low weak scattering, high optical transmission within absorption (pumping) and emission wavelengths, high solubility of dyes in it, high thermal and chemical stability, and high laser damage threshold. Inorganic glasses and polymers were over the years very popular candidates as host matrices for dyes (Costela *et al* 2003). In comparison to polymers, inorganic glasses used to have better photostability in combination with good laser efficiency (Rahn and King 1995, Faloss *et al* 1997), higher rate of heat dissipation, and, consequently, higher thermal damage threshold (O'Connell and Saito 1983, Barnes 1995, Nikogosyan 1997, Rahn and King 1998). However, with as time progressed, new modified polymers were synthesized with characteristics comparable to inorganic glasses and even better, such as, for example, chemical compatibility providing superior dye–polymer material homogeneity to achieve excellent optical properties (Duarte 1994, Rahn and King 1998) and to optimize polymer properties for particular application via chemical composition (Sastre and Costela 1995, Duarte and Pope 1995). Therefore, polymers became the most favorable host materials for organic dyes. As mentioned above, there exist different favorable host polymers for each particular dye (Costela *et al* 2003). However, we will not consider it in detail in this chapter; rather, we focus here on the tunability of dye lasers depending on the laser design (cavity layout and wavelength selectivity methods (Bornemann *et al* 2011)). Solid-state dye lasers can be pumped by other lasers, laser diodes, and LEDs (Yang *et al* 2008, Sakata and Takeuchi 2008, Zhao *et al* 2015, Burdukova *et al* 2017). As for tunability, one needs to make a clear distinction between the dynamic or 'real' option, where the lasing wavelength can be continually tuned by variation of certain cavity parameters, and the case where the wavelength is fixed for every particular laser device, fabricated or assembled, but can be adjusted via change of the cavity geometry at the stage of fabrication. In the following considerations, we will use notation 'fixed-tuned' for the second option.

10.2.1 VECSOLs

We start our consideration with dynamically tunable solid-state dye lasers (SSDL). One of the interesting representatives in this class is vertical external-cavity surface-emitting organic laser known as VECSOL firstly introduced by Rabbani-Haghighi *et al* (2010). Apparent advantages over other types of tunable SSDLs are its compact design with the cm-long resonator allowing insertion of intracavity elements (spatial or spectral filters (Rabbani-Haghighi *et al* 2010), nonlinear crystals (Forget *et al* 2011)), high output power with linear scalability (without increased photodegradation rate) via variation of pump spot size, excellent beam quality (Rabbani-Haghighi *et al* 2010, Forget *et al* 2011), and diffraction limited output.

The basic construction of the VECSOL is shown in figure 10.1. It is easy to notice that it resembles VECSELs (Okhotnikov 2010) structure consisting of an active layer located over a highly reflective surface-distributed Bragg reflector (DBR) (Forget *et al* 2011). In the VECSOLs configuration, an active layer is dye-doped polymer (DDP) film with thickness varying from several micrometers to centimeters (Rabbani-Haghighi *et al* 2010) for different devices, while DBR is represented as a standard multilayer structure with alternating layers of different refractive indices. An operating principle of DBR is based on Bragg reflection of light at a periodic structure (Westphal *et al* 2013). The DBR together with an external mirror (output-

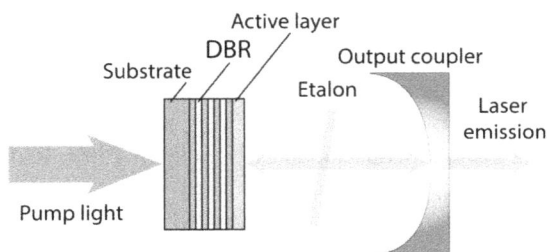

Figure 10.1. Generic schematics of VECSOL.

coupler) form a resonator. The output-coupler has partial transmission at emission wavelength, and in some cases can work as a dichroic mirror and block the pump light. Its choice is a compromise between the transmission losses (for emission output) and material gain saturation and, thus, is individual for each VECSOL design.

For optimized operation of VECSOLs, pump pulse duration and the resonator length (Rabbani-Haghighi *et al* 2010, 2011) have to be properly adjusted. Laser oscillations need time to build up a standing wave during the pump pulse. It means that pump pulse duration has to be long enough to create a stable population inversion that would produce amplification to overcome optical losses inside the cavity. If the pump pulse duration is shorter than the time needed for the laser oscillations to build up, then the laser cannot reach its steady state conditions. The slope efficiency is only 7% for short pump pulses, order of nanoseconds. For longer pump pulses (about 10 ns) capable of creating the steady state condition, the laser emission with efficient extraction of the pump energy and, thus, high conversion and slope efficiencies (52%) can be reached. It is worth noticing, that the threshold value of pump energy is weakly independent of the pulse duration because it is regulated by the optical losses in the system. Similar to the previous case, the cavity length is related to the time required for oscillation to build up. Thus variation of cavity length is equivalent to the change of pump pulse duration relative to the oscillation build-up time, thereby affecting laser efficiency (Rabbani-Haghighi *et al* 2011). However, talking about the pulse duration, one needs to remember that pulse duration and its raising edge are in intimate relation, i.e., for longer pump pulses (for example, microsecond scale) the raising time can be longer than the life time of dye molecules at excited state, and thus, the population inversion will not be enough to initiate amplification and the laser emission will not be created at all.

The VECSOL cavity design offers several options for either dynamical or external tuning of such lasers. The open cavity layout allows insertion of an intracavity etalon, as shown in figure 10.1, for continual tuning of the emission wavelength over tens of nanometers (Rabbani-Haghighi *et al* 2011, Mhibik *et al* 2013). The etalon acts as a Fabry–Perot resonator with the transmission periodically varying with the wavelength. The thickness of the etalon determines the wavelength comb propagating with minimum losses inside the cavity (figure 10.2(a)). Therefore, it works as a frequency filter for the modes generated inside the VECSOL's cavity. In a schematic

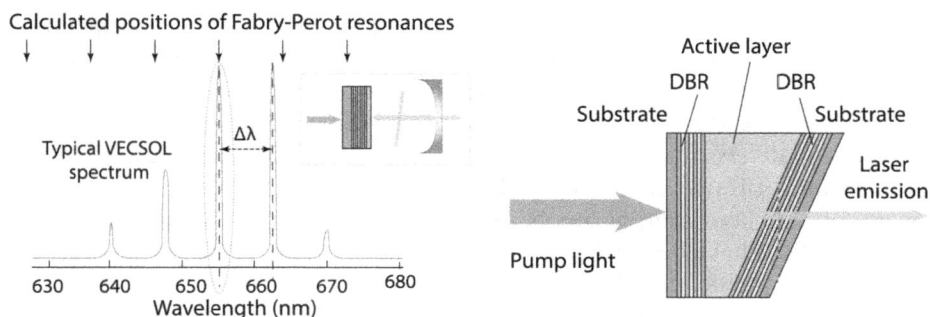

Figure 10.2. The schematics of VECSOLs tuning. (a) VECSOL with etalon inserted and (b) wedge-shaped VECSOL structure.

example shown in figure 10.2, among possible lasing modes formed from the dye gain profile and cavity standing modes (red line), only one frequency matches the etalon resonance (black arrows), thus only one emission wavelength highlighted by dashed blue will be generated at the laser output. The finesse of the etalon determines the spectral line width of the emitted laser line. Thus, VECSOL admits diffraction limited generation and excellent beam quality only for one TEM$_{00}$ laser mode.

With another design approach, a wedge-shaped structure of VECSOL (figure 10.2(b)), laser emission wavelength can be dynamically tuned by shifting the pump beam over the VECSOL side edge (Schütte *et al* 2008). The wedge-shaped design assumes gradually changing thickness of the active layer. Therefore, the cavity length determined as a distance between the resonator mirrors (two DBRs) is changed properly. These variations in the resonator length are reflected in the output spectrum of a VECSOL. In this design, one of the DBR mirrors works as an output-coupler for emitted light. It was also demonstrated (Rabbani-Haghighi *et al* 2011), that in a typical VECSOL design (figure 10.1) dye-polymer film itself might form a Fabry–Perot etalon which modifies emission spectrum without additional intra-cavity elements inserted. In this case, tuning of the laser emission spectra can be achieved by engineered variation of the film thickness. Again, by shifting the pump beam over the sample, the thickness of the formed Fabry–Perot etalon varies and output emission wavelength too (Rabbani-Haghighi *et al* 2011). There exist other examples of VECSOLs tuned by doping variation of the active layer (Kozlov *et al* 1998). However, this way demands the fabrication of new devices and strictly speaking cannot be considered as dynamically tunable, rather fixed-tuned devise, according to our previous definition. However, with the same approach and the same active material, it is possible to create lasers emitting at different wavelengths.

In addition to external geometry of the cavity, the VECSOL tunability is affected by inherent physical and chemical properties of broad-band absorption/emission spectra of dye in the active layers. They can be fabricated following different recipes: spin-coating of dye-doped polymer thin films (several microns) (Schülzgen *et al* 1998, Stagira *et al* 1998, Zavelani-Rossi *et al* 2003), radical polymerization (Garcia

Moreno *et al* 2009) or sol–gel technique (Canva *et al* 1995) giving the thickness of active layer in mm or thicker scales. However, two latter approaches are rather complicated, time-consuming and they demand samples to be polished to optical quality before use (Garcia Moreno *et al* 2009). Since typically the active layer thickness is relatively small, the lifetime resource of VECSOLs is quite limited due to photobleaching of dye molecules. The photobleaching effect leads to gradual decrease of the laser output upon operation until total cease of laser action. Significant efforts have been made to diminish the problem. There are some reported solutions to eliminate this drawback, such as separation of gain layer from the reflective element, developing disposable and replaceable dye capsules (Mhibik *et al* 2013), or careful adjustment of the chemical bonding between polymer and dye molecules to increase the material photostability, as was shown by Costela *et al* (2003).

Although a relatively new type of SSDL was introduced in 2010, the potential of use of VECSOLs as compact tunable organic solid-state lasers is still not fully explored and efficient solutions can be expected. However, based on already obtained achievements, we can conclude that VECSOLs will find application in different fields such as spectroscopy (Woggon *et al* 2010, Oki *et al* 2005), bio/chemistry sensing (Lu *et al* 2008, Yang *et al* 2010), polymer fiber optical communication (McGehee and Heeger 2000). One more possibility to utilize open cavity structure, apart from the wavelength tunability, is the use of nonlinear crystals to achieve SHG and to give VECSOLs an extended implementation in the blue and UV range.

10.2.2 DFB organic lasers

Another interesting class of SSDL offering the wavelength tunability range from near UV to near IR (Ubukata *et al* 2005, Spehr *et al* 2005, Riedl *et al* 2006, Ge *et al* 2010) is distributed feedback lasers (DFB). The concept was introduced by Kogelnik and Shank (1971) and later utilized for the creation of VECSOLs discussed above. In fact, DFBs' construction is similar to that for VECSOLs. The main differences in the design are the absence of an external mirror and alignment of Bragg reflector, longitudinal or transverse, respectively. Typically, the distributed feedback laser structure is formed by a deposited thin film of polymer–dye material over the distributed refractive index grating (Bragg grating) working as a waveguide, so the light travels along the active film (figure 10.3(a)) (Navarro-Fuster *et al* 2012). The optical feedback is realized by multiple Bragg reflections on the borders of layers (substrate/active layer) with different refractive indices, so providing the possibility of generation of standing waves distributed through the length of periodic structures. There exists an elegant version of DFB design (figure 10.3(b)) (Zhu *et al* 2000, Voss *et al* 2001, Tsutsumi and Fujihara 2005), where the periodic refractive index structure is created by constructive interference pattern of the coherent pump beams, as was first shown by Kogelnik and Shank (1971). Although the operation principle of such DFB structures is exactly the same as for DFB lasers with in-built

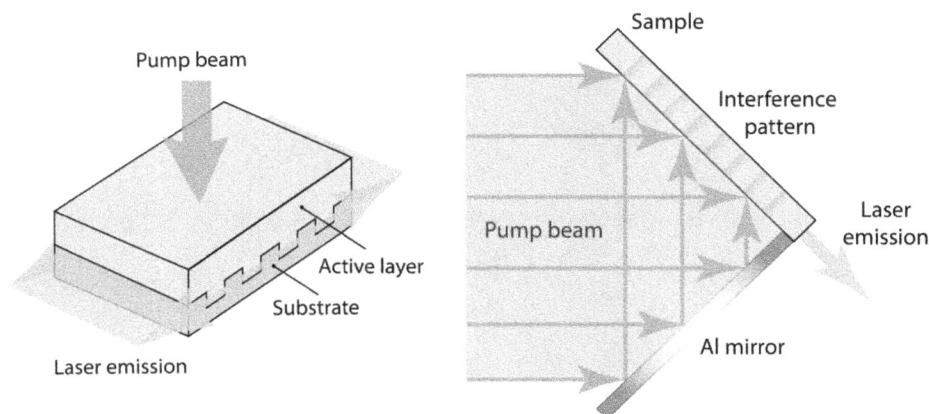

Figure 10.3. The schematic layout of DFB SSDL. (a) Longitudinal DFB profile and (b) interference formed DFB profile.

Bragg reflectors, it gives a possibility for tuning the output wavelength via varying the interference pattern.

The main component besides the active material in DFB lasers is the Bragg reflector providing the standing wave generation. The emission wavelength(s) experiencing the maximum coupling in both (forward and backwards) directions along the gain material plane is selected by Bragg condition (Tsutsumi and Fujihara 2005):

$$m\lambda_b = 2nA, \tag{10.1}$$

where m is integer number (the order of diffraction), λ_b is Bragg wavelength of the coupled light, n is effective refractive index of light, and A is grating period. Coupled mode theory (Kogelnik and Shank 1972) predicts that light with wavelength that exactly satisfies equation (10.1) cannot propagate in the film. A photonic gap (stop-band) centered at λ_b appears in the periodic structure as in the case of photonic crystals. Therefore, the lasing wavelengths are those located at the edges of the stop-band dip close to the Bragg wavelength. The spectral width of the photonic gap and, therefore, the spacing between this pair of wavelengths is determined by the strength of coupling which depends on refractive index contrast provided by the grating (Navarro-Fuster *et al* 2012). In general, the periodic structure generates the set of diffraction orders corresponding to m in equation (10.1) diffracted in the different directions. However, in the vicinity Bragg wavelength only a limited number of orders is in phase and they should be of significant amplitude (typically up to second order (Kogelnik and Shank 1972)). The output spectrum is the result of mutual influence of material gain spectrum and Bragg grating generating a certain number of modes and optical losses. It was demonstrated that DFB structures might work in single and multimode regimes (Schneider *et al* 2004a).

It is known, that geometrical parameters of the grating, such as shape (Karnutsch *et al* 2006), pitch size (Horn *et al* 2014), corrugation height (Ye *et al* 2007, Tsutsumi

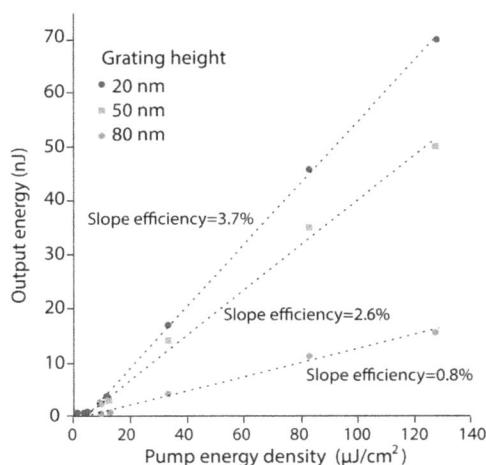

Figure 10.4. Influence of height of the Bragg grating on the DFB output performance.

et al 2008), and aspect ratio (Döring *et al* 2014) define a DFB laser performance. Optimization of these parameters besides corrugation period and the gain material is needed for optimal laser operation (Döring *et al* 2014). The output energy as well as slope efficiency increases with changing the corrugation height (figure 10.4), which could be explained by stronger coupling due to larger overlap of the grating profile and guided laser mode, while at the same time the threshold stays the same most probably due to residual imperfections of the grating (Döring *et al* 2014).

In the considered schemes, DFB lasers can be tuned in different ways: one option is to change the pitch of the Bragg grating (Schneider *et al* 2003, Schneider *et al* 2004b, Vannahme *et al* 2010). That can be done either by changing the interference pattern created by two overlapping coherent pump beams (Shank *et al* 1971, Schneider *et al* 2003) (figure 10.3(b)) or by fabrication of gratings with different pitch size (Schneider *et al* 2004c, Horn *et al* 2014). The photo-induced changes of refractive index with simultaneous profiling of the pumped regions by the interference pattern of two coherent beams can gradually change conditions for formation of different wavelengths. There exist various interfering schemes for DFB excitation, for example, with the use of a sophisticated interferometer, allowing better control of pump beam characteristics (Gale *et al* 1987) or so-called Lloyd's mirror (Maeda *et al* 1997, Tsutsumi and Fujihara 2005) (figure 10.3(b)). To tune the wavelength in the Lloyd-mirror configuration the angle of light incidence should be changed (Shank *et al* 1971, Kranzelbinder *et al* 2002), thus the efficient feedback is realized via spatial modulation of both refractive index or/and gain profile via creating intensively pumped regions.

To change the pitch size of the grating in its classical representation, a new laser chip must be fabricated. Strictly speaking, the fabrication of structures with different parameters of the grating cannot be considered as tunable. This is namely the case earlier defined as fixed-tuned, meaning that with the same approach and the same active material, it is possible to create laser emission at different wavelengths.

However, there is a possibility to manufacture the DFB structure with several different Bragg gratings at one chip (figure 10.5) (Döring *et al* 2014) or make the DFB layout with spatially and continuously changing grating period as in Schneider *et al* (2003). In this case, the output emission wavelength is tuned by shifting the pump beam over different gratings designed for corresponding emission wavelengths.

Another way to realize the wavelength tuning in DFB SSDLs is modification of effective refractive index in the waveguide (distributed grating) by changing the thickness of the active film (Kozlov *et al* 1998, Riechel *et al* 2001, Heliotis *et al* 2004a, 2004b, Xia *et al* 2005, Klinkhammer *et al* 2009, Calzado *et al* 2010, Ramirez *et al* 2011) or the depth of the grating (Navarro-Fuster *et al* 2012). Indeed, this approach is also quasi-tunable since it requires the fabrication of a new device with different thickness of the active layer or grating. However, if the wedge-shaped geometry similar to VECSOL (figure 10.2) is used, the DFB laser can be tuned dynamically by shifting the position of pump beam over the sample to illuminate a film of different thickness (figure 10.6) (Klinkhammer *et al* 2009). There have been

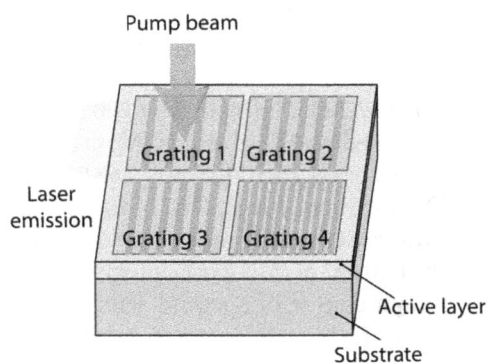

Figure 10.5. DFB laser geometry with several Bragg gratings.

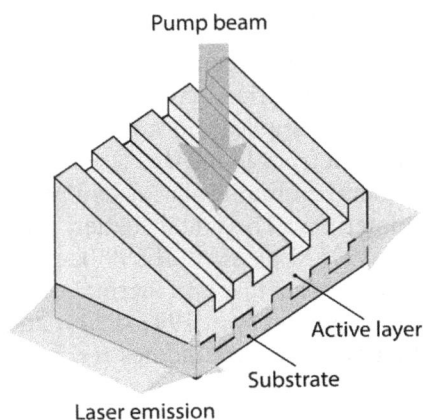

Figure 10.6. Wedge-shaped DFB laser layout.

reported several works indicating that the wavelength of generated light is red-shifted as a dependence of the film thickness and concentration of dye (McGehee *et al* 1998, Riechel *et al* 2001, Oki *et al* 2002a).

The third way mentioned in the literature is changing the emission wavelength of a DFB laser via the concentration of other active dopants in a host material. The main mechanism responsible for such a behavior is the Förster resonance energy transfer (FRET) from dopants to dye molecules (Schneider *et al* 2004a). By varying the concentration of dopants, the ASE is modified. However, using this approach the concentration of dopants has to be carefully controlled, because high concentrations can lead to quenching of the dye emission (Calzado *et al* 2006).

Summarizing, there is a variety of different methods to fabricate DFB SSDLs. The DFB structure has been realized in photopolymers doped with dyes, by either fabricating a permanent DFB structure (Kallinger *et al* 1998, Schneider *et al* 2003), or generating photo-induced gain modulation under illumination with pump pulses (Voss *et al* 2001, Zhu *et al* 2000). The required periodic structures for DFB in polymers can be realized with different fabrication methods such as selective etching (Hu and Kim 1976), holography (Oki *et al* 2002b, Tsutsumi *et al* 2006), nano-imprinting (Nilsson *et al* 2004), UV-lithography (Nilsson *et al* 2005), and electron/proton beam writing (Balslev *et al* 2005, Rao *et al* 2008), laser direct writing, or photolithography (Ubukata *et al* 2005, Nilsson *et al* 2005). Among others, it is worth mention that DFB can be shaped as 1D gratings (Voss *et al* 2001) or 2D photonic crystals (Baumann *et al* 2010), and can also be based on single-transverse mode nanoribbon waveguides (Deotare *et al* 2014). One of the obvious advantages of DFB structure is that the resonator can be easily integrated in a planar organic waveguide which makes possible fabrication of low-cost integrated photonics circuits with active components with a high degree of spectral selection for lab-on-chip integration. Nowadays, DFB lasers are broadly used in telecommunications (Clark and Lanzani 2010), and bio- and chemical-sensing (Lu *et al* 2008, Christiansen *et al* 2009).

10.2.3 Sol–gel silica organic lasers

Advances in the sol–gel technology opened the novel possibility of introducing organic dyes into inorganic host matrices. Although this type of organic laser does not have strong recognition and application nowadays, it was at one time a very important step for the transition from liquid dye lasers to solid-state ones. At earlier stages of dye laser development, inorganic glasses had better properties than organic polymers as dye host materials. They had better photostability, higher laser efficiency (Rahn and King 1995, Faloss *et al* 1997), greater thermal conductivity (which means not so strongly overheated) and thermal damage threshold (O'Connell and Saito 1983, Barnes 1995, Nikogosyan 1997, Rahn and King 1998). Therefore, inorganic glasses were the first choice candidates for fabrication of solid-state dye lasers. However, until sol–gel technology (Iler *et al* 1986) was introduced, it was extremely difficult to incorporate dye molecules into an inorganic glass matrix because the high temperature needed for the glass-making process would destroy

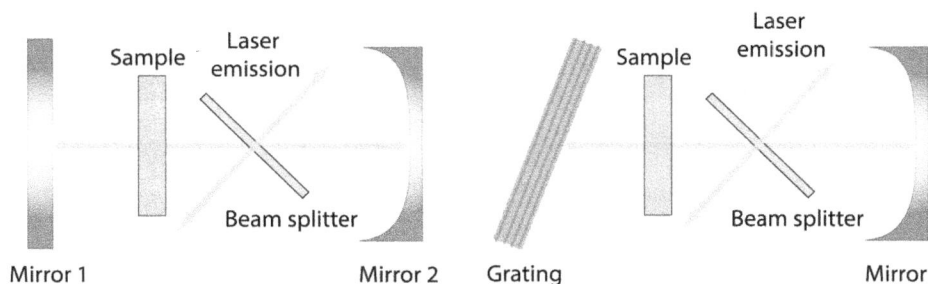

Figure 10.7. Typical layout of SSDL based on sol–gel silica.

organic molecules. There were numerous efforts for glass post-processing in attempts to incorporate dye into ready glass. One of the best known approaches is the use of porous glass (Dolotov *et al* 1992), but it did not become a popular solution because the quality of such glass was not enough for lasing properties— the dye impregnation process was very difficult and did not provide the homogeneity needed for laser quality material. Instead, the sol–gel method enabled the preparation of single- or multiple-component glasses. During the gel preparation, it is possible to embed the molecules of organic dye into the starting mixture before it is solidified. The incorporated dye is uniformly embedded in the host glass material. The final product of the process is a gel with embedded dye molecules that can be dried out and sintered into a glass (Zhu *et al* 2000).

For implementation with sol–gel gain material, the lasing samples are shaped as a thick disk, and the standard two-mirror cavity can be used for a laser design (figure 10.7(a)) (Salin *et al* 1989). Here, two mirrors have high reflectivity, and, therefore, a beam splitter is placed inside the resonator to extract the laser emission. With one optional modification, the emission wavelength tuning is realized by substitution of one mirror with a Littrow diffractive grating (figure 10.7(b)) (Salin *et al* 1989). The laser output wavelength can be tuned by rotating the grating. The slope efficiency (figure 10.8), however, is lower in this configuration in comparison to the two-mirror scheme, that can be attributed to low diffraction efficiency of the grating. Since this type of laser has an open cavity design, it could be possible to tune a laser by inserting a Fabry–Perot etalon (however, this is not demonstrated in the literature).

Some examples have been reported discussing implementation of fiber based structures with the sol–gel technique, and doped with dyes (Gvishi *et al* 1998). The tunability of such lasers was realized by the use of multiple dyes. However, the interest towards this approach was diminished due to significant progress in modification of polymers that are currently considered better hosts for dyes.

10.2.4 Multi-prism SSDL and nanoparticle technology

One of the interesting findings based on the mixture of organic dye (Rh6G) and nanoparticles was demonstrated in SSDL with intracavity tuning components (Dunn *et al* 1990). The system implemented dye-doped organically modified silicate

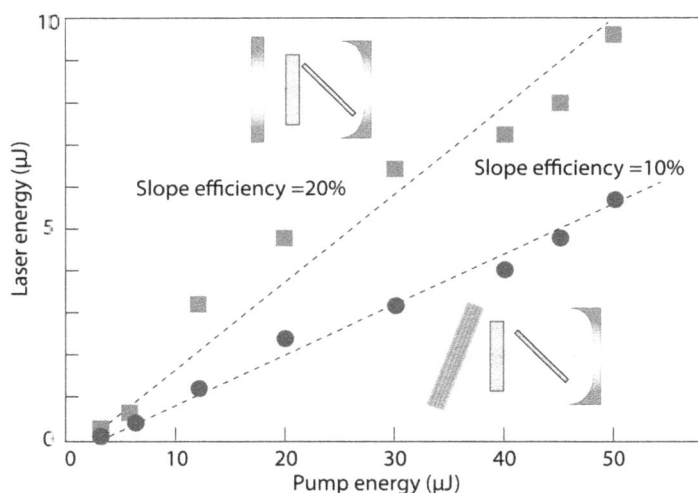

Figure 10.8. The output performance of SSDL with high reflective mirror design.

(Ormosil), or silica–polymer composites (Duarte and Pope 1995) and tetraethoxysilane (Duarte 1994, Dunn *et al* 1990). Originally, the doping of polymers with nanoparticles was introduced to overcome such impairments of early polymeric matrices as homogeneity and weak thermal dissipation. The former factor (homogeneity) significantly decreases the quality of the output laser emission, whereas the latter leads to quick degradation of the gain material due to overheating and following thermo-destruction of organic dyes. The first works reported highly homogeneous dye-doped polymers, but with strong thermal dependence of refractive index (that, for example, limits pump repetition frequency), and/or organic–inorganic media with improved thermal characteristics, but with a noticeable variation of a refractive index inside the combined material such that it caused internal interference and, thus, severe laser beam inhomogeneity. Duarte and James (2003) almost eliminated the laser beam inhomogeneity by nearly uniform distribution of nanoparticles in the dye-doped polymer matrices, so the refractive index variation was nearly negligible. The technology patented in 2002 (Duarte *et al* 2002) made it possible to realize an ultra-narrow-linewidth (Duarte 1994, Duarte *et al* 1997) single-longitudinal-mode emission (Duarte 1995, 1999) in multiple-prism grating oscillator with dye-doped polymer nanoparticle (DDNP) material (Duarte and James 2003).

DDPN based lasers do not have very specific features of a resonator such as, for example VECSOLs or DFB lasers, rather, they implement a modified version of a two-mirror resonator with an inserted etalon and advanced multi-prism scheme to provide the wavelength tunability. In such geometry, the finesse of an etalon determines the width of an emission line. However, the continuous tuning of a laser is usually complicated, since etalon principle of operation is based on a Fabry–Perot cavity and implies a set of output wavelengths. Duarte has demonstrated an alternative cavity design for continually tunable ultra-narrow emission line over

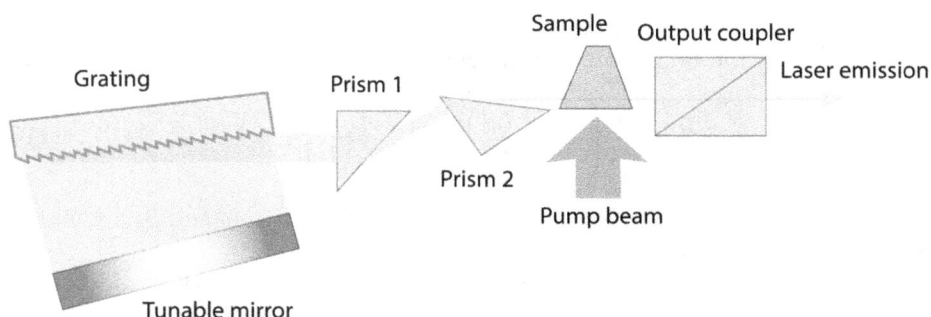

Figure 10.9. Multiple-prism grating oscillator with dye-doped nanoparticle gain material (Duarte 1997).

the spectral range provided by the gain material (1994). The suggested resonator includes diffraction grating allowing continual tuning and one or more prism. The example of a multi-prism geometry, hybrid multiple-prism grazing-incident (HMPGI) grating oscillator (Duarte 1997), is shown in figure 10.9. It includes a diffractive gating, multiple prism expander introduced by two prisms and Glan–Thompson polarizer to serve as an output-coupler (the inner window has antireflection coating; the outer window is partially reflective). The active medium was of wedge shape with sides polished to optical quality ($\lambda/4$). It was transversally pumped by a 532 pulsed laser (SHG of YAG:Nd) with the use of a cylindrical lens. In the case of Littrow configuration where incident and reflection angles are equal, the tunable mirror is not required (Duarte 1999), and the design is called multiple-prism Littrow grating oscillator.

An advantage of the use of prism expander is not only with the tunability of output emission, but also narrow-line width of laser irradiation and low beam divergence. Two latter parameters are directly related to the cavity design via the linewidth equation (Duarte 1990, 2001, 2015) (equation (4.1) in Duarte 2008)

$$\Delta\lambda = \frac{\Delta\theta}{MR}(\partial\theta/\partial\lambda \pm RK)^{-1}, \qquad (10.2)$$

where M is the total beam magnification obtained with the multiple-prism beam expander, R is the number of intracavity return passes, $\partial\theta/\partial\lambda$ is the grating dispersion, and K represents the return-pass multiple-prism dispersion which can be reduced by a proper design. Thus, in order to minimize $\Delta\lambda$ for the narrow emission line, the prism dispersion and beam-magnification factors should be increased, while the divergence of the beam should be decreased. Increasing the beam magnification is the main benefit of the geometry with multiple prisms. It provides a higher M-factor and lower prism losses defined by the prism position and dispersion. As earlier mentioned, the cavity design defines the grating dispersion. For example, for single-prism geometry in Littrow configuration the dispersion is (Duarte 2001, 2015):

$$\partial\theta/\partial\lambda = (2\tan\theta)/\lambda. \qquad (10.3)$$

The beam divergence (Duarte 2001, Duarte 2015) is given by the following expression: (equation (4.3) from Duarte 2008)

$$\Delta\theta \approx \frac{\lambda}{\pi w}\left(1 + \left(\frac{L_R}{B_R}\right)^2 + \left(\frac{LA_R}{B_R}\right)^2\right)^{-1},$$ (10.4)

where $L_R = \pi w^2/\lambda$ is the Rayleigh length, w is a beam waist, A_R and B_R are multiple-prism propagating matrix elements (Duarte 2001, 2015).

There are plenty of research reports with different designs where one or more prisms were utilized to expand the beam, whereas the one described above is the most efficient. In principle, the design of such cavities can be applied to any dye-based gain media with/without nanoparticles. While describing the basic operating principle of grating expander based resonators, we have not focused on other similar designs, as their pros and cons are considered in detail in chapter 5.

10.3 Fixed-tuned lasers

10.3.1 Fiber based dye lasers

As we agreed in the beginning of the chapter, the term 'tunability' can be also implemented for SSDLs with fixed wavelengths, which are predefined but changeable at the stage of design and fabrication. For this purpose, optical fibers are attractive options due to their beneficial features, such as light guiding, design flexibility, dispersion engineering, ease of handling and miniaturization. There have been plenty of optical components realized with optical fibers, for instance beam splitters, amplifiers, Bragg gratings, lasers, etc. Obviously, dye materials with their optical gain properties became interesting candidates for realization of fiber based dye lasers. There exist a vast set of examples of fiber based dye lasers: dye-doped step index polymer optical fibers (POF) laser (Shinzo *et al* 1988); dye doped sol–gel glass optical fiber (Gvishi *et al* 1996); graded index POF doped with a combination of several dyes (Kuriki *et al* 2000a); photonic gap fibers with integrated dye molecules (Shapira *et al* 2006), and others. Although the particular laser design can vary in a broad range, the basic principles are the same. Dye–polymer material is placed in the core of the fiber (middle part of optical fiber), and optical feedback is realized by different means either by an external resonator introduced by mirror(s) and/or grating (Gvishi *et al* 1998, Kuriki *et al* 2000b), or by photonic band gap (PBG) structure (Shapira *et al* 2006). Since external cavity fiber based dye lasers utilize classical principles known from the beginning of the laser era, we focus on a more interesting and sophisticated solution using PBG design.

The photonic band gap fiber laser shown in figure 10.10 (Shapira *et al* 2006) has the dye gain medium surrounded by the photonic band gap structure (multilayer) (Temelkuran *et al* 2002, Hart *et al* 2002, Kuriki *et al* 2004) working as a resonator. It is worth specifically noticing, that the PBG structure in this example is located inside the fiber cladding and has circular symmetry around the core, rather than commonly used fiber Bragg gratings (FBGs) placed inside the fiber core and affecting optical waves propagating along the fiber. In this case, the fiber works as a transverse laser cavity for light emitted by dye molecules along the whole surface area of the fiber

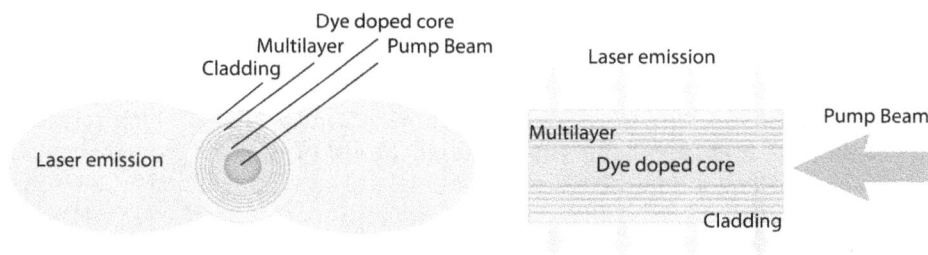

Figure 10.10. Photonic band gap fiber laser.

and, at the same time, as a waveguide for the pump light (figure 10.10(b)). The dye laser radiation is polarized and emitted as a dipole-like wave in the direction perpendicular to the fiber (figure 10.10(a)). In such a PBG fiber laser, there is a possibility of integration of different dyes within certain segments along the fiber, and to generate the output emission of pre-selected wavelengths. Strictly speaking, it is not a classical tuning according to conventional definition, but the structure can provide different outputs. There were reported 12 different dyes inserted in separate fibers with the band gaps corresponding to their emission peaks, and they were used to cover numerous wavelengths from UV till near IR (Shapira *et al* 2006).

In Sheeba *et al* (2007) authors used polymer optical fiber with the core doped by two dyes (Rh6G and RhB). The fiber operated as a series of connected microring resonators pumped along the fiber axis. Therefore, the mode spacing is established by the fiber diameter. Nonradiative energy transfer (Kumar *et al* 2002) from one dye (donor Rh6G) to another one (acceptor RhB) is the basic phenomenon. Its origin is described elsewhere (Dexter 1953, Főrster 1959, Bennett 1964). The probability of the energy transfer is steered by the overlap of emission spectrum of donor with absorption spectrum of acceptor. The energy transfer strongly depends on the dyes' concentration in relation to each other. Therefore, the emission wavelength can be controlled by the relative concentration of the components. This method requires the fabrication of new polymer optical fibers doped with different dye concentrations and, thus, is considered to be a fixed-tuned, or quasi-tunable, system. The elongation of the fiber introduces the red-shift of the emission spectrum due to dye reabsorption and reemission processes, which become more noticeable with longer fibers. There are other examples, where several dyes were combined together in different laser designs, so the emission wavelength could be tuned (at the fabrication stage), for example, organic thin films in DFB laser (Berggren *et al* 1997), sol–gel derived porous glass (Ruland *et al* 1996), whispering gallery mode laser (Vietze *et al* 1998), cascaded grazing incident grating cavity (Rana *et al* 2016), and others.

10.3.2 Whispering gallery mode lasers

In recent decades, micro-ring (-disc) resonators became the object of great interest in the micro-photonics area due to their unique combination of small sizes and high Q-factor. The main property of such resonators is the use of a selected area (or volume) of the material, where whispering gallery modes (WGM) can effectively propagate.

Due to high Q-factor, such devices are very suitable for implementation in laser design. In addition, they offer low power laser thresholds (Sandoghdar *et al* 1996) and compactness of design (Armani *et al* 2003). Apparently, dye gain medium was an excellent candidate for trying in micro-ring (disc) dye lasers, which resulted in vast reported achievements in this area. Initially, most of the data were dedicated to the dye solution flowing in the fiber capillary (that can be attributed to fiber lasers discussed above) to decrease dye photobleaching (Shopova *et al* 2007). However, there appeared several works devoted to solid state WGM dye lasers (Berggren *et al* 1997, Vietze *et al* 1998).

A typical design of a WGM laser is presented in figure 10.11. It consists of a micro-disk made of active dye–polymer material, a substrate, and a waveguide for extraction of laser emission. The pumping of a WPG laser can be realized by several means: with the use of a waveguide, from a side, or from the top of a micro-disk. In Berggren *et al* (1997) authors demonstrated an array of several micro-disks of 3 μm diameter, and separated by the distance of 20 μm. The active material was a mixture of dyes (Berggren *et al* 1997). The total spectrum is shown with several peaks (figure 10.12), caused by the slight differences of the micro-disks' diameters related by

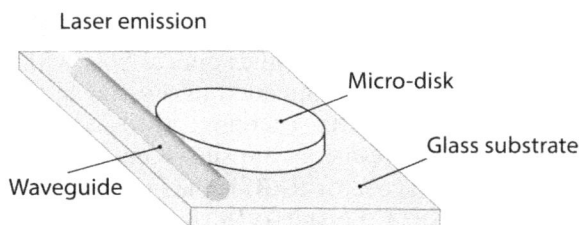

Figure 10.11. The schematic layout of whispering gallery mode dye laser.

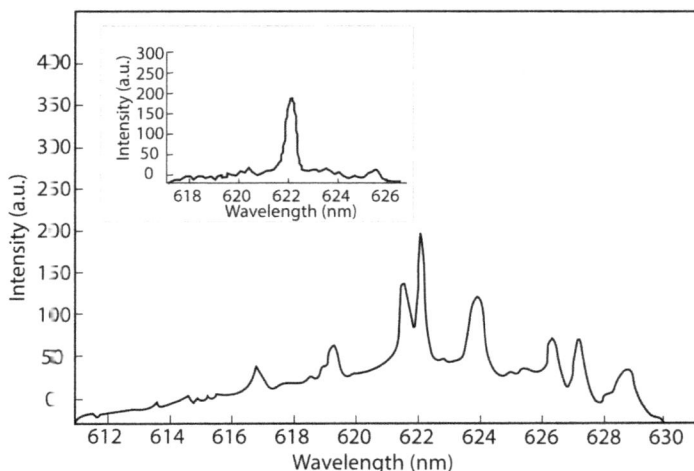

Figure 10.12. Emission spectrum of an array of WGM dye lasers. Inset—the emission spectrum of a single WGM laser from the array.

$$M\lambda = 2\pi Rn, \tag{10.5}$$

where R is the disk diameter, n is the effective refractive index, and M is the mode index. The inset in figure 10.12 represents the spectrum of one micro-disk. By varying the pump spot size, the number of illuminated disks can be controlled, so the output emission spectrum will have several peaks. By moving the pump spot over the structure and illuminating disks one by one, single emission wavelength can be obtained.

10.4 Transparent wood as novel laser media

10.4.1 Optically transparent wood material

With continually growing implementation of different organic materials, there was recently renewed strong interest in transparent wood (TW). Initially, TW was introduced as a method to investigate the inherent hierarchical structure of wood (Fink 1992). Later it was realized that together with excellent mechanical properties, TW can have an optical functionality after proper treatment. A unique feature of wood texture is that it possesses small- and large-scale ordering, although far from perfect, that makes this medium strongly anisotropic and quasi-ordered (Li *et al* 2018). Natural wood is not transparent because it contains an organically based colored component—lignin. To turn the wood into transparent form, certain processing is required. First, it should be decolorized via removing the lignin (or bleaching it) by chemical methods (Li *et al* 2018) or by deactivating the chromophores within lignin (Li *et al* 2017). However, wood scaffold consists of nanocellulose fibrils, fibers, vessels and rays filled with air that make the bleached wood a strongly scattering substance albeit having very low absorption in the visible range (Berglund Lars and Burgert 2018). To make the obtained structure transparent, one needs to eliminate strong light scattering via filling the sponge-like wood skeleton with index-matching material. Since the refractive index of the cellulose-based substances forming the wood is in the range of 1.5–1.6, one of most suitable and popular materials is polymethyl methacrylate (PMMA), with refractive index around 1.49 which can be varied by modification of chemical composition, polymerization technique, etc. After this two-stage process is completed, wood template becomes transparent in the visible range, and it looks like well-structured sponge consisting of three main inclusions: vessels, fibers, and rays filled with polymer (figure 10.13). Indeed, these components are tube-like voids of different size scale, rather than solid formations, where vessels are of 200–300 μm in diameter, and fibers and rays are of 20–30 μm (Borrega *et al* 2015). It is worth noticing that fibers have highest volume ratio inside the material and parallel to vessels, whereas rays are located perpendicularly to both. The two latter components occupy much less volume fraction than fibers.

The natural structure of wood has hierarchical structure scaling from nanometers up to microns. Also, it is not totally ordered, and, thus, all aforementioned formation elements have slightly different shape, diameter, lengths, and thickness of their walls, that prevents total polymer infiltration and prevents making the final

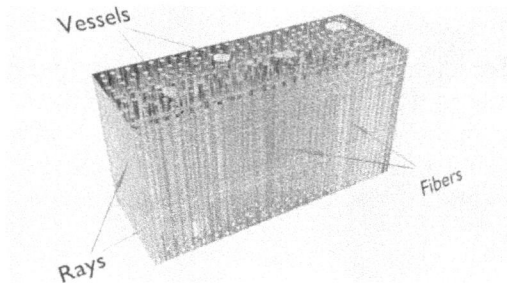

Figure 10.13. Schematic structures of transparent wood sample. Vessels, fibers and rays, and intermediate voids are infiltrated with a polymer.

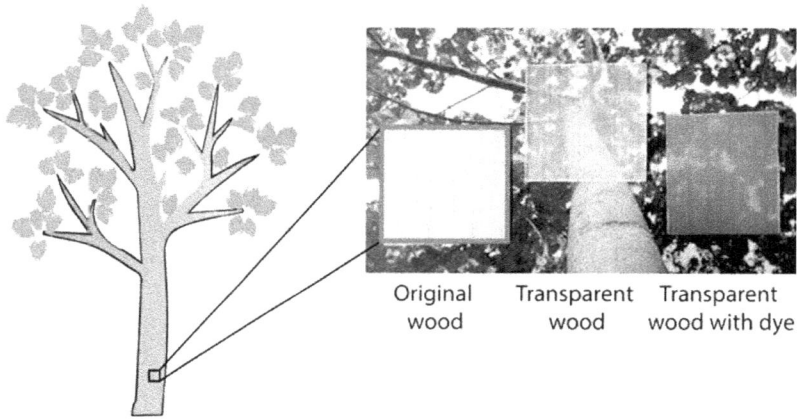

Original	Transparent	Transparent
wood	wood	wood with dye

Figure 10.14. Stages of fabrication of dye-doped TW material: delignification (or bleaching) of the natural wood; infiltration with pure or dye-doped polymer to obtain optical transparency.

material 100% transparent. Additionally, there is still some mismatch between the wood substance and polymer refractive index, and, as a result, noticeable residual light scattering at non-homogenous inclusions. Since the transparent wood, before polymer infiltration, is a porous medium, it was a natural idea to use it as a host material and to fill it with a polymer–dye solution with further solidification, similar to the technology tested with porous glasses at earlier stages of SSDL development. It proved to be a feasible approach, and TW-dye (Rh6G)-doped material demonstrated its lasing operation in 2017 (Vasileva *et al* 2017). Figure 10.14 shows the appearance of the TW-dye samples at different stages of fabrication.

10.4.2 Transparent wood dye laser

Recalling the structure of TW host material (figure 10.13) filled with dye-PMMA mixture, it is easy to see it as a collection of tubes of variable lengths and diameters containing optical gain media. With logical assumption to pump such material with a side-pump scheme (figure 10.15), it can be imagined as a set of simple resonators forming separate SSDLs pumped along the fiber direction. Since such tubes are

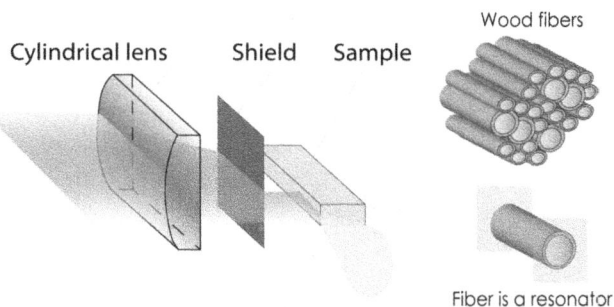

Figure 10.15. Layout of side-pump scheme with variable stripe realization. The black shield allows adjustment of the pumping length along fibers filled with PMMA and Rh6G.

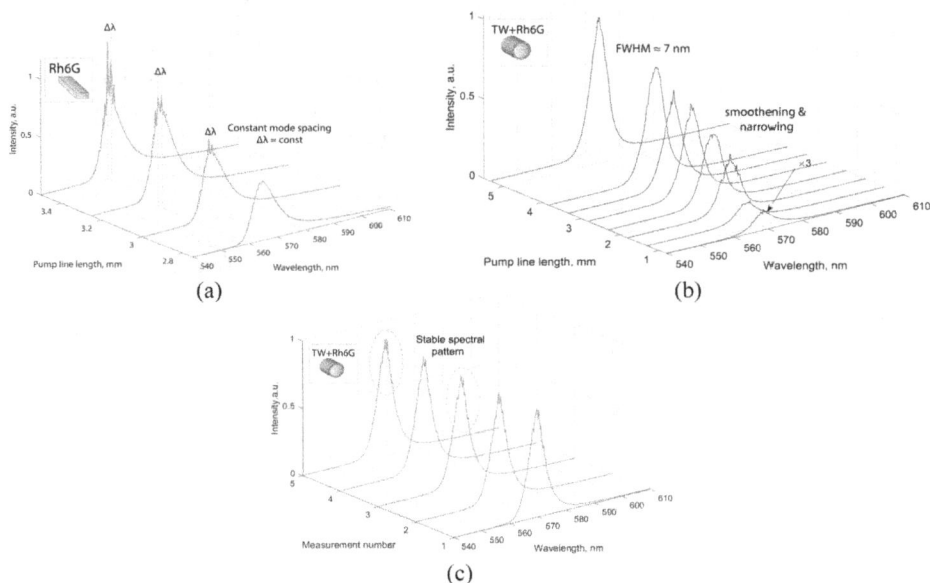

Figure 10.16. Output spectra of TW laser. (a) Change of the linewidth with extended pump line length dye-PMMA; (b) change of the linewidth with extended pump line length TW-dye-PMMA; and (c) pulse-to-pulse stability of the output spectral shape.

missing good quality output-couplers (mirrors), Q-factors of such resonators are rather low, but still demonstrate clear lasing performance (Vasileva *et al* 2017). The lasing linewidth of about 6–7 nm (figure 10.16) is comparable to the low quality laser cavity formed by a one-mirror cavity with PMMA based solid-state gain material demonstrated earlier (Popov 1998a, 1998b).

It is interesting to compare the lasing performance of such a structure to real Fabry–Perot cavities which are also rather simple, but have more traditional realization, for example made of a piece of solid PMMA-dye sample with both facets polished (or a mirror at one facet) to extract laser irradiation.

Comparison of output spectra for such samples (figure 10.16) reveals an interesting observation for the variable stripe pumping scheme. For monolithic samples (dye-PMMA), extending the pumping length (equivalent to a larger pump volume) results in narrowing the lasing linewidth (figure 10.16(a)) with lasing peaks distinguished over amplified spontaneous emission ASE, whereas for the TW piece (tube-bunch set), the longer pump line leads to a slightly broader lasing profile with separate peaks on the top, which finally are washed out with the increased pump volume (figure 10.16(b)). This can be explained by behavior of competing modes in the different samples. For the case of monolithic species, numerous modes finally degenerate to one (or very few) strongest, thus making the final spectral line narrow. In the TW material, modes are still maintained separately, and contribute to the output radiation incoherently, thus, broadening the spectral line while the total amount of fibers involved (figure 10.16(c)) in the lasing process becomes larger, and increasing the total intensity. Poor divergence of the TW lasing output and comparison of spatial coherence for these cases reasonably support such an assumption (Koivurova et al 2018). As was mentioned earlier, an internal structure of these simplified Fabry–Perot cavities is far from perfect, and light experiences rather strong and random scattering from internal walls. Thus, it could be appropriate to treat such TW SSDLs as a quasi-random structure since they occupy an intermediate position between well-defined laser cavities and random (without external cavity) lasers. Producing low coherent emission, but yet of high spectral brightness, they can find numerous applications where speckle-free sources with high intensity and narrow linewidth are required. Recent advances in transparent wood gain media for tunable organic lasers are discussed by Höglund et al (2023).

References

Armani D, Kippenberg T, Spillane S and Vahala K 2003 Ultra-high-Q toroid microcavity on a chip *Nature* **421** 925–8

Balslev S, Rasmussen T, Shi P and Kristensen A 2005 Single mode solid state distributed feedback dye laser fabricated by gray scale electron beam lithography on a dye doped SU-8 resist *J. Micromech. Microeng.* **15** 2456

Barnes N P 1995 *Tunable Lasers Handbook* (New York: Academic)

Bass M, Deutsch T and Weber M 1968 Frequency- and time-dependent gain characteristics of laser- and flashlamp-pumped dye solution lasers *Appl. Phys. Lett.* **13** 120–4

Bass M and Steinfeld J 1968 Wavelength dependent time development of the intensity of dye solution lasers *IEEE J. Quantum Electron.* **4** 53–8

Baumann K, Stöferle T, Moll N, Raino G, Mahrt R, Wahlbrink T, Bolten J and Scherf U 2010 Design and optical characterization of photonic crystal lasers with organic gain material *J. Opt.* **12** 065003

Bennett R 1964 Radiationless intermolecular energy transfer. I. Singlet$\rightarrow$ singlet transfer *J. Chem. Phys.* **41** 3037–40

Berggren M, Dodabalapur A and Slusher R 1997 Stimulated emission and lasing in dye-doped organic thin films with Forster transfer *Appl. Phys. Lett.* **71** 2230–2

Berglund Lars A and Burgert I 2018 Bioinspired wood nanotechnology for functional materials *Adv. Mater.* **30** 1704285

Borisevich N, Kalosha I and Tolkachev V 1973 Lasing of complex organic molecules in the gas phase *J. Appl. Spectrosc.* **19** 1646–47

Bornemann R, Thiel E and Bolívar P H 2011 High-power solid-state cw dye laser *Opt. Express* **19** 26382–93

Borrega M, Ahvenainen P, Serimaa R and Gibson L 2015 Composition and structure of balsa (*Ochroma pyramidale*) wood *Wood Sci. Technol.* **49** 403–20

Brock E, Csavinszky P, Hormats E, Nedderman H, Stirpe D and Unterleitner F 1961 Coherent stimulated emission from organic molecular crystals *J. Chem. Phys.* **35** 759–60

Burdukova O, Gorbunkov M, Petukhov V and Semenov M 2017 Diode pumped tunable dye laser *Appl. Phys.* B **123** 84

Calzado E M, Villalvilla J M, Boj P G, Quintana J A and Díaz-García M A 2006 Concentration dependence of amplified spontaneous emission in organic-based waveguides *Org. Electron.* **7** 319–29

Calzado E M, Villalvilla J M, Boj P G, Quintana J A, Postigo P A and Díaz-García M A 2010 Blue surface-emitting distributed feedback lasers based on TPD-doped films *Appl. Opt.* **49** 463–70

Canva M, Georges P, Perelgritz J-F, Brum A, Chaput F and Boilot J-P 1995 Perylene- and pyrromethene-doped xerogel for a pulsed laser *Appl. Opt.* **34** 428–31

Christiansen M B, Lopacinska J M, Jakobsen M H, Mortensen N A, Dufva M and Kristensen A 2009 Polymer photonic crystal dye lasers as optofluidic cell sensors *Opt. Express* **17** 2722–30

Clark J and Lanzani G 2010 Organic photonics for communications *Nat. Photonics* **4** 438–46

Costela A, García-Moreno I and Sastre R 2003 Polymeric solid-state dye lasers: recent developments *Phys. Chem. Chem. Phys.* **5** 4745–63

Deotare P B, Mahony T S and Bulovic V 2014 Ultracompact low-threshold organic laser *ACS Nano* **8** 11080–5

Deutsch T, Bass M, Meyer P and Protopapa S 1967 Emission spectrum of rhodamine B dye lasers *Appl. Phys. Lett.* **11** 379–81

Dexter D L 1953 A theory of sensitized luminescence in solids *J. Chem. Phys.* **21** 836–50

Dolotov S M, Koldunov M F, Aleksandr A M, Roskova G P, Sitnikov N M, Khaplanova N E and Tsekhomskaya T S 1992 Composite material based on a polymer and a porous glass for fabricating laser components *Sov. J. Quantum Electron.* **22** 1060

Döring S, Rabe T and Stumpe J 2014 Output characteristics of organic distributed feedback lasers with varying grating heights *Appl. Phys. Lett.* **104** 263302

Duarte F J 1990 Narrow-linewidth pulsed dye laser oscillators *Dye Laser Principles* (New York: Academic) pp 133–83

Duarte F J 1994 Solid-state multiple-prism grating dye-laser oscillators *Appl. Opt.* **33** 3857–60

Duarte F J 1995 Solid-state dispersive dye laser oscillator: very compact cavity *Opt. Commun.* **117** 480–4

Duarte F J 1997 Multiple-prism near-grazing-incidence grating solid-state dye-laser oscillator *Opt. Laser Technol.* **29** 513–6

Duarte F J 1999 Multiple-prism grating solid-state dye laser oscillator: optimized architecture *Appl. Opt.* **38** 6347–9

Duarte F J 2001 Multiple-return-pass beam divergence and the linewidth equation *Appl. Opt.* **40** 3038–41

Duarte F J 2008 *Tunable Laser Applications* (Boca Raton, FL: CRC)

Duarte F J 2015 *Tunable Laser Optics* (Boca Raton, FL: CRC)

Duarte F J, Costela A, Garcia-Moreno I, Sastre R, Ehrlich J J and Taylor T S 1997 Dispersive solid-state dye laser oscillators *Opt. Quantum Electron.* **29** 461–72

Duarte F J, Ehrlich J J, Davenport W E, Taylor T S and McDonald J 1993 A new tunable dye laser oscillator: preliminary report *Proc. Int. Conf. Lasers '92* **95** pp 293–6

Duarte F J and James R 2003 Tunable solid-state lasers incorporating dye-doped, polymer–nanoparticle gain media *Opt. Lett.* **28** 2088–90

Duarte F J, James R O and Rowley L 2002 Dye-doped polymer nanoparticle gain medium for use in a laser *US Patent* 20040120373A1

Duarte E and Pope E 1995 *Optical Inhomogeneities in Sol–Gel Derived Ormosils and Nanocomposites* (Westerville, OH: American Ceramic Society)

Dunn B S, Mackenzie J D, Zink J I and Stafsudd O M 1990 Solid-state tunable lasers based on dye-doped sol-gel materials (Defense Technical Information Center) (translated by International Society for Optics and Photonics) pp 174–82

Faloss M, Canva M, Georges P, Brun A, Chaput F and Boilot J-P 1997 Toward millions of laser pulses with pyrromethene-and perylene-doped xerogels *Appl. Opt.* **36** 6760–3

Fink S 1992 Transparent wood—a new approach in the functional study of wood structure *Holzforschung: Int. J. Biol. Chem. Phys. Technol. Wood* **46** 403

Forget S, Rabbani-Haghighi H, Diffalah N, Siove A and Chénais S 2011 Tunable ultraviolet vertically-emitting organic laser *Appl. Phys. Lett.* **98** 131102

Förster T 1959 10th Spiers Memorial Lecture. Transfer mechanisms of electronic excitation *Discuss. Faraday Soc.* **27** 7–17

Gale G, Ranson P and Denariez-Roberge M 1987 Coherent spectroscopy with a distributed feedback dye laser *Appl. Phys. B: Lasers Opt.* **44** 221–33

Garcia Moreno I, Costela A, Martin V, Pintado Sierra M and Sastre R 2009 Materials for a reliable solid state dye laser at the red spectral edge *Adv. Funct. Mater.* **19** 2547–52

Ge C, Lu M, Jian X, Tan Y and Cunningham B T 2010 Large-area organic distributed feedback laser fabricated by nanoreplica molding and horizontal dipping *Opt. Express* **18** 12980–91

Gvishi R, Narang U, Ruland G, Kumar Deepak N and Prasad Paras N 1998 Novel, organically doped, sol–gel derived materials for photonics: multiphasic nanostructured composite monoliths and optical fibers *Appl. Organomet. Chem.* **11** 107–27

Gvishi R, Ruland G and Prasad P N 1996 New laser medium: dye-doped sol-gel fiber *Opt. Commun.* **126** 66–72

Hart S D, Maskaly G R, Temelkuran B, Prideaux P H, Joannopoulos J D and Fink Y 2002 External reflection from omnidirectional dielectric mirror fibers *Science* **296** 510–3

Heliotis G, Xia R, Bradley D, Turnbull G, Samuel I, Andrew P and Barnes W 2004a Two-dimensional distributed feedback lasers using a broadband, red polyfluorene gain medium *J. Appl. Phys.* **96** 6959–65

Heliotis G, Xia R, Turnbull G A, Andrew P, Barnes W L, Samuel I D W and Bradley D D 2004b Emission characteristics and performance comparison of polyfluorene lasers with one and two dimensional distributed feedback *Adv. Funct. Mater.* **14** 91–7

Höglund M, Baitenov A, Berglund L A and Popov S 2023 Transparent wood biocomposite of well-dispersed content for fluorescence and lasing applications *ACS Appl. Opt. Mater.* **1** 1043–51

Horn W, Kroesen S and Denz C 2014 Two-photon fabrication of organic solid-state distributed feedback lasers in rhodamine 6 G doped SU-8 *Appl. Phys.* B **117** 311–5

Hu C and Kim S 1976 Thin film dye laser with etched cavity *Appl. Phys. Lett.* **29** 582–5

Iler R K, Hench L and Ulrich D 1986 *Science of Ceramic Chemical Processing* (New York: Wiley)

Kallinger C, Hilmer M, Haugeneder A, Perner M, Spirkl W, Lemmer U, Feldmann J, Scherf U, Müllen K and Gombert A 1998 A flexible conjugated polymer laser *Adv. Mater.* **10** 920–23

Karnutsch C, Gýrtner C, Haug V, Lemmer U, Farrell T, Nehls B, Scherf U, Wang J, Weimann T and Heliotis G 2006 Low threshold blue conjugated polymer lasers with first-and second-order distributed feedback *Appl. Phys. Lett.* **89** 201108

Klinkhammer S, Woggon T, Geyer U, Vannahme C, Dehm S, Mappes T and Lemmer U 2009 A continuously tunable low-threshold organic semiconductor distributed feedback laser fabricated by rotating shadow mask evaporation *Appl. Phys. B: Lasers Opt.* **97** 787–91

Kogelnik H and Shank C 1971 Stimulated emission in a periodic structure *Appl. Phys. Lett.* **18** 152–4

Kogelnik H and Shank C 1972 Coupled wave theory of distributed feedback lasers *J. Appl. Phys.* **43** 2327–35

Koivurova M, Vasileva E, Li Y, Berglund L and Popov S 2018 Complete spatial coherence characterization of quasi-random laser emission from dye doped transparent wood *Opt. Express* **26** 13474–82

Kozlov V, Bulovic V, Burrows P, Baldo M, Khalfin V, Parthasarathy G, Forrest S, You Y and Thompson M 1998 Study of lasing action based on Förster energy transfer in optically pumped organic semiconductor thin films *J. Appl. Phys.* **84** 4096–108

Kranzelbinder G, Toussaere E, Zyss J, Pogantsch A, List E, Tillmann H and Hörhold H-H 2002 Optically written solid-state lasers with broadly tunable mode emission based on improved poly (2, 5-dialkoxy-phenylene-vinylene) *Appl. Phys. Lett.* **80** 716–8

Kumar G, Thomas V, Jose G, Unnikrishnan N and Nampoori V 2002 Energy transfer in Rh 6G: Rh B system in PMMA matrix under cw laser excitation *J. Photochem. Photobiol. A: Chem.* **153** 145–51

Kuriki K, Kobayashi T, Imai N, Tamura T, Koike Y and Okamoto Y 2000a Organic dye doped polymer optical fiber laser *Polym. Adv. Technol.* **11** 612–6

Kuriki K, Kobayashi T, Imai N, Tamura T, Nishihara S, Nishizawa Y, Tagaya A, Koike Y and Okamoto Y 2000b High-efficiency organic dye-doped polymer optical fiber lasers *Appl. Phys. Lett.* **77** 331–3

Kuriki K, Shapira O, Hart S D, Benoit G, Kuriki Y, Viens J F, Bayindir M, Joannopoulos J D and Fink Y 2004 Hollow multilayer photonic bandgap fibers for NIR applications *Opt. Express* **12** 1510–7

Lempicki A and Samelson H 1996 Organic liquid lasers *Lasers: A Series of Advances* ed A L Levine (New York: Dekker) vol 1

Lempicki A and Samelson H 1967 Liquid lasers *Sci. Am.* **216** 80–90

Li Y, Fu Q, Rojas R, Yan M, Lawoko M and Berglund L 2017 Lignin retaining transparent wood *ChemSusChem* **10** 3445–51

Li Y, Vasileva E, Sychugov I, Popov S and Berglund L 2018 Optically transparent wood: recent progress, opportunities, and challenges *Adv. Opt. Mater.* **6** 1800059

Lo D, Parris J and Lawless J 1992 Multi-megawatt superradiant emissions from coumarin-doped sol-gel derived silica *Appl. Phys. B* **55** 365–7

Lu M, Choi S, Wagner C, Eden J and Cunningham B 2008 Label free biosensor incorporating a replica-molded, vertically emitting distributed feedback laser *Appl. Phys. Lett.* **92** 261502

Lu M, Choi S S, Irfan U and Cunningham B 2008 Plastic distributed feedback laser biosensor *Appl. Phys. Lett.* **93** 111113

Lupton J M 2008 Laser technology: Over the rainbow *Nature* **453** 459–60

Maeda R, Oki Y and Imamura K 1997 Ultrashort pulse generation from an integrated single-chip dye laser *IEEE J. Quantum Electron.* **33** 2146–9

Maslyukov A, Sokolov S, Kaivola M, Nyholm K and Popov S 1995 Solid-state dye laser with modified poly (methyl methacrylate)-doped active elements *Appl. Opt.* **34** 1516–8

McFarland B 1967 Laser second-harmonic-induced stimulated emission of organic dyes *Appl. Phys. Lett.* **10** 208–9

McGehee M D, Gupta R, Veenstra S, Miller E K, Díaz-García M A and Heeger A J 1998 Amplified spontaneous emission from photopumped films of a conjugated polymer *Phys. Rev. B* **58** 7035

McGehee M D and Heeger A J 2000 Semiconducting (conjugated) polymers as materials for solid state lasers *Adv. Mater.* **12** 1655–68

Mhibik O, Leang T, Siove A, Forget S and Chénais S 2013 Broadly tunable (440–670 nm) solid-state organic laser with disposable capsules *Appl. Phys. Lett.* **102** 041112

Murakawa S-i, Yamaguchi G and Yamanaka C 1968 Wavelength shift of dye solution laser *Japan. J. Appl. Phys.* **7** 681

Navarro-Fuster V, Vragovic I, Calzado E M, Boj P G, Quintana J A, Villalvilla J M, Retolaza A, Juarros A, Otaduy D and Merino S 2012 Film thickness and grating depth variation in organic second-order distributed feedback lasers *J. Appl. Phys.* **112** 043104

Nikogosyan D N 1997 *Properties of Optical and Laser-related Materials: A Handbook* (New York: Wiley) 388–91

Nilsson D, Balslev S, Gregersen M M and Kristensen A 2005 Microfabricated solid-state dye lasers based on a photodefinable polymer *Appl. Opt.* **44** 4965–71

Nilsson D, Nielsen T and Kristensen A 2004 Molded plastic micro-cavity lasers *Microelectron. Eng.* **73** 372–6

O'Connell R M and Saito T T 1983 Plastics for high-power laser applications: a review *Opt. Eng.* **22** 224393

Okhotnikov O G 2010 *Semiconductor Disk Lasers: Physics and Technology* (New York: Wiley)

Oki Y, Aso K, Zuo D, Vasa N J and Maeda M 2002a Wide-wavelength-range operation of a distributed-feedback dye laser with a plastic waveguide *Japan. J. Appl. Phys.* **41** 6370

Oki Y, Miyawaki S, Maeda M and Tanaka M 2005 Spectroscopic applications of integrated tunable solid-state dye laser *Optical Rev.* **12** 301–6

Oki Y, Yoshiura T, Chisaki Y and Maeda M 2002b Fabrication of a distributed-feedback dye laser with a grating structure in its plastic waveguide *Appl. Opt.* **41** 5030–5

Pacheco D P, Aldag H R, Itzkan I and Rostler P S 1988 A solid-state flashlamp-pumped dye laser employing polymer hosts *Proc. Int. Conf. Lasers '87*; F J Duarte (McLean, VA: STS) pp 330–7

Peterson O and Snavely B 1968 Simulated emission from flashlamp-excited organic dye in polymethyl methacrylate *Appl. Phys. Lett.* **12** 238–40

Peterson O, Tuccio S and Snavely B 1970 CW operation of an organic dye solution laser *Appl. Phys. Lett.* **17** 245–7

Popov S 1998a Dye photodestruction in a solid-state dye laser with a polymeric gain medium *Appl. Opt.* **37** 6449–55

Popov S 1998b Influence of pump repetition rate on dye photostability in a solid-state dye laser with a polymeric gain medium *Pure Appl. Optics: J. Eur. Opt. Soc. Part A* **7** 1379

Rabbani-Haghighi H, Forget S, Chénais S and Siove A 2010 Highly efficient, diffraction-limited laser emission from a vertical external-cavity surface-emitting organic laser *Opt. Lett.* **35** 1968–70

Rabbani-Haghighi H, Forget S, Siove A and Chenais S 2011 Analytical study of vertical external-cavity surface-emitting organic lasers *Eur. Phys. J. Appl. Phys.* **56** 34108

Rahn M D and King T A 1995 Comparison of laser performance of dye molecules in sol–gel, polycom, ormosil, and poly (methyl methacrylate) host media *Appl. Opt.* **34** 8260–71

Rahn M D and King T A 1998 High-performance solid-state dye laser based on peryleneorange-doped polycom glass *J. Mod. Opt.* **45** 1259–67

Ramirez M G, Boj P G, Navarro-Fuster V, Vragovic I, Villalvilla J M, Alonso I, Trabadelo V, Merino S and Díaz-García M A 2011 Efficient organic distributed feedback lasers with imprinted active films *Opt. Express* **19** 22443–54

Rana P, Sridhar G and Manohar K 2016 Characteristics of a cascaded grating multi wavelength dye laser *Opt. Laser Technol.* **86** 39–45

Rao S V, Bettiol A and Watt F 2008 Characterization of channel waveguides and tunable microlasers in SU8 doped with rhodamine B fabricated using proton beam writing *J. Phys. D: Appl. Phys.* **41** 192002

Rautian S and Sobel'Man I 1961 Remarks on negative absorption *Opt. Spectrosc.* **10** 65

Riechel S, Lemmer U, Feldmann J, Berleb S, Mückl A, Brütting W, Gombert A and Wittwer V 2001 Very compact tunable solid-state laser utilizing a thin-film organic semiconductor *Opt. Lett.* **26** 593–5

Riedl T, Rabe T, Johannes H-H, Kowalsky W, Wang J, Weimann T, Hinze P, Nehls B, Farrell T and Scherf U 2006 Tunable organic thin-film laser pumped by an inorganic violet diode laser *Appl. Phys. Lett.* **88** 241116

Ruland G, Gvishi R and Prasad P N 1996 Multiphasic nanostructured composite: Multi-dye tunable solid state laser *JACS* **118** 2985–91

Sakata H and Takeuchi H 2008 Diode-pumped polymeric dye lasers operating at a pump power level of 10 mW *Appl. Phys. Lett.* **92** 106

Salin F, Bagnall C, Zarzycki J, Le Saux G, Georges P and Brun A 1989 Efficient tunable solid-state laser near 630 nm using sulforhodamine 640-doped silica gel *Opt. Lett.* **14** 785–7

Samuel I D W and Turnbull G A 2007 Organic semiconductor lasers *Chem. Rev.* **107** 1272–95

Sandoghdar V, Treussart F, Hare J, Lefevre-Seguin V, Raimond J-M and Haroche S 1996 Very low threshold whispering-gallery-mode microsphere laser *Phys. Rev. A* **54** R1777

Sastre R and Costela A 1995 Polymeric solid state dye lasers *Adv. Mater.* **7** 198–202

Schäfer F, Schmidt W and Marth K 1967 New dye lasers covering the visible spectrum *Phys. Lett. A* **24** 280–81

Schäfer F P 2013 *Dye Lasers* (Berlin: Springer)

Schäfer F P, Schmidt W and Volze J 1966 Organic dye solution laser *Appl. Phys. Lett.* **9** 306–9

Schneider D, Hartmann S, Benstem T, Dobbertin T, Heithecker D, Metzdorf D, Becker E, Riedl T, Johannes H-H and Kowalsky W 2003 Wavelength-tunable organic solid-state distributed-feedback laser *Appl. Phys. B: Lasers Optics* **77** 399–402

Schneider D, Rabe T, Riedl T, Dobbertin T, Kröger M, Becker E, Johannes H-H, Kowalsky W, Weimann T and Wang J 2004a Laser threshold reduction in an all-spiro guest–host system *Appl. Phys. Lett.* **85** 1659–61

Schneider D, Rabe T, Riedl T, Dobbertin T, Kröger M, Becker E, Johannes H-H, Kowalsky W, Weimann T and Wang J 2004b Ultrawide tuning range in doped organic solid-state lasers *Appl. Phys. Lett.* **85** 1886–8

Schneider D, Rabe T, Riedl T, Dobbertin T, Werner O, Kröger M, Becker E, Johannes H-H, Kowalsky W and Weimann T 2004c Deep blue widely tunable organic solid-state laser based on a spirobifluorene derivative *Appl. Phys. Lett.* **84** 4693–5

Schülzgen A, Spiegelberg C, Morrell M, Mendes S, Kippelen B, Peyghambarian N, Nabor M, Mash E and Allemand P 1998 Near diffraction-limited laser emission from a polymer in a high finesse planar cavity *Appl. Phys. Lett.* **72** 269–71

Schütte B, Gothe H, Hintschich S, Sudzius M, Fröb H, Lyssenko V and Leo K 2008 Continuously tunable laser emission from a wedge-shaped organic microcavity *Appl. Phys. Lett.* **92** 153

Shank C, Bjorkholm J and Kogelnik H 1971 Tunable distributed-feedback dye laser *Appl. Phys. Lett.* **18** 395–6

Shapira O, Kuriki K, Orf N D, Abouraddy A F, Benoit G, Viens J F, Rodriguez A, Ibanescu M, Joannopoulos J D and Fink Y 2006 Surface-emitting fiber lasers *Opt. Express* **14** 3929–35

Sheeba M, Thomas K J, Rajesh M, Nampoori V P, Vallabhan C P and Radhakrishnan P 2007 Multimode laser emission from dye doped polymer optical fiber *Appl. Opt.* **46** 8089–94

Shinzo M, Akitoshi A, Osamu Y, Takayuki H and Hiroshi I 1988 Tunable laser using sheet of dye doped plastic fibers *Electron. Commun. Japan. (Part II: Electronics)* **71** 47–52

Shopova S I, Zhou H, Fan X and Zhang P 2007 Optofluidic ring resonator based dye laser *Appl. Phys. Lett.* **90** 221101

Snavely B, Peterson O and Reithel R 1967 Blue laser emission from a flashlamp-excited organic dye solution *Appl. Phys. Lett.* **11** 275–6

Soffer B and McFarland B 1967 Continuously tunable, narrow band organic dye lasers *Appl. Phys. Lett.* **10** 266–7

Sorokin P, Culver W, Hammond E and Lankard J 1966 End-pumped stimulated emission from a thiacarbocyanine dye *IBM J. Res. Dev.* **10** 401

Sorokin P, Lankard J, Moruzzi V and Hammond E 1968 Flashlamp pumped organic dye lasers *J. Chem. Phys.* **48** 4726–41

Sorokin P, Lankard J R, Hammond E C and Moruzzi V L 1967 Laser-pumped stimulated emission from organic dyes: experimental studies and analytical comparisons *IBM J. Res. Dev.* **11** 130–48

Sorokin P P and Lankard J 1966 Stimulated emission observed from an organic dye, chloro-aluminum phthalocyanine *IBM J. Res. Dev.* **10** 162–3

Spaeth M and Bortfeld D 1966 Stimulated emission from polymethine dyes *Appl. Phys. Lett.* **9** 179–81

Spehr T, Siebert A, Fuhrmann-Lieker T, Salbeck J, Rabe T, Riedl T, Johannes H, Kowalsky W, Wang J and Weimann T 2005 Organic solid-state ultraviolet-laser based on spiro-terphenyl *Appl. Phys. Lett.* **87** 161103

Stagira S, Zavelani-Rossi M, Nisoli M, DeSilvestri S, Lanzani G, Zenz C, Mataloni P and Leising G 1998 Single-mode picosecond blue laser emission from AQ solid conjugated polymer *Appl. Phys. Lett.* **73** 2860–62

Stepanov B, Rubinov A and Mostovnikov V 1967 Optic generation in solutions of complex molecules *JETP Lett.* **5** 117–9

Steyer B and Schäfer F P 1974 A vapor-phase dye laser *Opt. Commun.* **10** 219–20

Temelkuran B, Hart S D, Benoit G, Joannopoulos J D and Fink Y 2002 Wavelength-scalable hollow optical fibres with large photonic bandgaps for CO_2 laser transmission *Nature* **420** 650–53

Tsutsumi N and Fujihara A 2005 Tunable distributed feedback lasing with narrowed emission using holographic dynamic gratings in a polymeric waveguide *Appl. Phys. Lett.* **86** 061101

Tsutsumi N, Fujihara A and Hayashi D 2006 Tunable distributed feedback lasing with a threshold in the nanojoule range in an organic guest-host polymeric waveguide *Appl. Opt.* **45** 5748–51

Tsutsumi N, Takeuchi M and Sakai W 2008 All-plastic organic dye laser with distributed feedback resonator structure *Thin Solid Films* **516** 2783–7

Ubukata T, Isoshima T and Hara M 2005 Wavelength programmable organic distributed feedback laser based on a photoassisted polymer migration system *Adv. Mater.* **17** 1630–3

Vannahme C, Klinkhammer S, Christiansen M B, Kolew A, Kristensen A, Lemmer U and Mappes T 2010 All-polymer organic semiconductor laser chips: parallel fabrication and encapsulation *Opt. Express* **18** 24881–7

Vasileva E, Li Y, Sychugov I, Mensi M, Berglund L and Popov S 2017 Lasing from organic dye molecules embedded in transparent wood *Adv. Opt. Mater.* **5** 1700057

Vietze U, Krauss O, Laeri F, Ihlein G, Schüth F, Limburg B and Abraham M 1998 Zeolite-dye microlasers *Phys. Rev. Lett.* **81** 4628

Voss T, Scheel D and Schade W 2001 A microchip-laser-pumped DFB-polymer-dye laser *Appl. Phys. B: Lasers Optics* **73** 105–9

Westphal A, Klinkebiel A, Berends H-M, Broda H, Kurz P and Tuczek F 2013 Electronic structure and spectroscopic properties of mononuclear manganese(iii) Schiff base complexes: a systematic study on [Mn(acen)X] complexes by EPR, UV/vis, and MCD Spectroscopy (X = Hal, NCS) *Inorg. Chem.* **52** 2372–87

Woggon T, Klinkhammer S and Lemmer U 2010 Compact spectroscopy system based on tunable organic semiconductor lasers *Appl. Phys. B: Lasers Optics* **99** 47–51

Xia R, Heliotis G, Campoy-Quiles M, Stavrinou P, Bradley D, Vak D and Kim D-Y 2005 Characterization of a high-thermal-stability spiroanthracenefluorene-based blue-light-emitting polymer optical gain medium *J. Appl. Phys.* **98** 083101

Yang Y, Turnbull G and Samuel I 2008 Hybrid optoelectronics: A polymer laser pumped by a nitride light-emitting diode *Appl. Phys. Lett.* **92** 150

Yang Y, Turnbull G A and Samuel I D 2010 Sensitive explosive vapor detection with polyfluorene lasers *Adv. Funct. Mater.* **20** 2093–7

Ye C, Wong K, He Y and Wang X 2007 Distributed feedback sol–gel zirconia waveguide lasers based on surface relief gratings *Opt. Express* **15** 936–44

Zavelani-Rossi M, Lanzani G, Anni M, Gigli G, Cingolani R, Barbarella G and Favaretto L 2003 Organic laser based on thiophene derivatives *Synth. Met.* **139** 901–3

Zhao Z, Mhibik O, Nafa M, Chénais S and Forget S 2015 High brightness diode-pumped organic solid-state laser *Appl. Phys. Lett.* **106** 051112

Zhu X-L, Lam S-K and Lo D 2000 Distributed-feedback dye-doped solgel silica lasers *Appl. Opt.* **39** 3104–7

IOP Publishing

Organic Lasers and Organic Photonics (Second Edition)

F J Duarte

Chapter 11

Electrically-pumped organic semiconductor laser emission

F J Duarte and K M Vaeth

Electrically-pumped organic semiconductor coherent emission is discussed in detail. The first part of this chapter describes in detail the molecular chemistry and materials aspects of the organic semiconductors utilized whilst the second part describes and discusses the physics and the architecture of the interferometric emitters yielding near-Gaussian beams characterized by a divergence of $\Delta\theta \approx 1.09 \times (\lambda / \pi w)$ and interferograms with a visibility of $\mathcal{V} = 0.901 \pm 0.088$. Comments on available literature on this subject are also provided.

11.1 Introduction

Interest in direct electrical excitation of organic gain media dates back to the first decade of the organic dye laser (Steyer and Schäfer 1974, Marowsky *et al* 1976).

Partially motivated by keen interest in tunable miniature organic lasers integrated to chip-on-chip technology, the issue of *direct electrical excitation* of organic gain media has continued to generate widespread interest, varied efforts, and a large number of publications over many years. Since these publications are too numerous to include here, a number of representative reviews on this subject are mentioned (Kranzelbinder and Leising 2000, Baldo *et al* 2002, Karnutsch 2007, Samuel and Turnbull 2007, Grivas and Pollnau 2012).

In this chapter a *highly coherent* pulsed electrically-excited organic-semiconductor emitter powered by the laser dye coumarin tetramethyl 545 is described (Duarte *et al* 2005, Duarte 2007, 2008, 2010, 2016). The emission from this electrically-excited organic semiconductor sub-microcavity integrated interferometric source exhibits *single-transverse-mode* characteristics with a near-Gaussian profile. More specifically, its highly directional beam has a measured beam divergence which is nearly diffraction limited, as faithfully and permanently documented using silver-halide black and white film (Duarte *et al* 2005). Its interferometrically derived high-visibility ($\mathcal{V} \approx 0.9$)

emission linewidth is estimated to be in the $0.1 \leqslant \Delta\lambda \leqslant 1.5$ nm range. More to the point: for a cavity length of $L \approx 300$ nm, related to an intracavity mode spacing of $\delta\lambda \approx 486$ nm, this linewidth range is undoubtedly consistent with *single-longitudinal-mode* emission (Duarte 2010, 2016).

In this chapter a detailed description of the electrically-pumped tandem organic semiconductor giving origin to the observed coherent emission is provided. The physics of the integrated interferometric emitter used to record the spatially and spectrally coherent emission is described in detail. Furthermore, the recent literature (2009–2017) reporting on subsequent laser emission from organic semiconductor devices is also analyzed.

11.2 Organic semiconductors

The performance of organic light emitting diodes (OLEDs) has made significant progress since the initial report of efficient electroluminescence (Tang and VanSlyke 1987), leading to commercialization of the technology in mobile devices and large-screen displays. This revolution has been driven by improvements in OLED material and device design, which must be optimized as a system to achieve the target device efficiencies and operational lifetimes. In this section, the materials and device architectures used in typical OLEDs will be discussed, along with the advances that have enabled generation of highly efficient devices capable of exhibiting spatial and spectral coherence under electrical stimulation.

11.2.1 Energetics of organic semiconductors

The electronic properties of thin solid films used in inorganic microelectronic devices are typically controlled by finely tuning the material properties at the atomic level through doping. From electrical to optical properties, dopants in inorganic thin films alter the bandgap characteristics of the host material to achieve the desired conductivity, emission, and absorption properties of the film. A similar approach can be used with thin organic films used in OLEDs, but the implementation is different in practice, with doping occurring at the molecular level rather than the atomic level. This is mainly due to the amorphous nature of the organic films, and a tendency for localization of the electronic charge on the molecule.

Use of host-dopant systems in OLED devices was first demonstrated in 1989, when it was shown that doped solid films of electrically conductive organic materials with aromatic laser dyes allowed greater control over the emission properties of the device (Tang *et al* 1989, Adachi *et al* 1990). This allowed the device electrical and optical properties to be adjusted fairly independently, with the host optimized for charge transport and efficient electron–hole pair formation and transfer, and the dopant optimized for emission color and electroluminescence efficiency. This approach enabled substantial improvements in device efficiency, and led to a significant amount of work in the design and optimization of host-dopant combinations for OLED applications.

Organic host and dopant materials used in OLED devices are typically aromatic in structure, with delocalized π-orbitals that split into a highest occupied molecular

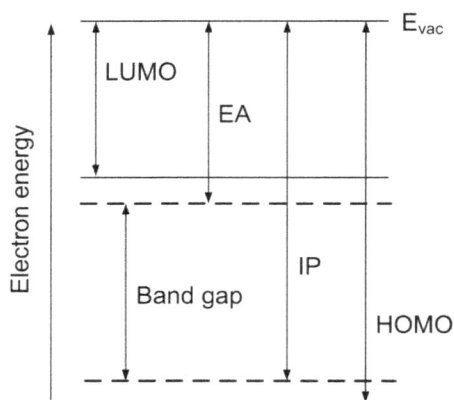

Figure 11.1. HOMO–LUMO energy diagram for organic semiconductors.

orbital (HOMO), lowest unoccupied molecular orbital (LUMO), and energy gap, all of which control the optical and electrical properties (see figure 11.1). The tendency for charge localization in the solid state makes the energy gap, HOMO, and LUMO levels of the isolated molecule good indicators of the characteristics observed in the solid film, particularly for the dopant, which is dispersed in the host layer at low concentrations (~0.5–5%). It should be noted that the energy gap of the host film, which is defined by the difference between the ionization potential (IP) and electron affinity (EA) of the layer, is often smaller than the difference between the HOMO and LUMO levels of the isolated molecule (figure 11.1), due to molecular interactions in the solid state brought on by Van der Waals forces (Djurovich *et al* 2009). Shown in table 11.1 are the electronic and optical properties of common organic used as host and charge transport in OLEDs. The properties of OLED dopant molecules, respectively, are listed in table 11.2.

When charges are injected into an OLED device, the electrons and holes diffuse through the layers and combine on the host to form an excited state. If a dopant is present in the film, the excited state can transfer to that molecule, either via non-radiative absorption, known as Föster resonance energy transfer, or by direct transfer of an excited electron, known as Dexter energy transfer (figure 11.2). In Föster resonance energy transfer (figure 11.2(a)), the excited state of the host is transferred to the dopant via resonant dipole–dipole interactions, followed by emission from the dopant. The transfer process is non-radiative, typically occurring on length scale of 1–10 nm, but efficient transfer requires overlap of the host fluorescent emission spectrum and dopant absorption spectrum. In Dexter energy transfer, the host and dopant molecules exchange electrons, resulting in transfer of the excited state to the dopant molecule, which can then recombine to produce a photon (figures 11.2(b) and (c)). Typical length scales for Dexter energy transfer are 1 nm, since a more intimate interaction between the molecules is required. In both Föster and Dexter energy transfer processes the host dopant system must be optimized for efficient photon generation, as there are many non-radiative recombination pathways that can compete with these processes to produce heat rather than

Table 11.1. OLED host and charge generation molecules.

Organic molecule[a]	Formula and molecular weight (amu)	HOMO [LUMO] (eV)	Application
Tris-(8-hydroxyquinoline) aluminum (Alq$_3$)	C$_{27}$H$_{18}$AlN$_3$O$_3$ 459.43	-5.7 [-2.9][b]	Electron transport and host
Bis(2-methyl-8-quinolinolate)-4-(phenyl phenolato) aluminum (BAlq)	C$_{32}$H$_{25}$AlN$_2$O$_3$ 512.53	-5.9 [-2.9][c]	Electron transport and host
2 Methyl 9,10-di (2 naphtyl) anthracene (tbADN)	C$_{35}$H$_{24}$ 444.56	-5.6 [-2.64][b,d]	Host
4,4′Bis(N-carbazolyl) biphenyl (CBP)	C$_{36}$H$_{24}$N$_2$ 484.59	-6.1 [-2.64][c]	Host
2,2′,2′-(1,3,5-Benzinetriyl)-tris (1-phenyl-1-H-benzimidazole) (TPBI)	C$_{45}$H$_{30}$N$_6$ 654.76	-6.2 [-2.7][e]	Electron transport and exciton blocking
2,9-Dimethyl-4,7diphenyl-1,10 phenanthroline (BCP)	C$_{26}$H$_{20}$N$_2$ 360.45	-6.7 [-3.2][e]	Electron transport and exciton blocking
4,4′-Bis-[N-(1-napthyl)-N-phenyl-amino] biphenyl (NPB)	C$_{44}$H$_{32}$N$_2$ 588.74	-5.4 [-2.3][e]	Hole transport
2,3,5,6-Tetrafluoro-7,7,8,8-tetracyano-quinodimethane (F4-TCNQ)	C$_{14}$F$_4$N$_4$ 276.15	-8.3 [-5.4][b]	Charge generation and interconnect layer
4,4′,4′-Tris[(3-methylphenyl) phenylamino]triphenylamine (m-TDATA)	C$_{57}$H$_{48}$N$_4$ 789.02	-5.1 [-2.4][f]	Charge generation and interconnect layer

[a] Abbreviated name given in parenthesis following the molecular name.
[b] Djurovich *et al* (2009).
[c] Yoon *et al* (2014).
[d] Kim *et al* (2010).
[e] Tao *et al* (2011).
[f] Goushi and Adachia (2012).

light. As shown in figure 11.3(a) the host IP is usually chosen to be larger than the dopant HOMO, so that there is little barrier for an excited state formed on the host to transfer to the dopant. Note also that the electrode materials, typically indium tin oxide (ITO) as the anode and magnesium stabilized with silver or aluminum doped

Table 11.2. OLED dopant molecules.

Dopant molecule[a]	Formula and molecular weight (amu)	HOMO [LUMO] eV	Function
4-(Dicyanomethylene)-2-*tert*-butyl-6-(1,1,7,7-tetramethyl julolidin-4-ylvinyl)-4*H*-pyran (DCJTB) Fluorescent	$C_{30}H_{35}N_3O$ 453.62	−5.26	[−3.11][b]
Coumarin 545 tetramethyl (C 545 T)	$C_{26}H_{26}N_2O_2S$ 430.56	−5.5 [−3.3][c]	Fluorescent[d]
5,6,11,12-Tetraphenyl tetracene (Rubrene) Fluorescent	$C_{42}H_{28}$ 532.70	−5.36	[−3.15][b]
Perylene Fluorescent	$C_{20}H_{12}$ 252.32	−5.20	[−2.37][e]
Tris[2-phenylpyridinato C^2,N] iridium(III) (Ir(ppy)$_3$)	$C_{33}H_{24}IrN_3$ 654.78	−5.2 [−2.8][f]	Phosphorescent
Bis[2-(4,6-difluorophenyl) pyridinato-C^2,N] (picolinato) iridium(III) (Blue Ir(ppy)) (FIrpic)	$C_{28}H_{16}F_4IrN_3O_2$ 694.66	−5.8 [2.9][g]	Phosphorescent
Platinum octaethylporphyrin (PtOEP)	$C_{36}H_{44}N_4Pt$ 729.86	−5.3 [−3.2][5]	Phosphorescent

[a] Abbreviated name given in parenthesis following the molecular name.
[b] Kan and Lee (2004).
[c] Liu *et al* (2010).
[d] C 545 T is an efficient laser dye (Duarte *et al* 2006).
[e] Djurovich *et al* (2009).
[f] Mäkinen *et al* (2002).
[g] Dong *et al* (2013).

with lithium as the cathode, are selected such that the work functions of the inorganic materials are well aligned with the IP and EA of the organic layers.

The dopant molecules are chosen to have short radiative lifetimes in the host molecule and high luminescent efficiencies, and it is worth noting that many OLED dopants were first used in optically-pumped organic dye lasers. Dopants can also impact the device lifetime, and so a balance must be made between electroluminescent efficiency and operational stability in order to achieve the overall desired performance (Hamada *et al* 1995, Shi and Tang, 1997, Aziz *et al* 1999).

Early OLED devices used fluorescent molecules for the host and dopant, producing emission via Föster resonance energy transfer (figure 11.2(a)) (Kalinowski, 1999). One of the drawbacks of these fluorescent materials is that photon emission is only allowed from the singlet excited state. This limits the theoretical internal electroluminescent efficiency of the device to 25% with these materials, since formation of electron–hole pairs by electrical injection is typically governed by statistics, with only ¼ of the injected charges forming the singlet states, and the other ¾ forming triplets. Discovery of organic phosphorescent dopants, that

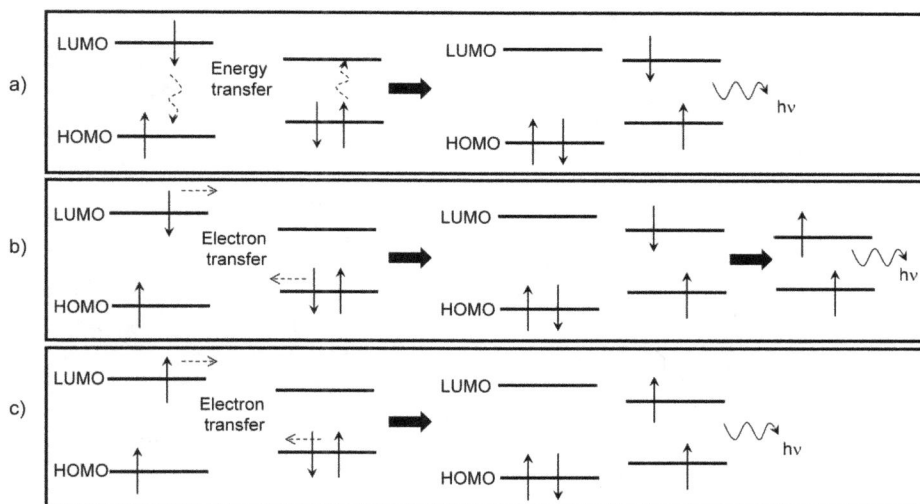

Figure 11.2. (a) Föster resonance energy transfer and (b) and (c) Dexter energy transfer.

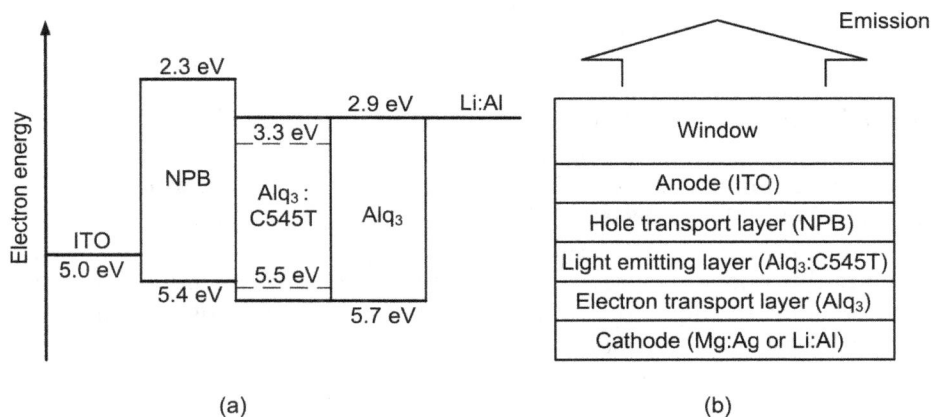

(a)   (b)

Figure 11.3. (a) Energy diagram for (b) single OLED device.

exhibit high electroluminescent efficiencies from the triplet state and fast rates of singlet–triplet intersystem crossing led to further improvements in device perform-ance as Dexter energy transfer could now be utilized to harvest singlet and triplet excited states from the host, to produce radiative triplet excited states on the dopant, see figures 11.2(b) and (c) (Baldo *et al* 1998, Baldo 1999). Internal device electro-luminescent efficiencies of close to 100% have been reported with these materials (Adachi *et al* 2001).

One of the drawbacks of phosphorescent dopants is the tendency to exhibit lower device operational stability than fluorescent materials, particularly for blue emis-sion. The dopant system chosen for a particular application will depend on the overall luminescence and stability requirements of the system, and both fluorescent

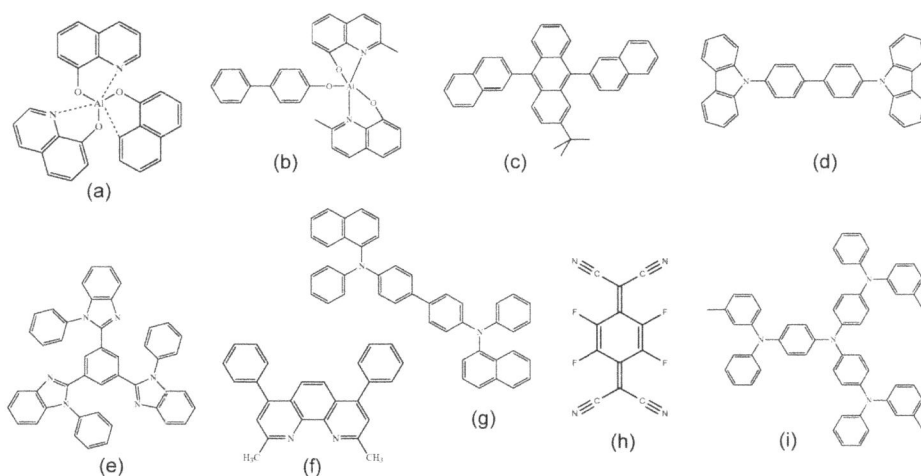

Figure 11.4. Molecular structures for (a) Tris-(8-hydroxyquinoline) aluminum (Alq$_3$), (b) bis(2-methyl-8-quinolinolate)-4-(phenyl phenolato) aluminum (BAlq), (c) 2 methyl 9,10-di (2 naphthyl) anthracene (tbADN), (d) 4,4′bis(*N*-carbazolyl)biphenyl (CBP), (e) 2,2′,2′-(1,3,5-benzinetriyl)-tris(1-phenyl-1-*H*-benzimidazole) (TPBI), (f) 2,9-dimethyl-4,7diphenyl-1,10 phenanthroline (BCP), (g) 4,4′-bis-[*N*-(1-napthyl)-*N*-phenyl-amino] biphenyl (NPB), (h) 2,3,5,6-tetrafluoro-7,7,8,8-tetracyano-quinodimethane (F4-TCNQ), (i) 4,4′,4′-tris[(3-methylphenyl)phenylamino]triphenylamine (*m*-TDATA).

and phosphorescent dopants have been used in commercial products. There has been a significant amount of work to further refine the host–dopant systems used in OLED devices, sometimes involving multiple dopants per layer, which allows even finer control over the processes of electron–hole pair formation and recombination (Endo *et al* 2009, 2011). State-of-the-art OLED devices often utilize several dopants to balance the processes of device electroluminescence and operational lifetime.

The structures of Alq$_3$ and other host and charge transport molecules are depicted in figure 11.4 while the molecular structures of the laser dye coumarin 545 tetramethyl and other OLED dopants are illustrated in figure 11.5.

11.2.2 High luminescence OLEDs

Early reports of organic electroluminescence were based on structures that consisted of a single layer of organic material sandwiched between an inorganic anode and cathode (Dresner 1965, Helfrich and Schneider 1965, Pott and Williams 1969, Williams and Schadt 1970). These devices exhibited electroluminescence at high voltages and were operationally unstable. A significant breakthrough in OLED architecture was reported in the late 1980s with the demonstration of a bilayered device, where two layers of organic material, one for hole transport and another for electron transport and luminescence, were used (Tang and VanSlyke 1987). An energy diagram of a simple OLED device is depicted in figure 11.3(a) and a cross section of the same device is shown in figure 11.3(b). This approach improved the luminescence efficiency and reduced the operational voltage of OLEDs significantly, and provided a path for improved operational stability. Indeed, modern OLEDs can

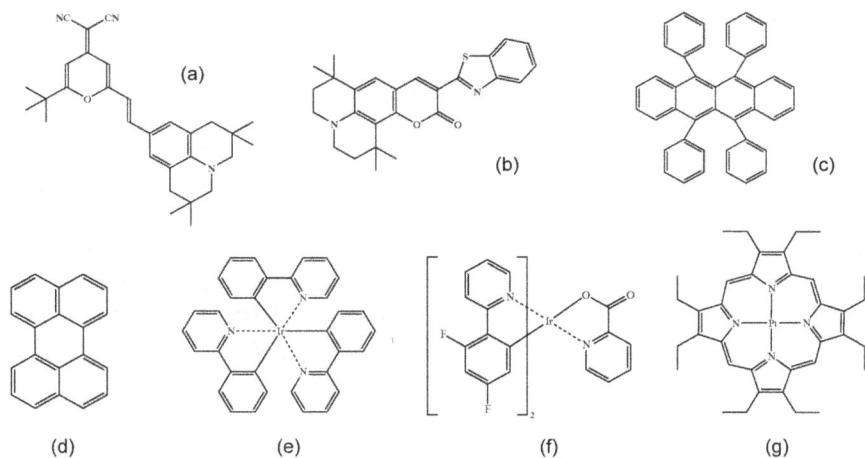

Figure 11.5. Molecular structure of various dopants: (a) 4-(dicyanomethylene)-2-*tert*-butyl-6-(1,1,7,7-tetra-methyljulolidin-4-ylvinyl)-4*H*-pyran (DCJTB), (b) 10-(2-benzothiazolyl)-2,3,6,7-tetrahydro-1,1,7,7,-tetra-methyl 1-1*H*,5*H*,11*H*-[1]benzopyrano [6,7,8-ij]quinolizin-11-one (C 545 T), (c) 5,6,11,12-tetraphenyl tetracene (Rubrene), (d) Perylene, (e) bis[2-(4,6-difluorophenyl)pyridinato-C2,N] (picolinato)iridium(III), (f) blue Ir(ppy) (FIrpic), (g) platinum octaethylporphyrin (PtOEP).

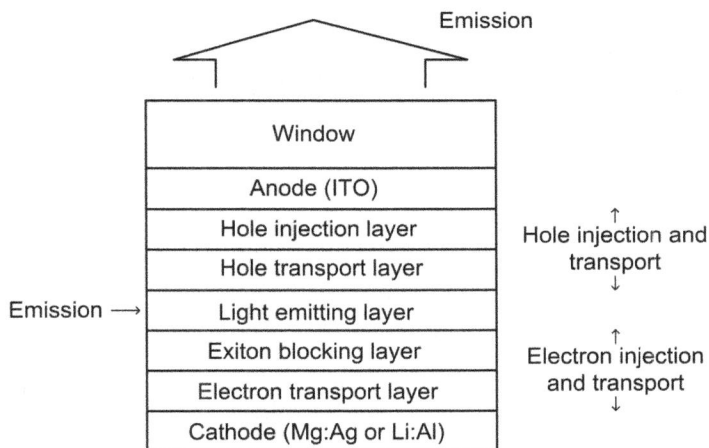

Figure 11.6. Generic OLED device layer diagram.

have as many as five or more organic layers to enable fine control over the processes of charge injection, charge transport, electron–hole pair (exciton) diffusion, and photon generation. As shown in figure 11.6, the layers in a single device stack typically group into three zones: electron injection and transport, hole injection and transport, and recombination and emission, the latter of which contains the host/dopant system (Adachi *et al* 1988). Use of this approach, combined with improvements made in transport, host, and dopant materials, has led to OLED devices exhibiting electroluminescent efficiencies suitable for flat panel display applications.

Another key attribute of OLED devices is the operational lifetime, and optimization of the device structure and material design is critical for achieving performance suitable for commercial use. The operational lifetime of an OLED is typically directly proportional to the amount of current that moves through it, due to Coulombic degradation, localized heating, and material decomposition (Aziz *et al* 1999, Liu *et al* 2003, Lin *et al* 2001, Meerheim *et al* 2006, Jiang *et al* 2007, Cester *et al* 2010). One way to address this failure mode is to maximize the device electroluminescence efficiency, which lowers the amount of current required for a particular target luminance. Dopant molecule design and host–dopant optimization have provided fruitful ground to achieve higher efficiencies, but device architecture can also play a role. Although more material can be added to the emissive layer, providing more recombination sites, in practice it is difficult to spatially control distribution of the excited states throughout the emissive layer leaving some dopant sites unsampled

A more successful approach for improving device efficiency came when OLED designers again turned to previous work in thin film inorganic device architecture (Lamorte and Abbott, 1979, Kim *et al* 1999, Guo *et al* 2001) demonstrating vertical integration of multiple OLED device electroluminescent units, each consisting of a hole transport emission, and electron transport zone, coupled in series by intermediate connectors that serve as charge generation layers (Kido *et al* 2003, Liao *et al* 2004) (see figure 11.7). When electrons and holes are injected into the device from the cathode and anode, respectively, electrons and holes are also generated at the interface of the n-type and p-type layers of the intermediate connector. These internally generated charges then migrate to populate the inner electroluminescent

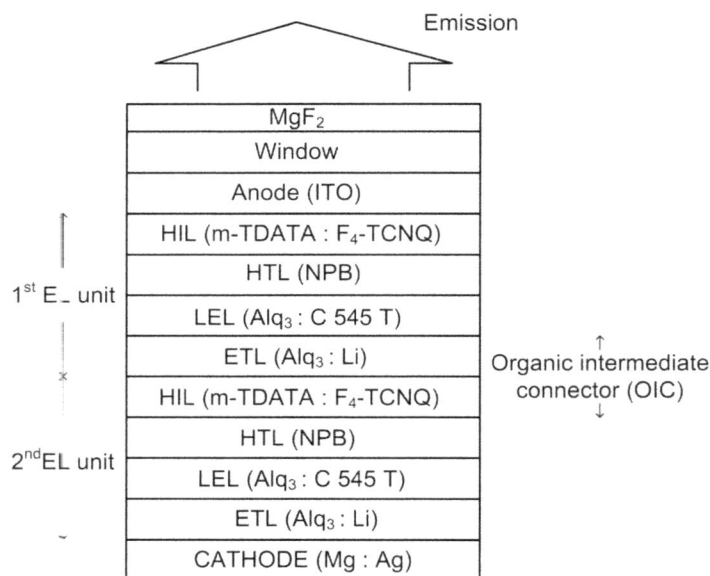

Figure 11.7. Tandem OLED device.

units, leading to generation of N electron–hole pairs in a device with N electroluminescent units for every externally injected electron–hole pair. This *tandem* device architecture achieves higher luminance for the same current density, or conversely, the same luminance at a lower current density, due to the multiple electroluminescent units incorporated in the device architecture. The electroluminescent efficiency of a tandem device with N electroluminescent units is approximately N times that of the single device, allowing longer device lifetimes to be achieved. The corresponding operating voltage with such a device is also higher by about a factor of N. In some cases, the device electroluminescence efficiency can be higher than N times, due to better charge balance throughout the layers of the device, as well as a reduction in the quenching of the electron–hole pairs via interaction with the metal electrodes, as the inner emission zones of the device are more removed from these regions due to the greater overall device thickness (Fung *et al* 2016). The tandem architectures are also critical for realizing electrically driven OLED lasers, as high luminescent efficiency is required for achieving self-sustained operation at relatively high current densities that can be withstood by the device (Duarte *et al* 2005, Duarte 2008).

Many different materials have been investigated as the intermediate connector layers in tandem OLED architectures, ranging from bilayers of thin evaporated metals, metal oxides, doped organics, and combinations thereof (Kido *et al* 2003, Liao *et al* 2004, Tsutsui and Terai 2004, Fung *et al* 2016). Some of the most efficient intermediate connectors consist of doped and undoped n- and p-type organic materials, as the materials provide good optical clarity for luminance output, can be deposited at lower temperatures without damaging the previously deposited layers in the device architecture, and do not quench the electron–hole pairs to as great a degree as do thin metal electrodes (see table 11.1).

11.3 Electrical excitation of the tandem organic semiconductor active region

The organic semiconductor gain region used in these experiments is a tandem OLED device with two device units, and therefore *two emitting regions in series* (Liao *et al* 2004, Duarte *et al* 2010). The emission medium in these active regions is a coumarin 545 tetramethyl (C 545 T) dye-doped tris (8-hydroxyquinolinato) aluminum, Al $(C_9H_6NO)_3$, or $C_{27}H_{18}AlN_3O_3$, matrix (Alq$_3$). The organic coumarin 545 tetramethyl (C 545 T) laser dye molecule used to dope the Alq$_3$ matrix is illustrated in figure 11.5. The architecture of the electrically excited organic semiconductor sub microcavity is depicted in figure 11.8, which is an alternative depiction of the tandem device illustrated, from a materials science perspective, in figure 11.7.

At this introductory stage it is important to mention explicitly that C 545 T has been shown to yield highly efficient laser emission tunable in the $501 \leqslant \lambda \leqslant 574$ nm range (Duarte *et al* 2006).

The multilayered laser-dye-doped Alq$_3$ gain medium has a length of $L \approx 300$ nm and is within a mirror–mirror asymmetrical cavity comprised of a high reflectivity back mirror M_1 ($R_1 \approx 0.9$) and a low reflectivity output coupler mirror M_2 ($R_2 \approx 0.1$).

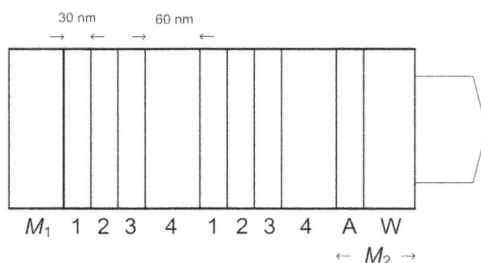

Figure 11.8. Basic structure of the electrically excited organic semiconductor sub-microcavity. The compositions of regions 1–4 are explained in the text. The light emitting regions are the two regions numbered as 2 and are composed of the Alq$_3$:C 545 T compound. M_1 is also the cathode and provides a reflectivity of $R_1 \approx 0.9$ while the output-coupler mirror M_2 is configured by the ITO coating on the output window which offers a reflectivity of at least $R_2 \approx 0.1$. It should be noticed that the active region is the same as the one depicted in figure 11.7.

In this configuration, M_1 has the double role of also performing as the excitation cathode and M_2 also doubles as the excitation anode. The reflectivity of the output coupler mirror M_2 is provided by a layer of ITO.

The OLED device is configured electrically and optically in series. There are two electro-luminescent (EL) units sandwiched between the semi-transparent indium tin oxide anode and the metal cathode mirror. Each EL unit consists of:

1. A 30 nm electron-transporting layer (ETL)
2. A 30 nm light-emitting layer (LEL)
3. A 30 nm hole-transporting layer (HTL), and
4. A 60 nm hole-injecting layer (HIL)

In reference to figure 11.8, the cathode (C) is comprised of Mg:Ag and the anode (A) is indium tin oxide (ITO). The regions numbered 1–4 correspond directly to the listing given above. The HIL is m-TDATA layer doped with 3 vol% F$_4$-TCNQ; the HTL is an NPB layer; the LEL is an Alq$_3$ layer doped with 1 vol% coumarin 545 T; and the ETL is an Alq$_3$ layer doped with 1.2% lithium. In this description, m-TDATA refers to 4,4′,4′-tris[(3-methylphenyl) phenylamino] triphenylamine; F$_4$-TCNQ refers to 2,3,5,6-tetrafluoro-7,7,8,8-tetracyano-quinodimethane; NPB refers to 4,4-bis[N-(1-naphthyl)-N-phenylamino]biphenyl. The interface between the ETL and of the first OLED unit and the HIL of the second OLED unit in the tandem device also serves as the intermediate connector, where the internal charge generation takes place during device operation. This tandem OLED device is fabricated on a glass window whose external surface is coated with an antireflection layer (MgF$_2$) to suppress possible intra-glass interference.

The excitation of this organic semiconductor sub-microcavity was performed using a pulsed variable voltage generator that provides pulses up to 100 V high with a ~5 ns rise time (Agilent 8114 A). The excitation pulse width is in the 500–1000 μs range and the repetition rate utilized was 67 Hz (Duarte *et al* 2005). Excitation was also performed with the transmission line of a nitrogen gas laser which yielded nanosecond pulses at a voltage approaching 10 kV at the organic semiconductor load (Duarte 2008).

11.4 Integrated interferometric coherent emitter (IICE)

Once the organic semiconductor sub microcavity is integrated to external interfero-metric components it becomes an *integrated interferometric coherent emitter* (IICE) or *integrated interferometric source* (IICS). There are two versions of this IICE. The first version, IICE-1, is depicted in figure 11.9 and consists of confinement of the emission aperture by an aperture of width $2w \approx 150$ μm, followed by a distance z_1, and a second aperture of width $2w \approx 150$ μm, which is in turn followed by a distance z_2, toward the detection plane d.

If only slits are involved, why designate this as an integrated interferometric emitter? The answer lies in the interferometric essence of diffraction. In this regard, the diffraction induced by a single slit can be accurately represented and predicted by the physics of interference as it applies to a single slit mathematically represented by a multitude of minute N-subslits. In other words, the single slit of width $\delta = 2w$ can be precisely quantified via the generalized interferometric equation given by Duarte (1991, 1993)

$$| \langle d|s \rangle |^2 = \sum_{j=1}^{N} \Psi(r_j)^2 + 2\sum_{j=1}^{N} \Psi(r_j) \left(\sum_{m=j+1}^{N} \Psi(r_m)\cos(\Omega_m - \Omega_j) \right) \qquad (11.1)$$

where N is the number of subslits comprising the single wider slit of width $2w$, and Ψ (r_j) are individual wave functions, of 'ordinary wave optics' (Dirac 1978), associated with propagation through each of the j slits in the N-slit array. For the sake of completeness, it should be noted that

$$\langle d|s \rangle = \langle d|j \rangle \langle j|s \rangle \qquad (11.2)$$

and

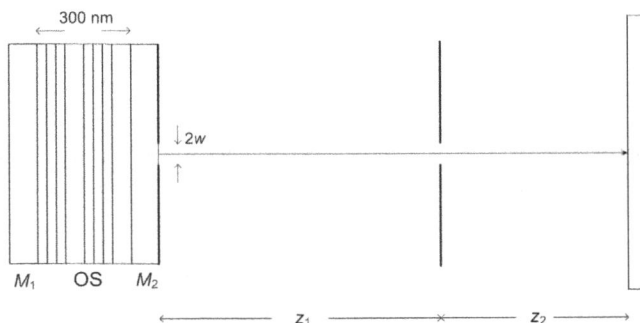

Figure 11.9. Integrated interferometric emitter configured with two sequential interferometric apertures for straight directional emission (IICE-1).

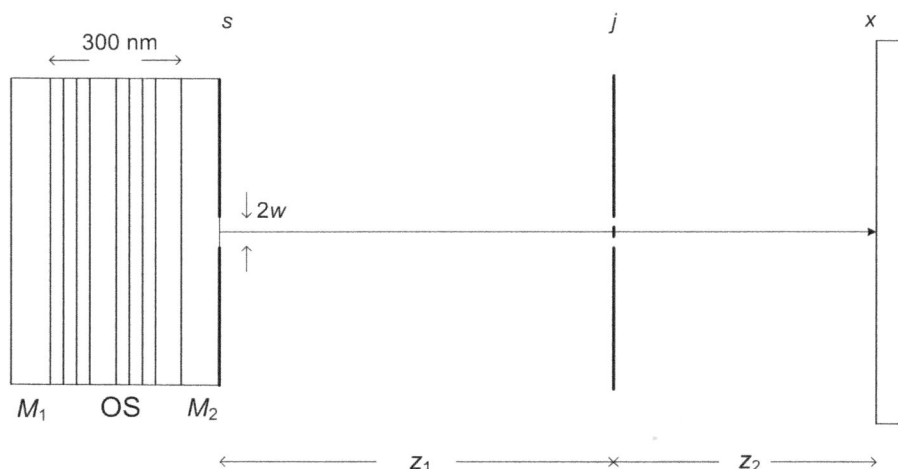

Figure 11.10. Integrated interferometric emitter configured with one interferometric aperture followed by a double-slit interferometric array. This IICE-2 configuration is used for determining the spectral coherence of the emission.

$$| \langle d|s \rangle |^2 = \langle x|s \rangle \langle x|s \rangle^* \qquad (11.3)$$

The reader can find the derivation of equation (11.1) explained in detail in Duarte (2014).

The second version of the IICE, that is IICE-2, is depicted in figure 11.10 and consists of confinement of the emission aperture by an aperture of width $2w \approx$ 150 μm, followed by a distance z_1, a double-slit configuration, which is in turn followed by a distance z_2, toward the detection plane x. In this particular configuration each slit is 50 μm wide whilst separated by a 50 μm space.

In reference to figure 11.10, the generalized interferometric equation now computes the probability, for a single photon, or an ensemble of indistinguishable photons, to propagate from the source (s) to the interference plane (x) via an array of j slits ($j = 1, 2$ in this case) and is also given by equation (11.1) (Duarte 1991, 1993, 2014).

11.5 Energetics of the organic semiconductor IICE

A graphical representation of the output intensity as a function of input current density, while using the IICE-1 configuration, shows an initial linear increase with a gradient of $\eta_1 \approx 0.66$ nW (A cm^{-2})$^{-1}$ followed by a subsequent linear increase of $\eta_2 \approx$ 1.54 nW (A cm^{-2})$^{-1}$. The discontinuity occurs at $\rho_1 \approx 0.8$ A cm^{-2} (Duarte *et al* 2005, Duarte 2007). The ratio of the two gradients is $(\eta_1/\eta_2) \approx 2.33$ which should be compared with a previously reported ratio of $(\eta_1/\eta_2) \approx 2.50$ for a conventional optically-pumped semiconductor laser (Rose *et al* 2005) and with $(\eta_1/\eta_2) \approx 3.12$ for an electrically excited inorganic semiconductor laser using an asymmetrical cavity (Lau 1993).

Beyond the comparison of slopes in asymmetrical cavities the situation becomes far more interesting once it is considered that the cavity length is in the sub-micrometer domain at $L \approx 300\text{nm}$ and emission wavelength is $\lambda \approx 540$ nm. This brings this sub-microcavity to the $L \leqslant \lambda$ regime where laser thresholds decrease dramatically. Indeed, De Martini and Jakobovitz (1988) report on a near 'zero-threshold-laser' for a cavity length of $L \approx \lambda/2$. This observation was made while experimenting on a multiple-transverse-mode optically-pumped dye laser.

The reference of De Martini and Jakobovitz (1988), which was overlooked in the 2005–08 Duarte publications, provides fairly compelling background evidence in support of threshold behavior at very low excitation densities for cavities in the $L \leqslant \lambda$ range. That is, a low threshold current density regime in the $0.8 \leqslant \rho \leqslant 0.9$ A cm^{-2} range would be consistent with what would be expected in a sub micrometer cavity where the conditions $L \leqslant \lambda$ are observed (Duarte 2010).

In regard to excitation at the ~10 kV level, it is instructive to mention the gain medium in the IICE-1 configuration emitted only for a short series of nanosecond pulses before undergoing irreversible destruction. This destructive excitation took place at current densities of $\rho \approx 190$ A cm^{-2} in the nanosecond regime (Duarte 2008).

11.6 Spatial coherence of the emission from the organic semiconductor IICE

The data and results presented and discussed in this section refer to the pulsed electrically-pumped organic semiconductor IICE-1 configuration illustrated in figure 11.9. The emission from the IICE-1 configuration is highly directional with its beam profile illustrated in figure 11.11. This black and white silver-halide photograph was captured at a distance of $z_2 = 340$ mm. At this distance, the full width of the beam is ~1.7 mm. The digital profile of this emission beam is nearly Gaussian, as illustrated in figure 11.12 (Duarte *et al* 2005).

Next, an analysis of the measurements presented above is done using classical Gaussian beam propagation theory as exposed in various previous reviews (Duarte 1990, 2015) and the presentation follows the style used in Duarte (2016). In this regard, the multiple-pass beam divergence is given by

$$\Delta\theta_R \approx \frac{\lambda}{\pi w}\left(1 + \left(\frac{L_{\mathcal{R}}}{B_R}\right)^2 + \left(\frac{L_{\mathcal{R}}A_R}{B_R}\right)^2\right)^{1/2} \tag{11.4}$$

where $L_{\mathcal{R}} = (\pi w^2/\lambda)$ is known as the Rayleigh length, while A_R and B_R are the corresponding multiple-pass elements of the propagation matrix (Duarte 2015). For a mirror–mirror cavity, for a single intracavity return-pass ($R = 1$), and in the absence of intracavity beam expansion, that is $A \approx 1$, equation (11.4) reduces to

$$\Delta\theta \approx \frac{\lambda}{\pi w}\left(1 + 2\left(\frac{L_{\mathcal{R}}}{B}\right)^2\right)^{1/2} \tag{11.5}$$

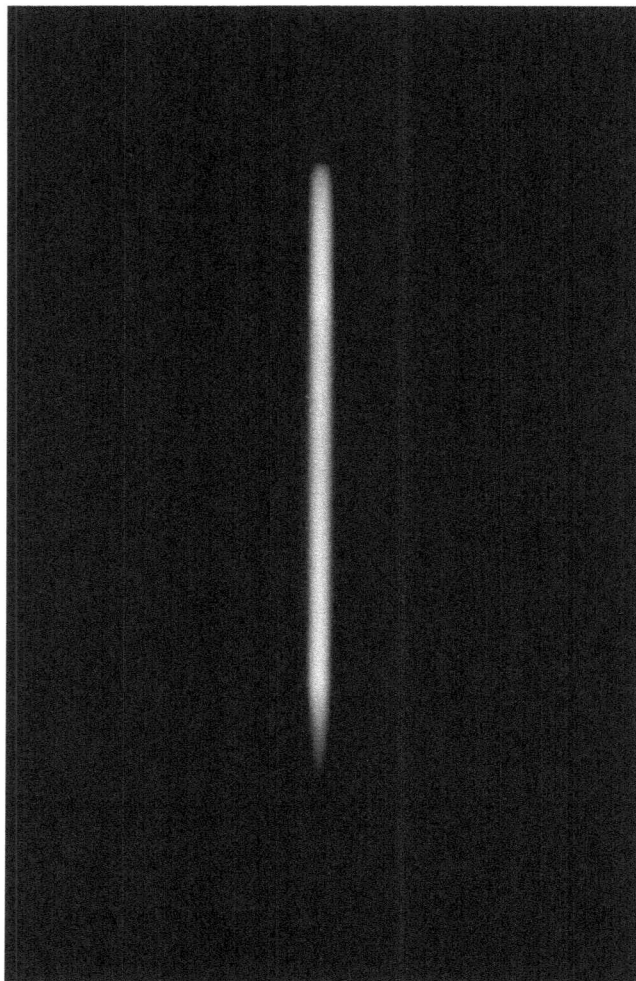

Figure 11.11. Photographic image of the cross section of the highly directional on-axis emission from the IICE-1 electrically excited organic semiconductor device at $z_2 = 340$ mm (from Duarte *et al* (2005); courtesy of the Optical Society).

For the sub-microcavity arrangement of figure 11.8, $2w \approx 150$ μm, $L_\mathscr{R} \approx 32.43$ mm, and $B \approx 2L$. Since the diminutive cavity length is only $L \approx 300$ nm, the single-pass divergence is, as expected, very large. However, a substantially different result is found for the arrangement of figure 11.9, where $z_1 \gg L$, so that $B \approx 2z_1$. Here, for $2w \approx 150$ μm, $L_\mathscr{R} \approx 32.43$ mm, and $z_1 \approx 130$ mm, equation (11.5) yields

$$\Delta\theta \approx 1.015 \frac{\lambda}{\pi w} \tag{11.6}$$

which is close to the diffraction limit. The measured divergence for the beam shown in figure 11.11 is $\Delta\theta_R = 2.53 \pm 0.13$ mrad, which for $2w \approx 150$ μm means that

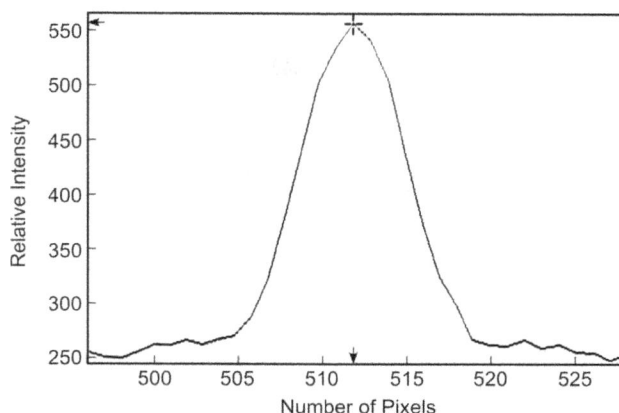

Figure 11.12. Digital profile of the cross section of the highly directional on-axis emission from the IICE-1 electrically excited organic semiconductor device $z_2 = 40$ mm. Each pixel is 25 μm (from Duarte *et al* (2005); courtesy of the Optical Society).

$$\Delta\theta_R \approx 1.09 \times \left(\frac{\lambda}{\pi w}\right) \qquad (11.7)$$

In other words, the measured beam divergence $\Delta\theta_R$ is ~1.09 × the diffraction limit.

This indicates that the double aperture configuration introduces an overwhelming discrimination in favor of the observed on-axis nearly diffraction-limited spatially-coherent beam.

In order to determine the number of return intracavity-passes R, it is necessary to measure the time delay between the leading edge of the excitation pulse and the leading edge of the coherent emission (Duarte 2001). This data is not available for these experiments. However, in the 5 ns interval corresponding to the rise-time of the excitation pulse, the number of intracavity return-passes for a cavity length of $L \approx$ 300 nm is enormous, $R \approx 2.5 \times 10^6$.

For completeness: the diffraction-limit of the beam divergence equation

$$\Delta\theta \approx \left(\frac{\lambda}{\pi w}\right) \qquad (11.8)$$

originates from the interferometric identity

$$\Delta\lambda \approx \left(\frac{\lambda^2}{\Delta x}\right) \qquad (11.9)$$

which itself can be derived either from the generalized interferometric equation (11.1), or Heisenberg's uncertainty principle

$$\Delta p \, \Delta x \approx h \qquad (11.10)$$

What is important here is that the observed spatial emission has a near Gaussian intensity profile and obeys the rules of Gaussian propagation. Furthermore, using the IICE-1 source it can be established that the observed emission corresponds to *single-transverse-mode* emission. Thus it can be concluded that the highly-directional *on-axis emission* from the pulsed electrically-pumped organic semiconductor interferometric emitter has a high degree of spatial coherence.

11.6.1 Interferometrically determined transverse-mode structures

Using the generalized interferometric equation (11.1) the transverse mode structure at the output aperture of IICE-1 is illustrated in figure 11.13, for $2w = 150$ µm, $z_1 = 2$ mm, and $\lambda = 545$ nm. In this figure, as expected, there is a strong multiple-transverse-mode structure.

For $2w = 150$ µm, $z_2 = 40$ mm, and $\lambda = 545$ nm, the calculated transverse mode structure is depicted in figure 11.14. The calculated profile shows a principal smooth emission mode with two adjacent satellite modes which do not show up in the

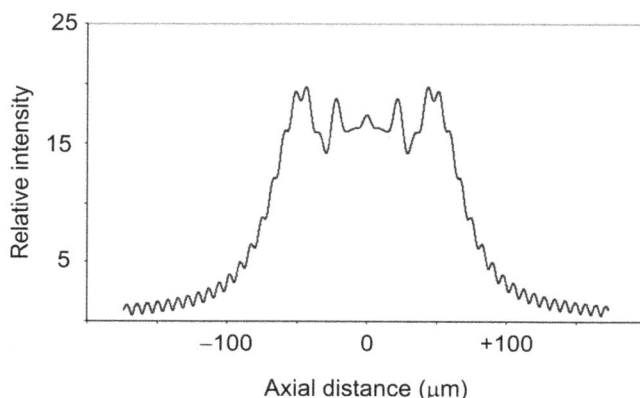

Figure 11.13. Calculated multiple-transverse-mode beam profile for the directional on-axis emission from the IICE-1 electrically excited organic semiconductor device, $z_1 = 40$ mm.

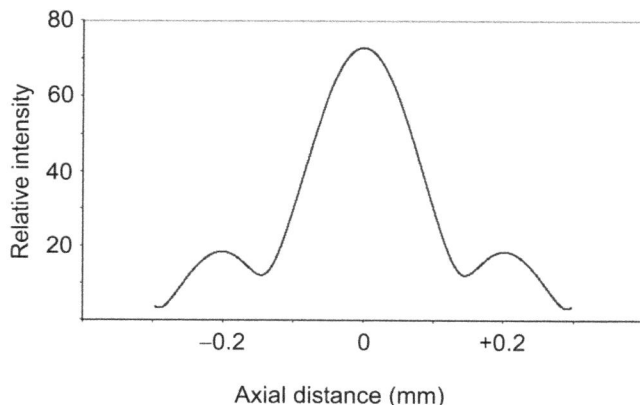

Figure 11.14. Calculated single-transverse-mode beam profile for the directional on-axis emission from the IICE-1 electrically excited organic semiconductor device $z_2 = 40$ mm.

measurement (figure 11.12) most likely due to transmission losses. This is typical of narrow-linewidth oscillator behavior (Duarte 2015). The calculated full-width of the near-Gaussian single-transverse-mode is 303 μm as compared to a measured near-Gaussian full-width of ~330 μm (see figure 11.12).

11.7 Spectral coherence of the emission from the organic semiconductor IICE

The data and results presented and discussed in this section refer to the pulsed electrically-pumped organic semiconductor IICE-2 configuration illustrated in figure 11.10. The double-slit ($N = 2$) interferogram from the emission of the IICE-2 configuration is illustrated in figure 11.15. This digital profile of the interferogram was captured at a distance of $z_2 = 50$ mm. The emission wavelength is $\lambda \approx 540$ nm and each slit has a width of $\delta = 50$ μm and they are separated by an interslit space of $\delta' = 50$ μm (Duarte *et al* 2005).

For direct comparison purposes a TEM$_{00}$ beam generated by the $3s_2$–$2p_{10}$ transition of a narrow-linewidth ($\Delta\lambda \approx 0.001$ nm) He–Ne laser at $\lambda = 543.30$ nm was slightly expanded to illuminate the IICE-2 configuration in the same manner as illustrated in figure 11.10. The double-slit ($N = 2$) interferogram thus generated is illustrated in figure 11.16. This digital profile of the interferogram was captured at a distance of $z_2 = 50$ mm. Again, the width of each slit is $\delta = 50$ μm separated by an interslit space of $\delta' = 50$ μm (Duarte *et al* 2005).

One further avenue of comparison using the identical arrangement for IICE-2 consists in displaying, side-by-side, the interferograms generated by the radiation of a high-power dye laser tuned at $\lambda \approx 540$ nm with the interferogram generated from the pulsed electrically-pumped organic semiconductor IICE-2. To this effect, the Glan–Thompson output-coupler polarizer Littrow-grating cavity powered by the C 545 T laser dye yielding a laser linewidth of $\Delta\lambda \approx 3$ nm at $\lambda \approx 540$ nm illuminates the

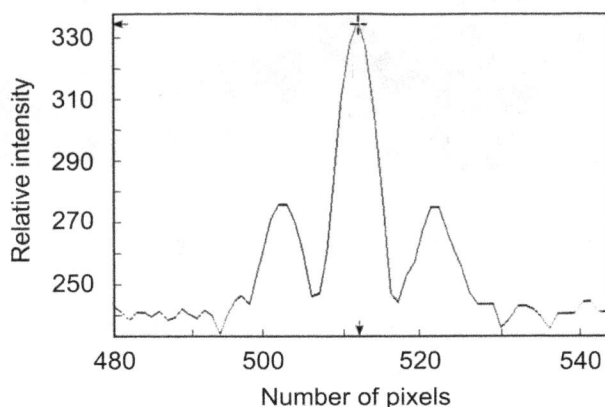

Figure 11.15. Digital image of the double-slit interferogram of the highly directional on-axis emission from the IICE-2 electrically excited organic semiconductor device, measured at $z_2 = 50$ mm. Each pixel is 25 μm (from Duarte *et al* (2005); courtesy of the Optical Society).

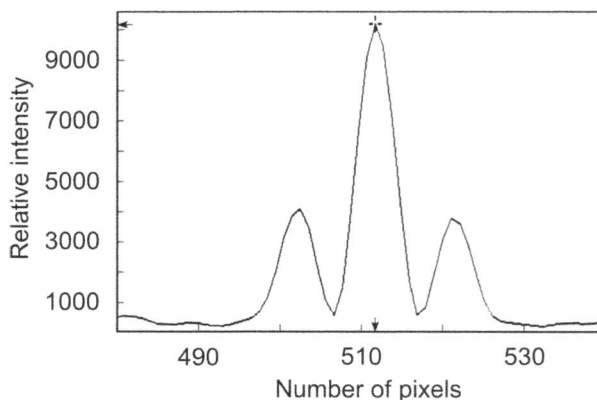

Figure 11.16. Digital image of the double-slit interferogram of the highly directional on-axis emission from the $3s_2–2p_{10}$ transition of the He–Ne laser using an identical IICE-2 configuration, measured at $z_2 = 50$ mm. Each pixel is 25 µm (from Duarte *et al* (2005); courtesy of the Optical Society).

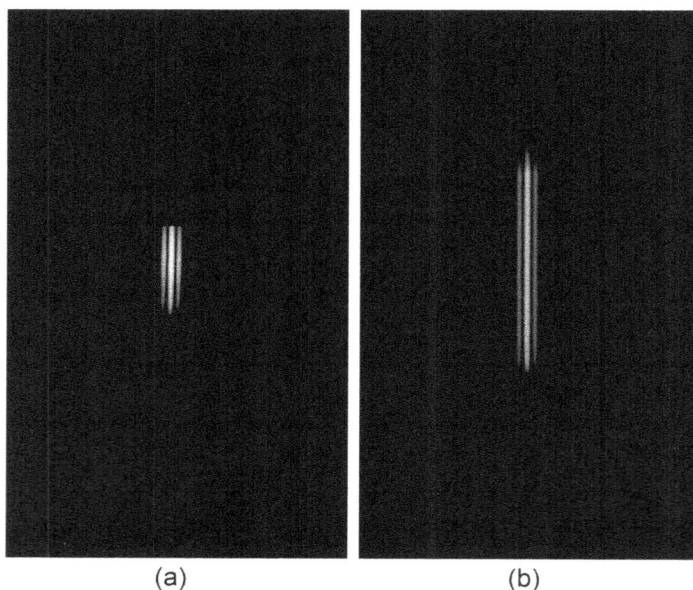

(a)   (b)

Figure 11.17. (a) Double-slit interferogram generated by a high-power C 545 T grating tuned dye laser using the IICE-2 configuration. (b) Double-slit interferogram generated by the electrically-pumped organic semiconductor in an identical IICE-2 configuration (from Duarte *et al* (2005); courtesy of the Optical Society).

IICE-2 interferometric configuration yielding the interferogram illustrated in figure 11.17(a) at $z_2 = 175$ mm. The interferogram generated from the pulsed electrically-pumped organic semiconductor IICE-2 at a wavelength of $\lambda \approx 540$ nm and a distance of $z_2 = 175$ mm is illustrated in figure 11.17(b). Again, the width of each slit is $\delta = 50$ µm separated by an interslit space of $\delta' = 50$ µm (Duarte *et al* 2005).

Table 11.3. Measured visibility via double-slit interference.[a]

Source	λ (nm)	$\mathcal{V}$	$\Delta\lambda$[b] (nm)	Reference
B Alq$_2$ OLED[c]	611	0.34		Xie *et al* (2016)
Alq$_3$ OLED[c]	520	~0.4	~100	Saxena *et al* (2006)
Hg lamp	~579	0.593		Thompson and Wolf (1957)
Dye ASE[d]	617	~0.65	~17	Dharmadhikari *et al* (2005)
sLED[e]	662	0.67		Deng and Chu (2017)
DPSS laser[f]	671	0.88		Deng and Chu (2017)
IICE-1[g]	540	~0.90	$0.1 \leqslant \Delta\lambda \leqslant 1.5$[h]	
He–Ne laser	543.30	~0.95	~0.001	Duarte (2007)

[a] Adapted from Duarte (2007).
[b] All linewidths are at half width or FWHM.
[c] Organic light emitting diode.
[d] Amplified spontaneous emission.
[e] Superluminescent light emitting diode.
[f] Diode pumped solid state.
[g] Integrated interferometric coherent emitter.
[h] Estimated linewidth range.

Using the visibility definition of Michelson (1927)

$$\mathcal{V} = \frac{I_1 - I_2}{I_1 + I_2} \tag{11.11}$$

it can be estimated that for the integrated interferometric emitter $\mathcal{V} = 0.901 \pm 0.088$ whilst the visibility for the He–Ne laser emitting at $\lambda \approx 543.30$ nm is $\mathcal{V} = 0.952 \pm 0.031$ (Duarte 2007).

To provide a broader comparative perspective the visibility results obtained for the IICE-2 are compared with the visibility results from other well-known semi-coherent and coherent sources in table 11.3.

The data presented in table 11.3 indicates that visibilities in the $0.34 \leqslant \mathcal{V} \leqslant 0.67$ range originate from radiation emanating from semi-coherent sources. On the other hand the last two entries in table 11.3, $\mathcal{V} \approx 0.9$, and $\mathcal{V} \approx 0.95$ belong exclusively in the visibility range $0.85 \leqslant \mathcal{V} \leqslant 1.00$, which is unequivocally associated in the literature with laser emission (Shimkaveg *et al* 1992, Trebes *et al* 1992, Ditmire *et al* 1996, Lucianetti *et al* 2004).

11.7.1 Emission linewidth from *N*-slit interferograms

It is not widely recognized in the optics profession that the interferometric equation

$$| \langle d|s \rangle |^2 = \sum_{j=1}^{N} \Psi(r_j)^2 + 2\sum_{j=1}^{N} \Psi(r_j)\left(\sum_{m=j+1}^{N} \Psi(r_m)\cos(\Omega_m - \Omega_j)\right)$$

can be applied, in conjunction with measured interferograms, to determine the wavelength λ and to estimate an upper limit for the linewidth $\Delta\lambda$ of the emission under observation (Duarte 2003, 2008, 2016). This is explained in detail in Duarte (2016).

Very succinctly, the interferometric equation applies to either single-quanta propagation or to the propagation of ensembles of *indistinguishable* photons. This kind of single-wavelength radiation produces very sharp, highly defined, interferograms. The phase term (in parenthesis) in the interferometric equation is a function of the exact geometry of the interferometer and the wavelength of the emission. Thus, if the emission of different laser wavelengths is observed, under identical geometrical conditions, then the interferograms will be uniquely a function of emission wavelength (Duarte 2003). It follows that since this interferometric method can yield information about λ it can also be applied to yield information about the linewidth $\Delta\lambda$ of the emission (Duarte 2016).

Based on the premise above, a spatial-graphical technique, for estimating $\Delta\lambda$, has been introduced that depends on the spatial contrast between the sharpest interferogram originating from very narrow-linewidth radiation and broader interferograms generated by wider linewidth radiation (Duarte 2007, 2008). An interferogram generated by pure narrow-linewidth radiation can be characterized directly by the original interferometric equation (11.1). On the other hand, for broadband emission a multitude of interferograms are integrated simultaneously at the detector and this cumulative interferometric signal can be characterized by (Duarte 2014)

$$\sum_{\lambda=\lambda_1}^{\lambda_n} |\langle d|s\rangle|_\lambda^2 = \sum_{\lambda=\lambda_1}^{\lambda_n} \left(\sum_{j=1}^{N} \Psi(r_j)_\lambda^2 + 2\sum_{j=1}^{N} \Psi(r_j)_\lambda \left(\sum_{m=j+1}^{N} \Psi(r_m)_\lambda \cos(\Omega_m - \Omega_j) \right) \right) \quad (11.12)$$

which generates wider interferograms with diminished sharpness. For the sake of completeness, at this stage, the reader should be reminded that the quantum intensity, that is, the measurable quantity, is proportional to the probability (Sargent *et al* 1974, Duarte 2014).

In this approximation a comparative *broadening factor* Δb is sought to yield information about the upper limit of $\Delta\lambda$. In this regard, either the full width, or half-width, of the interferogram under examination is defined. Next, the width of the narrow reference interferogram (W_r) and the corresponding width of the broader interferogram (W_m) are measured. The broadening factor Δb is defined as (Duarte 2008)

$$\Delta b = \frac{W_m - W_r}{W_r} \quad (11.13)$$

Using this definition, we obtain the broadening factor of the interferogram from the integrated interferometric emitter relative to the interferogram generated with the $3s_2 - 2p_{10}$ transition, of the He–Ne laser.

The adjoining part of this approach requires the generation of a calculated interferogram at the reference wavelength followed by a series of calculated interferograms at closely spaced wavelength increments above and below the reference wavelength. This approximation then uses graphical methods to relate the broadening factor Δb to $\Delta\lambda$. For the case at hand the broadening factor can be determined to be $\Delta b \approx 0.027$ that is related to a half-width linewidth of $\Delta\lambda \approx 1.5$ nm, which is an upper limit. Previously, the estimated linewidth was significantly overestimated as $\Delta\lambda \approx 10.5$ nm (Duarte 2008) mainly due the use of data at the base of the interferogram where the effect of electronic noise is at its maximum.

Although transparent from a theoretical perspective, in practice, the accuracy of this method depends on (a) the precise graphical generation of a series of theoretical interferograms and (b) the faithful recording of interferometric traces for comparison purposes. The latter requires the interferograms being recorded under identical geometrical conditions, which is difficult to achieve when using silver-halide film photographs. Accuracy also depends of the assumption of the calculated interferometric profiles being identical to the measured profiles which can provide an experimental challenge.

Given the caveats mentioned above, an alternative approach is to plot measured visibility values $\mathcal{V}$ as a function of measured linewidths $\Delta\lambda$, on a logarithmic scale (see table 11.3). For a measured visibility of $\mathcal{V} \approx 0.9$ this exercise yields a linewidth of the order of $\Delta\lambda \leqslant 0.01$ nm. Thus, the array of laser linewidths corresponding to $\mathcal{V} \approx 0.9$ is conservatively assigned to the $0.1 \leqslant \Delta\lambda \leqslant 1.5$ nm range.

11.8 On the origin of the coherent emission

The emission characteristics from the pulsed electrically-excited organic semiconductor integrated interferometric emitter can be succinctly summarized as

$$\Delta\theta_R \approx (1.09) \times \left(\frac{\lambda}{\pi w}\right)$$
$$\mathcal{V} \approx 0.9$$

First of all, it has been previously well established that organic dye lasers configured in microcavities have a 'zero-threshold' behavior (De Martini and Jakobovitz 1988). This means that the absence of a pronounced traditional threshold behavior does not preclude laser action.

The emission is highly directional along the propagation axis and is contained in a near Gaussian beam shape entirely consistent with single-transverse-mode emission. The beam divergence associated with this emission is $\times 1.09$ its diffraction limit. This means that in terms of directionality, beam shape, and beam divergence, this emission is as good as or even better than what would be expected from a highly optimized macroscopic laser oscillator (see, for example, Duarte 1999).

In terms of interferometric visibility the measured visibility of $\mathcal{V} \approx 0.9$ is right in the middle of the $0.85 \leqslant \mathcal{V} \leqslant 1.00$ range, which is explicitly associated in the literature with laser emission (Shimkaveg *et al* 1992, Trebes *et al* 1992, Ditmire *et al* 1996, Lucianetti *et al* 2004).

As far as the emission linewidth is concerned, the interferometrically derived range of $0.1 \leqslant \Delta\lambda \leqslant 1.5$ nm is a rather conservative and indirectly determined value which might appear tenuously wide at first glance. However, once the length of the sub-microcavity is considered, $L \approx 300$ nm, then it is immediately revealed that the corresponding intracavity mode spacing is $(\lambda^2/2L) \approx 486$ nm. This enormous intracavity-mode spacing implies that the interferometrically inferred linewidth corresponds unequivocally to *single-longitudinal-mode* emission.

The observed emission passes the threshold criteria, the spatial coherence criteria, and the spectral coherence criteria. Indeed, coherently speaking the emission from the electrically excited organic semiconductor integrated interferometric emitter considered here *is indistinguishable from broadband dye laser emission* with two important exceptions of distinction: it is single-transverse-mode and it is single-longitudinal mode.

But, is there an alternative explanation? From a spatial coherence perspective Collett–Wolf sources are partially-coherent sources known to produce Gaussian-like beams that are also directional (Collett and Wolf 1978). However, there is *no record* in the published literature of Collett–Wolf sources yielding interferometric visibilities in the laser range, that is, in the $0.85 \leqslant \mathcal{V} \leqslant 1.00$ range.

The physics imposed by the IICE geometry, illustrated in figure 11.9, implies that only strictly *on-axis emission* is utilized in the beam propagation measurements and in the interferometric measurements. Any emission that is not aligned in this highly unique and precise direction does not participate. In other words, the sequential narrow-slit arrangement imposes a geometrically highly-selective and discriminatory mechanism that allows only on-axis emission. This on-axis emission is also entirely orthogonal to the surface of the back total reflector and the output-coupler and thus highly likely to be the result of an enormous number of multiple return-pass emissions given the extremely short cavity length, $L \approx 300$ nm. This highly-selective scenario is totally consistent with laser emission.

An alternative scenario would have to invoke a subtle quantum effect yet to be discovered.

In conclusion, there are two possible explanations for the observed spatial and spectral coherence of the pulsed electrically excited organic semiconductor integrated interferometric emitter: it is either laser emission or it is a new type of yet-to-be-discovered spatially coherent and spectrally coherent radiation.

11.9 Perspective on the literature

As outlined in the first edition of this book, in this very section, the cited literature on electrically pumped organic semiconductor lasers became ambiguous and contradictory. So this section is being updated simply by briefly referring to what appears to be a straight forward report on regular lasing from an electrically excited incoherent organic semiconductor (OLED) optically pumping a DFB polymer laser (Yoshida et al 2023). This integrated pulsed device is said to emit at $\lambda \approx 542$ nm and to exhibit a beam divergence of $\Delta\theta = 2.4 \pm 0.2$ mrad.

11.10 Quantum coherence

In the first edition of this book the section on the origin of the coherent emission concluded with the thought that the observed coherent emission is *either laser emission or it is a new type of yet-to-be-discovered spatially coherent and spectrally coherent radiation.*

The sentence in italics becomes more accurate if the concept of yet-to-be-discovered is replaced by *yet-to-be-identified*. In fact, that is what was done by Duarte and Taylor (2022) in a paper on quantum coherence in electrically-pumped organic interferometric emitters. In that paper it was revealed that assembly states of indistinguishable quanta

$$|x\rangle_I = |x\rangle_1 |x\rangle_2 |x\rangle_3 \ldots |x\rangle_g \ldots \qquad (11.14)$$

$$|y\rangle_{II} = |y\rangle_1 |y\rangle_2 |y\rangle_3 \ldots |y\rangle_g \ldots \qquad (11.15)$$

are perfectly allowed by the boson physics of Dirac's identities. These states describe ensembles of indistinguishable quanta in a given state. This coherent emission, referred to as Diracian emission, is described in detail in chapter 16.

Assembly states of indistinguishable quanta exist at a very fundamental level, they perfectly describe coherent emission, and they are entirely compatible with laser emission. The main difference of laser emission and coherent emission from assembly states, of indistinguishable quanta, is the distinctive higher intensities associated with laser emission.

11.11 Miniaturization prospects

Initial numerical simulations on the performance of a coherent organic IICE device with minimized dimensions have also been performed for a device reduced by a factor of 10 (Duarte 2008). Further dimension reductions, while still maintaining coherence, are attainable. A second generation device configuration for the electrically-pumped tandem OLED coherent source introduced by Duarte *et al* (2005) is proposed in figure 11.18. In

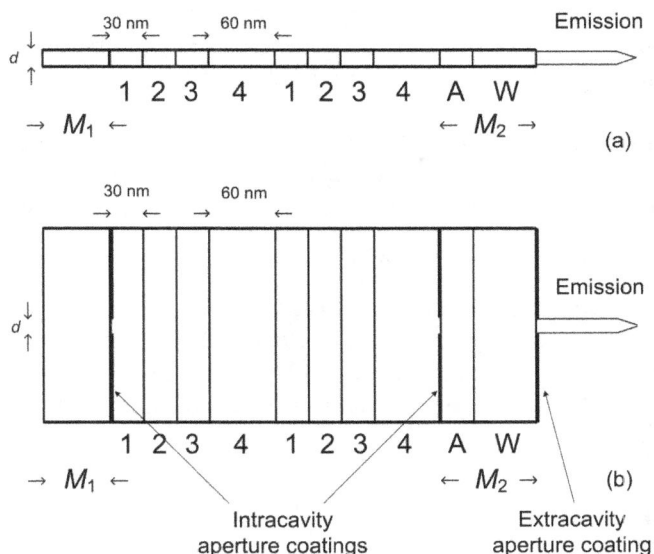

Figure 11.18. Proposed second generation interferometric configurations variants of the spatially and spectrally coherent OLED device introduced by Duarte *et al* (2005). (a) The whole tandem OLED device, with M_1 ($R_1 \approx 0.9$) and the output-coupler mirror M_2 ($R_2 \approx 0.1$), is reduced to a diameter in the range $10 \leqslant d \leqslant 150$ μm. (b) The configuration of (a) is replicated via the use of intracavity apertures plus an extracavity aperture in the $10 \leqslant d \leqslant 150$ μm range.

the first alternative the active geometry is reduced to a diameter in the $10 \leqslant d \leqslant 150$ μm region. In the second alternative the emission is confined to a diameter in the $10 \leqslant d \leqslant 150$ μm region through the introduction of intracavity apertures, formed via appropriate coatings, that confine the emission to a beam with a diameter in the $10 \leqslant d \leqslant 150$ μm range. The emission immediately following the output aperture will be multiple-transverse mode, similar to that depicted in figure 11.13. Discrimination in favor of the central TEM_{00} mode can be accomplished via the use of an extracavity aperture, as described in the schematics of IICE-1 coherent source.

11.12 Conclusion

Samuel and colleagues (Xie *et al* 2016) stated: 'Duarte *et al* reported an investigation of the spatial coherence of an electrically-driven OLED... the spatial distribution of the light beam after two spatial filters (two single slits) was modified dramatically, which contributed to a small beam divergence... and similar fringes to those from a He–Ne laser... these are interesting results but they do not directly measure the spatial coherence of the OLED.' Three mandatory observations are in order here:
  1. Duarte *et al* measured both *spatial and spectral coherence.*
  2. The interferometric slits do not create their own emission.
  3. The interferometric slits are there only to select and discriminate in favor of on-axis emission from the electrically-driven tandem OLED sub-microcavity.

In other words, the coherent emission is intrinsically present and the interfero-metric configuration's only function is to *reveal it.* The emission, in its entirety, is generated at the C 545 T-doped Alq_3 gain medium of the tandem device, and not at the slits.

The coherent emission reported by Duarte *et al* (2005) can be correctly described as either coherent emission with laser characteristics or as intrinsically coherent emission from assembly states of indistinguishable quanta, such as

$$|x\rangle_I = |x\rangle_1 |x\rangle_2 |x\rangle_3 ... |x\rangle_g ...$$

Given that the coherent emission observed from the electrically excited organic interferometric emitter exhibited a measured intensity at nW power levels, in the pulsed regime, it is our opinion that most likely this emission is coherent emission originating from intrinsic assembly quantum states of indistinguishable quanta, in other words, Diracian emission.

11.13 Problems

  - 11.1 Using the information in table 11.2, estimate the wavelength and emission color from the following OLED dopant materials: (a) Ir(ppy)3, (b) Perylene, (c) PtOEP.
  - 11.2 Using the information from Problem 11.1, sketch the emission spectra of an OLED made with Ir(ppy)3, Perylene, and PtOEP dopant emitters. Assume

all dopants with equal intensity from the device. What color will the emission appear to the observer?

- 11.3 For a tandem OLED structure such as the one shown in figure 11.7, describe the expected difference in brightness between devices with two and six EL units, when operated at the same current density.
- 11.4 Explain the main difference between fluorescent and phosphorescent OLED dopants.
- 11.5 In OLED devices, typical fluorescent dopant concentrations in the OLED host of the EL unit are in the range of 0.5%, while phosphorescent dopant concentrations are more typically 5%. Explain why this is the case.
- 11.6 Verify that equation (11.4) can be expressed as equation (11.5) for $A = 1$ and $R = 1$.
- 11.7 Verify that for $B \approx 2z_1$, $2w \approx 150\ \mu m$, $L_{\mathcal{R}} \approx 32.43$ mm, and $z_1 \approx 130$ mm, equation (11.5) reduces to equation (11.6).
- 11.8 Show that the interferometric identity given in equation (11.9) follows from Heisenberg's uncertainty principle $\Delta p \approx \Delta x \approx h$.
- 11.9 Show that Heisenberg's uncertainty principle $\Delta p \Delta x \approx h$ can be derived from the generalized interferometric equation given in equation (11.1). Hint: focus on the phase term.
- 11.10 Using equation (11.11), estimate the visibility $\mathcal{V}$ of the interferogram displayed in figure 11.16.

References

Adachi C, Baldo M A, Thompson M E and Forrest S R 2001 Nearly 100% internal phosphorescence efficiency in an organic light-emitting device *J. Appl. Phys.* **90** 5048–51

Adachi C, Tokito S, Tsutsui T and Saito S 1988 Electroluminescence in organic films with three-layer structure *Japan. J. Appl. Phys.* **27** 59–61

Adachi C, Tsutsui T and Saito S 1990 Blue light-emitting organic electroluminescent devices *Appl. Phys. Lett.* **56** 799–801

Aziz H, Popovic Z D, Hu N-X, Hor A-M and Xu G 1999 Degradation mechanism of small molecule-based organic light-emitting devices *Science* **283** 1900–02

Baldo M A 1999 Very high-efficiency green organic light-emitting devices based on electro-phosphorescence *Appl. Phys. Lett.* **75** 4–6

Baldo M A, O'Brien D F, You Y, Shoustikov A, Sibley S, Thompson M E and Forrest S R 1998 Highly efficient phosphorescent emission from organic electroluminescent devices *Nature* **395** 151–4

Baldo M A, Holmes R J and Forrest S R 2002 Prospects for electrically pumped organic lasers *Phys. Rev.* B **66** 035321

Cai Y-Y, Chen X, Li N, Li C-W and Wang Y-Q 2017 Electrically pumped photonic crystal laser constructed with organic semiconductors *Laser Phys.* **27** 035801

Cester A, Bari D, Framarin J, Wrachein N, Meneghesso G, Xia S, Adamovich V and Brown J J 2010 Thermal and electrical stress effects of electrical and optical characteristics of Alq₃/NPD OLED *Microelectron. Reliab.* **50** 1866–70

Collett E and Wolf E 1978 Is complete spatial coherence necessary for the generation of highly directional light beams? *Opt. Lett.* **2** 27–9

De Martini F and Jakobovitz J R 1988 Anomalous spontaneous-emission-decay phase transition and zero-threshold laser action in a microscopic cavity *Phys. Rev. Lett.* **60** 1711–4

Deng Y and Chu D 2017 Coherence properties of different light sources and their effect on the image sharpness and speckle of holographic displays *Sci. Rep.* **7** 5893

Dharmadhikari J A, Dharmadhikari A K and Kumar G R 2005 High contrast interference pattern of amplified spontaneous emission from dyes under transient grating excitation *Opt. Lett.* **30** 765–7

Dirac P A M 1978 *The Principles of Quantum Mechanics* 4th edn (London: Oxford University Press)

Ditmire T, Gumbrell E T, Smith R A, Tisch J W G, Meyerhofer D D and Hutchison M H R 1996 Spatial coherence measurements of soft x-ray radiation produced by high-order harmonic generation *Phys. Rev. Lett.* **77** 4756–9

Djurovich P I, Mayo E I, Forrest S R and Thompson M E 2009 Measurement of the lowest unoccupied molecular orbital energies of molecular organic semiconductors *Organic Electron.* **10** 515–20

Dong S C, Liu Y, Li Q, Cui L S, Chen H, Jian Z Q and Liao L S 2013 Spiro-annulated triarylamine-based hosts incorporating dibenzothiophene for highly efficient single-emitting layer white phosphorescent organic light-emitting diodes *J. Mater. Chem.* C **1** 6575–84

Dresner J 1965 Double injection electroluminescence in anthracene *RCA Rev.* **30** 322–34

Duarte F J 1990 Narrow-linewidth pulsed dye laser oscillators *Dye Laser Principles* ed F J Duarte and L W Hillman (New York: Academic) ch 4

Duarte F J 1991 Dispersive dye lasers *High Power Dye Lasers* ed F J Duarte (Berlin: Springer) ch 2

Duarte F J 1993 On a generalized interference equation and interferometric measurements *Opt. Commun.* **103** 8–14

Duarte F J 1999 Multiple-prism grating solid-state dye laser oscillator: optimized architecture *Appl. Opt.* **38** 6347–9

Duarte F J 2001 Multiple-return-pass beam divergence and the linewidth equation *Appl. Opt.* **40** 3038–41

Duarte F J 2003 *Tunable Laser Optics* (New York: Academic)

Duarte F J 2007 Coherent electrically excited organic semiconductors: visibility of interferograms and emission linewidth *Opt. Lett.* **32** 412–4

Duarte F J 2008 Coherent electrically-excited organic semiconductors: coherent or laser emission? *Appl. Phys.* B **90** 101–8

Duarte F J 2010 Electrically-pumped organic semiconductor coherent emission: a review *Coherence and Ultrashort Pulsed Laser Emission* ed F J Duarte (Rijeka: Intech) ch 1

Duarte F J 2014 *Quantum Optics for Engineers* (New York: CRC)

Duarte F J 2015 *Tunable Laser Optics* 2nd edn (New York: CRC)

Duarte F J 2016 Coherent electrically-excited organic semiconductors *Tunable Laser Applications* ed F J Duarte (New York: CRC) 3rd edn ch 12

Duarte F J, Liao L S and Vaeth K M 2005 Coherence characteristics of electrically excited tandem organic light-emitting diodes *Opt. Lett.* **30** 3072–4

Duarte F J, Liao L S, Vaeth K M and Miller A M 2006 Widely tunable green laser emission using the coumarin 545 tetramethyl dye as the gain medium *J. Opt. A: Pure Appl. Opt.* **8** 172–4

Duarte F J and Taylor T S 2022 Quantum coherence in electrically-pumped organic interfero-metric emitters *Appl. Phys.* B**128** *11*

Duarte F J, Vaeth K M and Liao L S 2010 Electrically excited organic light-emitting diodes with spatial and spectral coherence *US Patent* 7667391

Endo A, Ogasawara M, Takahashi A, Yokoyama D, Kato Y and Adashi C 2009 Thermally activated delayed fluorescence from Sn^{4+}–porphyrin complexes and their application to organic light emitting diodes—a novel mechanism for electroluminescence *Adv. Mater.* **21** 4802–6

Endo A, Sato K, Yoshimura K, Kai T, Kawada A, Miyazaki H and Adachi C 2011 Efficient up-conversion of triplet excitons into a singlet state and its application for organic light emitting diodes *Appl. Phys. Lett.* **98** 083302

Fung M, Li Y and Liao L 2016 Tandem organic light-emitting diodes *Adv. Mater.* **28** 10381–408

Grivas C and Pollnau M 2012 Organic solid-state integrated amplifiers and lasers *Lasers Photon. Rev.* **6** 419–62

Guo X, Shen G-D, Wang G-H, Zhu W-J, Du J-Y, Gao G and Zou D-S 2001 Tunnel-regenerated multiple-active-region light-emitting diodes with high efficiency *Appl. Phys. Lett.* **79** 2985–6

Goushi K and Adachia C 2012 Efficient organic light-emitting diodes through up-conversion from triplet to singlet excited states of exiplexes *Appl. Phys. Lett.* **101** 023306

Hamada Y, Sano T, Shibata K and Kuroki K 1995 Influence of the emission site on the running durability of organic electroluminescent devices *Japan. J. Appl. Phys.* **34** L824–6

Helfrich W and Schneider W G 1965 Recombination radiation in anthracene crystals *Phys. Rev. Lett.* **14** 229–31

Jiang X-Y, Zhang Z-L, Cao J, Khan M A, Haq K-U and Zhu W Q 2007 White OLED with high stability and low driving voltage based on a novel buffer layer MoOx *J. Phys. D: Appl. Phys.* **40** 5553–7

Kalinowski J 1999 Electroluminescence in organics *J. Phys. D: Appl. Phys.* **32** R179–250

Kan H Y and Lee C H 2004 Electroluminescence properties of organic light-emitting diodes with a red dye doped into Alq_3: rubrene mixed host *J. Korean Phys. Soc.* **45** 756–60

Karnutsch C 2007 *Low Threshold Organic Thin Film Laser Devices* (Göttingen: Cuvillier)

Kido J, Matsumoto T, Nakada T, Endo J, Mori K, Kawamura N and Yokoi A 2003 High efficiency organic EL devices having charge generation layers *SID Symp. Dig. Tech. Papers* **34** 964–5

Kim J K, Hall E, Sjölund O and Coldren L A 1999 Epitaxially-stacked multiple-active-region 1.55µm lasers for increased differential efficiency *Appl. Phys. Lett.* **74** 3251–3

Kim T G, Oh H S, Kim W Y and Ki Y H 2010 Study of deep blue organic light-emitting diodes using doped BCzVBi with various blue host materials *Trans. Elec. Electron. Mat.* **11** 85–8

Kranzelbinder G and Leising G 2000 Organic solid-state lasers *Rep. Prog. Phys.* **63** 729–62

Lamorte M F and Abbott D 1979 Analysis of AlGaAs-GaInAs cascade solar cell under AM 0-AM 5 spectra *Sol. State Electron.* **22** 467–73

Lau K Y 1993 Dyanamics of quantum well lasers *Quantum Well Lasers* ed P S Sory (New York: Academic) ch 4

Liao L S, Klubek K P and Tang C W 2004 High-efficiency tandem organic light-emitting diodes *Appl. Phys. Lett.* **84** 167–9

Lin K K, Wang W and Lim S F 2001 Influence of electrical stress voltage on cathode degradation of organic light-emitting devices *J. Appl. Phys.* **90** 976–9

Liu T H, Iou C Y and Chen C H 2003 Doped red organic electroluminescent devices based on a cohost emitter system *Appl. Phys. Lett.* **83** 5241–3

Liu Z, Helander M G, Wang Z and Lu Z 2002 Efficient single-layer organic light-emitting diodes based on $C545T\text{-}Alq_3$ system *J. Phys. Chem. C* **114** 11931 5

Lucianetti A, Janulewicz K A, Kroemer R, Priebe G, Tümmler J, Sandner W, Nickless P V and Redkorechev V I 2004 Transverse spatial coherence of a transient nickel like silver soft-x-ray laser pumped by a single picosecond laser pulse *Opt. Lett.* **29** 881–3

Mäkinen A J, Hill I G and Kafafi Z H 2002 Vacuum level alignment in organic guest-host systems *J. Appl. Phys* **92** 1598–603

Marowsky G, Schäfer F P, Keto J W and Tittel F K 1976 Flourescence studies of electron beam pumped POPOP dye vapor *Appl. Phys.* **9** 143–6

Meerheim R, Walzer K, Pfeiffer M and Leo K 2006 Ultrastable and efficient red organic light emitting diodes with doped transport layers *Appl. Phys. Lett.* **89** 061111

Michelson A A 1927 *Studies in Optics* (Chicago, IL: University of Chicago)

Pott G T and Williams D F 1969 Electron photoemission from anthracene crystals *J. Chem. Phys.* **51** 203–10

Rose A, Zhu Z, Madigan C F, Swager T M and Bulovic V 2005 Sensitivity gains in chemosensing by lasing action in organic polymers *Nature* **434** 876–9

Samuel I D W and Turnbull G A 2007 Organic semiconductor lasers *Chem. Rev.* **107** 1272–95

Samuel I D W, Mamdas E B and Turnbull G A 2009 How to recognize lasing *Nat. Photon.* **3** 546–9

Sargent M, Scully M O and Lamb W E 1974 *Laser Physics* (Reading, MA: Addison-Wesley)

Saxena K, Mehta D S, Srivastava R and Kamalasaman M N 2006 Spatial coherence properties of electroluminescence from Alq$_3$ based organic light emitting diodes *Appl. Phys. Lett.* **89** 061124

Shi J and Tang C W 1997 Doped organic electroluminescent devices with improved stability *Appl. Phys. Lett.* **70** 1665–7

Shimkaveg G M *et al* 1992 X-ray laser coherence experiments in neon-like yttrium *Proc. of the Int. Conf. on Lasers' 91* ed F J Duarte and D J Harris (Mc Lean, VA: STS) pp 84–92

Steyer B and Schäfer F P 1974 A vapor phase dye laser *Opt. Commun.* **10** 219–20

Tang C W and VanSlyke S A 1987 Organic electroluminescent diodes *Appl. Phys. Lett.* **51** 913–5

Tang C W, VanSlyke S A and Chen C H 1989 Electroluminescence of doped organic thin films *J. Appl. Phys* **65** 3610–6

Tao Y, Yang C and Qin J 2011 Organic host materials for phosphorescent organic light-emitting diodes *Chem. Soc. Rev.* **40** 2943–70

Thompson B J and Wolf E 1957 Two-beam interference with partially coherent light *J. Opt. Soc. Am.* **47** 895–902

Trebes J E *et al* 1992 Measurements of spatial coherence of a soft x-ray laser *Phys. Rev. Lett.* **68** 588–91

Tsutsui T and Terai M 2004 Electric field-assisted bipolar charge spouting in organic thin-film diodes *Appl. Phys. Lett.* **84** 440–2

Williams D F and Schadt M 1970 A simple organic electroluminescent diode *Proc. IEEE* **58** 476

Xie G, Chen M, Mazilu M, Zhang S, Bansal A K, Dhokalia K and Samuel I D W 2016 Measuring and structuring the spatial coherence length of organic light-emitting diodes *Laser Photon. Rev.* **10** 82–90

Yoon J A, Kim Y H, Kim N H, Yoo S I, Lee S Y, Zhu F R and Kim W Y 2014 Highly efficient blue organic light-emitting diodes using quantum well like multiple-emissive layer structure *Nanosci. Res. Lett.* **9** 191–7

Yoshida K, Gong J, Kanibolotsky A L, Skabara P J, Turnbull G A and Samuel I D W 2023 Electrically driven organic laser using integrated OLED pumping *Nature* **621** 746

IOP Publishing

Organic Lasers and Organic Photonics (Second Edition)

F J Duarte

Chapter 12

Organic photonics

F J Duarte

A brief handbook-like survey of organic photonics is given in terms of organic dyes, polymers, organic-inorganic composites, organic lasers, and organic quantum dots. In the area of applications emphasis is given to communications, medicine, industry, nonlinear optics, science, and sensors.

12.1 Introduction

As already articulated at the introduction of this book, *organic photonics* is a very large field that includes emission, transmission, detection, and processing of light via the use of organic materials. Chapters 5–11 focus on *organic sources of coherent radiation* and it is time to turn our attention to aspects of organic photonics in the materials and applications side of this field. In this regard, this chapter should be considered a companion chapter to chapters 2–4.

From the perspective of the subject matter already covered in this book one of the main polymer materials used in organic lasers, and organic photonics, is poly(methyl methacrylate) (PMMA) which was registered in the 1930s under the commercial name of Perspex. It is also known as Lucite, Acryle, and Pexiglass. Here, it is interesting to observe that the transparent polymer PMMA had been used in commercial and industrial illumination devices long before the word *photonics*, a post laser term, ever came into existence. Thus, part of the material in this chapter does relate to PMMA.

The other very important component, on the materials side, of organic photonics is the *organic dye* (see chapter 2) which was also used in light, and illumination, related applications long before the word photonics was coined.

The aim of this chapter is to provide a broad-range, fast-moving, perspective on organic photonics leaving the cited references to fill in the details. An effort has been made to cite references of chronological importance but, regrettably, that is not a consistent trend in this presentation.

12.2 Elements of organic photonics

In this section some basic elements of polymers and organic dyes for photonics applications are examined.

12.2.1 Organic dyes

The main article on this topic is in chapter 2 as related to *organic laser dyes*. Organic dyes applicable to medicine are considered in chapter 14. One class of organic laser dyes that have applications beyond the field of lasers are the coumarin tetramethyl dyes that given their methanol–water and methanol–water solubility have found applications in the lithography industry under UV laser excitation.

Table 12.1 lists some of the coumarin tetramethyl dyes. The molecular structures of coumarin 102 T and coumarin 153 T are depicted in figures 12.1 and 12.2, respectively.

Table 12.1. The organic laser dye coumarin tetramethyl family.

Organic dye	Molecular weight (amu)	Laser tuning range (nm)	Reference
Coumarin 102 T	311.32	$450 \leqslant \lambda \leqslant 510$	Duarte (1989a)
Coumarin 153 T	365.29	$508 \leqslant \lambda \leqslant 588$	Chen *et al* (1988)
Coumarin 314 T	369.35	$478 \leqslant \lambda \leqslant 525$	Chen *et al* (1988)
Coumarin 334 T	339.33	$500 \leqslant \lambda \leqslant 546$	Chen *et al* (1988)
Coumarin 338 T	397.41	$477 \leqslant \lambda \leqslant 526$	Chen *et al* (1988)
Coumarin 545 T	430.56	$501 \leqslant \lambda \leqslant 574$	Duarte *et al* (2006)

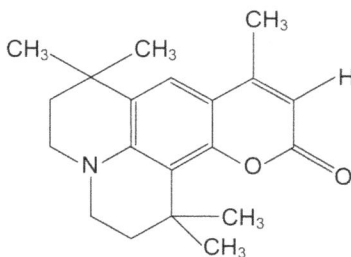

Figure 12.1. Molecular structure of coumarin 102 tetramethyl.

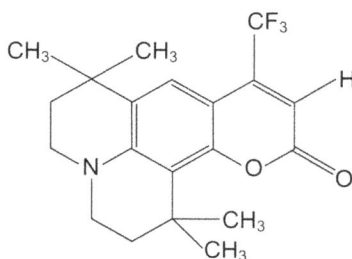

Figure 12.2. Molecular structure of coumarin 153 tetramethyl.

Organic dyes, such as C 545 T are used to dope Alq$_3$ in organic light emitting semiconductors commonly known as OLEDs (Chen and Tang 2001, Liao *et al* 2004, Duarte *et al* 2006), as described in chapter 11.

12.2.2 Polymers

Poly(methyl methacrylate), or PMMA (C$_5$O$_2$H$_8$), is probably the most emblematic, and widely employed, of the polymers utilized in photonic applications. For instance, it was with an organic laser dye-doped PMMA matrix that laser action in a solid-state organic laser was first reported (Soffer and McFarland 1967, Peterson and Snavely 1968). The molecular structure of PMMA is illustrated in figure 12.3.

In addition to PMMA, other photonically apt polymers include:

1. 2-Hydroxyethyl methacrylate methyl methacrylate (HEMA-MMA) (Costela *et al* 2016b)
2. 2-Hydroxyethyl methacrylate pentaerythritol tetraacrylate (HEMA-PETRA) (Costela *et al* 2016b)
3. 2-Hydroxyethyl methacrylate pentaerythritol triacrylate (HEMA-PETA) (Costela *et al* 2016b)
4. 2-Hydroxyethyl methacrylate trimethylolpropane (HEMA-TMPTMA) (Costela *et al* 2016b)
5. Cyclic olefin copolymer (James *et al* 2006)
6. Polyhydroxyethylmethacrylate, HEMA (C$_6$O$_{10}$H$_3$), see figure 12.4
7. Modified PMMA (MPMMA) (Maslyukov *et al* 1995)
8. Polycarbonate (James *et al* 2006)
9. Polystyrene (James *et al* 2006)
10. Polystyrene nanotubes (Dersch *et al* 2005)
11. Polysulfone (James *et al* 2006)
12. Polystyrene-poly(methyl methacrylate) (Paquet and Kumacheva 2008)

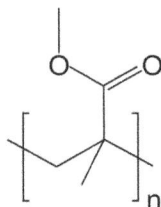

Figure 12.3. The molecular structure of the polymer PMMA.

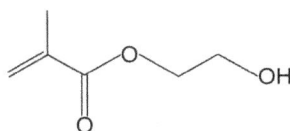

Figure 12.4. The molecular structure of the polymer HEMA.

13. Poly(ethyl acrylate) (Paquet and Kumacheva 2008)
14. Poly(styrene-*b*-isoprene) (Paquet and Kumacheva 2008).

The $\partial n/\partial T$ values for transparent polymers of interest are listed in table 12.2. The refractive index n and the $\partial n/\partial T$ values for dye-doped polymers of interest are listed in table 12.3.

Although solid-state dye lasers were introduced in the mid to late 1960s utilizing dye-doped PMMA (Soffer and McFarland 1967, Peterson and Snavely 1968), the field did not become firmly established mainly due to the absence of a laser-grade high-optical quality polymer. That situation changed rather spectacularly with the introduction of modified PMMA (MPMMA) (Maslyukov *et al* 1995). This form of poly(methyl methacrylate) was manufactured using a proprietary process that led to a superior class of PMMA.

As such this higher-class of PMMA provides the following characteristics (Duarte 1994):

1. Surface quality better than $\lambda/4$
2. Surface damage threshold approaching several J cm^{-2}
3. Laser dye bleaching threshold ~1 J cm^{-2}

These were the material characteristics that made possible narrow-linewidth oscillation in solid-state dye lasers (Duarte 1994) down to the $\Delta\nu\,\Delta t \approx 1$ limit (Duarte 1999).

12.2.3 Organic–inorganic composites

Sol–gel and organically modified silicates (ORMOSIL) have been known as laser and photonics materials since the late 1980s (Reisfeld *et al* 1989, Knobbe *et al* 1990, Dunn *et al* 1990, Rahn and King 1995).

Table 12.2. $\partial n/\partial T$ for clear optical polymers.

Polymer	$\partial n/\partial T$ (K^{-1})	Reference
Cyclic olefin copolymer	-1.02×10^{-4}	James *et al* (2006)
Polycarbonate	-1.14×10^{-4}	James *et al* (2006)
Poly(methyl methacrylate) (C$_5$O$_2$H$_8$)$_n$	-1.10×10^{-4}	Wunderlich (1989)
Polystyrene (C$_8$H$_8$)$_n$	-1.27×10^{-4}	James *et al* (2006)
Polysulfone	-1.00×10^{-4}	James *et al* (2006)

Table 12.3. Refractive index and $\partial n/\partial T$ for rhodamine 6 G-doped optical polymers.

Dye-doped polymer matrix	n^a	$\partial n/\partial T^b$ (K^{-1})	Reference
MPMMA	1.4943	$(-1.4 \pm 0.2) \times 10^{-4}$	Duarte *et al* (2000)
HEMA-MMA	1.5039	$(-1.3 \pm 0.2) \times 10^{-4}$	Duarte *et al* (2000)
Bz-MA HEMA-MMA		$(-1.4 \pm 0.3) \times 10^{-4}$	Duarte *et al* (2000)

[a] Measured at $\lambda = 593.93$ nm and $T = 297$K.
[b] Measured in the $297 \leqslant T$ 337 K range.

Table 12.4. Measured $\partial n/\partial T$ for DDP and PN solid-state matrices.[a]

Matrix	λ^b (nm)	$\partial n/\partial T$	Reference
PN[c,d] 30% SiO_2	632.82	-0.8840×10^{-4}	Duarte and James (2003)
PN[c,d] 50% SiO_2	632.82	-0.6484×10^{-4}	Duarte and James (2003)

[a] Adapted from Duarte and James (2003).
[b] Measurement wavelength.
[c] Polymer-nanoparticle.
[d] The SiO_2 nanoparticles used in these experiments had a diameter, on average, of ~10 nm.

Using a gain medium of rhodamine 6 G-doped ORMOSIL, in a multiple-prism grating cavity, linewidths of $\Delta\nu \approx 3$ GHz have been reported (Duarte *et al* 1993, Duarte 1994).

As already explained in chapter 9, some of these organic–inorganic materials did not possess the necessary refractive-index medium homogeneity to support the emission of homogeneous TEM_{00} laser beams (Duarte and Pope 1995). This meant a preference for pure polymer gain matrices which, as has been previously explained, exhibit unfavorable $\partial n/\partial T$ values that can lead to unnecessary higher beam divergences due to thermal lensing (Duarte and James 2003).

In order to improve $\partial n/\partial T$, beam quality, and beam divergence in these solid-state gain media, Duarte and James (2003) introduced organic dye-doped polymer nanoparticle gain media. The nanoparticles in this class of organic–inorganic gain media are SiO_2 nanoparticles, which are introduced into the polymeric volume in a nearly uniform manner that deny the occurrence of internal interferometric effects that lead to laser beam inhomogeneities (Duarte and James 2004). Moreover, the reduction in $\partial n/\partial T$ is rather significant, as it is evident via the values posted in table 12.4, as is the decrease in laser beam divergence (Duarte and James 2003). Please refer to chapter 9 for further details.

A detailed description on the synthesis of this laser-grade polymer-nanoparticle matrix is given by Duarte and James (2016).

A partial list of organic–inorganic materials applicable to the field of photonics includes:
1. HEMA-MMA methyltriethoxysilane (TRIEOS) (Costela *et al* 2016a)
2. HEMA-MMA tetraethoxylane (TEOS) (Costela *et al* 2016a)
3. HEMA 3-(trimethoxysilyl)propyl methacrylate (TMSPMA) (Costela *et al* 2016a)
4. MMA 3-(trimethoxysilyl)propyl methacrylate (TMSPMA) (Costela *et al* 2016a)
5. Organically modified ceramic (ORMOCER) (Serbin *et al* 2003)
6. ORMOSILs (Knobbe *et al* 1990)
7. PMMA-nanoparticle matrices (Duarte and James 2003)
8. Silica aerogel filled with MMA 2,2,2-trifluoroethyl methacrylate (TFMA) (Costela *et al* 2016a).

12.2.4 Organic lasers

Organic sources of coherent radiation, including organic lasers, offer the field of organic photonics interaction at three levels: level 1: at the low power regime where small-scale and miniaturized devices can find a plethora of applications in chemistry, biomedicine, medicine, and sensing. These devices can be of the traditional optically-pumped laser class, using semiconductor excitation capabilities, as already outlined in chapters 5 and 7, or electrically-pumped integrated interferometric emitters, as described in chapter 11. Level 2: at the high-average-power regime (tens of kW) for highly-selective applications such as atomic vapor laser isotope separation (AVLIS) applications (Akerman 1990, Webb 1991, Bass *et al* 1992) where low-noise tunable narrow-linewidth liquid organic lasers offer ample advantages (Duarte and Piper 1984, Bass *et al* 1992). Level 3: at the high-pulse energy regime (~1 kJ per pulse) regime where, for applications such as directed energy, liquid organic dye lasers have shown an unsurpassed advantage. For further details, please see chapter 8.

12.2.5 Quantum dots

Semiconductor nanocrystals, also known as quantum dots, become an additional component of the organic photonics armamentarium when they are encapsulated with an amphiphilic polymer (Wu *et al* 2003, Larson *et al* 2003). An example of such quantum dot is a CdSe–ZnS nanocrystal encapsulated by a polymer (Larson *et al* 2003). For a description on the physics of quantum dots please refer to Duarte (2015).

12.3 Organic optical elements

Polymer based optics is a pragmatic option, for the optical engineer, in an array of field instrumentation including: binoculars, cameras, periscopes, range finders, and telescopes. Polymer based optics is a necessity when the application demands low weight and flexibility. The issue of physical flexibility is particularly acute when considering medical instrumentation such as catheters for observation and catheters to transmit laser light to a particular internal tissue site, or organ, for therapeutical purposes. Polymer optics is also found in viewing screens and optical windows of scientific instrumentation and consumer electronics.

12.3.1 Organic lenses

Small polymer lenses are widely applied in consumer electronics, particularly in digital cameras and the in-built cameras of cell phones, or mobile phones, and computer screens. The work of James *et al* (2006) was, to a large degree, inspired to improve the $\partial n/\partial T$ for the focusing optics of commercial digital cameras and movie cameras. In summary, polymer, or organic–inorganic nanocomposite, lenses can play a central role in the optics of
1. Binoculars (Steiner and Lang 2003)
2. Digital cameras (James *et al* 2006)
3. Imaging systems (Duarte 1993a)
4. Laser range finders (available commercially)
5. Ophthalmic lenses (James *et al* 2006)
6. Movie cameras (James *et al* 2006)
7. Telescopic contact lenses (Arianpour *et al* 2015)
8. Transmission, or refraction, telescopes.

Both PMMA and HEMA are used in the manufacture of soft contact lenses (Refojo 1969).

12.3.2 Organic–inorganic prisms

As partially described by Duarte (2015) prisms and multiple-prism configurations can be found in a variety of devices, optical instruments and systems, including:

1. Beam expanding devices for optical processing (Lohmann and Stork 1989)
2. Beam expansion for imaging systems in astronomy (Sirat *et al* 2005)
3. Beam shaping optics for miniaturized diode lasers (Maker and Ferguson 1989)
4. Multiple-prism assemblies for Amici prisms for astronomy applications (Wynne 1997, Duarte 2013)
5. Multiple-prism assemblies for polarization rotators (Duarte 1989b)
6. Multiple-prism beam expansion for N-slit interferometry (Duarte 1987a, 1993a, 1993b)
7. One-dimensional beam expanders yielding *extremely elongated* Gaussian beams, or light sheets, for microscopy, or nanoscopy (Duarte 1987a, 1993a, 1993b)
8. One-dimensional beam expansion for application in laser printers (Duarte *et al* 2005)
9. Pulse compressors for ultrashort pulse lasers of the femtosecond class lasers (Dietel *et al* 1983, Fork *et al* 1984, Duarte 1987b).

In this regard, lightweight prism systems, and in particular achromatic lightweight multiple-prism arrays comprised of polymer organic–inorganic nanocomposite materials offer attractive alternatives to homologous systems made with traditional optical glasses. This is an area offering wide opportunities for material innovations in optical engineering.

12.3.3 Organic fibers

Organic fibers have been known in the open literature for a long time (Wente *et al* 1954). More specifically, polymeric optical fibers began to be noticed in the late 1980s (Koishi, *et al* 1989). Dye-doped sol–gel fibers, as a laser medium, were introduced by Gvishi *et al* (1996) and dye-doped poly(methyl methacrylate) fibers as gain media were demonstrated by Tagaya *et al* (1997).

An array of application of organic fibers include:

1. Chemical sensing (Narang *et al* 1996)
2. Laser gain media (Gvishi *et al* 1996, Tagaya *et al* 1997)
3. Medical scintillators (Allemand *et al* 1985, Therriault-Prouix *et al* 2013)
4. Medical catheters (Parker *et al* 1990)
5. Optical communications (Yuto and Kameo 1984)
6. Nonlinear optics (Borzo and Stuetz 1990)
7. Sensing in the automobile industry (Guerrero *et al* 1993).

12.4 Applications

Here, some salient application arenas for organic lasers, and organic photonics, are outlined. These include: communications, directed energy, industry, medicine, nonlinear optics, science, and sensors. Communications, directed energy, industry, and the science sections refer mainly to the utilization of organic dye lasers. The industrial sector also includes the use of OLEDs in the TV and cellular phone markets. On the other hand, medicine benefits from the use of organic dyes, organic lasers, organic fibers, and organic quantum dots while nonlinear optics utilizes organic fibers and organic–inorganic nanocomposites. Finally, the technology of organic sensors appears to mainly benefit from the use of organic fibers and OLED based techniques.

As an introduction, a tabular perspective of the field is outlined. Table 12.5 lists broad application categories for elements of organic photonics such as organic dyes, polymers, and organic–inorganic nanocomposites. Table 12.6 lists broad application categories for devices of organic photonics such as organic dye lasers, organic fibers, organic light emitting diodes (OLEDs), and organic lasers (second generation).

The description of organic lasers of the second generation refers in general to organic lasers that do not belong to the category of the *organic dye laser*, liquid or solid state. These second generation organic lasers are low-power optically-pumped lasers including organic conjugate polymer lasers, organic semiconductor lasers, organic optofluidic lasers, and organic VCSELs. This class of lasers is catalogued in section 7.5.

Table 12.5. Broad application categories for elements of organic photonics.

Element of organic photonics	Applications
Organic dye	Biology
	Flourescence light sheet microscopy
	Imaging
	Medicine
	Micro lithography
	Microscopy
	Optogenetics
	Organic lasers
	Organic fibers
Polymers: PMMA, HEMA	Ophthalmology
	Fibers
	Industrial manufacturing
	Optics
	Organic lasers
Organic–inorganic nanocomposites	Fibers
	Optics
	Organic lasers
	Nonlinear optics

Table 12.6. Broad application categories for devices of organic photonics.

Device of organic photonics	Applications
Organic dye lasers	Astronomy
	Atomic laser isotope separation
	Directed energy
	Materials diagnostics
	Medicine
	Optogenetics
	Photochemical industry
	Physics
	Spectroscopy
Organic fibers	Automobile industry
	Medicine
	Scintillators
	Sensors
Organic lasers (second generation)	Bioimaging
	Biosensors
OLEDs	Cellular phone industry
	Imaging display industry

A more nuanced description of this gigantic field, including research references, is given in the following subsections.

12.4.1 Communications

Tunable organic lasers are applicable to communications, however, this is a field heavily dominated by traditional inorganic semiconductor lasers (Killinger 2002) and other types of inorganic sources of coherent radiation such as optical parametric down converters (Yin, *et al* 2017). Perhaps diode-pumped solid-state organic lasers will make a future entrance in the field of communications. Despite some initial promise (Yuto and Kameo 1984) the field of organic fibers for optical communications also appears to remain overshadowed by their inorganic counterparts.

12.4.2 Directed energy

The main article on this subject is provided in chapter 8. For the sake of completeness here it is just mentioned that high-performance ruggedized liquid organic dye lasers have demonstrated exquisite tunable performance with low-divergence TEM_{00} beams and single-longitudinal-mode oscillation (Duarte *et al* 1991). At the same time, large pulse-energy liquid organic lasers have demonstrated single-pulse energy levels of hundreds of Joules per pulse (Baltakov *et al* 1974) and even at the *kilo Joule* level (Tang *et al* 1987, Klimek *et al* 1992).

12.4.3 Industry

Organic semiconductor coatings, more specifically OLED type coatings, are used in TV screens (Perry and Zorpette 2013, Cho and Daim 2016) and in the screens of cellular, or mobile, phones (Kim *et al* 2017). OLED displays are also applied to the automobile industry (Hack *et al* 2017) and the underwater illumination industry (Duarte 2017). The TV industry also utilizes other types of organic–inorganic hybrids in films incorporating organic dyes (Escribano *et al* 2007).

The photochemical industry is yet another area of productivity that has benefited from tunable organic lasers specifically in the area of atomic vapor laser isotope separation (AVLIS). For reviews on AVLIS readers are referred to Paisner and Solarz (1987), Akerman (1990), Bokhan *et al* (2006), and Duarte (2016). For laser aspects of AVLIS please refer to Duarte (1984), Broyer *et al* (1984), Webb (1991), Bass *et al* (1992), Singh *et al* (1994), and Sugiyama *et al* (1996).

A very detailed review on industrial and manufacturing applications of high-power organic dye lasers is given by Klick (1990). Although tailored specifically for dye lasers, the information provided by Klick should prove useful to researchers working on industrial applications of lasers in general.

12.4.4 Medicine

In chapter 14 the application of organic dyes and organic lasers in medicine is considered with a particular emphasis on photodynamic therapy. Recent advances in this field include the discovery of a whole new family of fine-tuned rhodamine dyes for live-cell and *in vivo* biological imaging (Grimm *et al* 2017). Besides demonstrating applications in cells and tissues, these authors utilized computational molecular modeling of the new dyes in addition to quantum yield measurements. Rhodamine dyes were fine-tuned using 3-substituted azetidines (Grimm *et al* 2017). It should be mentioned that fluorophores, like these fine-tuned rhodamines, are used in microscopy and in light-sheet fluorescence microscopy (see chapter 14).

Further biological and biomedical applications of organic dyes and organic–inorganic materials include:

1. Femtosecond lasers used to improve precision in optogenetics (Gunaydin *et al* 2010). This might open further application avenues for organic dye lasers emitting in the femtosecond regime. For optogenetics see chapter 13.
2. HEMA for soft contact lenses (Refojo 1969, Gasset and Kaufman 1970).
3. Lanthanide-doped organic–inorganic materials in biotechnology and clinical diagnosis (Escribano *et al* 2007).
4. Lanthanide-doped organic–inorganic materials in DNA fluorescence (Escribano *et al* 2007).
5. Organic dyes for fluorescence light sheet microscopy (Verveer *et al* 2007, Holekamp *et al* 2008). Also, see chapter 14.
6. Organic dyes for optogenetics (Luccardini *et al* 2009, Greb 2012). Also, see chapter 13.
7. Organic dyes for photodynamic therapy (PDT) (see chapter 14).

8. Organic fibers for medical scintillators (Allemand *et al* 1985, Therriault *et al* 2013)
9. Organic fibers for medical catheters (Parker *et al* 1990)
10. PMMA for soft contact lenses (Refojo 1969).
11. Quantum dots encapsulated within an amphiphilic polymer for imaging *in vivo* (Larson, *et al* 2003)
12. Quantum dots linked to immunoglobulin G as cancer markers (Wu *et al* 2003).

For a perspective on the application of organic lasers to medicine, please refer to chapter 14 and the following reviews on the subject: Goldman (1990), Jelinkova (2013), and Costela *et al* (2016b).

12.4.5 Nonlinear optics

As already indicated in the previous section, nonlinear optics was identified in conjunction with organic photonics via the use of polymers and silica glass-polymer composites including 2-methyl-4-nitroaniline (Borzo and Stuetz 1990). These researchers indicate that the nonlinear properties in these polymers are related to the existence of large numbers of delocalized π-electrons. A detailed study on the optical nonlinear properties of polymer-nanoparticle type composites, applicable to the composites developed by Duarte and James (2003, 2004) has been conducted by Dolgaleva *et al* (2009) and Dolgaleva and Boyd (2012). Particular attention in these studies was paid to local-field effects in nanocomposite materials. From a different perspective in this field, He *et al* (1995) and Bhawalkar *et al* (2013) discuss multiphoton processes in organic materials. Nonlinear behavior in organic dyes has been studied by Erande *et al* (2017). Two reviews on nonlinear organic photonics are given by Kajzar and Zyss (2012) and Norwood (2013).

12.4.6 Science

Undoubtedly, tunable narrow-linewidth organic lasers (Peterson *et al* 1970, Hänsch 1972, Littman and Metcalf 1978, Duarte and Piper 1984) brought a revolution and renaissance to analytical chemistry and optical spectroscopy (Duarte *et al* 1992). Among the scientific fields that have benefited enormously from high-performance tunable narrow-linewidth organic lasers are:
1. Astronomy: via the introduction of *laser guide stars* (Everett 1989, Primmerman *et al* 1991, Bass *et al* 1992) and via the introduction of laser stabilization techniques (Drever *et al* 1983) that ultimately made possible the detection of *gravitational waves* (Abbott *et al* 2016)
2. Physics: for example, atomic optics (Letokhov 1992), diffraction of atoms by light or the Kapitza–Dirac effect (Gould *et al* 1986), and parity non-conservation (Drell and Commins 1985)
3. Spectroscopy: atomic (Sneddon *et al* 1996) and molecular (Demtröder 2007, 2008, 2015).

12.4.7 Sensors

Perhaps one of the first elements of organic photonics to be associated with sensors were polymer fibers for medical applications, more specifically for hyperthermia (Wickersheim 1987).

A partial listing of sensory applications for elements of organic photonics include:

1. Organic fibers in the automobile industry: as speedometers, tachometers, and magnetometers (Guerrero *et al* 1993)
2. Organic fibers as chemical sensors (Lieberman and Brown 1989, Narang *et al* 1996)
3. Organic fibers as humidity sensors (Sadaoka *et al* 1991)
4. Organic fibers in medicine as hyperthermia sensors (Wickersheim 1987)
5. Organic optofluidic lasers as biosensors (Chen *et al* 2010, Choi *et al* 2013)
6. Organic optofluidic lasers for single molecule detection (Chen *et al* 2010)
7. Organic semiconductors as organic photodiodes (Arca *et al* 2013)
8. Organic semiconductors as detectors in artificial vision (Martino *et al* 2013)
9. Organic semiconductors as biotechnological sensors (Krujatz *et al* 2016)
10. Organic semiconductors in chemical sensors (Nalwa *et al* 2010).

12.5 Perspective

This broad and yet terse survey tends to indicate that organic lasers have made an impact mainly in medicine, the photochemical industry, and various branches of the sciences. As far as organic photonics materials is concerned, the impact has been mainly registered in biotechnology, medicine, nonlinear optics, and sensors. Emerging areas of interest include self assembly of organic crystals into hetero-structures (Xu *et al* 2023a), magnetically controlled assembly of hierarchical organic nanostructures (Xu *et al* 2023b), organic nanocrystal lasers (Shi *et al* 2023), and organic nanocrystals for integrated circuits (Pradeep *et al* 2023).

12.6 Problems

- 12.1 Perform a literature search and add at least one further optical instrument, or optical technique, that applies multiple-prism arrays to the list given in subsection 12.3.2.
- 12.2 Perform a literature search and add at least one further application for organic fibers to the list given in subsection 12.3.3.
- 12.3 Perform a literature search and add one further application for each organic element in table 12.5.
- 12.4 Perform a literature search and add at least one further application, of organic dyes in medicine, to the list given in subsection 12.4.4.
- 12.5 Perform a literature search and add at least one further application, of organic–inorganic composites in medicine, to the list given in subsection 12.4.4.
- 12.6 Perform a literature search and add at least one further application, of organic photonics in sensor technology, to the list given in subsection 12.4.7.

References

Abbot B P *et al* 2016 Observation of gravitational waves from a binary black hole merger *Phys Rev. Lett.* **116** 061102

Akerman M A 1990 Dye laser isotope separation *Dye Laser Principles* ed F J Duarte and L W Hillman (New York: Academic) ch 9

Allemand L-R, Calvet J, Cavan J C and Thevenin J-C 1985 Optical fibers with plastic core and polymer cladding *US Patent* 4552431

Arca F, Tedde F S, Sramek M, Rauh J, Lugli P and Hayden O 2013 Interface trap states in organic photodiodes *Sci. Rep.* **3** 1324

Arianpour A, Schuster G M, Tremblay E J, Stamenov I, Groisman A, Legerton J, Meyers W, Alonso Amigo G and Ford J E 2015 Wearable telescopic contact lens *Appl. Opt.* **54** 7195–204

Baltakov F N, Barikhin B A and Sukhanov L V 1974 400 J pulsed laser using a solution of rhodamine 6 G in ethyl alcohol *JETP Lett.* **19** 174–5

Bass I L, Bonano R E, Hackel R H and Hammond P R 1992 High-average-power dye laser at Lawrence Livermore National Laboratory *Appl. Opt.* **31** 6993–7006

Bhawalkar J D, He G S and Prasad P N 2013 Nonlinear multiphoton processes in organic and polymeric materials *Prog. Quantum Electron.* **59** 1041–70

Bokhan P A, Buchanov V V, Fateev N V, Kalugin M M, Kazaryan M A, Prokhorov A M and Zakrevskii D E 2006 *Laser Isotope Separation in Atomic Vapor* (Weinheim: Wiley-VCH)

Borzo M and Stuetz D E 1990 Organic nonlinear optical media *US Patent* 4898691

Broyer M, Chevaleyre J, Delacretaz G and Woste L 1984 CVL-pumped dye laser for spectroscopic application *Appl. Phys.* B **35** 31–6

Chen C H, Fox J L, Duarte F J and Ehrlich J J 1988 Lasing characteristics of new coumarin-analog dyes: broadband and narrow-linewidth performance *Appl. Opt.* **27** 443–5

Chen C H and Tang C W 2001 Efficient green organic light-emitting diodes with stericly hindered coumarin dopants *Appl. Phys. Lett.* **79** 3711–3

Chen Y, Lei L, Zhang K, Shi J, Wang L, Li H, Zhang X M, Wang Y and Chan H L W 2010 Optofluidic microcavities: dye lasers and biosensors *Biomicrofluidics* **4** 043002

Cho Y and Daim T 2016 OLED TV technology forecasting using technology mining and the Fisher–Pry diffusion model *Foresight* **18** 117–37

Choi E Y, Mager L, Cham T T, Dorkenoo K D, Fort A, Wu J W, Barsella A and Ribierre J-C 2013 Solven-free fluid organic dye lasers *Opt. Express* **21** 11368–75

Costela A, García-Moreno I and Gómez C 2016a Medical applications of organic dye lasers *Tunable Laser Applications* ed F J Duarte (New York: CRC) 3rd edn ch 8

Costela A, García-Moreno I and Sastre R 2016b Solid-state organic dye lasers *Tunable Laser Applications* ed F J Duarte (New York: CRC) 3rd edn ch 3

Demtröder W 2007 *Laserspektroscopie: Grundlagen und Techniken* (Berlin: Springer)

Demtröder W 2008 *Laser Spectroscopy: Basic Principles* 4th edn (Berlin: Springer)

Demtröder W 2015 *Laser Spectroscopy 1* 5th edn (Berlin: Springer)

Dersch R, Steinhart M, Boudriot U, Greiner A and Wendorff J H 2005 Nanoprocessing of polymers: applications in medicine, sensors, catalysis, photonics *Poly. Adv. Tech.* **16** 276–81

Dietel W, Fontaine J J and Diels J-C 1983 Intracavity pulse compression with glass: a new method of generating pulses shorter than 60 fs *Opt. Lett.* **8** 4–6

Dolgaleva K, Boyd R W and Milonni P 2009 The effect of local fields on laser gain for layered and Maxwell Garnett composite materials *J. Opt. A: Pure Appl. Opt.* **11** 024002

Dolgaleva K and Boyd R W 2012 Local field effects in nanostructures photonic materials *Adv. Opt. Photon.* **4** 1–77

Drell P S and Commins E D 1985 Parity nonconservation in atomic thallium *Phys. Rev.* A **32** 2196–210

Drever R W P, Hall J L, Kowalski F V, Hough J, Ford G M, Munley A J and Ward H 1983 Laser phase and frequency stabilization using an optical resonator *App. Phys.* B **31** 97–105

Duarte F J 1987a Beam shaping with telescopes and multiple-prism beam expanders *J. Opt. Soc. Am.* A **4** 30

Duarte F J 1987b Generalized multiple-prism dispersion theory for pulse compression in ultrafast dye lasers *Opt Quantum Electron.* **19** 223–9

Duarte F J 1989a Ray transfer matrix analysis of multiple-prism dye laser oscillators *Opt. Quantum Electron.* **21** 47–54

Duarte F J 1989b Optical device for rotating the polarization of a light beam *US Patent* 4822150

Duarte F J 1993a On a generalized interference equation and interferometric measurements *Opt. Commun.* **103** 8–14

Duarte F J 1993b Electro-optical interferometric microdensitometer system *US Patent* 5255069

Duarte F J 1994 Solid-state multiple-prism grating dye-laser oscillators *Appl. Opt.* **33** 3857–60

Duarte F J 1999 Multiple-prism grating solid-state dye laser oscillator: optimized architecture *Appl. Opt.* **38** 6347–9

Duarte F J 2013 Tunable laser optics: applications to optics and quantum optics *Prog. Quantum Opt.* **37** 326–47

Duarte F J 2015 *Tunable Laser Optics* 2nd edn (New York: CRC)

Duarte F J 2016 Tunable laser atomic vapor laser isotope separation *Tunable Laser Applications* ed F J Duarte (New York: CRC) 3rd edn ch 11

Duarte F J 2017 Diffractive and prismatic OLED wireless and LED wireless underwater pool light sources *US Patent* 2017/0167717 A1

Duarte F J, Costela A, García-Moreno I and Sastre R 2000 Measurements of $\partial n/\partial T$ in solid-state dye-laser gain media *App. Opt.* **39** 6522–3

Duarte F J, Davenport W E, Ehrlich J J and Taylor T S 1991 Ruggedized narrow-linewidth dispersive dye laser oscillator *Opt. Commun.* **84** 310–6

Duarte F J, Ehrlich J J, Davenport W E, Taylor T S and McDonald J C 1993 A new tunable dye laser oscillator: preliminary report *Proc. of the Int. Conf. on Lasers '92* ed C P Wang (McLean, VA: STS) pp 293–6

Duarte F J and James R O 2003 Tunable solid-state lasers incorporating dye- doped polymer-nanoparticle gain media *Opt. Lett.* **28** 2088–90

Duarte F J and James R O 2004 Spatial structure of dye-doped polymer nanoparticle laser media *Appl. Opt.* **43** 4088–90

Duarte F J and James R O 2016 Organic dye-doped polymer-nanoparticle tunable lasers Tunable Laser Applications 3rd edn ed F J Duarte (New York: CRC Press) pp 177–201

Duarte F J, Liao L S, Vaeth K M and Miller A M 2006 Widely tunable green laser emission using the coumarin 545 tetramethyl dye as the gain *J. Opt. A: Pure Appl. Opt.* **8** 172–4

Duarte F J, Paisner J A and Penzkofer A 1992 Dye lasers: introduction by the feature editors *Appl. Opt.* **31** 6977–8

Duarte F J and Piper J A 1984 Narrow-linewidth high-prf copper laser-pumped dye laser oscillators *Appl. Opt.* **23** 1391–4

Duarte F J and Pope E J A 1995 Optical inhomogeneities in sol–gel derived ORMOSILS and nanocomposites *Ceram. Trans.* **55** 267–73

Dunn B S, Mackenzie J D, Zink J I and Stafsudd O M 1990 Solid-state tunable lasers based on dye-doped sol–gel materials *SPIE Proc.* **1328** 174–82

Erande Y, Warde U and Sekar N 2017 Investigation of NLO properties of fluorescent BORICO dyes: a comprehensive experimental and theoretical approach *J. Fluores.* **27** 2253–62

Escribano P, Julián-López B, Planelles-Aragó J, Cordoncillo E, Viana B and Sanchez C 2007 Photonic and nanophotonic properties of lanthalide-doped hybrid organic–inorganic materials *J. Mat. Chem.* **18** 23–40

Everett P N 1989 300-Watt dye laser for field experimental site *Proc. the Int. Conf.on Lasers '88* ed R C Sze and F J Duarte (McLean, VA: STS) pp 404–9

Fork R L, Martinez O M and Gordon J P 1984 Negative dispersion using pairs of prisms *Opt. Lett.* **9** 150–2

Gasset A R and Kaufman H E 1970 Therapeutic uses of hydrophilic contact lenses *Am. J. Ophthalmol.* **69** 252–9

Goldman L 1990 Dye lasers in medicine *Dye Laser Principles* ed F J Duarte and L W Hillman (New York: Academic) ch 10

Gould P L, Ruff G A and Pritchard D E 1986 Diffraction of atoms by light: the near resonant Kapitza–Dirac effect *Phys. Rev. Lett.* **56** 827–30

Greb C 2012 *Fluorescent Dyes* (Wetzlar: Leyca Microsystems)

Grimm J B *et al* 2017 A general method to fine- tune fluorophores for live-cell and *in-vivo* imaging *Nat. Meth.* **14** 987–94

Guerrero H, Zoido J, Escudero J L and Bernabeu E 1993 Characterization and sensor applications of polycarbonate optical fibers *Fiber Integr. Opt.* **12** 257–68

Gunaydin L A, Yizhar O, Berndt A, Sohal V S, Deisseroth K and Hegemann P 2010 Ultrafast optogenic control *Nat. Neurisci.* **13** 387–93

Gvishi R, Ruland G and Prasad P N 1996 New laser medium: dye-doped sol gel fiber *Opt. Commun.* **126** 66–72

Hack M, Weaver M S, Thompson N J and Adamovich V 2017 High resolution low power consumption OLED display with extended lifetime *US Patent* 2017/0207281 A1

Hänsch T W 1972 Repetitively pulsed tunable dye laser for high-resolution spectroscopy *Appl. Opt.* **11** 895–8

He G S, Gvishi R, Prasad P N and Reinhardt B A 1995 Two-photon absorption based optical limiting and stabilization in organic molecule-doped solid materials *Opt. Commun.* **117** 133–6

Holekamp T F, Diwakar T and Holy T E 2008 Fast three dimensional fluorescence imaging of activity in neural populations by objective-coupled planar illumination microscopy *Neuron* **57** 661–72

James R O, Rowley L A, Hurley D F and Border J 2006 Core shell nanocomposite optical plastic article *US Patent* 7091271B2

Jelinkova H 2013 *Lasers for Medical Applications: Diagnostics, Therapy, and Surgery* (Oxford: Woodhead)

Kajzar F and Zyss J 2012 Organic nonlinear optics: historical survey and current trends *Nonlinear Opt. Quantum Opt.* **43** 31–95

Killinger D 2002 Free space optics for laser communications through the air *Opt. Photon. News* **13** 36–42

Kim B, Kim H and Kim Y 2017 Cellular phone *US Patent* D778865S

Klick D 1990 Industrial applications of dye lasers *Dye Laser Principles* ed F J Duarte and L W Hillman (New York: Academic) ch 8

Klimek D E, Aldag H R and Russell J 1992 High-energy high-power flashlamp- pumped dye laser *Conference on Lasers and Electro Optics* (Washington DC: Optical Society of America) p 332

Knobbe E T, Dunn B, Fuqua P D and Nishida F 1990 Laser behavior and photostability characteristics of organic dye doped silicate gel materials *Appl. Opt.* **29** 2729–33

Koishi T, Tanaka I, Yasumura T and Nishikawa Y 1989 Optical fiber having polymethylmethacrylate core and fluoro-copolymer cladding *US Patent* 4687295

Krujatz F, Hild O R, Fehse K, Jahnel M, Werner A and Bley T 2016 Exploiting the potential of OLED-based photo-organic sensors for biotechnological applications *Chem. Sci.* J **7** 1000134

Larson D R, Zipfel W R, Williams R M, Clark S W, Bruchez M P, Wise F W and Webb W W 2003 Water soluble quantum dots for mutiphoton fluorescence imaging *in vivo Science* **300** 1434–6

Letokhov V S 1992 Atomic optics with tunable dye lasers *Dye Lasers: 25 Years* ed M Stuke (Berlin: Springer) ch 11

Liao L S, Klubeck K B and Tang C W 2004 High-efficiency tandem organic light-emitting diodes *Appl. Phys. Lett.* **84** 167–9

Lieberman R A and Brown K E 1989 Intrinsic fiber optics chemical sensor based on two-stage fluorescence coupling *SPIE Proc.* **990** 104–10

Littman M G and Metcalf H J 1978 Spectrally narrow pulsed dye laser without beam expander *Appl. Opt.* **17** 2224–7

Lohmann A W and Stork W 1989 Modified Brewster telescopes *Appl. Opt.* **28** 1318–9

Luccardini C *et al* 2009 Measuring mitochondrial and cytoplasmic Ca^{2+} in EGFP expressing cells with a low affinity calcium ruby and its dextran conjugate *Cell Calcium* **45** 275–83

Maker G T and Ferguson A I 1989 Frequency-modulation mode locking of a diode-pumped Nd: YAG laser *Opt. Lett.* **14** 788–90

Martino N, Ghezzi D, Benfenati F, Lanzani G and Antognazza M R 2013 Organic conductors for artificial vision *J. Mat. Chem.* B **1** 3768–80

Maslyukov A, Sokolv S, Kailova M, Nyholm K and Popov S 1995 Solid-state dye laser with modified poly(methyl methacrylate)-doped active elements *Appl. Opt.* **34** 1516–8

Nalwa K S, Cai Y, Thoeming A L, Shinar J, Shinar R and Chaudhary S 2010 Polythiophene-fullerene based photodetectors: tuning of spectral response and application in photoluminescence based (bio)chemical sensors *Adv. Mat.* **22** 4157–61

Narang U, Gvishi R, Bright F V and Prasad P N 1996 Sol–gel-derived micron scale optical fibers for chemical sensing *J. Sol-Gel Sc. Technol.* **6** 113–9

Norwood R A 2013 Organic photonics: ready for prime time *Opt. Photon. News* **24** 40–7

Paisner J A and Solarz R W 1987 Resonance photoionization spectroscopy *Laser Spectroscopy and Its Applications* ed L J Radziemski, R W Solarz and J A Paisner (New York: Marcel Dekker) ch 3

Paquet C and Kumacheva E 2008 Nanostructured polymers for photonics *Mater. Today* **11** 48–56

Parker T L, Pedersen D R and Mossely J D 1990 Plastic optical fiber for *in vivo* use having a biocompatible polyurethane cladding *US Patent* 4893897

Perry T S and Zorpette G 2013 *IEEE Spectrum* **50** 46–7

Peterson O G and Snavely B B 1968 Stimulated emission from a flashlamp-excited organic dyes in polymethyl methacrylate *Appl. Phys. Lett.* **12** 238–40

Peterson O G, Tuccio S A and Snavely B B 1970 CW operation of an organic dye solution laser *Appl. Phys. Lett.* **17** 245–7

Pradeep V V, Chosenyah M, Mamonov E and Chandrasekar R 2023 Crystals photonics foundry: geometrical shaping of molecular single crystal into next generation optical cavities *Nanoscale* **15** 12220–6

Primmerman C A, Murphy D V, Page D A, Zollars B G and Barclay H T 1991 Compensation of atmospheric optical distortion using a synthetic beacon *Nature* **353** 141–3

Rahn M D and King T A 1995 Comparison of laser performance of dye molecules in sol–gel, polycom, ormosil, and poly(methyl methacrylate) host media *Appl. Opt.* **34** 8260–71

Refojo M F 1969 Artificial membranes for corneal surgery *J. Biomed. Mat. Res.* **3** 333–47

Reisfeld R, Brusilovsky D, Eyal M, Miron E, Burstein Z and Ivri J 1989 A new solid-state tunable laser in the visible *Chem. Phys. Lett.* **160** 43–4

Sadaoka Y, Matsugushim M and Sakai Y 1991 Optical fiber humidity sensor using nafion^r-triphenylcarbinol composity *J. Electrochem Soc.* **138** 614–5

Serbin J, Egbert A, Ostendorf A, Chichkov B N, Houbertz R, Domman G, Schulz J, Cronauer C, Fröhlich L and Popall M 2003 Femtosecons laser-induced two-photon polymerization of inorganic-organic hybrid materials for applications in photonics *Opt. Lett.* **28** 301–3

Shi Y-L, Hu Y, Lv Q, Yang W-Y and Wang X-D 2023 Construction of organic micro/nanocrystal lasers: from molecules to devices *Mater. Chem. Front.* **7** 3922–36

Singh S, Dasgupta K, Kumar S, Manohar K G, Nair L G and Chatterjee U K 1994 High-power high-repetition-rate capper-vapor-pumped dye laser *Opt. Eng.* **33** 1894–904

Sirat G Y, Wilner K and Neuhauser D 2005 Uniaxial crystal interferometer: principles and forecasted applications to imaging astrometry *Opt. Express* **13** 6310–22

Sneddon J, Thiem T L and Lee Y-I (ed) 1996 *Lasers in Analytical Atomic Spectroscopy* (New York: Wiley-VCH)

Soffer B H and McFarland 1967 Continuously tunable narrow-band organic dye lasers *Appl. Phys.* **10** 266–7

Steiner R F and Lang A J 2003 Binocular lens systems *US Patent* 6537317 B1

Sugiyama A, Nakayama T, Kato M, Maruyama Y and Arisawa T 1996 Characteristics of a pressure-tuned single-mode dye laser pumped by a copper vapor lasers *Opt. Eng.* **35** 1093–7

Tagaya A, Teramoto S, Nihei E, Sasaki K and Koike Y 1997 High-power high- gain optical fiber amplifiers: novel techniques for preparatio and spectral investigation *Appl. Opt.* **36** 572–8

Tang K Y, O'Keefe T, Treacy B, Rottler R and White C 1987 Kilojoule-output XeCl dye laser: optimization and analysis *Proceedings: Dye Laser/Laser Dye Technical Exchange Meeting* ed J H Bentley (Redstone Arsenal, Al: U. S. Army Missile Command) pp 490–502

Therriault-Prouix F, Beaulie L, Archambault L and Beddar S 2013 On the nature of the light produced within PMMA optical guide guides in scintillation fiber-optics dosimetry *Phys. Med. Biol.* **58** 2073–84

Verveer P J, Swoger J, Pampaloni F, Greger K, Marcello M and Stelzer E H K 2007 High-resolution three dimensional imaging of large specimens with light sheet based microscopy *Nature Meth.* **4** 311–3

Webb C E 1991 High-power dye lasers pumped by copper vapor lasers *High Power Dye Lasers* ed F J Duarte (Berlin: Springer) ch 5

Wente V A, Boone E L and Fluharty D C 1954 *Manufacture of Superfine Organic Fibers* (Washington, DC: Naval Research Laboratory)

Wickersheim K A 1987 A new fiberoptic thermometry system for use in medical hyperthermia *SPIE Proc.* **713** 150

Wu X, Liu H, Liu J, Haley K N, Treadway J A, Larson J P, Ge N, Peale F and Bruchez M P 2003 Immunofluorescent labeling of cancer marker Her2 and other cellular targets with semiconductor quantum dots *Nat. Biotech.* **21** 41–6

Wunderlich W 1989 Physical constants of poly(methyl methacrylate) *Polymer Handbook* (NewYork: John Wiley) pp 77–80

Xu C F 2023a Directed self-assembly of organic crystals into chip-like heterostructures for signal processing *Sci. China Mater.* **66** 733–9

Xu L, Jia H, Yin B and Yao J 2023b Magnetically controlled assembly: a new approach to organic integrated photonics *Chem. Sci.* **14** 8723–42

Wynne C G 1997 Atmospheric dispersion in very large telescopes with adaptive optics *Mon. Not. R. Astron. Soc.* **285** 130–4

Yin J *et al* 2017 Satellite based entanglement distribution over 1200 kilometers *Science* **356** 1140–4

Yuto M and Kameo Y 1984 Coated plastic optical fiber *US Patent* 4458986

IOP Publishing

Organic Lasers and Organic Photonics (Second Edition)

F J Duarte

Chapter 13

Organic dyes in optogenetics

Alfons Penzkofer, Peter Hegemann and Suneel Kateriya

13.1 Introduction

Optogenetics is a technique of the life sciences where photoreceptor proteins are expressed and delivered into cells to trigger and monitor biological action by light exposure. The name optogenetics developed with the expression of rhodopsins in neurons for photostimulation of selective cells in large cellular networks (Boyden *et al* 2005, Nagel *et al* 2005, Li *et al* 2005, Ishizuka *et al* 2006, Deisseroth *et al* 2006). The name was first used by Deisseroth *et al* (2006). In the meantime the name optogenetics is not restricted to photo-activation of neurons with rhodopsins but includes all genetically expressed photoreceptors delivered into cells or tissues to trigger and monitor functional actions of proteins with light and optical methods (Yizhar *et al* 2011, Knöpfel and Boyden 2012, Hegemann and Sigrist 2013, Yawo *et al* 2015, Ilango and Lobo 2015, Mathes and Kennis 2016, Kianianmomeni 2016, Appasani 2017, Rost *et al* 2017, Stroh 2018, O'Banion and Lawrence 2018). Optogenetics allows fast temporal and spatial localized control of accurately defined events of selected cells in complex biological systems such as neuron activation, switching or inhibition, RNA transcription, enzyme catalysis, and cell motility.

Optogenetic tools consist of a light sensor protein domain generally with incorporated organic dye cofactor that absorbs light and an effector protein domain that exerts biological activity like photocurrent response, cellular second messenger activation, catalysis (Ziegler and Möglich 2015). Various light sources may be used for sensor excitation like lamps, light emitting diodes (LEDs), lasers, light sources with fiber-optic delivery systems (Doronina *et al* 2015, Pisanello *et al* 2017). Besides single-photon excitation, two-photon-excitation techniques have been employed (Andrasfalvy *et al* 2010, Papagiakoumou *et al* 2010, Prakash *et al* 2012). Also photoreceptor excitation has been applied by near-infrared light pumped rare earth ion upconversion nanoparticles (Chen *et al* 2014, Zhou *et al* 2015a, Chen *et al* 2018). Often a fluorescent protein domain is coexpressed with the light-sensitive actuator for its localization in cellular networks or even within a single cell (Tantama *et al* 2012, Shcherbakova *et al* 2015, Harada *et al* 2017, Park *et al* 2017).

Most natural photoreceptors (Batschauer 2003, Briggs and Spudich 2005, Schmidt and Cho 2015) have both a light sensor domain and an effector domain (occasionally unified in a single protein). They sense light between 280 and 750 nm and regulate all kinds of processes in the biological world from prokaryotic cells, via unicellular eukaryotes, to plants, fungi, animals and humans. Such natural photoreceptor classes are microbial and animal rhodopsins (Ernst *et al* 2014), photosensitive flavoproteins (Müller 1991a, Silva and Edwards 2006, Conrad *et al* 2014) with the sub-families of LOV domains (Pudasaini *et al* 2015), BLUF domains (Masuda 2013, Penzkofer *et al* 2016b, Park and Tame 2017), and cryptochromes (Chaves *et al* 2011, Park *et al* 2017), xanthopsins (photoactive yellow proteins, Imamoto and Kataoka 2007, Gao *et al* 2016), phytochromes (Auldidge and Forest 2011, Li *et al* 2011), cobalamin (vitamin B_{12}) binding domains (Cheng *et al* 2016, Fang and Bauer 2017, Padmanabhan *et al* 2017), plant UV-B receptor UVR8 (Heijde and Ulm 2012, Tilbrook *et al* 2013, Jenkins 2014b), cyanobacterial orange carotenoid protein and its homologs (Bao *et al* 2017, Kerfeld *et al* 2017 and references therein), and phototransformable fluorescent proteins (Day and Davidson 2009, Tantama *et al* 2012, Shcherbakova *et al* 2015, Harada *et al* 2017, Park *et al* 2017, Chernov *et al* 2017). For optogenetic applications the photoreceptors—natural ones, mutated ones, and engineered ones—are synthesized by genetic methods in transgenic cells, tissues, or organisms and brought to targeting cells and tissues for their light triggered action.

In neuroscience mainly genetically encoded proteins from the families of rhodopsins, flavoproteins (BLUF, LOV, cryptochrome), phytochromes, UVR8, and fluorescent proteins are used (Kim *et al* 2017), but all of the above introduced photoreceptor families have already been applied for optical control of protein function in a wide variety of cells.

The organic dyes functionalized as light sensors (chromophores) in photoreceptor proteins are primarily natural organic dyes. These dyes are retinals (for rhodopsins), the flavins FMN and FAD (for LOV domains, BLUF proteins, cryptochromes), *p*-coumaric acid (for xanthopsins), linear tetrapyrroles (for phytochromes), cyclic tetrapyrrole vitamin B_{12} (for cobalamin-based photoreceptors), and tryptophan (for UVR8). For optogenetic tool localization various fluorescent proteins are tagged to the genetic encoded sensor and effector domains. This task may also be taken over by fluorescent quantum dots (Rowland *et al* 2015, Huang *et al* 2016). Quantum dots closely located at voltage-gated ion channels may also be used for stimulation of neurons by photo-induced changes of electric dipole moments (Lugo *et al* 2012, Huang *et al* 2016). Upconversion nanoparticles, which absorb near infrared light and emit UV–Vis light, allow the use of NIR light with deep penetration depth into tissue to excite visible light absorbing photoreceptors (Huang *et al* 2016, Wu *et al* 2016, Chen *et al* 2018).

In this chapter the applied organic dyes are introduced and spectroscopic characterized (absorption, emission, and photo-dynamics). Then the action of the organic dyes in photoreceptors of optogenetic tools will be briefly discussed. Some applications in neuroscience and cell biology will be presented.

13.2 Characterization of organic dyes applied in optogenetics

Optogenetic tools are genetically encoded proteins with photoreceptor domain and intrinsic effector function or additional biological effector domain. The photoreceptor domains require a light absorbing chromophore (light-sensitive cofactor) for light take-up which stimulates/triggers the effector domain. In most cases the photoreceptor in the signaling state recovers thermally back to its original receptor state (dark-adapted state) and closes the photocycle from where it might be reactivated by a subsequent photon. In some cases, the photoreceptor in the signaling state absorbs a second photon to convert back to the original dark state or at least accelerates the recovery by many orders of magnitude (for example in phytochromes, many invertebrate rhodopsins, or histidine kinase rhodopsins). In other cases the photoexcitation may cause permanent chemical change of the receptor without recovery in the dark (one-way stimulation/switch, for example in cobalamin-based photoreceptors).

The organic dyes in the various photoreceptors applied in optogenetics are characterized in the following.

13.2.1 Retinals

Retinals are organic dyes covalently bound to opsin trans-membrane proteins in microbes (type-I rhodopsins) and animals (type-II rhodopsins) (Kandori *et al* 2001, Wand *et al* 2013, Ernst *et al* 2014). The binding occurs by lysine Schiff base formation.

Some structural formulae of retinals are shown in figure 13.1. They consist of a β-ionone ring and a polyene chain. The π double-bond structure of the polyene part fixes the spatial almost planar shape. The arrows indicate deviation from planarity. The orientation of the double bonds determines the spatial isomer structure. In the top part of figure 13.1 retinal is shown in its all-*trans* form, and in the row below in its 13-*cis* isoform. In the third row 11-*cis*,12-*s-trans* retinal is depicted. In the fourth and fifth row all-*trans*,15-*anti* retinal Schiff base and all-*trans*,15-*syn* retinal Schiff base are shown. In rhodopsins R is the lysine residue that covalently binds retinal to the opsin protein. In rhodopsin models R is *n*-butyl (i.e., retinal Schiff base = *n*-butylamine retinal Schiff base). The sixth and seventh rows show 13-*cis*,15-*anti* retinal Schiff base and 13-*cis*,15-*syn* retinal Schiff base. In the last row the structural formula of protonated all-*trans*,15-*anti* retinal Schiff base is shown. In microbial rhodopsins (type-I rhodopsins) the dark-state retinal conformation generally is protonated all-*trans* retinal Schiff base and the primary product of photoexcitation is protonated 13-*cis* retinal Schiff base. In animal rhodopsins (type-II rhodopsins) the dark-state retinal conformation is protonated 11-*cis*,12-*s-trans* retinal Schiff base and the primary product of photoexcitation is protonated all-*trans* retinal Schiff base (Bassolino *et al* 2015).

In figure 13.2 some retinal absorption cross-section spectra are shown. The solid curve shows the absorption cross-section spectrum of all-*trans* retinal in ethanol (from Hubbard and Wald 1952, Robeson *et al* 1955). Its peak absorption cross-section is $\sigma_a(383 \text{ nm}) = 1.65 \times 10^{-16} \text{ cm}^2$ and its spectral half-width is

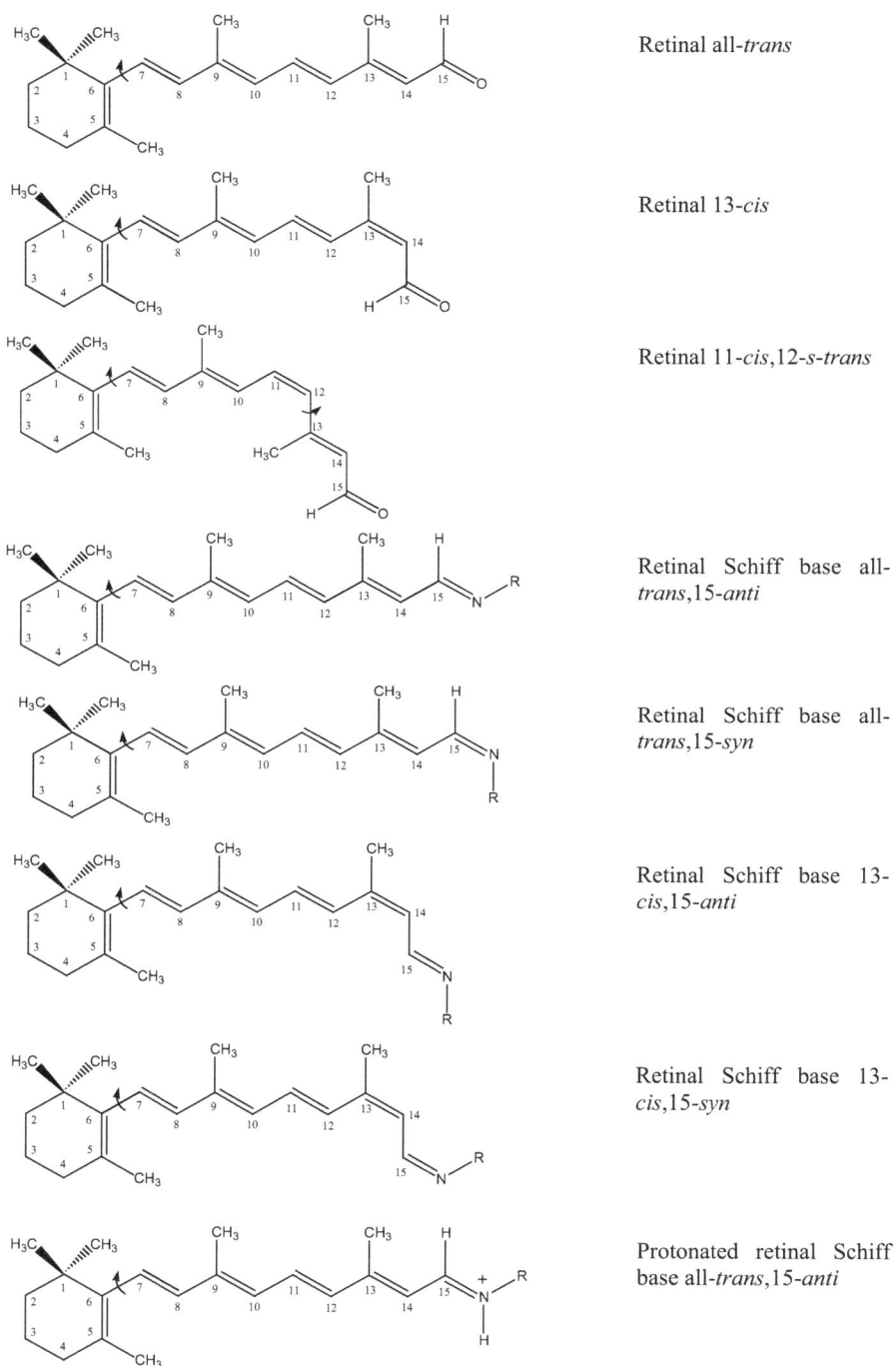

Retinal all-*trans*

Retinal 13-*cis*

Retinal 11-*cis*,12-*s-trans*

Retinal Schiff base all-*trans*,15-*anti*

Retinal Schiff base all-*trans*,15-*syn*

Retinal Schiff base 13-*cis*,15-*anti*

Retinal Schiff base 13-*cis*,15-*syn*

Protonated retinal Schiff base all-*trans*,15-*anti*

Figure 13.1. Structural formulae of some retinals.

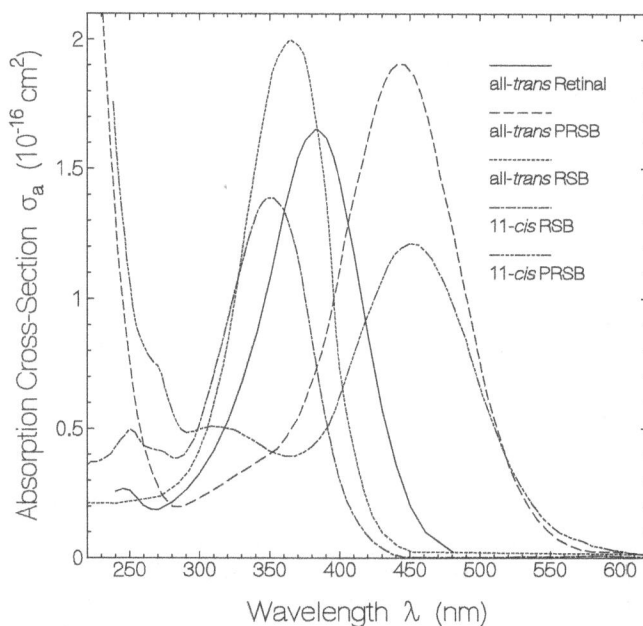

Figure 13.2. Absorption cross-section spectra of all-*trans* retinal in ethanol (solid curve, from Hubbard and Wald 1952 with calibration to Robeson *et al* 1955), unprotonated all-*trans* retinal n-butylamine Schiff base in methanol (dotted curve, from Becker and Freedman 1985), protonated all-*trans* retinal n-butylamine Schiff base in methanol (dashed curve, from Becker and Freedman 1985), unprotonated 11-*cis* retinal n-butylamine Schiff base in acetonitrile (dash-dotted curve, from Becker and Freedman 1985), and protonated 11-*cis* retinal n-butylamine Schiff base in acetonitrile (dashed-triple dotted curve, from Becker and Freedman 1985).

$\Delta \tilde{\nu}_a = 5840$ cm^{-1}. The dotted curve shows the absorption cross-section spectrum of all-*trans* retinal *n*-butylamine Schiff base (RSB) in methanol (from Becker and Freedman 1985). Its peak absorption cross-section is $\sigma_a(364$ nm$) = 2.0 \times 10^{-16}$ cm^2 and its spectral half-width is $\Delta \tilde{\nu}_a = 5470$ cm^{-1}. The dashed curve shows the absorption cross-section spectrum of all-*trans* protonated retinal *n*-butylamine Schiff base (PRSB) in methanol (from Becker and Freedman 1985). Its peak absorption cross-section is $\sigma_a(445$ nm$) = 1.9 \times 10^{-16}$ cm^2 and its spectral half-width is $\Delta \tilde{\nu}_a = 5080$ cm^{-1}. The protonation of retinal Schiff base causes a spectral red-shift of the first absorption band. Protonated all-*trans* retinal Schiff base in rhodopsin is further red shifted compared to solution (opsin shift, see Rajamani *et al* 2010; for example in Bacteriorhodopsin the wavelength position of the first absorption maximum is at 568 nm, see Kikukawa *et al* 2012). All-*trans* PRSB is the receptor state chromophore in microbial rhodopsins (Zhang *et al* 2011, Ernst *et al* 2014).

The dash-dotted curve in figure 13.2 shows the absorption cross-section spectrum of 11-*cis* retinal *n*-butylamine Schiff base in acetonitrile (CH$_3$CN) (from Becker and Freedman 1985). Its peak absorption cross-section is σ_a (350 nm) $= 1.39 \times 10^{-16}$ cm^2 and its spectral half-width is $\Delta \tilde{\nu}_a = 6320$ cm^{-1}. The 11-*cis* RSB likely consists of a 50% to 50% mixture of the 12-*s-cis* and 12-*s-trans* conformers (Rowan *et al* 1974). The dashed-triple dotted curve in figure 13.2 shows the absorption cross-section spectrum of protonated 11-*cis* retinal *n*-butylamine Schiff base in acetonitrile (from Becker and Freedman 1985). Its peak absorption cross-section is σ_a (452 nm) = 1.21 $\times 10^{-16}$ cm^2 and its spectral half-width is $\Delta \tilde{\nu}_a = 5220$ cm^{-1}. The 11-*cis* PRSB likely consists of a 50% to 50% mixture of the conformers 12-*s-cis* and 12-*s-trans* (Rowan *et al* 1974). Again the protonation of retinal Schiff base causes a spectral red-shift of the first absorption band. Protonated 11-*cis* retinal Schiff base in rhodopsin is further red shifted compared to solution (opsin shift, for example in bovine visual rhodopsin the wavelength position of the first absorption maximum is at 498 nm, Kikukawa *et al* 2012). 11-*cis*,12-*s-trans* PRSB is the receptor state chromophore in animal visual rhodopsins (Ernst *et al* 2014, Bassolino *et al* 2015).

Retinal, unprotonated retinal Schiff base, and protonated retinal Schiff base in liquid solution exist in various thermally stable isomeric forms (Rowan *et al* 1974). Photoexcitation causes a partial change of the isomeric composition due to excited-state isomerization (photoisomerization). Studies on the photoisomerization of retinal in various solvents were reported by Kropf and Hubbard (1970), Birge *et al* (1975), Larson *et al* (1997), Takeuchi and Tahara (1997), Yamaguchi and Hamaguchi (1998), and Jen *et al* (2015). Differences of the relaxation dynamics of retinal in unpolar and polar solvents were discussed. The photoisomerization dynamics of unprotonated retinal Schiff base in solutions was studied by Becker *et al* (1982) and Becker and Freedmann (1985). The photoisomerization dynamics of protonated retinal Schiff base in solutions was investigated by Becker *et al* (1982), Becker and Freedmann (1985), Logunov *et al* (1996), Hamm *et al* (1996), Zgrablić *et al* (2005), Bismuth *et al* (2007), Zgrablić *et al* (2012), Sovdat *et al* (2012), Bassolino *et al* (2014, 2015), and Mališ *et al* (2017).

The photoexcitation of a specific ground-state isomer in liquid solution generally leads to the formation of several different isomers (different relaxation paths along the multi-dimensional excited-state potential energy surface). These formed different isomers are generally thermally stable. The speed of isomerization depends on the shape of the excited-state potential energy surface (barrierless relaxation and barrier-involved relaxation). The quantum yield of photoisomerization depends on the branching of excited molecules from the locally excited state to isomerization paths and non-isomerization paths, and it depends on the forward isomerization to product isomers and on the backward isomerization to the original conformation at the internal conversion position of excited state and ground state (Sovdat *et al* 2012, Zgrablić *et al* 2012, Bassolino *et al* 2015). The situation is illustrated in figure 13.3(a) for barrierless photoisomerization, and in figure 13.3(b) for barrier-involved photoisomerization. The schemes include a non-reactive path (non-photo-isomerization path), where the excited retinal molecules deactivate to the original ground-state conformation without excited-state isomerization. The fraction of excited molecules following this non-reactive path is denoted by the quantum efficiency ϕ_{nr}. The fraction of excited molecules following the reactive path is $\phi_r = 1 - \phi_{nr}$. Along the non-reactive path and the reactive path some molecules recover by light emission (fluorescence lifetimes $\tau_{F,nr}$ and $\tau_{F,r}$, path-related fluorescence quantum yields $\phi_{F,nr}$ and $\phi_{F,r}$, total fluorescence quantum yield $\phi_F = \phi_{nr}\phi_{F,nr} + \phi_r\phi_{F,r}$). The excited retinal molecules along the reactive path may branch into different isomerization paths

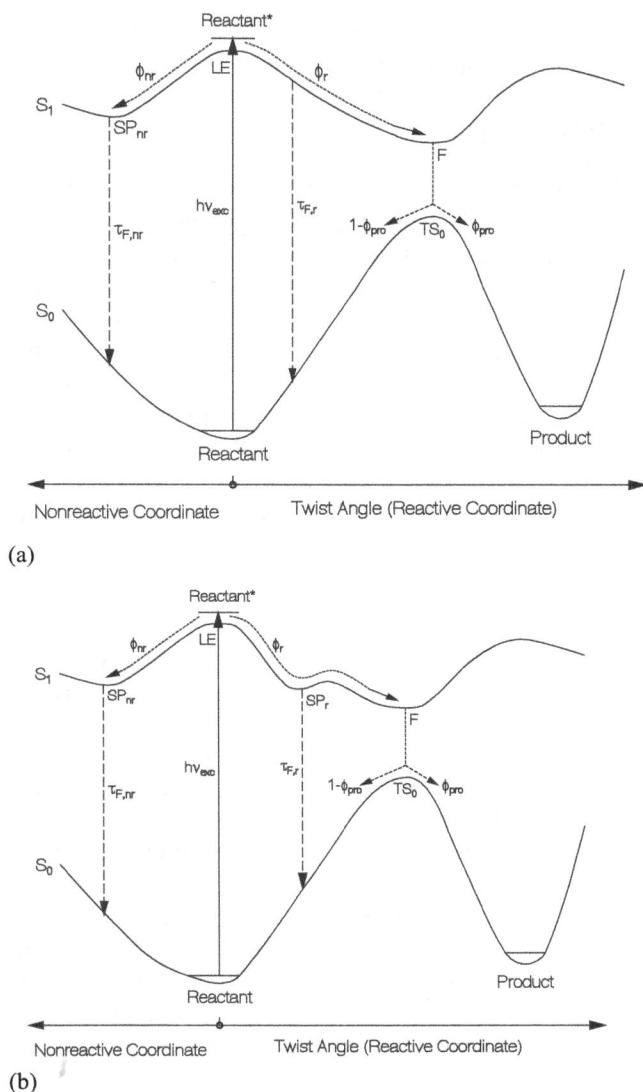

Figure 13.3. Reaction coordinate schemes of barrierless (a) and barrier-involved photoisomerization of retinals. Reactive coordinate indicates path of isomerization. Non-reactive coordinate indicates path of excited state recovery to the ground-state reactant without isomer formation. Reactant is initial *cis* or *trans* isomer. Product is formed *trans* or *cis* isomer caused by photoisomerization. LE: local excited state. F: funnel state of conical intersection. SP_r: stationary point along reactive coordinate. SP_{nr}: stationary point along non-reactive coordinate. TS_0: transition state on S_0 potential energy surface.

with different product isomers (for example photoisomerization of all-*trans* retinal partly to 13-*cis* retinal and partly to 11-*cis* retinal). The schemes of figure 13.3 show only one reactive coordinate (isomerization coordinate) to a product isomer. The reactive coordinate leads to an excited state funnel F (position of conical intersection, position of twisted internal conversion) where S_1 to S_0 internal conversion from funnel F to the transition state TS_0 takes place. At TS_0 a fraction ϕ_{pro} of molecules relaxes forward to the product conformation and a fraction $\phi_{backward} = 1 - \phi_{pro}$ recovers back to the original conformation. The quantum yield of photoisomerization ϕ_{iso} is $\phi_{iso} = \phi_r(1 - \phi_{s,r})\phi_{pro}$.

For retinals (Ret), retinal Schiff bases (RSB) and protonated retinal Schiff bases (PRSB) in liquid solution generally the non-reactive relaxation path dominates (Zgrablić *et al* 2012), while for protonated retinal Schiff base in rhodopsin the non-reactive relaxation path is less dominant (Hontani *et al* 2017) or even absent (Zgrablić *et al* 2012, Bassolino *et al* 2015). Ret, RSB and PRSB isomers in solution photoisomerize to a mixture of product isomers (Freedman and Becker 1986, exception: 11-*cis* PRSB photoisomerizes to all-*trans* PRSB, see Bassolino *et al* 2015), while in rhodopsins PRSB photoisomerizes to an almost pure single PRSB product isomer (Kandori *et al* 2001, Kikukawa *et al* 2012, Wand *et al* 2013, Ernst *et al* 2014: all-*trans* PRSB to 13-*cis* PRSB in microbial rhodopsins, 11-*cis* PRSB to all-*trans* PRSB in animal rhodopsins; 15-*syn* and 15-*anti* isomerization not considered here).

13.2.2 Flavins (isoalloxazine derivatives)

Alloxazines and isoalloxazines are a group of organic compounds based on pteridine, formed by the tricycle of benzene, pyrazine, and pyrimidine. Structural formulae of some alloxazines and isoalloxazines are shown in figure 13.4(a) and (b), respectively. Isoalloxazine is the tautomeric form of alloxazine by transferring a proton from N_1 to N_{10} with accompanying change of a double bond. In figure 13.4(a) the structural formulae of the parent alloxazine ($R_7 = R_8 = H$) and two derivatives, lumichrome ($R_7 = R_8 = CH_3$) and 8-amino-lumichrome ($R_7 = CH_3$, $R_8 = NH_2$), are shown. In figure 13.4(b) the structural formulae of the parent isoalloxazine ($R_7 = R_8 = R_{10} = H$) and of important isoalloxazine derivatives for optogenetics are depicted. The isoalloxazine derivatives are named flavins (the name is derived from common yellow color of isoalloxazine derivatives originating from the Latin word *flavus* for yellow). The most relevant flavin cofactors in optogenetics are FMN (= flavin mononucleotide) and FAD (= flavin adenine dinucleotide).

13.2.2.1 Redox states and ionic states of flavins
The flavins, Fl, exist in three redox states, which are the fully oxidized flavoquinones Fl_{ox}, the half-reduced flavosemiquinones Fl_{sq}, and the fully reduced

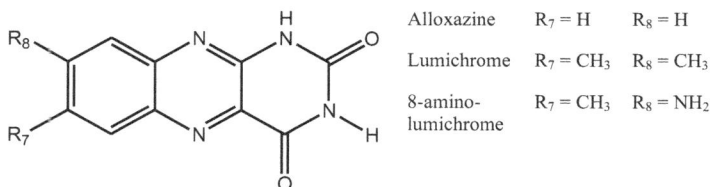

Alloxazine	$R_7 = H$	$R_8 = H$
Lumichrome	$R_7 = CH_3$	$R_8 = CH_3$
8-amino-lumichrome	$R_7 = CH_3$	$R_8 = NH_2$

Figure 13.4a. Structural formulae of some alloxazines

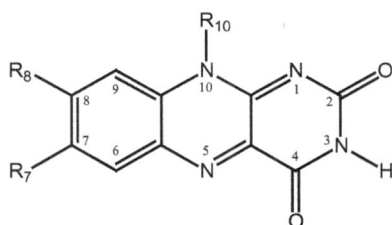

	R_7	R_8	R_{10}
Isoalloxazine	H	H	H
Lumiflavin	CH_3	CH_3	CH_3
Riboflavin	CH_3	CH_3	ribityl
FMN (= flavin mononucleotide)	CH_3	CH_3	ribityl-monophosphate
FAD (= flavin adenine dinucleotide)	CH_3	CH_3	ribityl-adenosine-diphosphate
Roseoflavin	CH_3	$N(CH_3)_2$	ribityl
8-amino-riboflavin	CH_3	NH_2	ribityl

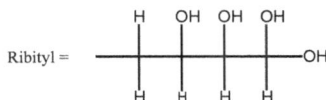

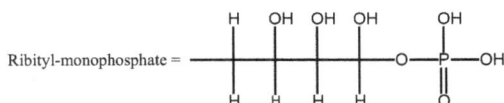

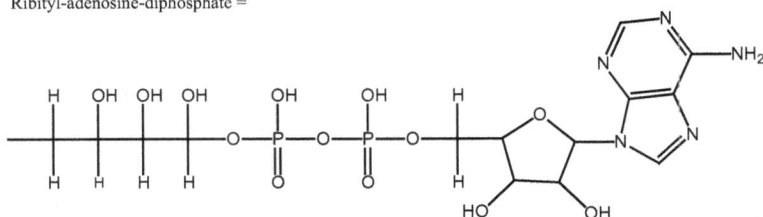

Figure 13.4b. Structural formulae of some isoalloxazines (flavins).

flavohydroquinones Fl_{red}. Depending on the pH of the aqueous solution each of the flavin redox species exist in cationic, neutral, and anionic forms (Müller 1991b, p 26). The structural formulae of the flavins in the different redox states and their different ionic states are displayed in figure 13.4(c). The pK_c indicates the pH value of aqueous solution where the concentration of cationic flavin $[Fl^+]$ is equal to the concentration neutral flavin $[Fl]$. The pK_a indicates the pH value of aqueous

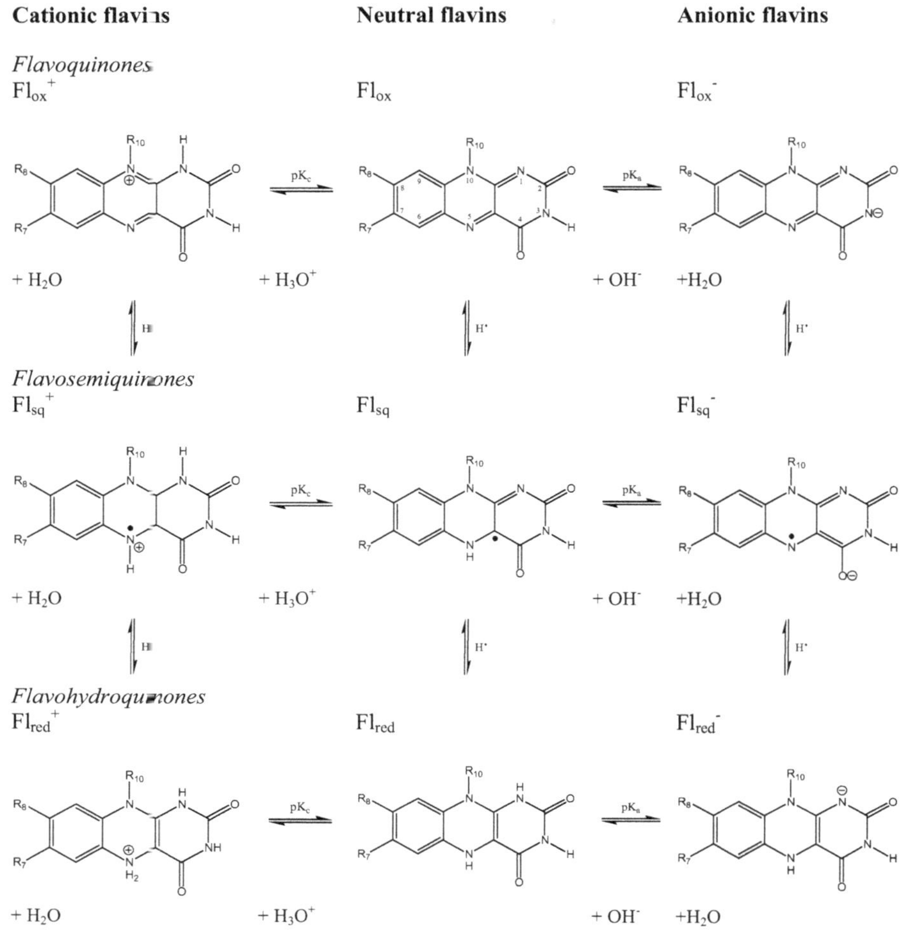

Cationic flavins	Neutral flavins	Anionic flavins

Flavoquinones
Fl_{ox}^{+} Fl_{ox} Fl_{ox}^{-}

$+ H_2O$ pK_c $+ H_3O^+$ pK_a $+ OH^-$ $+H_2O$

Flavosemiquinones
Fl_{sq}^{+} Fl_{sq} Fl_{sq}^{-}

$+ H_2O$ pK_c $+ H_3O^+$ pK_a $+ OH^-$ $+H_2O$

Flavohydroquinones
Fl_{red}^{+} Fl_{red} Fl_{red}^{-}

$+ H_2O$ pK_c $+ H_3O^+$ pK_a $+ OH^-$ $+H_2O$

Figure 13.4c. Structural formulae of neutral, cationic, and anionic forms of the different redox states of flavins (from Müller 1991b, p 26).

solution where the concentration of anionic flavin [Fl⁻] is equal to the concentration neutral flavin [Fl].

In aerated aqueous solutions the flavins are stable in oxidized form (Fl_{ox}^{+}, Fl_{ox}, Fl_{ox}^{-} depending on pH) (Müller 1991b, Stevenson *et al* 1997, Song *et al* 2007b). They may be reduced chemically with dithionite (Mayhew and Massey 1973, Lambeth and Palmer 1973), borohydride (Müller and Massey 1971), hydrogen (Ghisla *et al* 1974, Heizmann *et al* 1973) or natural substrates (Ghisla *et al* 1974, Ghisla 1980). Photoexcitation of oxidized flavins leads to transient reduction under aerobic conditions and to more permanent reduction under anaerobic conditions and the presence of reducing agents (Song *et al* 2007b and references therein). Transient photo-reduction of oxidized flavins in starch films was studied by Penzkofer (2012). The reduction of Fl_{ox} to Fl_{sq} is thought to occur by electron transfer from an external electron donor or a reductive amino acid residue of a flavoprotein, resulting in flavin quinone radical anion formation $Fl_{ox}^{\cdot-}$, and subsequent proton transfer (in total one hydrogen radical $H^{\cdot}$ transfer). Reduction of Fl_{sq} to Fl_{red} is thought to occur by a second electron transfer (flavin semi-quinone radical anion formation $Fl_{sq}^{\cdot-}$) followed by a subsequent proton transfer (in total one hydrogen radical $H^{\cdot}$ transfer, Penzkofer 2012). The structural formulae of flavin quinone radical anion $Fl_{ox}^{\cdot-}$ and flavin semiquinone radical anion $Fl_{sq}^{\cdot-}$ with their mesomeric structures (Fl_{sq}^{-} and Fl_{red}^{-}) are shown in figure 13.4(d).

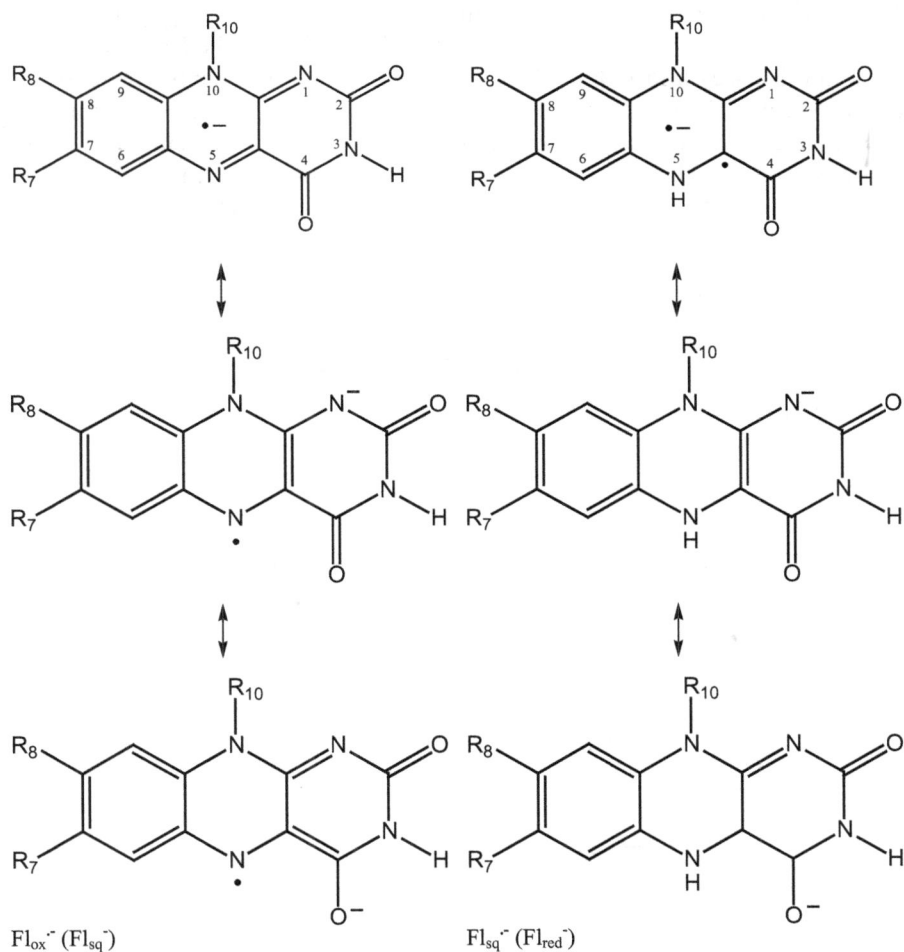

Figure 13.4d. Structural formulae of flavin quinone radical anion $Fl_{ox}^{\cdot-}$ with mesomeric structures (anionic flavin semiquinones Fl_{sq}^{-}), and flavin semiquinone radical anion $Fl_{sq}^{\cdot-}$ with mesomeric structures (anionic flavin hydroquinones Fl_{red}^{-}).

13.2.2.2 Absorption cross-section spectra of some flavins and alloxazines

Absorption cross-section spectra of some flavins are shown in figure 13.5. The top part of figure 13.5 displays the absorption cross-section spectra of fully oxidized riboflavin in neutral, cationic, and anionic form (from Drössler *et al* 2002). The absorption spectra of neutral and anionic riboflavin are similar with first absorption maximum (S_0–S_1 transition) around 445 nm. The cationic riboflavin has its first absorption maximum at about 400 nm. The absorption cross-section spectra of FMN and FAD are similar to those of riboflavin. In the bottom part of figure 13.5 absorption cross-section spectra are shown of fully reduced neutral FAD_{red} (dotted curve, from Müller 1991b, p 44), of anionic FAD_{red}^- (dash-dotted curve, from Müller 1991b, p 44), of neutral FMN_{sq} (solid curve, from Song *et al* 2007b), and of oxidized radical anion $FAD_{ox}^{\cdot-}$ (= anionic semiquinone FAD_{sq}^-, dashed curve, from Shirdel *et al* 2008, see also Berndt *et al* 2007 (there called red anionic FAD radical), and Massey and Palmer 1966 (there called red anionic flavin semiquinone radical)). Anionic flavin radicals $Fl_{ox}^{\cdot-}$ (= Fl_{sq}^-) and neutral flavin semiquinones Fl_{sq} are generally obtained only by photoexcitation of flavoproteins.

In the top part of figure 13.6 absorption cross-section spectra of alloxazine and 7,8-dimethyl-alloxazine (lumichrome, figure 13.4(a)) are shown. Alloxazine at pH 4

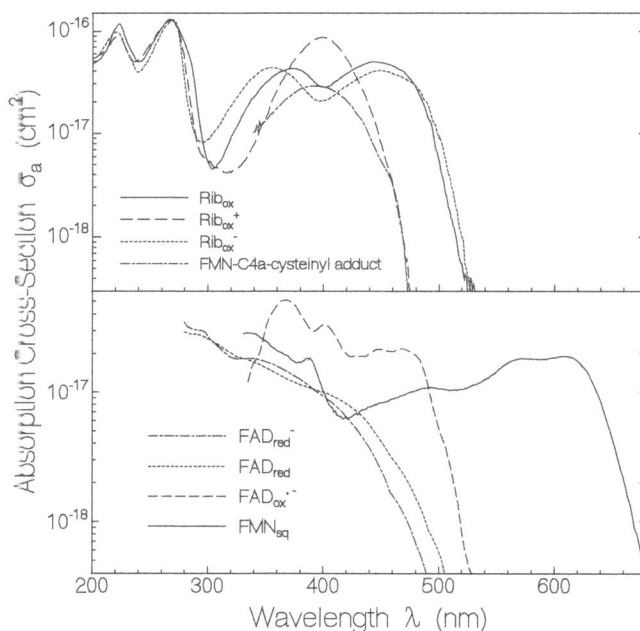

Figure 13.5. Top part: absorption cross-section spectra of neutral, cationic, and anionic fully oxidized riboflavins Rib_{ox}, Rib_{ox}^+, and Rib_{ox}^- (from Drössler *et al* 2002) and absorption cross-section spectrum of FMN-C4a cysteinyl adduct (from Holzer *et al* 2005a). Bottom part: absorption cross-section spectra of neutral and anionic fully reduced flavin adenine dinucleotide FAD_{red}^- and FAD_{red} (from Müller 1991b p 44), anionic flavin adenine dinucleotide radical $FAD_{ox}^{\cdot-}$ (from Shirdel *et al* 2008) and neutral flavin mononucleotide semiquinone FMN_{sq} (from Song *et al* 2007b).

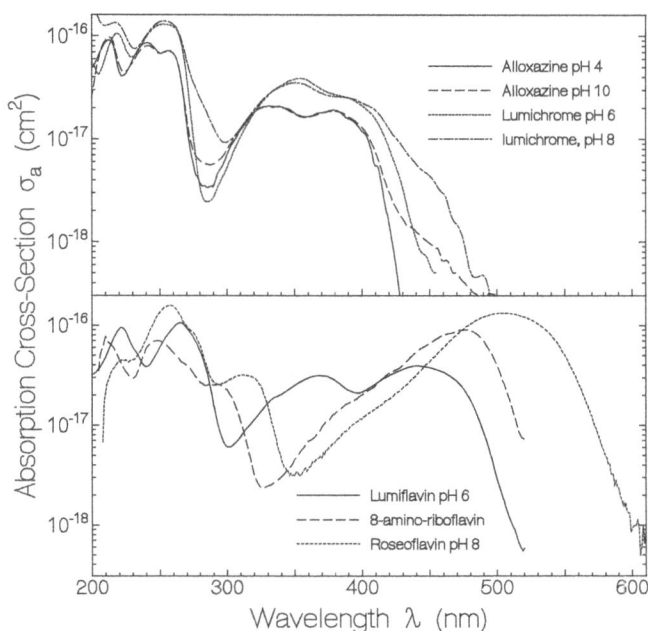

Figure 13.6. Top part: absorption cross-section spectra of aqueous solutions of alloxazine at pH 4 and pH 10 (from Penzkofer 2016), and of lumichrome at pH 6 and pH 8 (from Tyagi and Penzkofer 2011). Bottom part: absorption cross-section spectra of aqueous solutions of lumiflavin at pH 6 (from Tyagi and Penzkofer 2010a), of roseoflavin at pH 8 (from Tyagi *et al* 2010c), and of 8 amino-riboflavin in Millipore water (from Tyagi *et al* 2009c)

(from Penzkofer 2016) and lumichrome at pH 6 (from Tyagi and Penzkofer 2011) show the typical alloxazine-like spectrum with first absorption band around 380 nm. For alloxazine at pH 10 and for lumichrome at pH 8, absorption build-up around 450 nm is seen due to the tautomerization of the alloxazine-like structure to the isoalloxazine-like structure at high pH.

In the bottom part of figure 13.6 absorption cross-section spectra are shown of the isoalloxazine derivatives lumiflavin at pH 6 (from Tyagi and Penzkofer 2010a), of 8-amino-riboflavin in unbuffered neutral Millipore water (from Tyagi *et al* 2009c) and of roseoflavin at pH 8 (Tyagi *et al* 2010c). The change from $R_8 = CH_3$ (lumiflavin) to $R_8 = NH_2$ (8-amino-riboflavin) to $R_8 = N(CH_3)_2$ (roseoflavin) caused a red shift of the first absorption maximum with increasing absorption strength and lowering of the absorption strength of the second absorption band.

13.2.2.3 Photoexcitation dynamics of flavins

13.2.2.3.1 pH dependent photoexcitation dynamics of fully oxidized flavins
The photoexcitation and first excited-state relaxation dynamics of fully oxidized neutral flavin Fl_{ox} ($pK_c < pH < pK_a$), cationic flavin Fl_{ox}^+ ($pH < pK_c$), and anionic flavin Fl_{ox}^- ($pH > pK_a$) is illustrated in figure 13.7 (adapted from Tyagi and

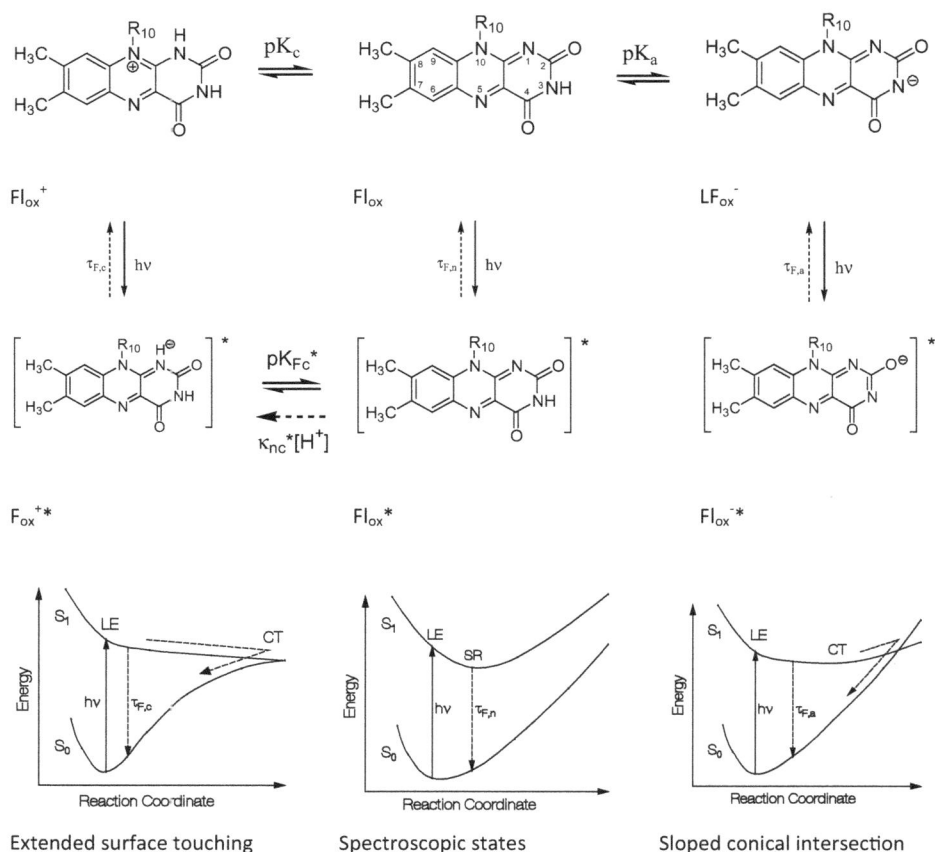

Figure 13.7. Photoexcitation dynamics of fully oxidized flavins (flavoquinones) in cationic, neutral, and anionic states. Top row: structural formulae of Fl_{ox}^+, Fl_{ox}, and Fl_{ox}^- with thermal equilibrium paths (pK_c and pK_a). Middle row: possible structural formulae of the excited flavoquinones Fl_{ox}^{+*}, Fl_{ox}^*, and Fl_{ox}^{-*} with excited-state relaxations ($\tau_{F,c}$, $\tau_{F,n}$, $\tau_{F,a}$) and excited-state intermolecular proton transfer $H_3O^+ + Fl_{ox}^* \rightarrow Fl_{ox}^{+*} + H_2O$ (shift of pK_c to $pK_{Fc}^* > pK_c$). Bottom row: illustrative S_0 and S_1 potential energy curves for Fl_{ox}^+, Fl_{ox}, and Fl_{ox}^- with excited-state relaxation dynamics. Left figure: barrier-less extended surface touching diabatic relaxation; middle figure: spectroscopic states adiabatic relaxation; right figure: barrier-slowed sloped conical intersection diabatic relaxation. LE: locally excited state. SR: vibrational and solvatation relaxed state. CT: charge transfer state. Adapted from Tyagi and Penzkofer 2010a.

Penzkofer 2010a). The pK values given by Müller (1991b, p 26) are $pK_c = 0$ and $pK_a = 10$.

The photo-dynamics of Fl_{ox} is shown by the middle column of figure 13.7. The structural formulae of Fl_{ox} (upper row of figure 13.7) and Fl_{ox}^* (middle row of figure 13.7) are the same. The relaxation behavior of Fl_{ox}^* for pH > pK_{Fc}^* is determined by photo-physical relaxation (fluorescence emission, internal conversion and intersystem crossing). The S_1-state potential energy surface is determined by vibronic relaxation and solvatation due to excited-state dipole moment changes (Simon 1990). One speaks of adiabatic optical electron transfer (Chen *et al* 2006).

A potential energy curve diagram is shown in the middle part of the bottom row of figure 13.7. The potential curves are called spectroscopic states according to Turro *et al* 2009. Franck–Condon excitation occurs from the S_0 ground-state to the locally excited-state position LE of the S_1 potential energy surface. It follows fast relaxation to the S_1-state potential energy surface minimum SR by vibrational relaxation and solvatation relaxation. From there fluorescence emission to the S_0 potential energy surface occurs with fluorescence quantum yield $\phi_{F,n}$ and fluorescence lifetime $\tau_{F,n}$. The relaxation of $Fl_{ox}*$ at low pH (pH $<$ $pK_{Fc}*$, (high H_3O^+ concentration) is enhanced by excited-state intermolecular proton transfer, $Fl_{ox}* + H_3O^+ \rightarrow Fl_{ox}^{+}* + H_2O$, converting $Fl_{ox}*$ to $Fl_{ox}^{+}*$ with the rate constant $\kappa_{nc}*[H^+]$ (see Drössler *et al* 2002 for riboflavin, Tyagi and Penzkofer 2010a for lumiflavin, and Tyagi and Penzkofer 2011 for lumichrome).

The photo-dynamics of Fl_{ox}^+ is shown by the left column of figure 13.7. The structural formulae of Fl_{ox}^+ and $Fl_{ox}^{+}*$ are thought to be different due to excited-state intra-molecular charge transfer. The fluorescence emission of $Fl_{ox}^{+}*$ is quenched (very small fluorescence quantum yield $\phi_{F,c}$, very short fluorescence lifetime $\tau_{F,c}$) by barrier-less intra-molecular charge transfer between locally excited state LE and charge transfer state CT (diabatic electron transfer according to Chen *et al* 2006). A potential energy curve diagram is shown at the lower left part of figure 13.7. A S_1–S_0 extended surface touching situation is sketched according to Turro *et al* (2009), but barrier-less conical intersection cannot be excluded.

The photo-dynamics of Fl_{ox}^- is shown by the right column of figure 13.7. The structural formulae of Fl_{ox}^- and Fl_{ox}^-* are thought to be different due to excited-state intra-molecular charge transfer. The fluorescence emission of Fl_{ox}^-* is quenched (small fluorescence quantum yield $\phi_{F,a}$, short fluorescence lifetime $\tau_{F,a}$) by barrier-slowed-down intra-molecular charge transfer between locally excited state LE and charge transfer state CT (diabatic electron transfer according to Chen *et al* 2006). A potential energy curve diagram is shown at the lower right part of figure 13.7 where an S_1–S_0 sloped conical intersection situation is sketched according to Turro *et al* (2009).

13.2.2.3.2 *Fluorescence and phosphorescence dynamics of flavins*

The fluorescence and phosphorescence dynamics of flavins Fl (Fl_{ox}, Fl_{sq}, Fl_{red}) is illustrated in figure 13.8. Fluorescence quenching by quenchers Q_s (e.g. Förster-type energy (excitation) transfer to acceptors (Förster 1951), Dexter-type energy (excitation) transfer to acceptors (Dexter 1953), photo-induced reductive or oxidative electron transfer and charge recombination (Hopfield 1974, Jortner 1976, Marcus 1993, Kuznetsov and Ulstrup 1999, Balzani 2001)), and phosphorescence quenching by quenchers Q_T (e.g. oxygen quenchers, see Parker 1968, McGlynn *et al* 1969, Birks 1970, Turro *et al* 2009) are schematically included. The excitation light (frequency ν_{exc}) excites molecules from the ground-state to a Franck–Condon position in the S_1 state with some excess energy. From there the molecules relax to a thermal equilibrium position (potential energy minimum) in the S_1 state by vibronic relaxation (VR). From the S_1 state thermal equilibrium position there occurs (i) S_1–S_0 fluorescence emission (quantum yield ϕ_F), (ii) equi-potential S_1–S_0 internal

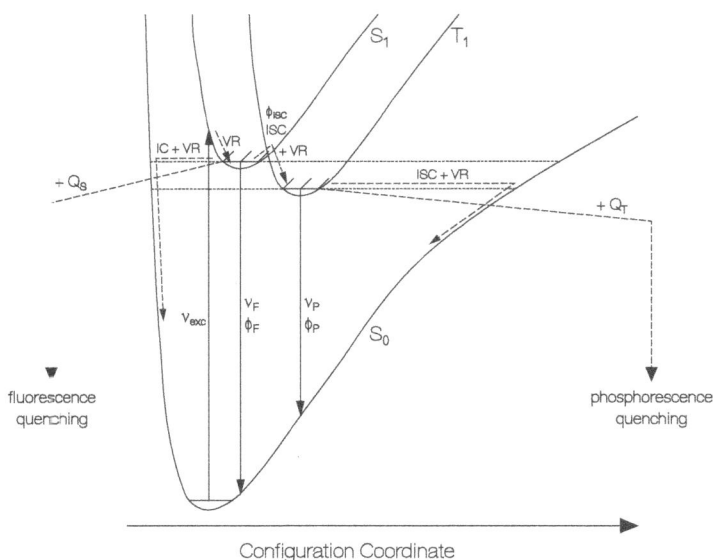

Figure 13.8. Fluorescence and phosphorescence of flavins illustrated in a singlet and triplet potential energy curve diagram. In the diagram is included fluorescence quenching by fluorescence quenchers Q_S (for example causing excited-state electron transfer) and phosphorescence quenching by phosphorescence quenchers Q_T (for example T_1 to S_0 non-radiative relaxation by collisional $O_2(^3\Sigma_g^-)$ to $O_2(^1\Delta_g)$ excitation). IC = internal conversion. ISC = single-triplet intersystem crossing. VR = vibrational relaxation. ϕ_F = fluoresence quantum yield. ϕ_P = phosphorescence quantum yield. ϕ_{ISC} = quantum yield of singlet–triplet intersystem-crossing (equal to quantum yield of triplet formation ϕ_T). Adapted from Tsuboi *et al* 2011.

conversion (IC) with subsequent S_0 state vibronic relaxation (VR), (iii) S_1–T_1 intersystem-crossing (ISC, quantum yield ϕ_{ISC}) with subsequent vibronic relaxation, and (iv) possible chemical reaction (fluorescence quenching). The molecules transferred to the T_1 triplet state thermalize there and leave the triplet state by (i) phosphorescence emission (frequency ν_P, quantum yield ϕ_P), (ii) equi-potential T_1–S_0 intersystem-crossing with subsequent vibronic relaxation (VR), and (iii) chemical reaction (phosphorescence quenching).

The fluorescence quantum yields of lumiflavin, riboflavin, and FMN in neutral aqueous solution at room temperature are of the order of 0.2–0.3 (see Drössler *et al* 2002, Tyagi and Penzkofer 2010a, Holzer *et al* 2005b and references therein). For FAD the fluorescence quantum yield in neutral aqueous solution at room temperature is reduced to about 0.033 because of stacked and un-stacked conformation of the isoalloxazine part and the adenosine part of FAD and reductive electron transfer from adenosine to excited isoalloxazine for the stacked FAD conformation (see figure 13.9(b), Islam *et al* 2003b).

Quantum yields of intersystem-crossing (triplet state formation) in aqueous solution at room temperature have been reported to be ϕ_{ISC}(lumiflavin, pH 6) = 0.67 (Grodowski *et al* 1977), ϕ_{ISC}(lumiflavin, pH 7) = 0.43 (Fritz *et al* 1987), ϕ_{ISC}(riboflavin, pH 6) = 0.67 (Grodowski *et al* 1977), ϕ_{ISC}(riboflavin, pH 6.1) = 0.51 (Fife and Moore 1979), ϕ_{ISC}(riboflavin, pH 7) = 0.375±0.05 (Islam *et al* 2003a),

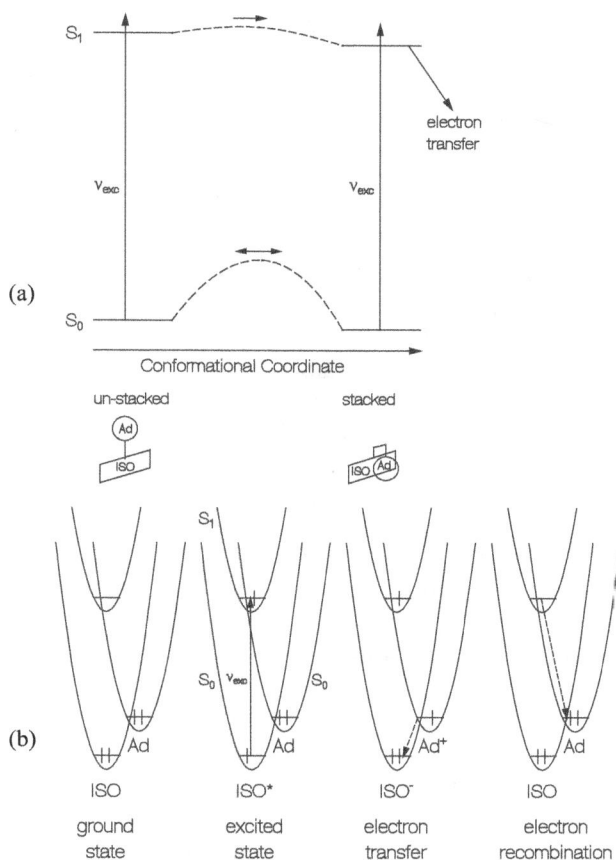

Figure 13.9. Isomerization and photo-induced electron-transfer situation of FAD consisting of reductive electron accepting isoalloxazine part (ISO) and electron donating adenosine part (Ad). (a): Ground-state and first excited-state potential energy diagram of un-stacked and stacked FAD (ISO-Ad) with bi-directional isomerization between un-stacked and stacked form in the ground-state and uni-directional unstacked to stacked transformation in the first excited state. (b): Illustration of reductive photo-induced electron transfer from Ad to ISO in stacked FAD. Adapted from Islam *et al* 2003b.

$\phi_{ISC}(\text{FMN}) \geqslant 0.50$ (Losi *et al* 2002). For FAD in neutral aqueous solution the quantum yield of triplet formation is expected to be low because only about $\beta_{ust} = 0.2$ of the molecules are in the un-stacked conformation and the fluorescence lifetime of the un-stacked conformation of 3 ns is a factor of $\kappa_{Fl} = 0.6$ of the fluorescence lifetime of riboflavin (Islam *et al* 2003b). We estimate $\phi_{ISC}(\text{FAD, pH 7}) \approx \beta_{ust} \times \kappa_{Fl} \times \phi_{ISC}(\text{riboflavin, pH 7}) \approx 0.045$.

The quantum yield of phosphorescence (ϕ_P = number of emitted photons from triplet state/number of singlet excited molecules) of flavins in aerobic aqueous solution is very low ($\phi_P < 10^{-4}$, see Penzkofer 2012) because of the long radiative lifetime of the triplet state and the efficient triplet state quenching by molecular diffusion allowing collisional bimolecular de-excitation (Penzkofer *et al* 2012) including triplet–triplet annihilation (P-type delayed fluorescence, see Parker 1968, McGlynn *et al* 1969), triplet–singlet thermal activated electron transfer (E-type delayed fluorescence, see Turro *et al* 2009, Valeur 2002), and impurity quenching (mainly by dissolved molecular oxygen, Schweitzer and Schmidt 2003, Redmond and Gamlin 1999). For lumiflavin in starch film $\phi_{ISC} = 0.415 \pm 0.03$ and $\phi_P = 0.0094$, and for riboflavin in starch film $\phi_{ISC} = 0.30 \pm 0.03$ and $\phi_P = 0.0025$ were measured (Penzkofer 2012).

13.2.2.3.3 Isomerization and photo-induced electron transfer in FAD

In FAD the photo-dynamics includes, besides the fluorescence and phosphorescence dynamics of figure 13.8, the conformational structuring of un-stacked arrangement and stacked arrangement of the isoalloxazine part and the adenosine part of the molecule, and the photo-induced reductive electron transfer dynamics between the adenosine and isoalloxazine part. The FAD isomerization and electron transfer situation is illustrated in figure 13.9.

Figure 13.9(a) shows the conformational structuring situation (Islam *et al* 2003b). The lower part sketches the un-stacked (open) isoalloxazine-adenosine ISO-Ad conformation (left side) and the stacked (closed) ISO-Ad conformation (right side). In the S_0 ground-state there exists a thermal equilibrium between the un-stacked and stacked conformation. In the pH range from 4 to 8 about 80% of the molecules are in the stacked conformation and about 20% are in the un-stacked conformation. In the S_1 excited state there occurs a unidirectional transfer on the excited-state potential energy surface from the un-stacked to the stacked conformation and subsequent S_1-state deactivation by intra-molecular electron transfer. Figure 13.9(b) explains the reductive photo-induced electron transfer situation (Valeur 2002) for stacked FAD (van den Berg *et al* 2002, Stanley and MacFarlane 2000, Islam *et al* 2003b). The left part in figure 13.9(b) shows the S_0 ground-state level population of the isoalloxazine part ISO and the adenosine part Ad of FAD (ISO-Ad). The next potential energy scheme shows the energy level population after photoexcitation of the isoalloxazine part (ISO*-Ad). The empty electron state in the HOMO level of ISO* opens the channel for reductive electron transfer from the HOMO level of Ad to the HOMO level of ISO* resulting in the zwitter-ionic FAD$^{\pm}$ (ISO$^-$-Ad$^+$, third potential energy scheme). The zwitter-ionic molecule ISO$^-$-Ad$^+$ recovers non-radiatively to the ground-state electron level population by electron recombination from the LUMO position of ISO$^-$ to the HOMO position of Ad$^+$ (right part of figure 13.9(b)).

13.2.2.3.4 Photo-induced intramolecular charge transfer in roseoflavin

The photoexcitation dynamics of roseoflavin (figure 13.10) is thought to be determined by fast excited-state intra-molecular charge transfer from the dimethy-lamino electron donor group to the pteridin carbonyl electron acceptor group (zwitter-ion formation) followed by intra-molecular charge recombination to the original neutral ground-state conformation (Zirak *et al* 2009, Merz *et al* 2011). The dynamics in terms of structural formulae is illustrated in figure 13.10(a): S_0–S_1 photoexcitation changes the ground-state S_0 conformation to the locally excited state LE S_I conformation. From there occurs either planar intra-molecular charge transfer PICT to a zwitter-ionic charge-separated quinoid form (Shinkai *et al* 1985) or twisted intra-molecular charge transfer TICT to a zwitter-ionic intra-molecular charge-transfer state (Grabowski *et al* 2003). The zwitter-ionic conformations relax to the initial ground state by charge recombination. A fast fluorescence component (frequency $\nu_{F,LE}$) is observed due to direct locally-excited-state emission, and a slow

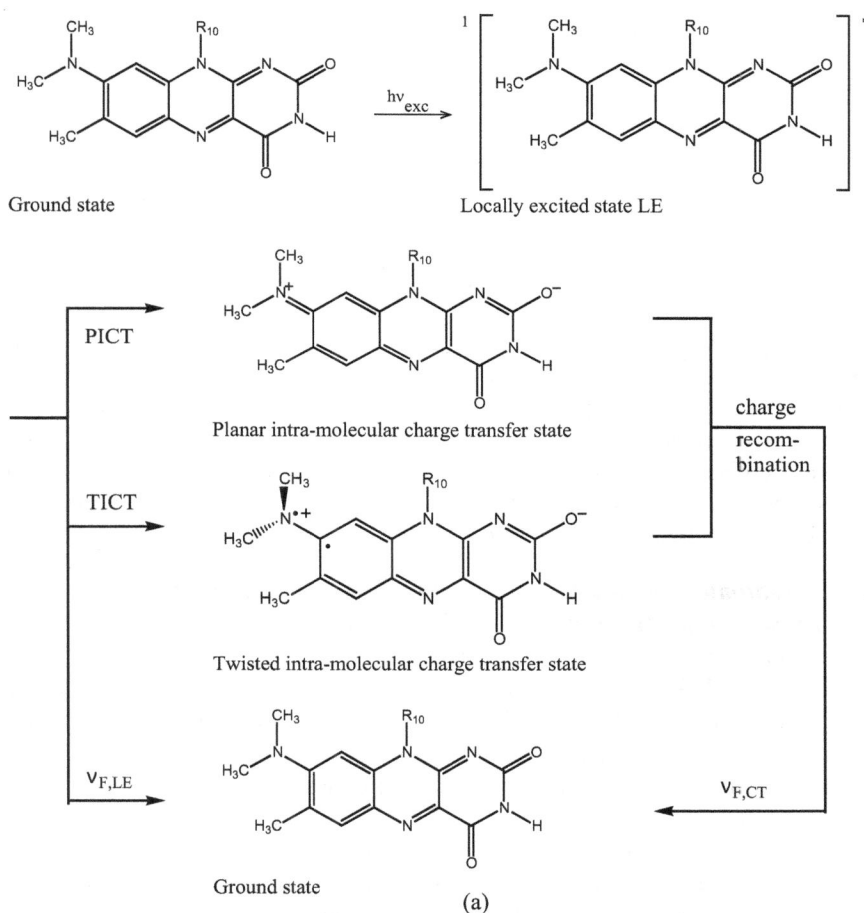

Figure 13.10. Photoexcitation dynamics of roseoflavin (R_{10} = ribityl). (a) Reaction scheme in terms of structural formulae of ground-state, locally excited state LE, formation of planar intra-molecular charge transfer state (PICT), formation of twisted intra-molecular charge transfer state (TICT), and recovery to the initial ground from the locally excited state (fluorescence frequency $\nu_{F,LE}$) and from the charge transfer states (charge recombination and fluorescence emission $\nu_{F,CT}$). (b) Reaction coordinate scheme of S_0 - S_1 photo-excitation to locally excited state LE, from there radiative ($\nu_{F,LE}$) and non-radiative relaxation to the ground state, as well as intra-molecular charge-transfer (ICT) to the PICT and/or TICT charge transfer states (CT), followed by charge recombination (CR) and fluorescence emission relaxation ($\nu_{F,CT}$) to the ground-state. Adapted from Zirak *et al* 2009.

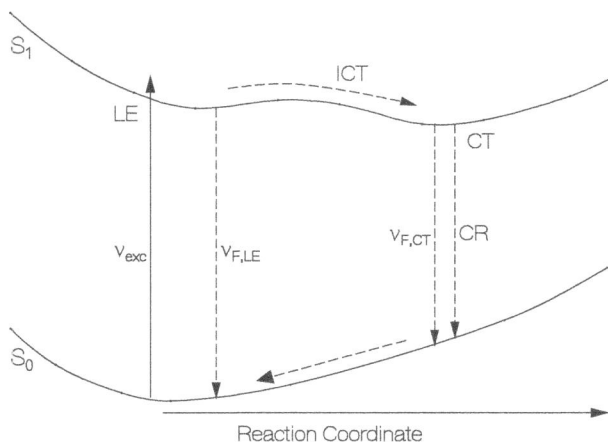

(b)

Figure 13.10. (Continued.)

fluorescence component is observed due to charge transfer state emission (frequency $\nu_{F,CT}$). In figure 13.10(b) the photoexcitation dynamics of roseoflavin is illustrated in a S_0–S_1 reaction coordinate scheme. It involves the photoexcitation (ν_{exc}), the fast fluorescence emission ($\nu_{F,LE}$), the intra-molecular charge transfer ICT from the locally excited state LE to the charge transfer state CT, the charge recombination CR and the slow fluorescence emission ($\nu_{F,CT}$).

13.2.2.3.5 *Tautomerization situation of alloxazines and isoalloxazines*

Alloxazines with R_{10} empty and R_1 = H (figure 13.4(a)) and isoalloxazines with R_{10} = H and R_1 empty (figure 13.4(b)) are tautomers. The tautomerization situation was studied at pH 4 and pH 10 for alloxazine ($R_7 = R_8$ = H) and lumichrome ($R_7 = R_8 = CH_3$, Penzkofer 2016). The pH dependent situation for lumichrome was studied by Tyagi and Penzkofer (2011). In figure 13.11(a) a ground-state and excited-state tautomerization reaction coordinate system is shown (from Penzkofer 2016).

For alloxazine and lumichrome at pH 4 in the ground state, only the alloxazine structure was found to be present in absorption measurements. For alloxazine at pH 10 a fraction of 9% of the molecules changed to the isoalloxazine-structure; and for lumichrome at pH 10 a fraction of 25.5% of the molecules changed to the isoalloxazine structure (Penzkofer 2016).

Photoexcitation of alloxazine at pH 4 and pH 10 caused within the S_1-state lifetime the excited-state tautomerization of about 17% of alloxazine* to isoalloxazine*. For lumichrome both at pH 4 and at pH 10 no photo-tautomerization was observed (Penzkofer 2016).

The pH dependent ground-state tautomerization equilibrium between lumichrome structure (LC) and lumichrome-10H-tautomer structure (LCH-10H) was studied by Tyagi and Penzkofer (2011) (see absorption, fluorescence quantum

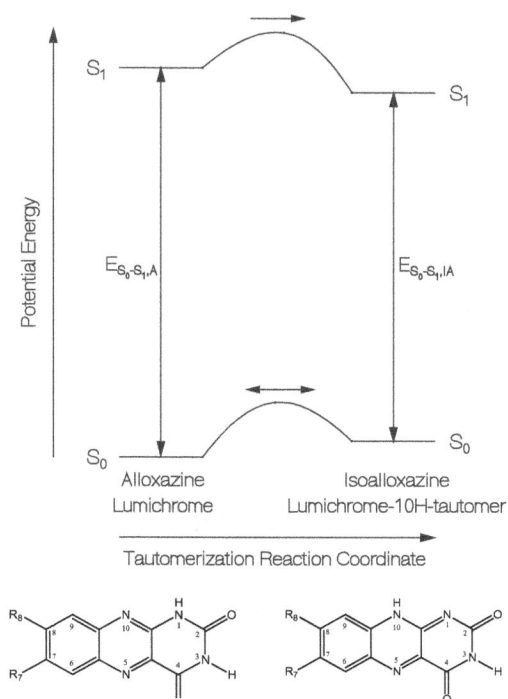

Figure 13.11a. Ground-state S_0 and first excited-state S_1 tautomerization situation for alloxazine ($R_7 = R_8 = H$) and lumichrome ($R_7 = R_8 = CH_3$). In the ground-state exists a pH dependent equililibrium between alloxazine-like N_1H structure and isoalloxazine-like $N_{10}H$ structure. In the first excited-state occurs photo-induced proton transfer from the alloxazine-like structure to the isoalloxazine-like structure for alloxazine. Adapted from Penzkofer 2016.

distribution and fluorescence lifetime curves there). Up to pH 7 only the alloxazine structure of lumichrome (LC) was observed. In the range from pH 7 to pH 9 ground-state tautomerization was observed. The content of LCT-10H increased from 0 to about 25%. In the range from pH 10 to pH 12 no further rise of the LCT-10H content was observed. In the range from pH 12 to pH 14 the alloxazine-like absorption behavior changed over to more and more complete isoalloxazine-like absorption behavior. The complex tautomerization behavior is thought to be caused by the anion formation of lumichrome (LC-1$^-$) and lumichrome-tautomer (LCT-10$^-$) with pK $\approx$ 10, and the dianion formation of lumichrome (LC-1$^-$3$^-$) and lumichrome-tautomer (LCT-10$^-$3$^-$) with pK $\approx$ 12.5 (see Marchena *et al* 2011, Prukala *et al* 2012). For the LC-1$^-$ and LCH-10$^-$ anions the tautomer content seems to be stable at about 25.5%. For the LC-1$^-$3$^-$ and LCH-10$^-$3$^-$ dianions the tautomerization seems to increase with rising pH towards complete conversion of LC-1$^-$3$^-$ to LCH-10$^-$3$^-$. This tautomerization/ionization situation is illustrated in figure 13.11(b).

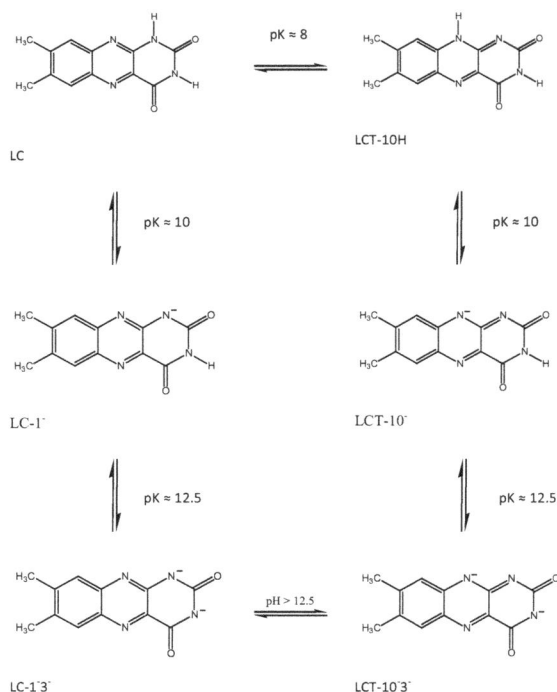

Figure 13.11b. Scheme of pH dependent tautomerization and anion formation of lumichrome (after Tyagi and Penzkofer 2011). LC = lumichrome. LCT lumichrome tautomer.

13.2.2.3.6 Photo-induced ground-state $n \rightarrow \pi$ electron transfer with elongation of radiative lifetime for alloxazines and isoalloxazines

In the alloxazines and isoalloxazines (flavins) the highest occupied molecular n orbital (n-HOMO) and the highest occupied molecular π orbital (π-HOMO) may be energetically close together (Salzmann and Marian 2009 and references therein). Depending on the specific alloxazine and isoalloxazine derivative and the surrounding solvent n-HOMO may be energetically below π-HOMO ($E_{n\text{-HOMO}} < E_{\pi\text{-HOMO}}$) and vice versa ($E_{n\text{-HOMO}} > E_{\pi\text{-HOMO}}$, energy difference: $E_{n\pi} = E_{n-\text{HOMO}} - E_{\pi-\text{HOMO}}$). The broad first alloxazine and first flavin absorption band is determined by the strongly absorbing π-HOMO $-$ π*-LUMO (ππ*) transition (LUMO = lowest unoccupied molecular orbital). The weakly absorbing n-HOMO $-$ π*-LUMO (nπ*) transition is hidden in the broad first absorption band. After ππ* photoexcitation an electron transfer from the full occupied n-HOMO to the half-emptied π-HOMO occurs with a fraction of thermal equilibrated n-HOMO to π-HOMO population transfer ρ_{eT} of

$$\rho_{eT} = 1 - \frac{\exp(-E_{n\pi}/(k_B\vartheta))}{1 + \exp(-E_{n\pi}/(k_B\vartheta))} \tag{13.1}$$

where k_B is the Boltzmann constant and ϑ is the temperature. The radiative rate constant k_{rad} of LUMO to HOMO relaxation becomes

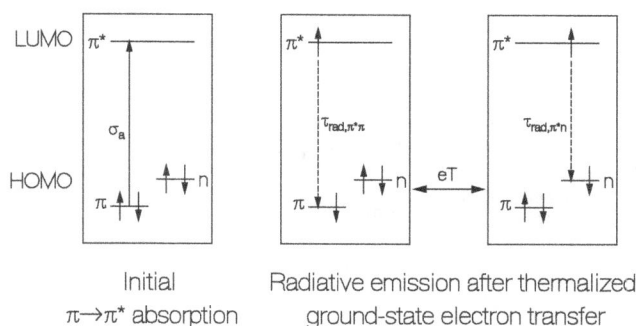

Initial
$\pi \to \pi^*$ absorption

Radiative emission after thermalized
ground-state electron transfer

Figure 13.12. Molecular orbital illustration of photo-dynamics of flavins (alloxazines and isoalloxnazines) with energetically overlapping HOMO-LUMO ($\pi\pi^*$) and ($n\pi^*$) transitions. Left: ground-state molecular orbital populations with indicated $\pi \to \pi^*$ photoexcitation (absorption cross-section σ_a). Right: molecular orbital population after $\pi\pi^*$ photoexcitation and thermalized ground-state $n\pi$ electron transfer between n-HOMO and π-HOMO orbitals. The left box indicates radiative repopulation of π-HOMO from π^*-LUMO (radiative lifetime $\tau_{rad,\pi^*\pi}$). The right box indicates radiative repopulation of n-HOMO from π^*-LUMO (radiative lifetime τ_{rad,π^*n}). eT: ground-state electron transfer from fully occupied n-HOMO to half-occupied π-HOMO.

$$k_{rad} = (1 - \rho_{eT})k_{rad,\,\pi^*\pi} + \rho_{eT}k_{rad,\,\pi^*n} \approx (1 - \rho_{eT})k_{rad,\,\pi^*\pi} \qquad (13.2)$$

and the corresponding radiative lifetime $\tau_{rad} = k_{rad}^{-1}$ becomes

$$\tau_{rad} \approx \frac{\tau_{rad,\,\pi^*\pi}}{1 - \rho_{eT}} \qquad (13.3)$$

This photo-dynamics of flavins and alloxazines with overlapping HOMO-LUMO ($\pi\pi^*$) and ($n\pi^*$) transitions is illustrated in figure 13.12.

In Penzkofer (2016) elongation of the radiative lifetime (equation (13.3)) by photo-induced n-HOMO $\to$ π-HOMO electron transfer was observed for isoalloxazine (alloxazine-10H tautomer), lumichrome, and lumichrome-10H tautomer. It was not observed for alloxazine (energy level of n-HOMO well below energy level of π-HOMO, $\rho_{ET} \to 0$, see equation (13.1)). For riboflavin in neutral aqueous solution no radiative lifetime elongation was observed by Drössler $et\ al$ (2002) indicating that the energy level of n-HOMO is well below the energy level of π-HOMO for Rib_{ox}.

13.2.3 Folates

Folic acid (pteroylglutamic acid, FA) is a widely distributed vitamin (called vitamin B_c, vitamin M, or vitamin B_9) (Rabinowitz 1960, Blakley 1969, Blakley and Benkovic 1984). FA is made up of the building blocks pterin, p-amino-benzoic acid and glutamic acid. Folate is the deprotonated form of folic acid and folic acid derivatives. (6R,S)-5,10-Methenyltetrahydrofolate (MTHF) is an important derivative of folate. It is the light harvesting cofactor in DNA repairing photolyases (energy transfer from excited MTHF to anionic form of flavin adenine dinucleotide hydroquinone FAD_{red}^- (Sancar 2003)). It is present as second cofactor besides FAD_{ox} in cryptochromes (Sancar 2003). Outside photolyases and cryptochromes in

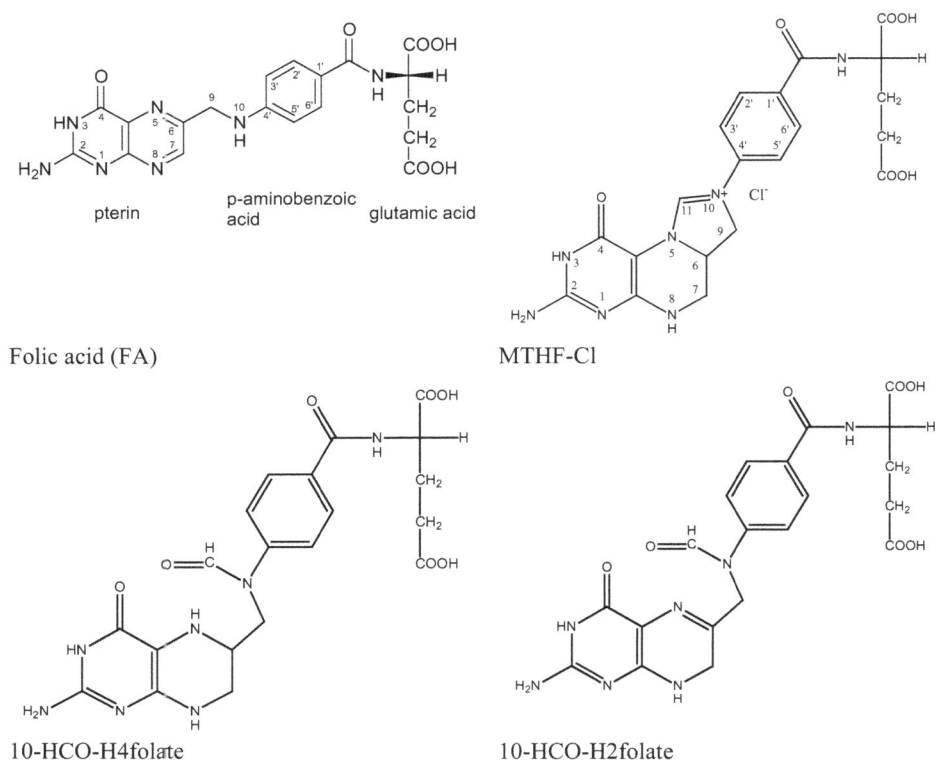

Figure 13.13. Structural formulae of folates. Folic acid (FA) = pteroylglutamic acid. MTHF-Cl = (6R,S)-5,10-methenyltetrahydrofolate chloride. 10-HCO-H4folate = (6R,S)-10-formyltetrahydrofolate. 10-HCO-H2folate = 10-formyldihydrofolate.

aqueous solution at room temperature the thermal stability of MTHF is limited. It hydrolyses to (6R,S)-10-formyltetrahydrofolate (10-HCO-H4folate), which oxidizes to 10-formyldihydrofolate (10-HCO-H2folate) under aerobic conditions (Tyagi *et al* 2009a).

In figure 13.13 the structural formulae of folic acid, MTHF chloride (MTHF-Cl), 10-HCO-H4folate, and 10-HCO-H2folate are shown. For folic acid the pterin, *p*-aminobenzoic acid, and glutamic acid parts are indicated.

In figure 13.14 the absorption cross-section spectra of folic acid at pH 6 (Tyagi and Penzkofer 2010b), and of MTHF-Cl, 10-HCO-H4folate, and 10-HCO-H2folate at pH 8 are shown (Tyagi *et al* 2009a, 2009b). The long-wavelength absorption (>300 nm) is caused by the pterin part. The pterin part and the *p*-aminobenzoic acid part contribute to the absorption in the range from 250 to 300 nm.

Fluorescence quantum yields of the folates at room temperature were determined to be ϕ_F(FA, pH 6, $\lambda_{F,exc}$ = 350 nm) $\approx$ 5 × 10^{-4} (Tyagi and Penzkofer 2010b), ϕ_F(MTHF, pH 8) = 9 × 10^{-4}, ϕ_F(10-HCO-H4folate, pH 8) $\approx$ 1.1 × 10^{-4}, and ϕ_F(10-HCO-H2folate, pH 8) = 4.5 × 10^{-3} (Tyagi *et al* 2009b). In all four folates the low fluorescence quantum yields are thought to be due to fluorescence quenching by

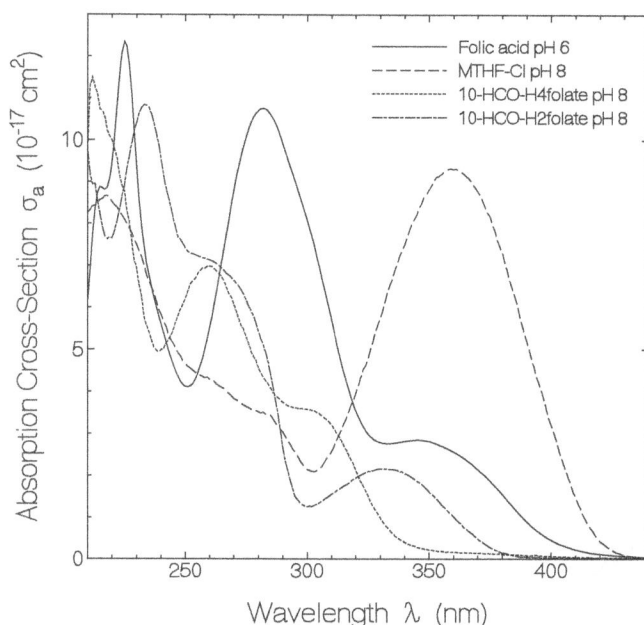

Figure 13.14. Absorption cross-section spectra of some folates in aqueous solutions. Solid curve: folic acid at pH 6 (from Tyagi and Penzkofer 2010b). Dashed curve: MTHF-Cl at pH 8 (from Tyagi *et al* 2009b). Dotted curve: 10-HCO-H4folate at pH 8 (from Tyagi *et al* 2009a and 2009b). Dash-dotted curve: 10-HCO-H2folate at pH 8 (from Tyagi *et al* 2009a and 2009b).

thermal activated photo-induced reductive intra-molecular electron transfer (Kavarnos 1993, Valeur 2002, Tyagi *et al* 2009b, Tyagi and Penzkofer 2010b) from the HOMO level of the benzoyl-glutamate subunit to the single occupied HOMO (SOMO) level of the pterin subunit via a ground-state potential energy barrier with subsequent charge recombination (situation similar to the stacked FAD of figure 13.9).

13.2.4 *p*-coumaric acid

p-coumaric acid (= *p*-hydroxycinnamic acid = 4-hydroxycinnamic acid) is the cofactor of the xanthopsin blue-light photoreceptor protein family, the photoactive yellow proteins (Kort *et al* 1996, Hellingwerf *et al* 2003, Van der Horst *et al* 2005a, 2005b).

In figure 13.15 structural formulae are shown of *trans-p*-coumaric acid (*p*-CAH$_2$), double anionic *cis-p*-coumaric acid (*cis-p*-CA^{2-}), single anionic *trans-p*-coumaric acid (*p*-CAH$^-$), double anionic *trans-p*-coumaric acid (*p*-CA^{2-}), anionic thiomethyl *trans-p*-coumaric acid (TM-*p*-CA$^-$), anionic thiophenyl *trans-p*-coumaric acid (TP-*p*-CA$^-$), anionic *p*-coumaric acid covalently bound to cysteine residue in photoactive yellow protein (PYP) in 7-*trans*-9,*s-cis* conformation (7-*trans*,9-*s-cis* Cys-*p*-CA$^-$, and anionic *p*-coumaric acid covalently bound to cysteine residue in photoactive yellow protein (PYP) in *7-trans,9-s-cis* conformation (7-*cis*,9-*s-trans* Cys-*p*-CA$^-$).

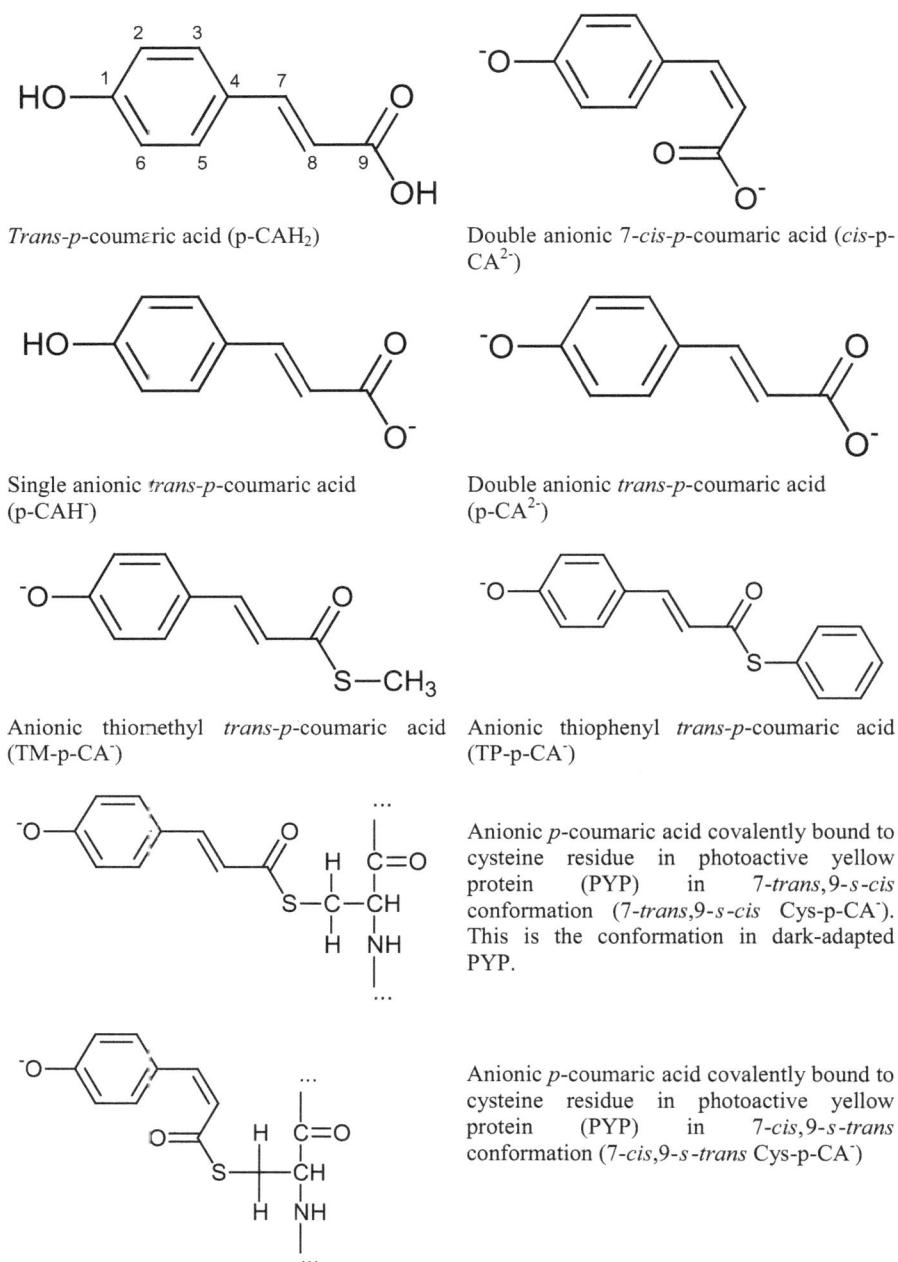

Trans-p-coumaric acid (p-CAH₂)

Double anionic 7-cis-p-coumaric acid (cis-p-CA²⁻)

Single anionic trans-p-coumaric acid (p-CAH⁻)

Double anionic trans-p-coumaric acid (p-CA²⁻)

Anionic thiomethyl trans-p-coumaric acid (TM-p-CA⁻)

Anionic thiophenyl trans-p-coumaric acid (TP-p-CA⁻)

Anionic p-coumaric acid covalently bound to cysteine residue in photoactive yellow protein (PYP) in 7-trans,9-s-cis conformation (7-trans,9-s-cis Cys-p-CA⁻). This is the conformation in dark-adapted PYP.

Anionic p-coumaric acid covalently bound to cysteine residue in photoactive yellow protein (PYP) in 7-cis,9-s-trans conformation (7-cis,9-s-trans Cys-p-CA⁻)

Figure 13.15. Structural formula of some p-coumaric acids

In figure 13.16 absorption cross-section spectra are shown for the molecules depicted in figure 13.15 except that of the photoisomer 7-cis,9-s-trans Cys-p-CA⁻ of PYP (Putschögl et al 2008, Changenet-Baret et al 2001, 2002, Larsen et al 2003,

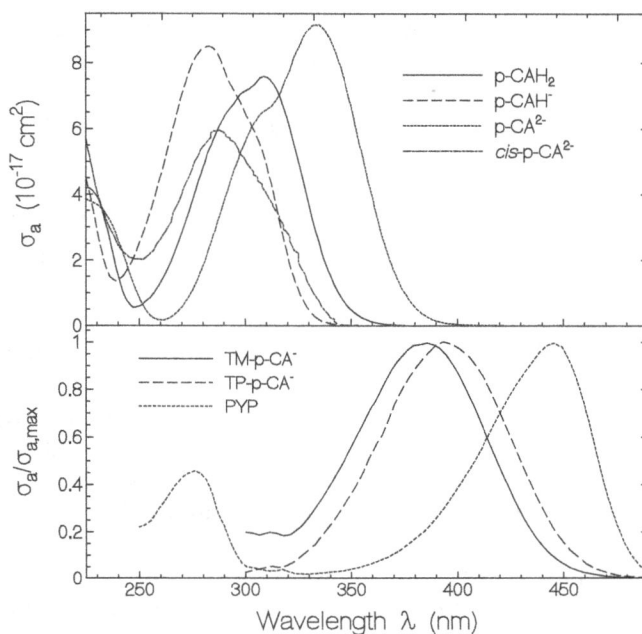

Figure 13.16. Absorption cross-section spectra of *p*-coumaric acids. Top part: absolute absorption cross-section spectra. Bottom part: normalized absorption cross-section spectra. *p*-CAH$_2$ = *trans-p*-coumaric acid (from Putschögl *et al* 2008); *p*-CAH$^-$ = single anionic *trans-p*-coumaric acid (from Putschögl *et al* 2008); *p*-CA^{2-} = double anionic *trans-p*-coumaric acid (from Putschögl *et al* 2008); *cis-p*-CA^{2-} = double anionic *cis-p*-coumaric acid (from Changenet-Baret *et al* 2001); TM-*p*-CA$^-$ = anionic thiomethyl *trans-p*-coumaric acid (from Larsen *et al* 2003); TP-*p*-CA$^-$ = anionic thiophenyl *trans-p*-coumaric acid (from Changenet-Baret *et al* 2002). PYP = photoactive yellow protein from *Halorhodospira halophila* (from Imamoto and Kataoka 2007).

Imamoto and Kataoka 2007). In the top part absolute absorption cross-section spectra are shown, and in the bottom part normalized absorption cross-section spectra are shown ($\sigma_{a,max}$(TP-*p*-CA$^-$) = 1.1 × 10^{-16} cm^2, see Changenet-Baret *et al* 2002, $\sigma_{a,max}$(7-*trans*,9-*s-cis* Cys-*p*-CA$^-$ of PYP from *Halorhodospira halophila*) = 1.7 × 10^{-16} cm^2, see Hellingwerf *et al* 2003). The wavelength positions of the first absorption maxima $\lambda_{a,max}$ are at 310 nm for *p*-CAH$_2$ (S$_0$–S$_1$), 283 nm for *p*-CAH$^-$ (S$_0$–S$_2$), 334 nm for *p*-CA^{2-} (S$_0$–S$_1$), 287 nm for *cis-p*-CA^{2-} (S$_0$–S$_2$), 386 nm for TM-*p*-CA$^-$ (S$_0$–S$_1$), 394 nm for TP-*p*-CA$^-$ (S$_0$–S$_1$), and 446 for 7-*trans*,9-*s-cis* Cys-*p*-CA$^-$ of PYP (S$_0$–S$_1$). The absorption position depends on the *cis* or *trans* structure, the ionization stage, the thio-ester binding, the solvent (here aqueous solution), and the protein embedding in the case of *p*-CA$^-$ covalently bound to a Cys residue in PYP. The molecule ionic state and the ion-solvent interaction change the ground-state and excited-state potential energy surfaces and thereby cause spectral shifts of the absorption spectra. The large red-shift of the S$_0$-S$_1$ absorption band of 7-*trans*, 9-*s-cis p*-CA$^-$ in PYP is caused by the covalent binding to Cys, the counter ion effect (Arg52 in *H. halophila*), the hydrogen-bonding effect (Tyr42, Glu46, and Thr50 in *H. halophila*), and the medium effect of the protein matrix (Yodo *et al* 2001).

Fluorescence quantum yields of *trans-p*-coumaric acids in aqueous solutions at room temperature were determined to be $\phi_F(p\text{-CAH}_2) = \phi_F(p\text{-CA}, \text{pH } 1, \lambda_{F,exc} = 311 \text{ nm}) \approx 1.4 \times 10^{-4}$ (Putschögl *et al* 2008), $\phi_F(p\text{-CAH}^-) = \phi_F(p\text{-CA}, \text{pH } 7, \lambda_{F,exc} = 311 \text{ nm}) \approx 1.4 \times 10^{-4}$ (Putschögl *et al* 2008), and $\phi_F(p\text{-CA}^{2-}) = \phi_F(p\text{-CA}, \text{pH } 12, \lambda_{F,exc} = 311 \text{ nm}) \approx 1.31 \times 10^{-3}$ (Putschögl *et al* 2008).

The photo-dynamics of *p*-coumaric acids was studied by Changenet-Baret *et al* 2001, 2002, Larsen *et al* 2003, Espagne *et al* 2006, 2007a, 2007b, Putschögl *et al* 2008, Rocha-Rinza *et al* 2010. If the electron donor–acceptor character of the end groups is weak ($p\text{-CA}^{2-}$, $\text{pH} > \text{pK}_{aa} = 9.35$) then *trans-cis* photoisomerization is observed. If it is strong ($p\text{-CAH}_2$, $p\text{-CAH}^-$, TP-p-CA$^-$) then photo-induced intra-molecular charge transfer occurs with ultrafast charge recombination before the occurrence of photoisomerization. The photo-dynamics is illustrated in figure 13.17 for $p\text{-CAH}_2$ (pH $< \text{pK}_{na} = 4.9$, photo-induced intra-molecular charge transfer to zwitter-ionic planar quinoid $p\text{-CAH}_{2, PICT}$ or zwitter-ionic twisted radical $p\text{-CAH}_{2, TICT}$ with charge recombination time $\tau_{CR} \approx \tau_F \approx 1.2$ ps, Putschögl *et al* 2008), $p\text{-CAH}^-$ (pK$_{na} = 4.9 < \text{pH} < \text{pK}_{aa} = 9.35$, photo-induced intra-molecular charge transfer to planar quinoid $p\text{-CAH}^-_{quinoid}$ with charge recombination time $\tau_{CR} \approx \tau_F \approx 1.6$ ps, Putschögl *et al* 2008), and $p\text{-CA}^{2-}$ (pH $> \text{pK}_{aa} = 9.35$, photo-induced acivation-barrier-slowed-down *trans-cis* isomerization to thermally stable *cis-p*-CA^{2-} with fluorescence lifetime $\tau_F \approx 9$ ps, Putschögl *et al* 2008).

13.2.5 Linear tetrapyrroles (phytochrome chromophores)

Tetrapyrroles are chemical compounds that contain four pyrrole or pyrrole-like rings. The pyrroles or pyrrole derivatives are linked by = CH– or –CH$_2$– units in either linear or cyclic fashion. Cyclic tetrapyrroles (porphyrines) form the basic structure of chlorophylls (Mg central metal) and hemes (Fe central metal). Several linear tetrapyrroles are products of the heme catabolism. Some linear tetrapyrroles are the chromophores of the phytochrome photoreceptors and the cyanobacterio-chrome photoreceptors (Gärtner and Braslavsky 2003, Tu and Lagarias 2005, Rockwell *et al* 2006, 2014, 2015, 2017, Ikeuchi and Ishizuka 2008, Auldridge and Forest 2011, Ziegler and Möglich 2015). They are phytochromobilin in plant and algal phytochromes; phycocyanobilin in cyanobacteria, algal phytochromes and cyanobacteriochromes; biliverdin IXα in bacterial phytochromes and fungal phyto-chromes; and phycoviolobilin in cyanobacteriochromes.

In figure 13.18 the structural formulae of phycocyanobilin (PCB), phytochromo-bilin (PΦB), biliverdin IXα (BV), and phycoviolobilin (PVB) are shown. They are presented as triple *Z,syn* isomers (ZZZsss, Cahn–Ingold–Prelog nomenclature is used where *Z* stands for Zusammen and *E* stands for Entgegen; configuration change Z ↔ E occurs by opening of C = C double bond, isomerization and new formation of double bond; conformation change *syn* ↔ *anti* occurs by rotation around C–C single bond, Dugave 2006).

In figure 13.19 the absorption behavior of phytochrome/cyanobacteriochrome chromophores in solution is shown. The top part shows absorption cross-section spectra of biliverdin dimethyl ester in methanol (wavelength position of maximum

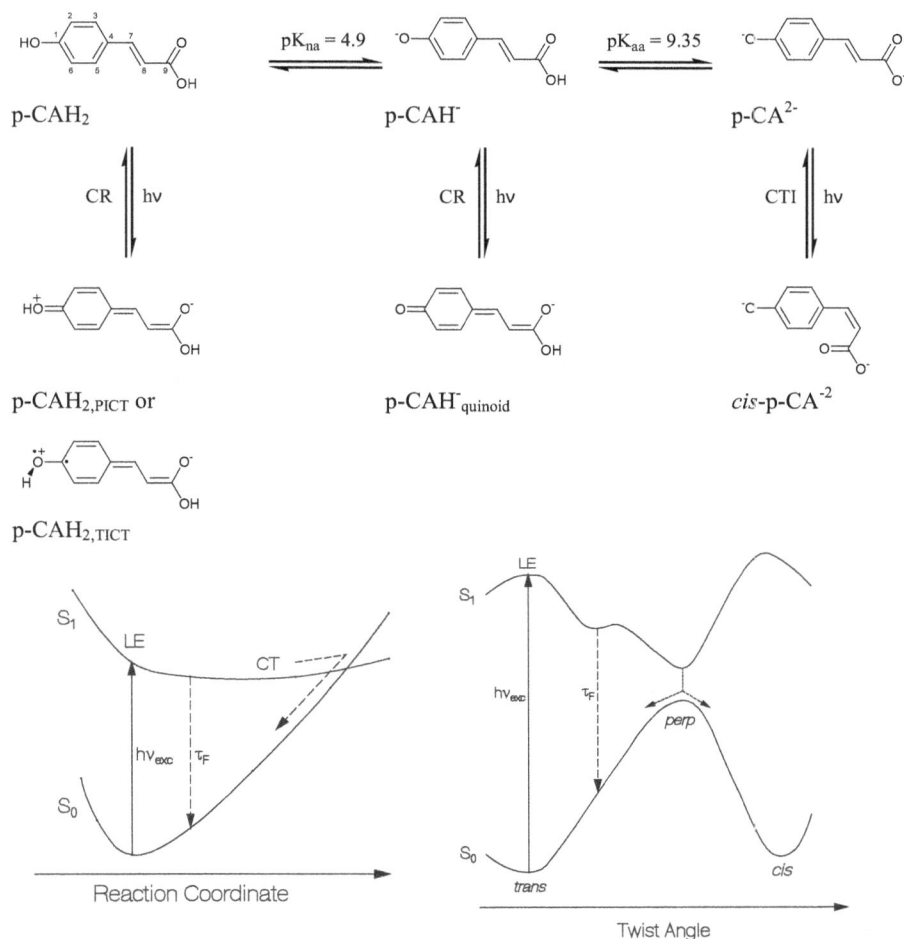

Figure 13.17. Photodynamics of *p*-coumaric acid. Top row: structural formulae of pH dependent neutral, single anionic and double anionic forms of *p*-coumaric acid. Middle row: possible structural formulae after photoexcitation. Bottom row: illustrative potential energy curves for ground state and first excited state with indicated transitions, where left part indicates photo-induced charge transfer (for *p*-CAH$_2$ and *p*-CAH$^-$) and right part indicates *trans-cis* photosomerization (for *p*-CA^{2-}). *p*-CAH$_{2,PICT}$: zwitter-ionic planar quinoid form of *p*-CAH$_2$ (PICT = planar intra-molecular charge transfer). *p*-CAH$_{2,TICT}$: zwitter-ionic twisted radical form of *p*-CAH$_2$ (TICT = twisted intra-molecular charge transfer). CR = charge recombination. CTI = *cis* to *trans* isomerization. LE = locally excited state. CT = intra-molecular charge transfer state.

S$_0$–S$_1$ absorption $\lambda_{a,max} \approx 665$ nm, from Holzwarth *et al* 1978), of phycocyanobilin dimethyl ester in methanol ($\lambda_{a,max} \approx 611$ nm, from Gossauer and Hirsch 1974), and of phytochromobilin dimethyl ester in methanol ($\lambda_{a,max} \approx 626$ nm, from Weller and Gossauer 1980). Biliverin has the longest absorption wavelength position because of the additional double bond between C2 and C3. Phytochromobilin absorbs at longer wavelength than phycocyanobilin because of the CH$_2$ = CH– group at C18 compared to CH$_3$–CH$_2$–. The absorbance of the denatured protein His-TePixJ-GAF from the

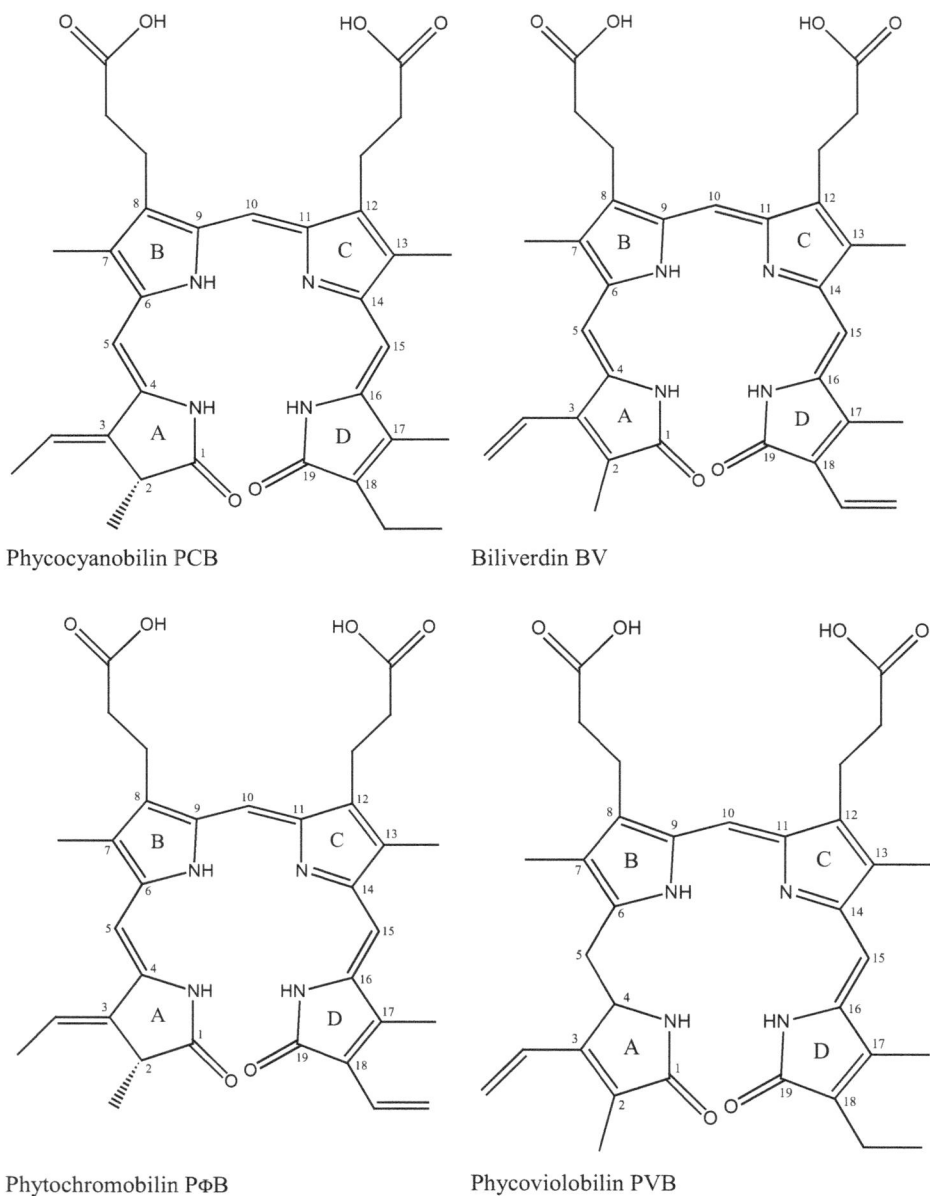

Phycocyanobilin PCB

Biliverdin BV

Phytochromobilin PΦB

Phycoviolobilin PVB

Figure 13.18. Structural formulae of phytochrome chormophors. Phycocyanobilin (PCB, $C_{33}H_{38}N_4O_6$, M_{mol} = 586.68 g mol^{-1}, from Rockwell *et al* 2006). Biliverdin (BV, $C_{33}H_{34}N_4O_6$, M_{mol} = 582.65 g mol^{-1}, from Rockwell *et al* 2006). Phytochromobilin (PΦB, $C_{33}H_{36}N_4O_6$, M_{mol} = 584.66 g mol^{-1}, from Rockwell *et al* 2006). Phycoviolobilin (PVB, $C_{33}H_{38}N_4O_6$, M_{mol} = 586.68 g mol^{-1}, from Rockwell *et al* 2017).

thermophilic cyanobacterium *Thermosynechoccus elongatus* strain BP-1 displayed in the bottom part of figure 13.19 (Ishizuka *et al* 2006) is determined by the chromophore phytoviolobilin. Its S_0–S_1 absorption maximum is located at $\lambda_{a,max} \approx 560$ nm.

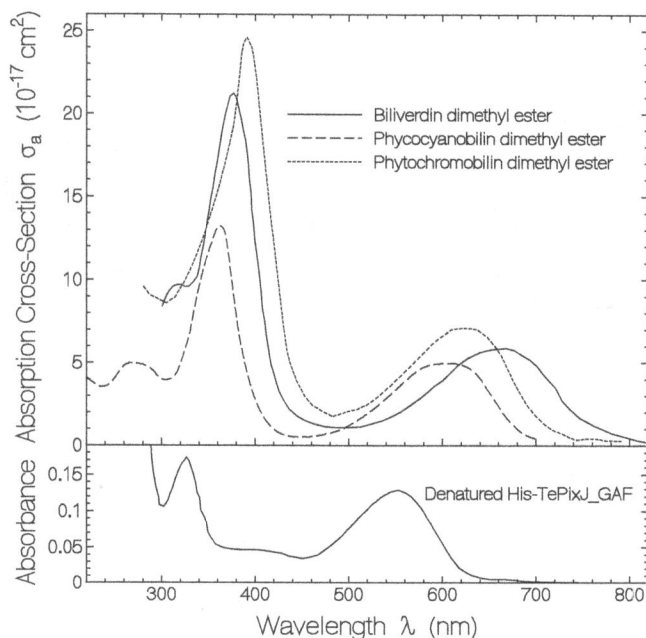

Figure 13.19. Top part: Absorption cross-section spectra of phytochrome chromophores in solution. Solid curve: biliverin dimethyl ester in methanol (from Holzwarth *et al* 1978). Dashed curve: phycocyanobilin dimethyl ester in methanol (from Gossauer and Hirsch 1974). Dotted curve: phytochromobilin dimethyl ester in methanol (from Weller and Gossauer 1980). Bottom part: absorbance of denatured His-TePixJ_GAF from thermophilic cyanobacterium *Thermosynechoccus elongatus* strain BP-1with phycoviolobilin chromophore (from Ishizuka *et al* 2006).

This blue-shift is due to the absence of the double bond between C4 and C5. The S_0–S_1 absorption bands of all four chromophores are rather broad since they are a racemic mixture of spatial isomers in solution.

The photodynamics of biliverdin dimethyl ester BVE in different solvents was studied by Braslavsky *et al* (1978, 1983) and Holzwarth *et al* (1978). The fluorescence quantum yield of BVE in ethanol at room temperature was found to be $\phi_F = 1.1 \times 10^{-4}$ (Braslavsky *et al* 1978). The small fluorescence quantum yield is caused by photoisomerization and intramolecular proton transfer (Holzwarth *et al* 1978, Braslavsky *et al* 1983).

13.2.6 Corrinoid-based cyclic tetrapyrroles (chromophores of cobalamin-based photoreceptors)

Cobalamins (vitamin B_{12} group members) are derivatives of the corrin nucleus, which contains three dihydropyrrole rings and a tetrahydropyrrole ring joined in a macrocycle by three =CH– groups and one direct carbon–carbon bond (see structural formula in lower part of figure 13.20). They are found in all living systems. They have biological functions in fatty acid and folate metabolism (Padmanabhan *et al* 2017).

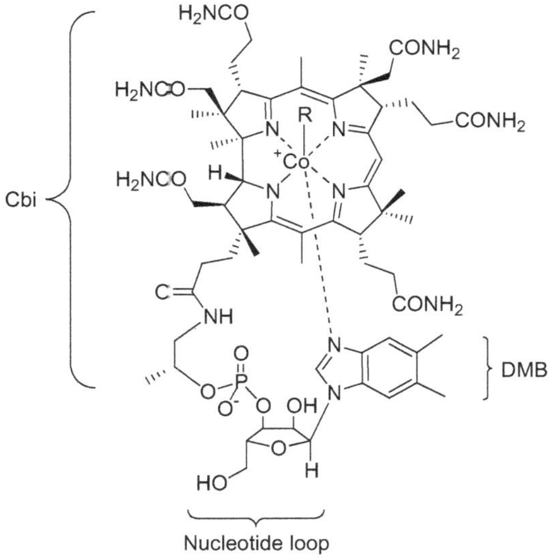

R =

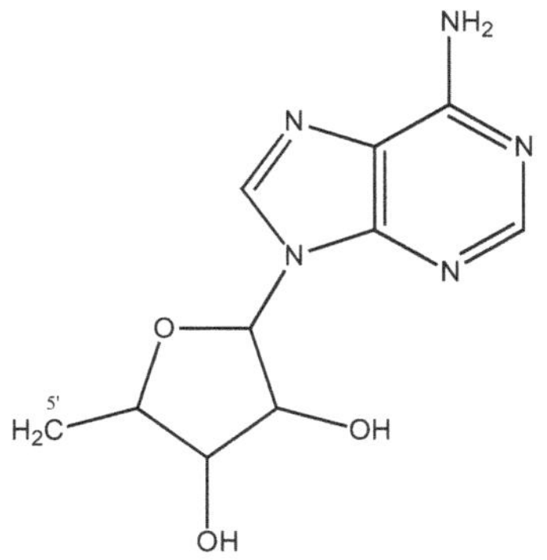

Adenosylcobalamin, coenzyme B_{12}, AdoCbl, $AdoB_{12}$

CH_3
Methylcobalamin, MeCbl, MeB_{12}

CN
Cyanocobalamin, vitamin B_{12}, CNCbl, CNB_{12}

OH
Hydroxocobalamin, OHCbl, OHB_{12}, vitamin B_{12b}

H_2O
Aquocobalamin, H_2OCbl, H_2OB_{12}, vitamin B_{12a}

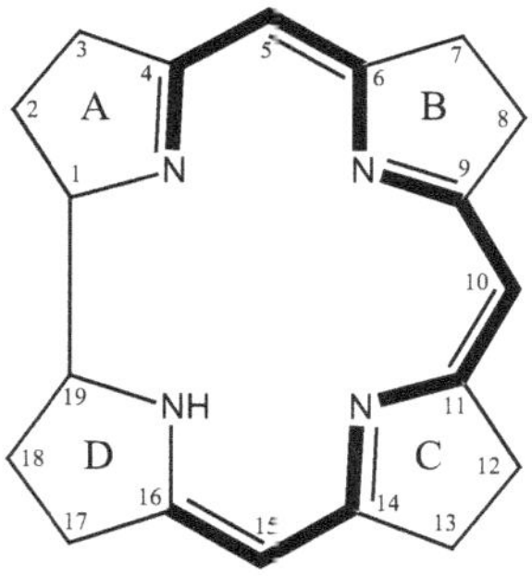

Corrin

Figure 13.20. Structural formulae of cobalamins and of the corrin core. The specific name of the cobalamines is determined by the specific upper axial ligand R. DMB = 5,6-dimethylbenzimidazole. Cbi =cobinamide (nucleotide loop including DMB is absent, and a water molecule occupies the lower axial position). Corrin consists of three pyrroline (dihydropyrrole) groups and a pyrrolidine (tetrahydropyrrole) group. The conjugated double-bond chain of corrin is indicated in bold.

Structural formulae of some members of the cobalamin family are depicted in the top part of figure 13.20. Cyanocobalamin (CNCbl, CNB$_{12}$) is vitamin B$_{12}$. Adenosylcobalamin (AdoCbl, AdoB$_{12}$) is coenzyme B$_{12}$. AdoCbl is the cofactor (chromophore) in the photoreceptor CarH (B$_{12}$-dependent repressor of carotenogenesis) of the obligate aerobic delta proteobacterium *Myxococcus xanthus* (Ortiz-Guerrero *et al* 2011), of the gram-negative eubacterium *Thermus thermophilus* (Takano *et al* 2011, Kutta *et al* 2015a), and the gram-positive, endospore-forming soil bacterium *Bacillus megaterium* (Takano *et al* 2015). Methylcobalamin (MeCbl) and/or AdoCbl are the cofactors (chromophores) in the photoreceptor AerR (aerobic repressor of photosynthesis) of the purple bacterium *Rhodobacter capulatus* (Cheng *et al* 2014). In the absence of other upper ligands, cobalamin in aqueous solution ligates a water molecule to form hydroxocobalamin (OHCbl) or aquocobalamin (H$_2$OCbl), which predominates at physiological pH (above pH 8 OHCbl dominates). Regardless of the upper axial ligand, free cobalamin forms are generally found in the DBM-on (base-on) conformation at physiological pH, as shown in figure 13.20. Protonation of DMB (5,6-dimethylbenzimidazole) at low pH leads to unligation of DMB from Co(III) and lower-axial water ligation to Co(III) yielding the base-off (DMB-off) conformation. The B$_{12}$-dependent photoreceptors and many B$_{12}$-using enzymes bind to cobalamins at the lower axis with a His side chain replacing the DMB ligand, giving a B$_{12}$-binding mode known as base-off/His-on coordination (Padmanabhan *et al* 2017).

Absorption cross-section spectra of some corrinoid compounds are displayed in figure 13.21. These compounds may be grouped as cobalt-free corrinoid (descobalt B$_{12}$), non-alkyl cobalamins where the upper axial (β face) ligand has no alkyl group connecting to the central Co transition metal, and alkyl-cobalamins where the upper axial ligand is connected to Co by a CH$_2$ group (Rury *et al* 2015). The upper part of figure 13.21 shows the absorption cross-section spectrum of the cobalt-free corrinoid compound descobalt B$_{12}$ from the photosynthetic bacterium *Chromatium* (Toohey 1965) and 'typical' cobalamin absorption cross-section spectra of the non-alkyl cobalamins CNCbl (vitamin B$_{12}$), H$_2$OCbl (vitamin B$_{12a}$) and OHCbl (vitamin B$_{12b}$). They have a typical rather sharp and strong γ-absorption band in the wavelength region around 350 nm. The absorption cross-section spectrum of the Co-free corrinoid complex descobalt B$_{12}$ is blue-shifted compared to the Co-containing corrinoid complexes. The longer wavelength absorption of the cobalamins compared to descobalt B$_{12}$ is thought to be due to mixing of MLCT (metal-to-ligand charge transfer), LMCT (ligand-to-metal charge transfer), and σ-bonds to the ππ* transitions (Lodowski *et al* 2011, 2015). The lower part of figure 13.21 shows the 'unique' cobalamin spectra (Stich *et al* 2003) of the alkyl-cobalamin photoreceptor chromophores adenosylcobalamin and methylcobalamin. For these corrinoid complexes the sharp γ-band is lost to a broad structured absorption in the 300–380 nm region.

The photoexcitation dynamics of corrinoid compounds is different for (i) cobalt-free corrinoids, (ii) non-alkyl cobalamins, and (iii) alkyl cobalamins.

    (i) Cobalt-free corrinoid compounds: photoexcitation of cobalt-free corrinoid compounds (descobalt B$_{12}$ from *Chromatium* (Toohey 1965), synthetic

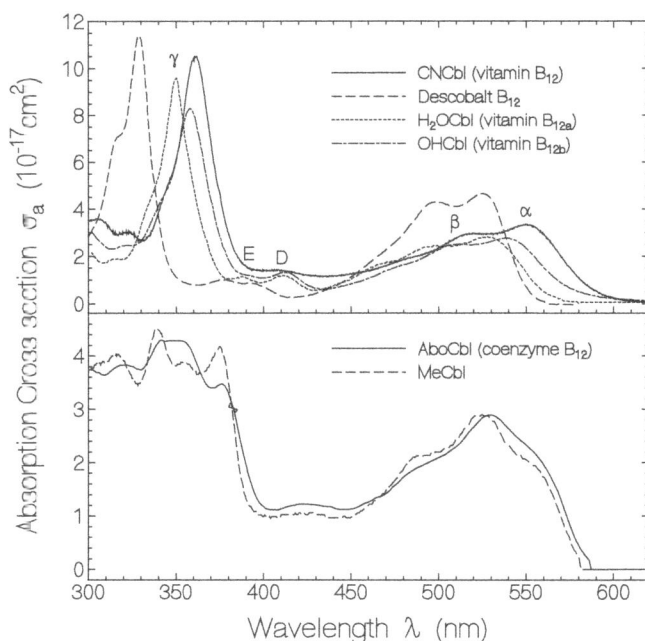

Figure 13.21. Absorption cross-section spectra of cobalamins (numbering of bands α, β, γ, D, and E follows Stich *et al* 2003 and Chemaly *et al* 2004). Top part: 'typical' cobalamin spectra (with no alkyl group at upper (β) face, see Stich *et al* 2003). Solid curve: CNCbl = vitamin B_{12} = cyanocobalamin (from PhotochemCAD ID: R06). Dashed curve: Descobalt B_{12} = cobalt –free corrinoid compound from photosynthetic bacterium *Chromatium* (from Toohey 1965). Dotted curve: H_2OCbl = Aquocobalamin (upper axial CN group of vitamin B12 replaced by H_2O) (redrawn from Stich *et al* 2003, calibrated to Walker *et al* 1998). Dash-dotted curve: OHCbl = Hydroxocobalamin (upper axial CN group of vitamin B12 replaced by OH) (taken from Cheng *et al* 2014, calibrated to Walker *et al* 1998). Bottom part: 'unique' cobalamin spectra (with alkyl group at upper (β) face (see Stich *et al* 2003). Solid curve: AboCbl = coenzyme B_{12} = 5′-deoxyadenosylcobalamin = adenosylcobalamim (from Stich *et al* 2003 calibrated to Moore *et al* 2014). Dashed curve: MeCbl = methylcobalamin (from Stich *et al* 2003, calibrated to Johnson *et al* 1963).

1,2,2,7,7,12,12-heptamethyl-15-cyanocorrin (Eschenmoser 1970)) leads to S_1–S_0 fluorescence emission, and non-radiative deactivation by singlet–triplet intersystem-crossing with T_1–S_0 decay, and S_1–S_0 internal conversion (Thomson 1969, Gardiner and Thomson 1974, Fugate *et al* 1976). The fluorescence quantum yield of 1,2,2,7,7,12,12-heptamethyl-15-cyanocorrin in ethanol was found to be $\phi_F = 0.08$ (Gardiner and Thomson 1974). Descobalt B_{12} from *Chromatium* (red corrin) is not very photostable converting to a yellow product I (Toohey 1965).

(ii) Non-alkyl cobalamins: non-alkyl cobalamins (CNCbl, H_2OCbl, and OHCbl) are generally photostable. Photoexcitation leads to a sequence of internal conversion processes repopulating the ground electronic state on a picosecond time scale involving $\pi^* \rightarrow d$ LMCT state relaxation (ligand-to-metal charge transfer) for CNCbl, and $d \rightarrow \pi^*$ MLCT state relaxation (metal-to-ligand charge transfer) for OHCbl. The photodynamics is described by Rury *et al* (2015) and references therein.

(iii) Alkyl cobalamins: photoexcitation of alkyl cobalamins generally results in rapid internal conversion to the lowest electronic excited state followed by cleavage of the upper axial C–Co bond with near unit quantum yield on a time scale ranging from ca. 10 to 100 ps. Geminate recombination limits the ultimate photolysis yield. MeCbl exhibits a wavelength dependent quantum yield for band cleavage: Excitation at 400 nm results in a branching between prompt bond cleavage and internal conversion to a low-lying excited state with a nanosecond lifetime. Excitation at 520 nm eliminates access to the prompt dissociation channel. For AboCbl excitation at 400 nm or 520 nm results in rapid internal conversion to the lowest electronic state and cleavage of the upper axial C–Co bond on a 100 ps timescale. The nature of the S_1 state of AdoCbl observed in experiments is dependent on the environment of AdoCbl. The detailed photodynamics is described by Rury *et al* 2015 and references therein.

13.2.7 Tryptophan (UVR8 chromophore)

The amino acid tryptophan (Trp, W) acts as chromophore of the UV-B photo-receptor UVR8 (= UV RESISTANCE LOCUS 8) from *Arabidopsis thaliana* (Rizzini *et al* 2011, Christie *et al* 2012b).

In figure 13.22 structural formulae of different ionic forms of Trp in aqueous solution are shown. For pH $\leqslant$ pK_c = 2.46 the cationic form of tryptophan (Trp$^+$) is dominantly present. In the pH range pK_c = 2.46 $\leqslant$ pH $\leqslant$ pK_a = 9.41 the zwitter-ionic form of tryptophan (Trp^{+-}) dominates (normally simply written as Trp). In the pH range pK_a = 9.41 $\leqslant$ pH $\leqslant$ pK_{a2} $\approx$ 12.1 anionic tryptophan (Trp$^-$) is dominating. For pH > pK_{a2} $\approx$ 12.1 a bi-anionic form of tryptophan (Trp^{2-}) is mainly present.

The absorption cross-section spectrum of Trp in water at pH 7 (0.1 M phosphate buffer) is shown by the solid curve in figure 13.23. The S_0–S_1 absorption maximum occurs at 278 nm with an absorption cross-section of $\sigma_a = 2.13 \times 10^{-17}$ cm^2. The absorption is determined by the indole part of tryptophan.

The photoexcitation dynamics of Trp was studied by Bent and Hayon (1975), Robbins *et al* (1980), Grossweiner (1984), Chen (1990), Blancafort *et al* (2002), Igarashi *et al* 2007, and others (see references in cited articles). The temporal fluorescence decay, the fluorescence quantum yield, the quantum yield of triplet formation, the quantum yield of intra-molecular proton transfer, and the quantum yield of photo-ionization (solvated electron formation) were studied as a function of pH and temperature by Robbins *et al* 1980.

For Trp$^+$ (pH $\ll$ pK_c) the fluorescence is thought to be quenched by excited-state barrierless intra-molecular charge transfer similar to the situation of cationic flavins Fl$_{ox}$$^{+*}$ of figure 13.7. For Trp^{+-} (dominant present for pK_c < pH < pK_a, presence of C9NH$_3$$^+$ and C10OO$^-$) the fluorescence signal decays bi-exponential with the fluoresence lifetimes $\tau_{F,1} = \tau_F(\text{Trp}^{+-}_{syn}) = 430$ ps (fraction 0.19), $\tau_{F,2} = \tau_F(\text{Trp}^{+-}_{anti}) = 3.32$ ns, and the fluorescence quantum yield $\phi_F = 0.137$

Cationic tryptophan Trp$^+$

$pK_c = 2.46$

Zwitter-ionic tryptophan Trp^{+-}

$pK_a = 9.41$

Anionic tryptophan Trp$^-$

$pK_{a2} = 12.1$

Bi-anionic tryptophan Trp^{2-}

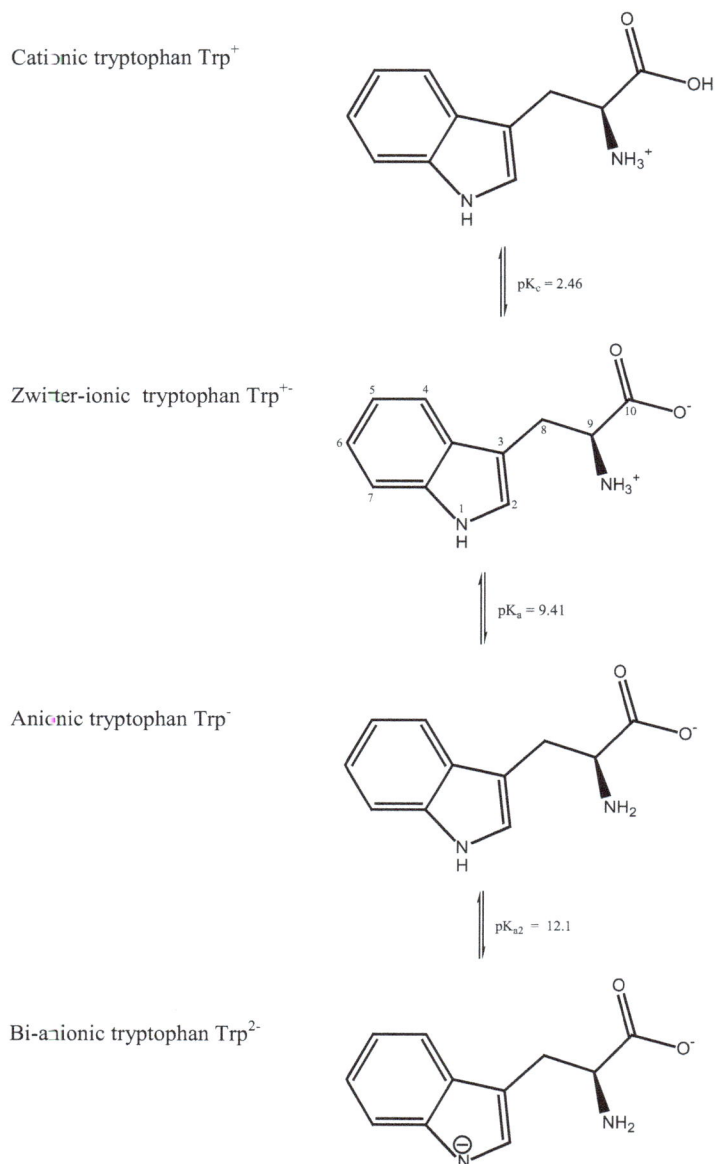

Figure 13.22. Structural formulae of different ionic forms of tryptophan. The pK values pK_c and pK_a are taken from Voet and Voet 2004, p 66. pK_{a2} is extracted from Bent and Hayon 1975, and Robbins *et al* 1980.

(Robbins *et al* 1980). This behavior is thought to be due to the presence of two isomeric tryptophan conformations (rotamers), Trp$_{syn}$ and Trp$_{anti}$ (see figure 13.24), with different efficiencies of intra-molecular proton transfer (C9NH$_3^+$ → 1NH to give C9NH$_2$ and 1NH$_2^+$). The fluorescence efficiency is reduced due to the simultaneous occurrence of intra-molecular proton transfer (IPT), intersystem crossing (ISC), and

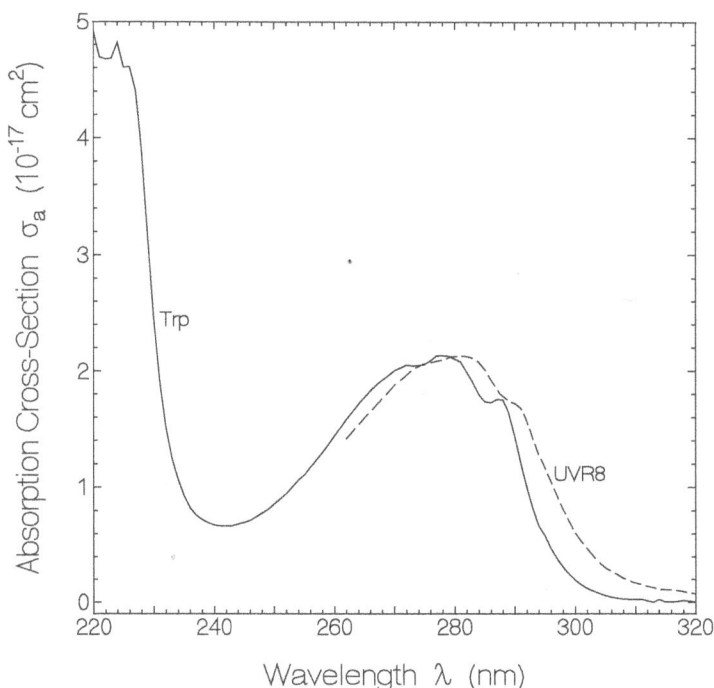

Figure 13.23. Absorption cross-section spectra of tryptophan in water, 0.1 M phosphate buffer, pH 7 (from Lindsey 2005) and UVR8 dimer (from Mathes *et al* 2015). The peak absorption of UVR8 was normalized to the peak value of Trp.

photo-ionization (PI, electron ejection, cationic tryptophan radical $\mathrm{Trp}^{\cdot+}$ and e^-_{aq} formation, see figure 13.24). For Trp^- (dominant in pH range from $\mathrm{pK}_a = 9.41$ to $\mathrm{pK}_{a2} = 12.1$) no fluorescence quenching by intra-molecular proton transfer is possible (1NH, C9NH$_2$, C10OO$^-$ is present). Therefore, the fluorescence decay is single-exponential with lifetime $\tau_F = \tau_F(\mathrm{Trp}^-_{syn}) = \tau_F(\mathrm{Trp}^-_{anti}) = 9.1$ ns and fluorescence quantum yield $\phi_F = 0.36$ (Robbins *et al* 1980). For pH > $\mathrm{pK}_{a2} = 12.1$ bi-anionic tryptophan Trp^{2-} (1 N$^-$, C10OO$^-$, and C9NH$_2$) is dominantly present with short fluorescence lifetime and low fluorescence quantum yield. The S_1 state relaxation is thought to occur by diabatic relaxation due to excited-state intra-molecular charge transfer and sloped S_1–S_0 conical intersection similar to anionic flavins Fl_{ox}^- of figure 13.7.

A schematic of the photoexcitation dynamics of Trp^{+-} is depicted in figure 13.25. This scheme is valid for both $\mathrm{Trp}^{\pm}_{syn}$ and $\mathrm{Trp}^{\pm}_{anti}$. Singlet ground-state $^1\mathrm{Trp}$ is photo-excited to $^1\mathrm{Trp}^*$. The $^1\mathrm{Trp}^*$ lifetime is $\tau_{F,1}$ for Trp_{syn} and $\tau_{F,2}$ for Trp_{anti}. $^1\mathrm{Trp}^*$ relaxes to Trp_{IPT} by intra-molecular proton transfer IPT (see first and second row of figure 13.24), to $^3\mathrm{Trp}^*$ by singlet-triplet intersystem crossing ISC, to the radical cation $\mathrm{Trp}^{\cdot+}$ by photo-ionization PI due to excited-state electron ejection ($-\mathrm{e}^-_{aq}$, see bottom row of figure 13.24), and to the singlet ground-state by fluorescence emission (fluorescence lifetimes $\tau_{F,1}$ and $\tau_{F,2}$, combined fluorescence quantum yield ϕ_F). For

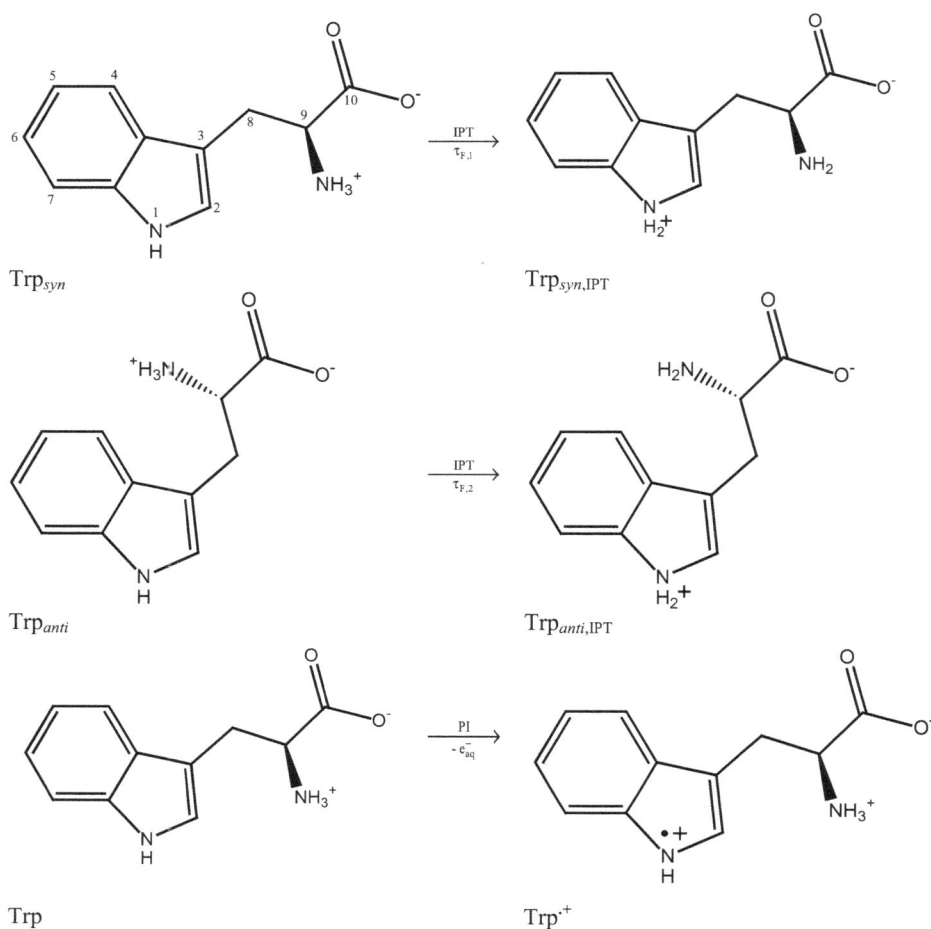

Figure 13.24. Illustration of photo-induced intra-molecular proton transfer (IPT) of *syn* and *anti* form of tryptophan, and photo-induced tryptophan radical cation formation by excited state electron ejection (photo-ionization PI).

Trp in aqueous solution at pH 7 and temperature of 25 °C the quantum yields of ^{1}Trp* relaxation were determined to be $\phi_{IPT} = 0.65$ (intra-molecular proton transfer), $\phi_T = 0.10$ (triplet formation), $\phi_{PI} = 0.12$ (electron ejection), and $\phi_F = 0.13$ (Robbins *et al* 1980). The triplet state lifetime was in the range from 11 μs to 16 μs, and the lifetime of Trp^{+} was ≈1 μs (Bent and Hayon 1975). The Trp$_{IPT}$ lifetime is thought to be ≈ 45 ns (Robbins *et al* 1980).

13.2.8 Fluorescent probes

Organic fluorescent dyes are applied as fluorescent probes in biological and medical research to detect particular components of complex biomolecular assemblies including live cells. They are applied as clinical diagnostic assays, biosensors and

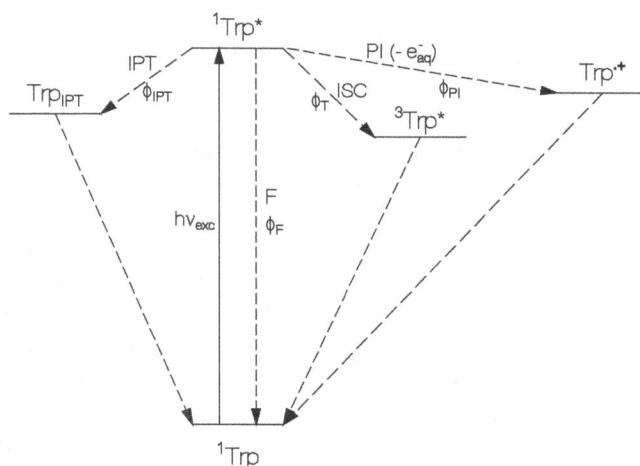

Figure 13.25. Photoexcitation dynamics of tryptophan in aqueous solution. F: fluorescence emission. ϕ_F: fluorescence quantum yield. ISC: intersystem crossing. ϕ_T: quantum yield of triplet formation. IPT: intra-molecular proton transfer. ϕ_{IPT}: quantum yield of intra-molecular proton transfer. PI: photo-ionization (electron ejection). e_{aq}^-: solvated electron. ϕ_{PI}: quantum yield of photo-ionization.

nanoscopic probes for monitoring cellular and subcellular activity. These fluoro-phores can be modified to reactive versions which can be bound to specific functional proteins, cell membranes and organelles for imaging and diagnostic purposes. Detailed information on fluorescent dyes and probes is found by Haugland (2002) and Sabnis (2015).

13.2.9 Fluorescent proteins

Fluorescent proteins are members of a structurally homologous class of proteins that share the unique property of being self-sufficient to form a visible wavelength chromophore from a sequence of three amino acids within their own polypeptide sequence. It is common research practice for biologists to introduce a gene (or a gene chimera) encoding an engineered fluorescent protein into living cells and subse-quently visualize the location and dynamics of the gene product using fluorescence microscopy (Campbell 2008, Jung 2012a, 2012b).

The fluorescent component in the bioluminescent organs of the *Aequorea victoria* jellyfish was identified as a fluorescent protein (Shimomura *et al* 1962, Shimomura 1979). The complete primary sequence of the 238 amino acids of *Aequorea victoria* green fluorescent protein AvGFP was determined by cloning and sequencing of its cDNA (Prasher *et al* 1992). AvGFP could be functionally expressed in worms and bacteria (Chalfie *et al* 1994, Inouye and Tsuji 1994). The mutagenesis of AvGFP resulted in a wide range of genetically encoded fluorescent proteins in the spectral range from blue to yellow (for reviews see Tsien 1998, Matz *et al* 2002, Campbell 2008, Day and Davidson 2009, Chudakov *et al* 2010, Remington 2011, Acharya *et al* 2017). The spectrum was extended to red fluorescence by cloning the fluorescent protein DsRed from the reef anthozoa *Discosoma* sp. mushroom anemone

(Matz *et al* 1999, 2002) and genetically engineering mutants thereof (Matz *et al* 2002, Campbell 2008, Day and Davidson 2009, Chudakov *et al* 2010, Remington 2011, Acharya *et al* 2017).

The fluorescent proteins have a β-can fold consisting of an 11-stranded β-sheet wrapped into a cylindrical β-barrel of ≈4 nm in height and ≈3 nm in diameter. The chromophore is located near the center of the protein which threads through the center of the β-can along its long axis (Yang *et al* 1996). There exist natural fluorescent proteins with at least five different chromophores. To them belong *Aequorea victoria* green fluorescent protein AvGFP (chromophore formed from Ser65, Tyr66 and Gly67 (Zhang *et al* 2006)), *Discosoma* sp red fluorescent protein DsRed (chromophore formed from Gln66, Tyr67 and Gly68 (Gross *et al* 2000)), *Zoanthus* yellow fluorescent protein zFP538 (chromophore formed from Lys66, Tyr67 and Gly68 (Remington *et al* 2005)), *Anemonio sulcata* kindling fluorescent protein asFP595 (chromophore formed from Met63, Tyr64, and Gly65 (Tretyakova *et al* 2007)), and *Trachyphyllia geoffroyi* red fluorescent protein Kaede (chromophore formed from His62, Tyr63, and Gly64 (Mizuno *et al* 2003)). An enormous variety of engineered fluorescent proteins have been expressed from the archetypes *Aequorea* green fluorescent protein and anthozoan *Discosoma* red fluorescent protein (Tsien 1998, Matz *et al* 2002, Campbell 2008, Day and Davidson 2009, Chudakov *et al* 2010, Remington 2011, Acharya *et al* 2017).

The spontaneous formation of the *Aequorea victoria* green fluorescent protein chromophore from Ser65, Tyr66 and Gly67 within the folded β-can protein structure involves cyclization between Ser65 and Gly67, dehydration (loss of H_2O), and oxidation with molecular oxygen (thereby removal of H_2O_2) (Zhang *et al* 2006). The wild-type chromophore of AvGFP exists as an equilibrium mixture of neutral tyrosine 66 (absorption maximum at 397 nm) and anionic tyrosinate 66 (absorption maximum at 475 nm) (Patterson *et al* 1997). The spontaneous formation of the *Discosoma* sp. DsRed chromophore from Gln66, Tyr67 and Gly68 involves besides cyclization, dehydration, and oxidation a second oxidation to extend the conjugated system by two double bonds (Gross *et al* 2000).

In figure 13 26 structural formulae are shown of the chromophores of some engineered widely applied monomeric fluorescent proteins spanning the spectral range of fluorescence emission from the blue to the red. They are EBFP (enhanced blue fluorescent protein), ECFP (enhanced cyan fluorescent protein), EGFP (enhanced green fluorescent protein), and EYFP (enhanced yellow fluorescent protein) derived from wild-type GFP of the jellyfish *Aequorea victoria* (Shimomura *et al* 1962), and mOrange and mCherry derived from DsRed (originally called drFP583) of the sea anemone *Discosoma* sp. (Matz *et al* 1999).

In figure 13.27 the absorption cross-section spectra of the fluorescent proteins of figure 13.26 are displayed. Their absorption strength is high as expected for organic dyes. The fluorescence quantum yield of the selected fluorescent proteins in figures 13.26 and 13.27 is high in the range from $\phi_F = 0.22$ for mCherry to $\phi_F = 0.69$ for mOrange (values included in figure 13.26).

The photodynamics of a wide range of fluorescent proteins, as those shown in figures 13.26 and 13.27, is similar to that of normal fluorescent organic dyes, as

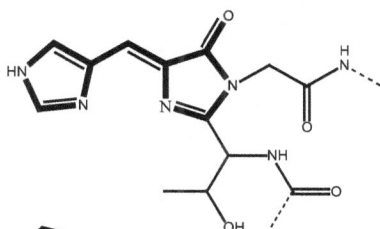

EBFP
F64L/S65T/Y66H/Y145F
$_{a,max}$ = 383 nm
$_{F,max}$ = 445 nm
$_F$ = 0.31

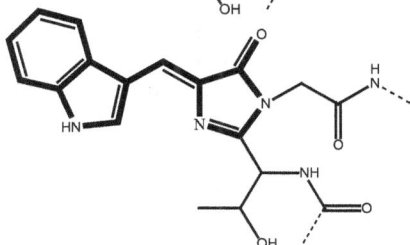

ECFP
F64L/S65T/Y66W/N146I/M
153T/V163A
$_{a,max}$ = 439 nm
$_{F,max}$ = 476 nm
$_F$ = 0.40

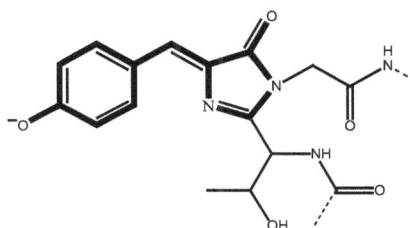

EGFP
F64L/S65T
$_{a,max}$ = 488 nm
$_{F,max}$ = 507 nm
$_F$ = 0.60

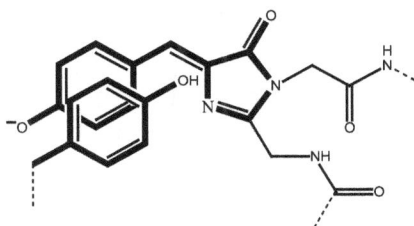

EYFP
S65G/ V68L/S72A/T203Y
$_{a,max}$ = 514 nm
$_{F,max}$ = 527 nm
$_F$ = 0.61

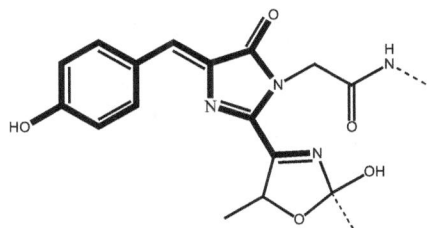

mOrange
Q66T/T147S/V7I/M182K/T1
95V/T41F/L83F
$_{a,max}$ = 548 nm
$_{F,max}$ = 562 nm
$_F$ = 0.69

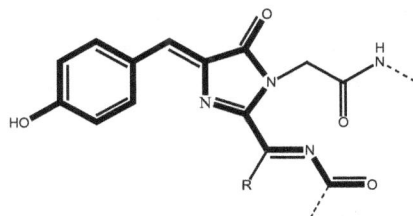

mCherry
Q66M/T147S/V7I/M182K/M
163Q/T195V
$_{a,max}$ = 587 nm
$_{F,max}$ = 610 nm
$_F$ = 0.22

Figure 13.26. Structural formulae of chromophores of some fluorescent proteins. The conjugated ring structure is indicated in bold drawing. Structures are redrawn from Sample *et al* 2009. EBFP, ECFP, EGFP, and EYFP are derived from wild-type GFP of the jellyfish *Aequorea victoria* (Shimomura *et al* 1962). The amino acid mutations relative to wild-type GFP are listed (taken from Patterson *et al* 2000). mOrange and mCherry are derived from DsRed (originally called drFP583) of the sea anemone *Discosoma* sp. (Matz *et al* 1999). The amino acid mutations relative to DsRed are listed (taken from Shaner *et al* 2004). EBFP: enhanced blue fluorescent protein (Patterson *et al* 1997). ECFP: enhanced cyan fluorescent protein (Shaner *et al* 2004). EGFP: enhanced green fluorescent protein (Patterson *et al* 1997). EYFP: enhanced yellow fluorescent protein (Spiess *et al* 2005). mOrange: monomeric orange fluorescent protein (Shaner *et al* 2004). mCherry: monomeric red fluorescent protein (Shaner *et al* 2004). $\lambda_{a,max}$: wavelength position of first absorption maximum. $\lambda_{F,max}$: wavelength position of fluorescence maximum. ϕ_F: fluorescence quantum yield. Values of $\lambda_{a,max}$, $\lambda_{F,max}$ and ϕ_F are found in Day and Davidson 2009.

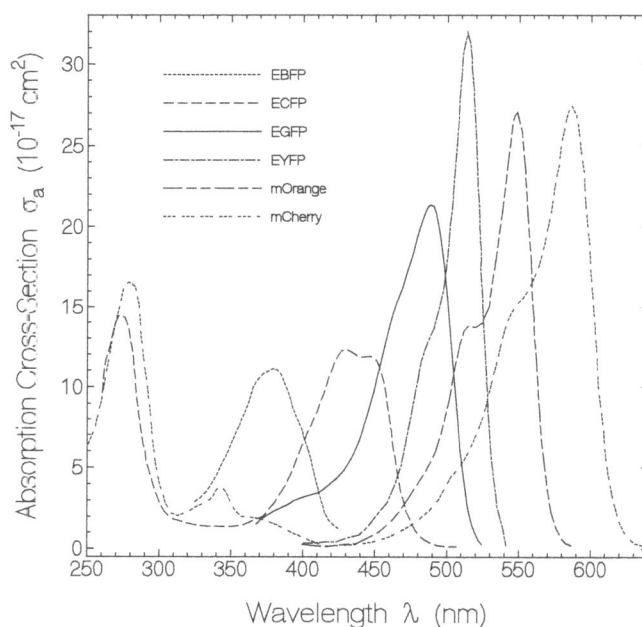

Figure 13.27. Absorption cross-section spectra of some fluorescent proteins. Dotted curve: EBFP (from Patterson *et al* 1997) Dashed curve: ECFP (from Fredj *et al* 2012). Solid curve: EGFP (from Patterson *et al* 1997) Dash-dotted curve: EYFP (from Spiess *et al* 2005, and AAT Bioquest: https://www.aatbio.com/spectrum/EYFP). Thick long dashed short double dashed curve: mOrange (from Shaner *et al* 2004). Thick triple dotted curve: mCherry (from Shaner *et al* 2004).

illustrated in the middle column of figure 13.7, for neutral fully oxidized flavins and in the potential energy curve diagram of figure 13.8. The wavelength position of the first absorption maximum and the wavelength position of the corresponding fluorescence emission (S_0–S_1 transition) is determined by the extension of the conjugated π-electron system and is influenced by the protonation stage (neutral phenol or anionic phenolate group of Tyr), the stereoisomer conformation (E or Z configuration), and the protein micro- environment that surrounds the chromophore.

Besides the above described fluorescent proteins, a variety of optical *highlighter* fluorescent proteins have been developed with various induced phenomena as photoactivation, positive and negative photoswitching, and photoconversion. They apply light-induced chromophore transformations. They open new possibilities in bioimaging and super-resolution microscopy. Detailed reviews on the photochemistry and photophysics of these optical *highlighter* fluorescent proteins are found in Day and Davidson (2009), Nienhaus and Wiedenmann (2009), Chudakov *et al* (2010), Remington (2011), Shcherbakova and Verkhusha (2014), and Acharya *et al* (2017).

Photoactivatable fluorescent proteins are capable of being activated from very low to bright fluorescent emission upon illumination with ultraviolet or violet light (for example in PA-GFP with T203H mutation, UV light exposure causes irreversible photochemical Glu 222 decarboxylation with changing the chromophore protonation state from neutral to anionic with strong cyan absorption and green fluorescence emission (Henderson *et al* 2009).

Photoswitchable fluorescence proteins (reversibly photoswitchable fluorescent proteins rsFPs) have an emission characteristics that can be alternatively turned on or off with specific illumination in a reversible manner. They involve photo-induced *cis-trans* isomerization accompanied with protonation/deprotonation changes and chromophore-protein interaction changes (Grotjohann *et al* 2011, Bourgeois and Adam 2012, Zhou and Lin 2013, Acharya *et al* 2017, Pennacchietti *et al* 2018).

Photoconversion fluorescent proteins (pcFPs), or also named irreversible photoswitchable fluorescent proteins, can be converted irreversibly from one fluorescence emission band to another one when illuminated by specific wavelengths. In green-to-red photoconvertible fluorescent proteins 405 nm photoexcitation causes a chromophore change by cleavage of the His 62 N^α–C^α bond and formation of a C^α–C^β double bond in the His 62 side chain which extends the delocalized π-electron system and changes green emitting fluorescent protein to a red emitting fluorescent protein (Kim *et al* 2015, Turkowyd *et al* 2017, Wachter 2017). In primed conversion green to red fluorescent proteins the combined illumination at 488 nm and 730 nm caused the chromophore change in a two-step mechanism from green emission to red emission (Turkowyd *et al* 2017).

13.2.10 Nanomaterials

Nanomaterials are solid matter in the size of typically 1–100 nm. To them belong insulators (large energy gap between conduction band and valence band, transparent in the visible spectral region), semiconductors (medium energy gap between conduction band and valence band, absorbing in the visible spectral region like organic dyes), and metals (conduction band partly filled with electrons, Fermi energy level lying in the conduction band, electrons around Fermi energy level are thermally free moving, plasmonic interaction with light in the visible spectral region involving inter-band transitions). The nanomaterial physical behavior begins to deviate from the bulk material behavior when the particle size becomes less than the exciton Bohr radius in insulators and semiconductor and less than the electron mean free path in metals (Gaponenko 2010, De Mello Donegá 2014).

The exciton Bohr radius a_B^* of semiconductors and insulators is given by (De Mello Donegá 2014)

$$a_B^* = a_0 \varepsilon_r m_e \left(\frac{1}{m_e^*} + \frac{1}{m_h^*} \right), \tag{13.4}$$

where $a_0 = 0.0529$ nm is the Bohr radius of the electron in the hydrogen atom, ε_r is the relative dielectric constant, m_e is the free electron mass, m_e^* is the electron reduced mass, and m_h^* is the hole reduced mass. For particles of diameter $d < 2a_B^*$ the bandwidths of the conduction band and the valence band become narrower and thereby the energy gap between the conduction band energy minimum and the valence band energy maximum becomes larger with decreasing cluster size (polymer to oligomer to octomer, heptamer, hexamer, pentamer, tetramer, trimer, dimer, monomer) approaching the HOMO–LUMO energy gap of a single molecule (monomer). This increase in energy gap with decreasing particle size is called quantum confinement.

In metals quantum confinement (decrease of first absorption band wavelength with decreasing metal cluster size, Zheng *et al* 2007) occurs for particle radius less than the Fermi wavelength λ_{Fermi} (de Broglie wavelength)

$$\lambda_{Fermi} = \frac{2\pi}{k_{Fermi}} = \frac{h}{p_{Fermi}} = \frac{h}{(2m_e^* E_{Fermi})^{1/2}}, \tag{13.5}$$

where k_{Fermi} is the Fermi wave vector, p_{Fermi} is the Fermi momentum, and E_{Fermi} is the Fermi energy. λ_{Fermi} is less than 2 nm for all metals (De Mello Donegá 2014). Changes of absorption bandwidth and absorption strength are occurring if the metal particle size becomes smaller than the electron mean free path l_{Fermi} determined by the Fermi velocity υ_{Fermi} and the inverse of the damping rate Γ

$$l_{Fermi} = \upsilon_{Fermi} \Gamma^{-1} = \hbar k_{Fermi} / (m_e^* \Gamma) \tag{13.6}$$

Typical values of l_{Fermi} are 40–50 nm (Gaponenko 2010).

13.2.10.1 Fluorescent semiconductor nanocrystals (quantum dots)
Quantum dots versus organic dyes are compared in the review of Resch-Genger *et al* (2008). Semiconductor quantum dot bioconjugates play an important role in biosensing with versatile energy transfer and electron transfer applications (for review see Hildebrandt *et al* 2017). Generally the quantum dots are built up of a semiconductor core, a semiconductor shell, and a coating for biomolecule conjugation (Jamieson *et al* 2007). In Type I quantum dots the energy gap of the core is less than the energy gap of the shell (an example: CdSe/ZnS), and the electron and hole tend to localize in the core. In reverse Type I quantum dots the energy gap of the core is larger than the energy gap of the shell (an example ZnS/CdSe), and the electron and hole tend to localize in the shell. In Type II quantum dots the energy

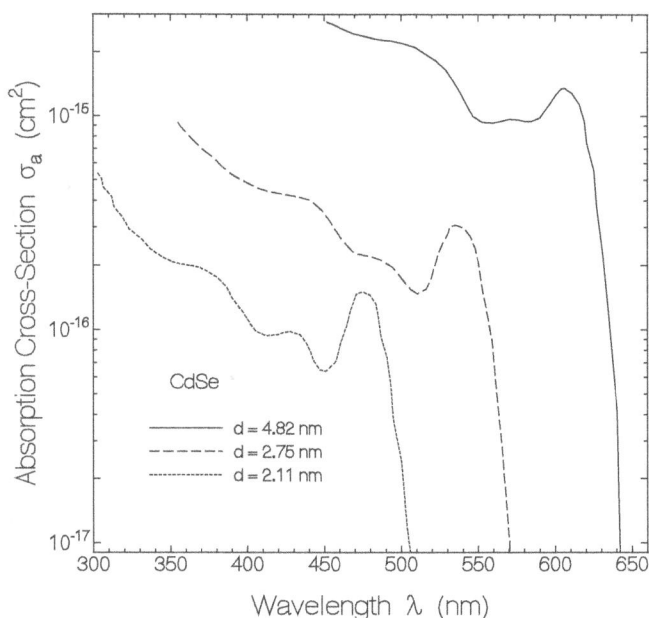

Figure 13.28. Absorption cross-section spectra $\sigma_a(\lambda)$ of three CdSe quantum dots of different size (diameter d given in figure). Curves are adapted from Yu *et al* 2003.

gaps of the semiconductors are asymmetric and one carrier is localized in the shell and the other carrier is localized in the core (an example: CdSe/ZnSe).

The absorption behavior of CdSe quantum dots is shown in figure 13.28 (curves adapted from Yu *et al* 2003). In this case the bulk energy gap is $E_g = 1.74$ eV giving an absorption edge wavelength of $\lambda_g = hc_0/E_g = 713$ nm (h: Planck constant, c_0: vacuum light velocity). The exciton Bohr radius is $a_B^* = 4.9$ nm (Madelung 2004). The CdSe molecule radius a_{CdSe} is

$$a_{CdSe} = \left(\frac{3}{4\pi} V_{CdSe}\right)^{1/3} = \left(\frac{3}{4\pi} \frac{M_{CdSe}}{N_A \rho_{CdSe}}\right)^{1/3}, \tag{13.7}$$

giving a value of $a_{CdSe} \approx 0.235$ nm (V_{CdSe} is volume of a CdSe molecule, $M_{CdSe} = 191.39$ g mol^{-1} is molar mass of CdSe, N_A is Avogadro constant, $\rho_{CdSe} = 5.816$ g cm^{-3} is density of CdSe). The number of CdSe molecules within a CdSe particle of exciton Bohr radius is $N_B = (a_B^*/a_{CdSe})^3 \approx 9000$.

In figure 13.28 the wavelength position of the first absorption maximum decreases with decreasing particle diameter (decreasing width of conduction band and valence band, center wavelength positions of conduction band and valence band approach the LUMO and HOMO wavelength positions of the CdSe molecule). The absorption cross-sections at the peak wavelength positions decrease with decreasing particle size since less CdSe molecules are present, which contribute to the absorption.

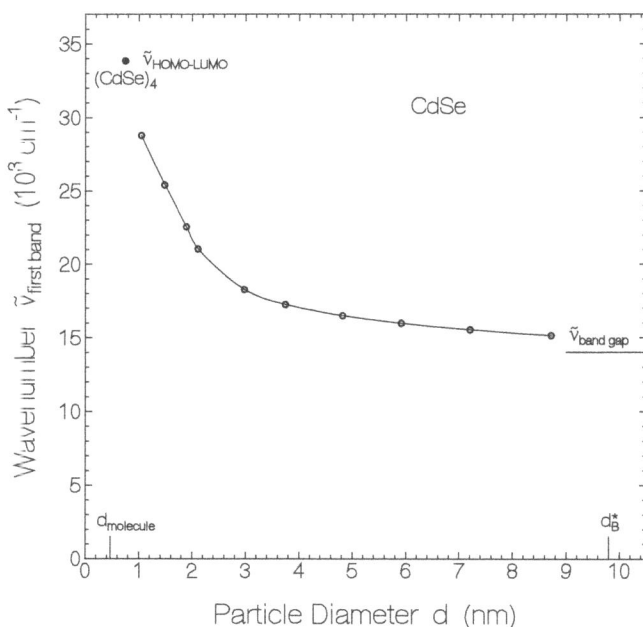

Figure 13.29. Wavenumber position of first absorption peak of CdSe quantum dots. Solid line connected circles are taken from Yu *et al* 2003. Dot is taken from Soloviev *et al* 2001. $\tilde{\nu}_{\text{band gap}}$ indicates energy band gap of CdSe bulk crystal ($E_g = 1.74$ eV). $\tilde{\nu}_{\text{HOMO-LUMO}}$ indicates energy difference between LUMO and HOMO energy levels of CdSe molecule. $d_{molecule}$ is diameter of a CdSe molecule. d_B^* is exciton Bohr diameter (value taken from Madelung 2004).

In figure 13.29 the wavenumber position $\tilde{\nu}_{first\ band}$ of the first absorption peak of CdSe quantum dots versus the particle diameter d is displayed (data taken from Yu *et al* 2003 and Soloviev *et al* 2001). For $d > d_B^* = 2a_B^*$ the wavenumber $\tilde{\nu}_{first\ band}$ approaches the bulk CdSe band gap $\tilde{\nu}_{band\ gap}$ between conduction band and valence band. For d approaching $d_{molecule}$ of a single CdSe molecule $\tilde{\nu}_{first\ band}$ approaches the discrete HOMO–LUMO wavenumber difference $\tilde{\nu}_{HOMO-LUMO}$.

The absorption strength of semiconductor quantum dots is high (see figure 13.28), their absorption wavelength position and their corresponding fluorescence wavelength position are tunable by the particle size (see figure 13.29), and the fluorescence quantum yield is high (especially in Type I quantum dots, Grabolle *et al* 2008, 2009, Pilla *et al* 2012). These properties make them attractive for biosensing and applications towards optogenetics (Huang *et al* 2016). Photoexcitation of quantum dots leads to charged exciton formation (photo-ionization) with change of the electric dipole moment of the quantum dots. Location of the quantum dots to cell membranes allows ion channel activation by the electric field of the photo-excited quantum dots (Lugo *et al* 2012, Rowland *et al* 2015).

13.2.10.2 Lanthanide upconversion nanoparticles
Photon upconversion is a nonlinear optical phenomenon known as anti-Stokes emission in which the sequential absorption of two or more low-energy photons

leads to high-energy luminescent emission (Auzel 2004, Wang and Liu 2009, Haase and Schäfer 2011). Upconversion nanocrystals are composed of an inorganic crystalline host matrix and trivalent lanthanide ions (Ln^{3+}) embedded in the host lattice (inner shell 4f electron transitions are involved in optical excitation of lanthanides, principle quantum number is 4, orbital quantum number is 3). For efficient upconversion intermediate states between the ground state and the emitting state are required to act as energy reservoirs. The $4f^n$ (n is number of f electrons of the considered lanthanide ion) energy levels are separated by symmetry splitting, Coulomb interaction, spin–orbit interaction, and crystal field interaction (Stark splitting) (Dieke 1968, Penzkofer 1988 and references therein). The f–f transitions are parity forbidden for electric dipole emission. They are parity allowed for magnetic dipole and electric dipole radiation. The radiative transitions are mainly due to field-induced electric dipole interactions in crystals without inversion symmetry, and due to magnetic dipole and phonon-induced electric dipole inter-action in crystals with inversion symmetry and in glasses. The oscillator strength of the transitions is low, leading to long-lived excited states with radiative lifetimes in the range of 50 μs to 10 ms (Moulton 1982). The f–f transitions involve inner shells. The crystal field is shielded by outer s and p orbitals. The free ion spectra are only slightly disturbed by the crystal host.

Upconversion nanoparticles are reviewed by Haase and Schäfer (2011), Chen *et al* (2014), in the Photon Upconversion Nanomaterials issue of *Chemical Society Reviews* edited by Liu *et al* (2015), and by Zhou *et al* (2015a). The nanoparticle generally consists of a crystalline core doped with infrared light absorbing sensitizer f-ions S and visible light emitting activator f-ions A excited to the emitting level by energy transfer upconversion processes. The lanthanide ion doped core is sur-rounded by an undoped crystalline shell for reducing surface quenching effects of the up-converted activator ions. For the nanoparticle core and shell crystalline host often hexagonal-phase sodium yttrium fluoride $NaYF_4$ is used. As sensitizer in most cases Yb^{3+} is used (absorption around 980 nm). As upconversion activator ions, Tm^{3+}, Er^{3+}, and Ho^{3+} are most frequently used. In most cases energy transfer upconversion is applied for anti-Stokes emission from the activator by sensitizer excitation. In a first step energy transfer from excited sensitizer to activator ground-state g takes place populating an excited state e1. In a second step continued sensitizer excitation causes energy transfer from the still populated activator excited state e1 to a higher excited state e2 during the e1 state lifetime. State e2 may relax to a lower lying state e2′. From state e2′ or e2 emission to the activator ground-state may occur, giving upconversion photoluminescence $UCPL_1$ (efficiency proportional to excitation laser intensity). In a third step continued sensitizer excitation causes energy transfer from the excited sensitizer to the activator in excited level e2′ which gets elevated to level e3 with possible non-radiative relaxation to level e3′. From e3′ or e3 to g occurs upconversion photoluminescence $UCPL_2$ (efficiency proportional to square of excitation intensity). This process may proceed in a third upconversion energy transfer process.

In figure 13.30 a scheme of a core/shell upconversion nanoparticle is shown. The excitation light of frequency ν_{exc} excites the sensitizer lanthanide ion S. Sensitizer S

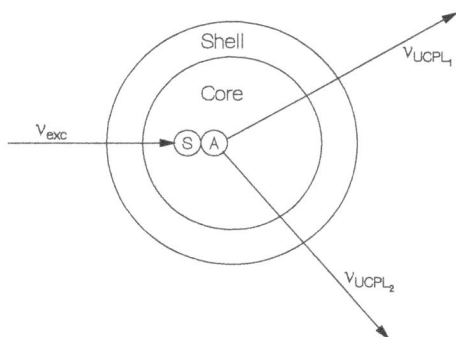

Figure 13.30. Scheme of a core–shell upconversion nanoparticle. Sensitizer S excitation with light of frequency ν_{exc} and activator A emission at the anti-Stokes frequencies ν_{UCPL_1} and ν_{UCPL_2} is indicated.

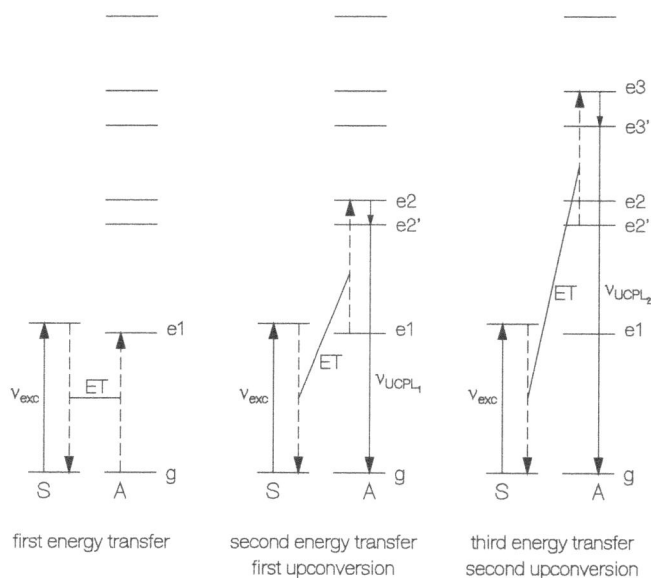

first energy transfer

second energy transfer
first upconversion

third energy transfer
second upconversion

Figure 13.31. Illustration of energy-transfer upconversion. An example of two-step photon upconversion is presented involving three energy transfer processes from sensor lanthanide S to activator lanthanide A with upconversion photoluminescence emission after the first upconversion step (frequency ν_{UCPL_1}) and after the second upconversion step (frequency ν_{UCPL_2}). Some non-radiative relaxation to lower energetic levels is indicated after the first upconversion step (e2 → e2') and after the second upconversion step (e3 → e3'). ν_{exc}: excitation frequency. ET: phonon-assisted energy transfer from S to A.

to activator A multi-step energy transfer causes activator upconversion photo-luminescence indicated by ν_{UCPL_1} and ν_{UCPL_2}.

In figure 13.31 the energy transfer upconversion mechanism is illustrated for a two-step photon upconversion involving three energy transfer processes from sensitizer S to activator A first in ground-state g, then in excited state e1, and then in relaxed excited state e2'. Upconversion photoluminescence emission of activator A from e2' to g (UCPL$_1$) and from e3' to g (UCPL$_2$) is indicated.

Organic dye sensitized lanthanide-doped upconversion nanoparticles have been developed to overcome the weak and narrow absorption of lanthanide ion sensitizers (Yb^{3+} and also Nd^{3+}). In these configurations, organic dyes with spectrally broad strong absorption bands are anchored to the surface of upconversion nanoparticles to harvest the excitation light energy and transfer it to the lanthanides via Förster and/or Dexter excitation transfer across the organic/inorganic interface (see Wu *et al* 2015, Chen *et al* 2015, and tutorial review by Wang *et al* 2017b). A list of applied organic dyes for dye-sensitized upconversion is given in Wang *et al* 2017b.

13.3 Action of the organic dyes in photoreceptors

13.3.1 Rodopsins

Rhodopsins consist of an opsin protein and a covalently bound retinal. The opsin apoprotein has the architecture of seven transmembrane α-helices (TM) with N-terminus facing out of the cell and C-terminus facing inside the cell. Retinal is attached by a Schiff base linkage to the ε-amino group of a lysine side chain in the middle of TM7. The retinal Schiff base (RSB) is protonated (PRSB) in most cases except in UV-rhodopsins (as they occur for example in night-active rodents, Salcedo *et al* 2003, Tsutsui *et al* 2007). Photoexcitation of rhodopsins starts rhodopsin specific photocycles which generally involve photoisomerization, opsin counter ion repositioning, PRSB deprotonation to RSB with protein conformation changes to signaling/effector action, and subsequent reprotonation and back isomerization to the initial receptor state (Kandori *et al* 2001, Zhang *et al* 2011, Kikukawa *et al* 2012, Wand *et al* 2013, Ernst *et al* 2014). One distinguishes two classes of rhodopsins: the microbial rhodopsins (type-I rhodopsins) and the animal rhodopsins (type-II rhodopsins). Both classes have already found application as optogenetic tools in neuroscience (Li *et al* 2005, Yizhar *et al* 2011).

The microbial rhodopsins (Grote *et al* 2014, Ernst *et al* 2014, Gushchin and Gordeliy 2018) include (i) ion pumps (H$^+$ outward pumping bacteriorhodopsins BR (Lanyi 2004) and proteorhodopsins PR (Béja *et al* 2000, Bamann *et al* 2014); Na$^+$ outward pumping *Krokinobacter eikastus* rhodopsin 2 KR2 (Kandori 2015, Kato *et al* 2016); Cl$^-$ inward pumping halorodopsins HR (Pal *et al* 2013)), (ii) cation conducting channelrhodopsins ChRs (Nagel *et al* 2003, Klapoetke *et al* 2014, Deisseroth and Hegemann 2017), (iii) anion-conducting channelrhodopsins ACRs (Govorunova *et al* 2015), (iv) sensory rhodopsins SR (SRI and SRII from archeon *Halobacter salinarum* (Hoff *et al* 1997) and SRI from *Salinibacter ruber* (Sudo *et al* 2014) mediate color-sensitive phototaxis response; *Anabena* SR (Wand *et al* 2011) functions as a photochromic sensor), and (v) enzymerhodopsins (photochromic histidine kinase rhodopsin HKR1 from *Chlamydomonas reinhardtii* (Luck *et al* 2012, Penzkofer *et al* 2014d) plays a role in the adaptation of behavioral responses in the presence of UVA light; rhodopsin–guanylyl cyclases (Avelar *et al* 2014, Gao *et al* 2015, Penzkofer *et al* 2016c, Penzkofer *et al* 2017, Scheib *et al* 2018) enable photo-orientation by light-activated second messenger cyclic guanosine monophosphate cGMP formation). In addition to natural microbial rhodopsins many engineered variants (primarily ChRs) with altered absorption wavelength, kinetics, and ion selectivity are widely applied in neurosciences (Wietek *et al* 2014, 2017, Schneider *et al* 2015, Berndt *et al* 2016).

Dark-adapted bacteriorhodopsin from *Halobacterium halobium* (later named *salinarum*) contains all-*trans*,15-*anti* PRSB (BR$_{AT}$, wavelength of first absorption maximum $\lambda_{a,max}$ = 568 nm) and 13-*cis*,15-*syn* PRSB (BR$_{13C}$, $\lambda_{a,max}$ = 548 nm) in a ratio of 1:1.5 (Harbison *et al* 1984). The primary photoisomerization of BR$_{AT}$ is all-*trans*,15-*anti* → 13-*cis*,15-*anti* (K-intermediate, $\lambda_{a,max}$ = 590 nm) formation with sub-picosecond time constant (Dobler *et al* 1988, Haake *et al* 2002). The photoexcitation of BR$_{13C}$ partly converts it to BR$_{AT}$ which is called 'light adaptation' (Kawanabe *et al* 2007). In the photocycle of BR$_{AT}$ follows K-intermediate → L-intermediate counterion repositioning (1 µs timescale, $\lambda_{a,max}$ = 550 nm), L-intermediate → M-intermediate proton release (50 µs timescale, $\lambda_{a,max}$ = 410 nm), M-intermediate → N-intermediate reprotonation (3 ms timescale, $\lambda_{a,max}$ = 560 nm), N-intermediate → O-intermediate 13-*cis*,15-*anti* → all-*trans*,15-*anti* isomerization (5 ms timescale, $\lambda_{a,max}$ = 640 nm), and O-intermediate → ground-state restructuring (10 ms timescale, $\lambda_{a,max}$ = 568 nm) (Lanyi 2004, 2006, Kawanabe *et al* 2007, Hirai and Subramaniam 2009, Ernst *et al* 2014, Wickstrand *et al* 2015). The equilibration to the initial dark-adapted isomer composition of 40% all-*trans*,15-*anti* and 60% 13-*cis*,15-*syn* retinal occurs with a half-time of 180 min in the dark at 20 °C (Balashov *et al* 1996).

Dark-adapted proteorhodopsin contains 95% all-*trans* and 5% 13-*cis* PRSB (Imasheva *et al* 2005). Photoexcitation causes primary all-*trans* → 13-*cis* isomerization on a sub-picosecond and picosecond timescale (Imasheva *et al* 2005, Rupenyan *et al* 2008, Neumann *et al* 2008).

Dark-adapted halorhodopsin HR from *Halobacterium halobium* (later named *salinarum*) contains 30% all-*trans* PRSB ($\lambda_{a,max}$ = 578 nm in the presence of Cl⁻, $\lambda_{a,max}$ = 568 nm without the presence of Cl⁻) and 70% 13-*cis* PRSB ($\lambda_{a,max}$ = 568 nm with and without the presence of Cl⁻) (Lanyi 1986). Only the all-*trans* form exhibits the full photocycle and gives the light-driven Cl⁻ transport (Lanyi 1986). Photoexcitation causes primary all-*trans* → 13-*cis* isomerization on a picosecond timescale (Polland *et al* 1985, Arlt *et al* 1995, Peter *et al* 2006) with a quantum yield of 34% (Oesterhelt *et al* 1985). Light-adapted HR recovered to the dark-adapted isomer composition with a half-time of 24 h at 22 °C and pH 7 (Kamo *et al* 1985). Halorhodopsin exhibits some photochromic behavior: blue illumination shifts the isomeric composition more towards all-*trans* while red illumination of blue-adapted samples shifts it more towards 13-*cis* (Lanyi 1986).

Initial dark-adapted channelrhodopsin-2 ChR2 from *Chlamydomonas reinhardtii* contains 100% all-*trans*,15-*anti* PRSB ($\lambda_{a,max}$ = 480 nm) (Bruun *et al* 2015, Becker-Baldus *et al* 2015). Photoexcitation causes primary all-*trans*,15-*anti* → 13-*cis*,15-*syn* isomerization on a sub-picosecond timescale with subsequent conformational rearrangements on a picosecond timescale, proton release on a microsecond timescale, reprotonation within 3 ms (13-*cis*,15-*anti* PRSB) and re-isomerization to the initial all-*trans*,15-*anti* PRSB in a tri-exponential manner (6 ms, 33 ms, 3.4 s) (Verhoefen *et al* 2010). Initial light exposure causes a high intrinsic conductance for cations. Continued light exposure changes the high cation conductance to low cation conductance by light-adaptation (Ritter *et al* 2013, Bruun *et al* 2015, Schneider *et al* 2015, Penzkofer *et al* 2017). The conversion dynamics of initial dark-adapted ChR2 to light-adapted ChR2 depended on the wavelength of light-exposure, pH, and membrane voltage (Schneider *et al* 2015).

Microbial rhodopsins were shown to function as fluorescent voltage reporters (Kralj *et al* 2011a and 2011b). The endogeneous retinal chromophore shows weak near infrared fluorescence when a proton resides on the Schiff base linking the retinal to the protein core (protonated retinal Schiff base). Changes in membrane potential shift the local electrochemical potential of protons and thereby tune the fluorescence determining acid-base equilibrium. Directed evolution identified mutations of Archaerhodopsin 3 (Arch) from *Holorubrum sodomense* resulted in fluorescent genetically encoded voltage indicators (GEVIs) with imroved brightness, kinetics, and voltage sensitivity with eliminated proton-pumping photocurrent (McIsaac *et al* 2014, Hochbaum *et al* 2014, Gong *et al* 2015, St-Pierre *et al* 2015, Xu *et al* 2017, Piatkevich *et al* 2018).

13.3.2 Flavoproteins

Photosensitive flavoproteins are natural photosensors (Müller 1991a, Kritsky *et al* 2010, Conrad *et al* 2014), and genetically engineered flavoproteins play an important role in optogenetics (Christie *et al* 2012a, Glantz *et al* 2016, Fraikin *et al* 2016). The sub-families of the photosensitive flavoproteins are the LOV domains (Christie *et al* 2012a, Masuda 2013, Pudasaini *et al* 2015, Glantz *et al* 2016, Fraikin *et al* 2016), the BLUF proteins (Christie *et al* 2012a, Fraikin *et al* 2016, Penzkofer *et al* 2016b, Park and Tame 2017), the DNA photolyases and the cryptochromes (Chaves *et al* 2011, Park *et al* 2017).

13.3.2.1 LOV Proteins
LOV (light-oxygen-voltage) domains are a subset of the family of Per-Arnt-Sim (PAS) group of environmental sensors (Henry and Crosson 2011) that contain a non-covalently bound flavin cofactor FMN or FAD (for reviews see Zoltowski and Gardner 2011, Losi and Gärtner 2011, 2012, 2017, Herrou and Crosson 2011, Christie *et al* 2012a, Masuda 2013, Correa *et al* 2013, Conrad *et al* 2014, Pudasaini *et al* 2015, Glantz *et al* 2016, Fraikin *et al* 2016). They are conserved in bacteria, archaea, algae, plants and fungi. They were discovered in plant phototropins (Phot proteins) which consist of two LOV domains (LOV1 and LOV2) each with a FMN cofactor and a downstream serine/threonine kinase (Christie *et al* 1998, Zoltowski and Gardner 2011, Losi and Gärtner 2011, 2012). In LOV-based natural photo-receptors and optogenetic devices the blue-light sensory LOV domain is coupled to downstream signal transduction effector domains like ATP-binding histidine kine-ases HK (Swartz *et al* 2007, Alexandre *et al* 2010, Rivera-Cancel *et al* 2014), helix–turn–helix HTH DNA binding domains (Strickland *et al* 2008, Nash *et al* 2011, Zoltowski *et al* 2013), phosphodiesterases (Cao *et al* 2010), sulphate transporter/antisigma-factor antagonists STAS (Avila-Pérez *et al* 2009), sporulation stage II

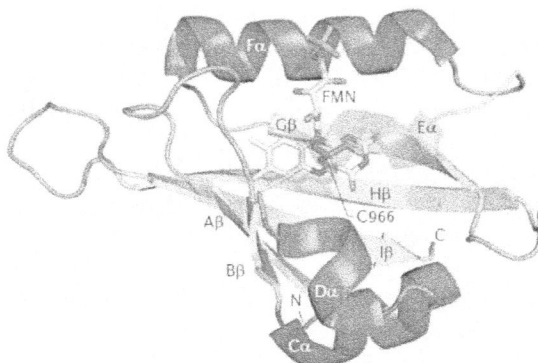

Figure 13.32a. Topology of a LOV domain (adapted from Herrou and Crosson 2011). Reprinted by permission from Springer Nature, copyright 2011.

protein E SpoIIE (Carniol *et al* 2005), and adenylyl cyclases (Raffelberg *et al* 2013, Chen *et al* 2014).

The topology of a LOV domain is shown in figure 13.32(a) (adapted from Herrou and Crosson 2011). Eα and Fα pack against the β-sheet to form a pocket to bind the flavin isoalloxazine ring. The Cys residue that forms the flavin–cysteinyl adduct in the photocycle resides on Eα within the conserved sequence GXNCRFLQ where X is any amino acid (Herrou and Crosson 2011, Conrad *et al* 2014).

Blue-light photoexcitation of LOV domains causes a temporary flavin–C4a cysteinyl adduct formation (covalent linkage between flavin C4a position and thiol moiety of the active site cysteine (Salomon *et al* 2000, Swartz *et al* 2001). The absorption cross-section spectrum of FMN–C4a cysteinyl adduct is included in the top part of figure 13.5 (from Holzer *et al* 2005a). LOV photocycle schemes have been developed (Swartz *et al* 2001, Crosson and Moffat 2001, Crosson *et al* 2003, Kottke *et al* 2003, Kernis *et al* 2003, Schleicher *et al* 2004, Dittrich *et al* 2005, Sato *et al* 2005, Alexandre *et al* 2009, Kutta *et al* 2015b). In figure 13.32(b) a photocycle scheme is shown, which is valid for LOV1 and LOV2 from *Chlamydomonas reinhardtii* (Holzer *et al* 2004, Holzer *et al* 2005a). The structural formulae of FMN and Cys involved in the photocycle of figure 13.32(b) are shown in figure 13.32(c) (Cys partly denoted as HS-Cys). The blue-light photoexcitation of FMN in LOV changes FMN from the S_0 ground-state to the S_1 excited state (indicated by ^{1}LOV*). From there occurs on a nanosecond timescale intersystem crossing to the T_1 state of FMN (indicated by ^{3}LOV*) and reductive electron transfer eT from adjacent ground-state cysteine residue to the single electron occupied HOMO of ^{1}FMN* forming the FMN$^{\cdot-}$-Cys$^{\cdot+}$ radical ion pair (LOV charge-transfer complex LOV$_{CT}$). The formed ^{3}LOV* converts within the microsecond lifetime of the ^{3}FMN* state to LOV$_{CT}$ again by reductive electron transfer from adjacent ground-state cysteine residue to the single electron occupied HOMO of ^{3}FMN*. The electrostatic force acting between the radical anion FMN$^{\cdot-}$ and the radical cation Cys$^{\cdot+}$ seems to be responsible for the covalent FMN–C4a cysteinyl adduct formation. The formed FMN-C4a–Cys adduct is in most LOVs thermally

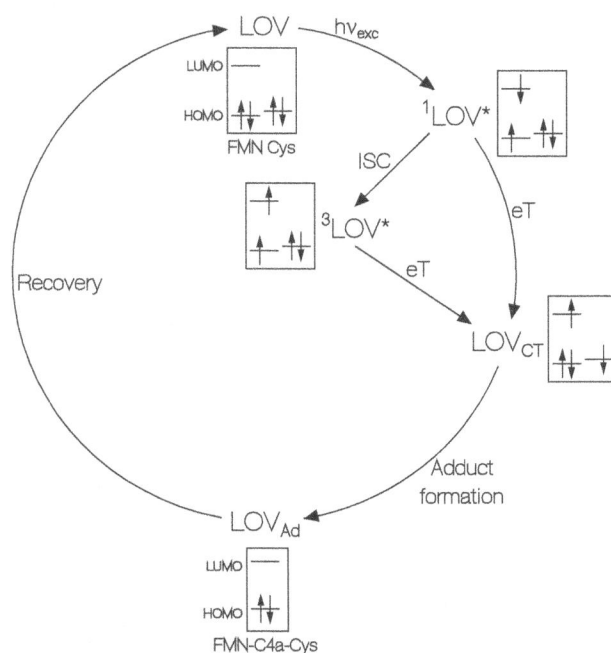

Figure 13.32b. Photocycle scheme of LOV domain in *Chlamydomonas reinhardtii* (from Holzer *et al* 2004, 2005a).

not stable and recovers back to the original unbound FMN and Cys-protein structure on a second-to-minute timescale for phototropin. However, photo-activated LOV-PAS-histidine kinases from *Brucella* spp. show only negligible back reaction to the dark state with almost infinite kinase activity (Swartz *et al* 2007, Rinaldi *et al* 2012). The ^{1}LOV* to LOV$_{CT}$ electron transfer channel besides the ^{3}LOV* to LOV$_{CT}$ electron transfer channel is concluded from the reduced FMN fluorescence quantum yield, the shortened fluorescence lifetime (Holzer *et al* 2002, Holzer *et al* 2004), and the reduced quantum yield of triplet formation (Islam *et al* 2003a) compared to FMN in aqueous solution.

The flavin–C4a cysteinyl adduct formation causes changes of the flavin structure and the LOV domain structure. Especially the light-induced restructuring of the Jα helix connecting the LOV domain to the effector domain (LOV domain-effector domain unfolding, tilting, rotation, dimerization) causes effector domain action (Herrou and Crosson 2011, Pudasaini *et al* 2015). Depending on the specific domain architecture of LOV domain containing proteins there is a range of flavin–C4a cysteinyl adduct stability (flavin–C4a cysteinyl adduct signaling state recovery in the dark to noncovalently bound flavin in the LOV domain pocket in the time range from seconds up to a day at room temperature). The recovery back from the adduct state to the non-covalently bound receptor state requires flavin N5 deprotonation, covalent bond breakage between Cys–S and flavin–C4a, and Cys–S protonation to Cys–SH. The rate of adduct decay depends on solvent access to the active site, the

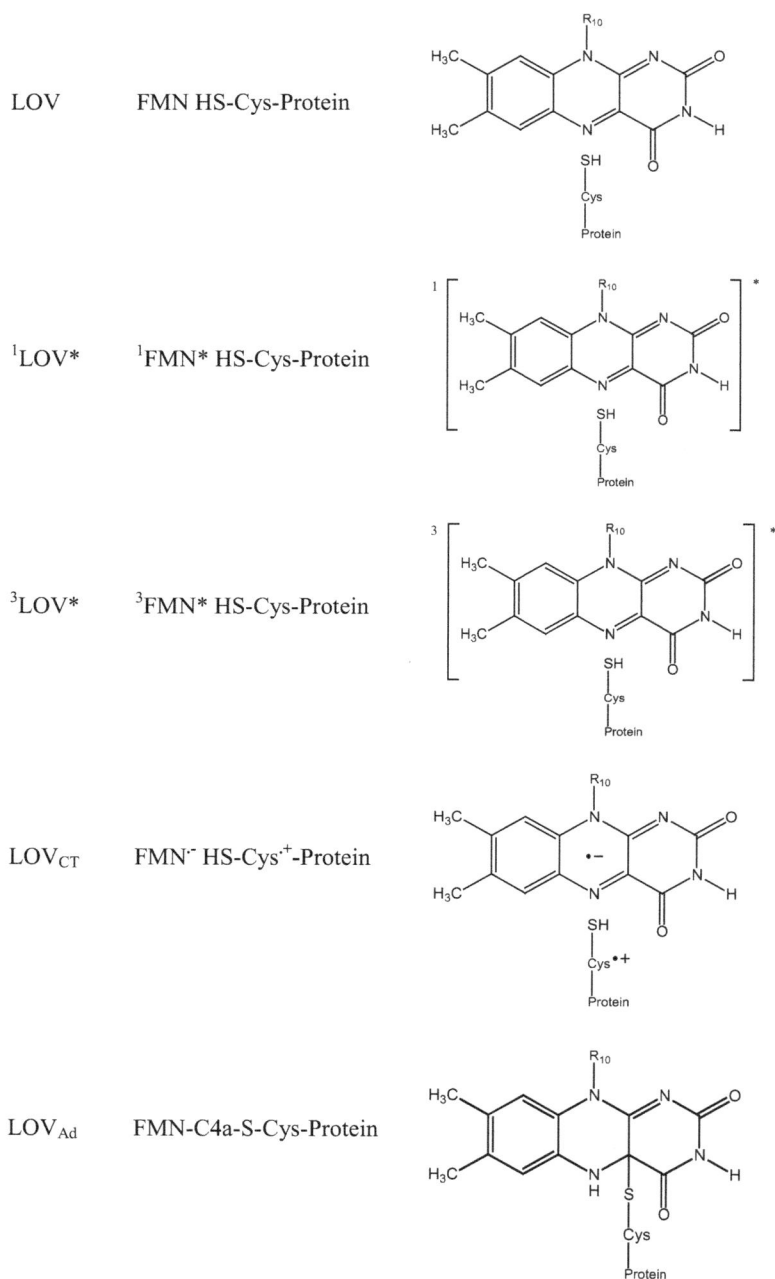

LOV	FMN HS-Cys-Protein
^{1}LOV*	^{1}FMN* HS-Cys-Protein
^{3}LOV*	^{3}FMN* HS-Cys-Protein
LOV$_{CT}$	FMN$^{•-}$ HS-Cys$^{•+}$-Protein
LOV$_{Ad}$	FMN-C4a-S-Cys-Protein

Figure 13.32c. Structural formulae of FMN and Cys involved in LOV photocycle. R_{10} = ribityl-mono-phosphate. SH group of Cys is explicitly shown.

hydrogen bonding network to the flavin cofactor, and the electronic environment surrounding the flavin (Conrad *et al* 2014, Pudasaini *et al* 2015). During violet light exposure of the LOV domain adduct state it is converted to the receptor state by adduct state photoexcitation (Kennis *et al* 2004, Penzkofer *et al* 2005, Losi *et al* 2013).

FMN in LOV domains loses its fluorescence property when photo-converted to FMN-C4a cysteinyl adduct in the photocycling process. Mutated LOV proteins where the active cysteine needed for adduct formation was replaced by a non-reactive serine (Song *et al* 2007b) or aspartic acid (Kopka *et al* 2017) changed to FMN semiquinone photo-cycling with loss of FMN fluorescence emission in the FMN semiquinone light-adapted state (FMN_{sq} is only weakly fluorescent according to Kao *et al* 2008). In other cases the replacement Cys with alanine led to no photocycling and hindered fluorescence quenching opening the application of these mutated LOV domains as flavin-based fluorescent proteins (FbFP) (for reviews see Mukherjee and Schroeder 2015, Buckley *et al* 2015). Fluorescent protein applications were achieved for the flavin based bacterial blue-light photoreceptor YtvA from *Bacillus subtilis* (BsFbFP), its *Escherichia coli* codon-optimized counterpart (EcFbFP), for sensory box protein SB2 from *Pseudomonas putida* (PpFbFP) (Drepper *et al* 2007), and for photoreversible LOV proteins with improved fluorescent properties derived from the LOV2 domain of phototropin 2 of *A. thaliana* (named iLOV, Chapman *et al* 2008, Christie *et al* 2012c). Compared to classical fluorescent proteins of section 13.2.9 the mutated LOV domain based fluorescent proteins have the advantage of smaller size, their applicability under aerobic and anaerobic conditions, and their high pH, thermal and photochemical stability (Chapman *et al* 2008, Christie *et al* 2012c).

13.3.2.2 Photolyases and cryptochromes

Photolyases are a class of flavoproteins with the blue/ultraviolet light active non-covalently bound U-shaped cofactor FAD_{red}^- and the non-covalently bound light-harvesting antenna chromophore 5,10-methenyltetrahydrofolate MTHF or 8-hydroxydeazaflavin (8-HDF). They use blue light to repair two types of ultraviolet light-induced DNA damages, namely cyclobutane pyrimidine dimer and pyrimidine-pyrimidone (6-4) photoproduct (Sancar 2003, Liu *et al* 2015, Zhang *et al* 2017). Cryptochromes are composed of an N-terminal photolyase homology region (PHR) with non-covalently bound FAD_{ox} and MTHF cofactors and a cryptochrome C-terminal extension CCE acting as an effector domain. They act as blue-light photoreceptors, but have in most cases lost the DNA repair ability of the photolyases (Öztürk *et al* 2007, Chaves *et al* 2011).

The photolyase/cryptochrome superfamily contains eight major clades (Öztürk *et al* 2007, Zhang *et al* 2017, Ozturk 2017): (i) plant cryptochromes, (ii) class III photolyases, (iii) class I photolyases, (iv) single strand DNA photolyases (previously classified as CRY with CRY-DASH designation), (v) animal type I cryptochromes (insect Cry1), (vi) (6-4) photolyases, (vii) animal type II cryptochromes (insect Cry2 and vertibrate cryptochromes), and (viii) class II photolyases. Further sub-grouping of animal cryptochromes is found in Haug *et al* (2015).

Class I, class II and class III photolyases have been found in archaea, bacteria, plants, insects, and vertebrates (Hitomi *et al* 2000). They repair UV-induced DNA lesions in form of cyclobutane pyrimidine dimers (CPD T<>T, they are called CPD photolyases). (6-4) Photolyases repair UV-induced DNA lesions in form of pyrimidine (6-4) pyrimidone photoproducts (6-4PPs). They have been found in some higher eukaryotes (Hitomi *et al* 2000). Single strand DNA photolyase (CRY-DASH, cry 3 in *A. thaliana*) repair CPDs specifically in single-stranded DNA (ssDNA) and in loop structures of double-stranded DNA (for review see Chaves *et al* 2011). The active FAD form for DNA photo-repair by photolyases is FAD_{red}^-. Whenever FAD is not in this state, photolyase may reduce FAD to FAD_{red}^- via another light-induced reaction called photoactivation (Martin *et al* 2017).

Plant cryptochromes have U-shaped FAD_{ox} in their receptor state. They regulate the circadian clock and control photomorphogenesis, photoperiodic flower induction, and other plant growth and development processes (Ahmad 2016, Yang *et al* 2017b). Animal cryptochromes have divergent roles in light perception, circadian timekeeping and beyond (Michael *et al* 2017). Type I animal cryptochromes (in invertebrates, also called *Drosophila*-type cryptochromes) are directly light-sensitive and act as circadian photoreceptor (in receptor state FAD_{ox} is U-shaped). Type II animal cryptochromes (found in vertebrates, there called vertebrate-type cryptochromes, and in invertebrates) have no light-dependent function and act as core clock proteins. Animal type II cryptochromes predominately do not bind FAD *in vivo* and should be seen as vestigal flavoproteins (Kutta *et al* 2017).

Photoexcitation of cryptochromes is expected to cause magnetic field-dependent sensing effects in plants, insects, and birds (Ritz *et al* 2010, Gegear *et al* 2010, Nießner *et al* 2013, Wiltschko and Wiltschko 2014, Hore and Mouritsen 2016, Qin *et al* 2016, Mouritsen 2018, Zeng *et al* 2018). Geomagnetic field impacts on cryptochrome and phytochrome plant signaling was observed (Agliassa *et al* 2018). Under blue light, the geomagnetic field regulation of gene expression in *Arabidopsis thaliana* was found to partly depend on cryptochrome activation.

The dynamics of the repair of the DNA cyclobutane pyrimidine dimer (T<>T) damage by CPD photolyase is described in Liu *et al* (2015). The reaction is shown by Scheme 1. Photoexcitation of FAD_{red}^- directly and via excitation energy transfer (ET) from photo-excited MTHF (or 8-HDF) causes an electron transfer (eT) from FAD_{red}^- to T<>T leading to FAD_{sq} and $T - T^-$. $T - T^-$ separates to $T + T^-$. Finally, electron back-transfer (electron return eR) occurs from T^- to FAD_{sq} giving FAD_{red}^- and the repaired thymine dinucleotide T + T.

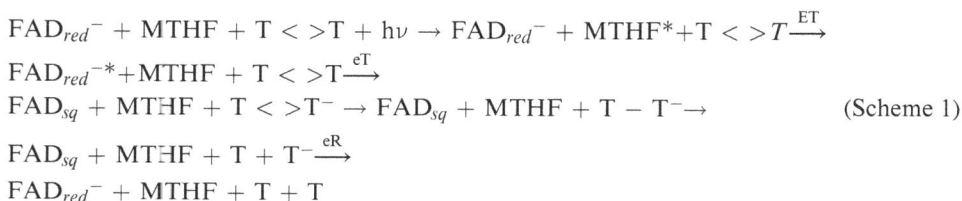

$FAD_{red}^- + MTHF + T <> T + h\nu \rightarrow FAD_{red}^- + MTHF^* + T <> T \xrightarrow{ET}$

$FAD_{red}^{-*} + MTHF + T <> T \xrightarrow{eT}$

$FAD_{sq} + MTHF + T <> T^- \rightarrow FAD_{sq} + MTHF + T - T^- \rightarrow$ (Scheme 1)

$FAD_{sq} + MTHF + T + T^- \xrightarrow{eR}$

$FAD_{red}^- + MTHF + T + T$

The dynamics of the repair of the DNA pyrimidine-pyrimidone (6-4) photo-product (6-4PP) is also described by Liu *et al* (2015). The reaction is shown in Scheme 2. Direct photoexcitation of FAD_{red}^- or via energy transfer from excited MTHF (or 8-HDF) causes an electron transfer (eT) from FAD_{red}^- to 6-4PP leading to FAD_{sq} and 6-4PP·$^-$. It follows a proton transfer (pT) from a neighboring histidine residue to change 6-4PP·$^-$ to 6-4PPH·. Then bond rearrangement occurs followed by proton and electron return (pR + eR) to FAD_{red}^- and the repaired thymine dinucleotide T + T.

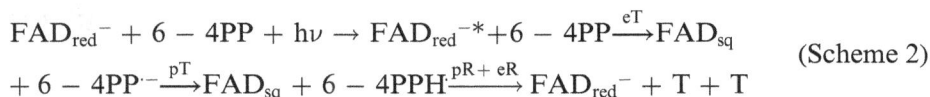

$$FAD_{red}^- + 6 - 4PP + h\nu \rightarrow FAD_{red}^{-*} + 6 - 4PP \xrightarrow{eT} FAD_{sq}$$
$$+ 6 - 4PP^{\cdot-} \xrightarrow{pT} FAD_{sq} + 6 - 4PPH \xrightarrow{pR + eR} FAD_{red}^- + T + T \quad \text{(Scheme 2)}$$

A more detailed description of the CPD photocycle and the (6-4) photolyase photocycle including intra-molecular photo-induced electron transfer from the isoalloxazine part to the adenine part of FAD_{red}^- is found in Zhang *et al* 2017.

The CRY-DASH cryptochromes (single strand DNA photolyases) retain the photolyase repair activity for single-stranded DNA and loops of double-stranded DNA (Chaves *et al* 2011). The cofactors of *A. thaliana* CRY-DASH, named *At*cry3, were identified to be FAD and MTHF (Göbel *et al* 2017). The photocycle dynamics of *At*cry3 was studied by Song *et al* (2006) and Zirak *et al* (2009). In *At*cry3 FAD was found to be present in oxidized form (FAD_{ox}), semiquinone form (FAD_{sq}) and anionic hydroquinone form (FAD_{red}^-). The photocycle dynamics induced by FAD_{ox} excitation is shown in scheme 3 (Zirak *et al* 2009). In a primary photocycle photoexcitation of FAD_{ox} to FAD* caused the formation of $FAD_{ox}^{\cdot-}$ (= FAD_{sq}^-) by intermolecular electron transfer from an adjacent Trp in a Trp triad (for theoretical description see Firmino *et al* 2016) and back recovery to FAD_{ox} in the dark on a 10 s timescale (Scheme 3a). Continued light exposure caused $FAD_{ox}^{\cdot-}$ excitation to $FAD_{ox}^{\cdot-*}$ and recovery to FAD_{ox} by electron release (Scheme 3a). The primary photocyle caused protein re-conformations changing to a secondary photocycle dynamics where FAD_{ox} excitation caused the formation of FAD_{sq} by electron and proton transfer. The excitation of the formed FAD_{sq} caused further reduction to FAD_{red}^- by a further electron transfer. After light switch-off a new equilibrium distribution between FAD_{red}^-, FAD_{sq} and FAD_{ox} established on a minute-to-hour timescale (Scheme 3b).

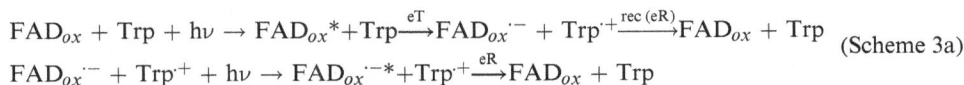

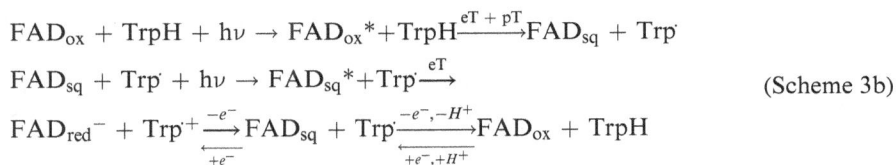

$$FAD_{ox} + Trp + h\nu \rightarrow FAD_{ox}^* + Trp \xrightarrow{eT} FAD_{ox}^{\cdot-} + Trp^{\cdot+} \xrightarrow{rec(eR)} FAD_{ox} + Trp \quad \text{(Scheme 3a)}$$
$$FAD_{ox}^{\cdot-} + Trp^{\cdot+} + h\nu \rightarrow FAD_{ox}^{\cdot-*} + Trp^{\cdot+} \xrightarrow{eR} FAD_{ox} + Trp$$

$$FAD_{ox} + TrpH + h\nu \rightarrow FAD_{ox}^* + TrpH \xrightarrow{eT + pT} FAD_{sq} + Trp^{\cdot}$$
$$FAD_{sq} + Trp^{\cdot} + h\nu \rightarrow FAD_{sq}^* + Trp^{\cdot} \xrightarrow{eT} \quad \text{(Scheme 3b)}$$
$$FAD_{red}^- + Trp^{\cdot+} \underset{+e^-}{\overset{-e^-}{\rightleftharpoons}} FAD_{sq} + Trp^{\cdot} \underset{+e^-,+H^+}{\overset{-e^-,-H^+}{\rightarrow}} FAD_{ox} + TrpH$$

The photocycle dynamics of plant cryptochromes (especially the *A. thaliana* cryptochromes Atcry1 and Atcry2) was studied by Procopio *et al* (2016) and Ahmad (2016). The photoexcitation of FAD_{ox} leads to the formation of FAD_{sq} by electron and proton transfer. After light-switch-off, there occurs back-recovery to FAD_{ox} (primary photocycle, Scheme 4a). The photoexcitation of the formed FAD_{sq} causes reduction to FAD_{red}^-, which recovers to FAD_{ox} in the dark (secondary photocycle, Scheme 4b)

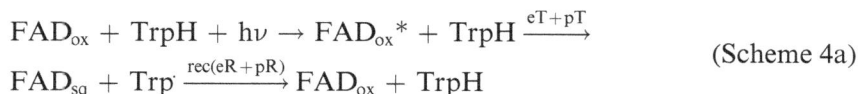

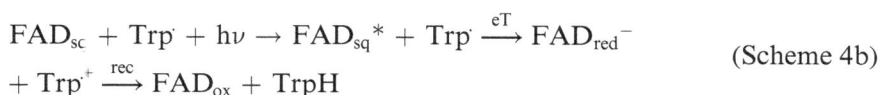

$$FAD_{ox} + TrpH + h\nu \rightarrow FAD_{ox}^* + TrpH \xrightarrow{eT+pT}$$
$$FAD_{sq} + Trp \xrightarrow{rec(eR+pR)} FAD_{ox} + TrpH$$
(Scheme 4a)

$$FAD_{sc} + Trp + h\nu \rightarrow FAD_{sq}^* + Trp \xrightarrow{eT} FAD_{red}^-$$
$$+ Trp^+ \xrightarrow{rec} FAD_{ox} + TrpH$$
(Scheme 4b)

In Scheme 4a and b tryptophan is written as TrpH.

The animal type I photocycle dynamics of the fruit fly *Drosophila melanogaster* dcry was studied by Shirdel *et al* (2008). The dynamics is shown in Scheme 5. The photoexcitation of FAD_{ox} caused reductive electron transfer from an adjacent Trp to FAD_{ox} changing it to FAD quinone radical anion $FAD_{ox}^{\cdot-}$ (= anionic FAD semiquinone FAD_{sq}^-). In the dark $FAD_{ox}^{\cdot-}$ recovered back to FAD_{ox} with a time constant of $\tau_{rec} = 1.6$ min (Scheme 5a). The formed $FAD_{ox}^{\cdot-}$ absorbs in the same wavelength region as FAD_{ox} (see figure 13.5). Therefore, during light exposure the formed $FAD_{ox}^{\cdot-}$ got photo-excited and returned back to FAD_{ox} by electron transfer from $FAD_{ox}^{\cdot-*}$ to the adjacent $Trp^{\cdot+}$ (Scheme 5b).

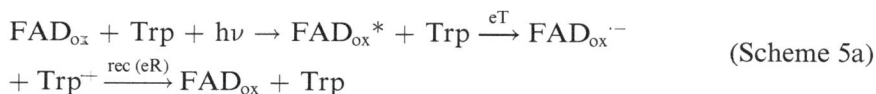

$$FAD_{ox} + Trp + h\nu \rightarrow FAD_{ox}^* + Trp \xrightarrow{eT} FAD_{ox}^{\cdot-}$$
$$+ Trp^+ \xrightarrow{rec(eR)} FAD_{ox} + Trp$$
(Scheme 5a)

$$FAD_{ox}^{\cdot-} + Trp^+ + h\nu \rightarrow FAD_{ox}^{\cdot-*} + Trp^+ \xrightarrow{eR} FAD_{ox} + Trp \quad \text{(Scheme 5b)}$$

Optogenetic applications involving plant cryptochromes have been carried out. Genetically encoded *A. thaliana* CRY2 and *A. thaliana* CIB1 (cryptochrome-interacting basic-helix-loop-helix 1) dimerize due to blue-light exposure with sub-second time resolution and subcellular spatial resolution (Kennedy *et al* 2010, Bugaj *et al* 2013, Wend *et al* 2014). In particular, this system was used to blue-light induced protein translocation, transcription, and Cre recombinase-mediated DNA recombination (Kennedy *et al* 2010). The light-induced AtCRY2 homo-oligomerization and AtCRY2-AtCIB1 heterodimerization for optogenetic manipulation in mammalian cells was characterized by Che *et al* (2015). Important determinants for efficient AtCRY2 clustering were found, and the introduction of a CRY2 module with a newly identified nine residues long peptide tag and diverse fluorescent proteins, named CRY2clust, allowed blue-light induced rapid and efficient homo-oligomerization of target proteins capable of regulating fine cellular signaling in living cells in response to light (Park *et al* 2017). An *A. thaliana* CRY2 E490G mutant, named CRY2olig, was expressed in mammalian cells (Taslimi *et al* 2014), and every expressed cell illuminated with blue light underwent rapid, reversible, and robust clustering within seconds, redistributing a majority of cytosolic proteins into clusters. The CRY2olig optogenetic module was used to interrogate protein interaction dynamics in live cells, to induce and reversibly control diverse cellular processes with spatial and temporal resolution (Taslimi *et al* 2014).

13.3.2.3 BLUF proteins

BLUF (= sensor for Blue Light Using FAD) domains (Gomelsky and Klug 2002) were reviewed by Losi (2007), Losi and Gärtner (2011, 2012), Christie *et al* (2012a), Masuda 2013, Park and Tame (2017), Fujisawa and Masuda (2018). They adopt a ferredoxin-like $\beta_1\alpha_1\beta_2\beta_3\alpha_2\beta_4$ fold (Wu *et al* 2008, Wu and Gardner 2009) made up of about 95 amino acids that envelop the isoalloxazine ring and the ribityl chain of the non-covalently bound flavin (FAD and also FMN and riboflavin (Laan *et al* 2004)). The topology of a BLUF domain is shown in figure 13.33(a) (adapted from Winkler *et al* 2014). BLUF domains have been found in bacteria and lower eukaryotes.

Blue-light photoexcitation of BLUF domains causes a temporary spectral red shift of about 10 nm of the flavin S_0–S_1 absorption band due to photo-induced reductive electron transfer from an adjacent Tyr to the excited flavin with subsequent protein conformational changes and hydrogen bond restructuring including flavin and the neighboring tyrosine and glutamine (Christie *et al* 2012a, Masuda 2013, Mathes and Götze 2015, Park and Tame 2017, and references in these publications). In the dark the BLUF domain is in its receptor state $BLUF_r$

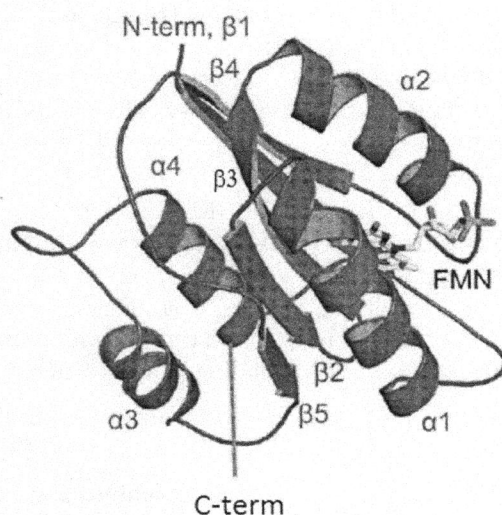

Figure 13.33a. Topology of a BLUF domain. Adapted from Winkler *et al* (2014). Copyright 2014. This article is available under a CC BY 3.0 licence https://creativecommons.org/licenses/by/3.0/.

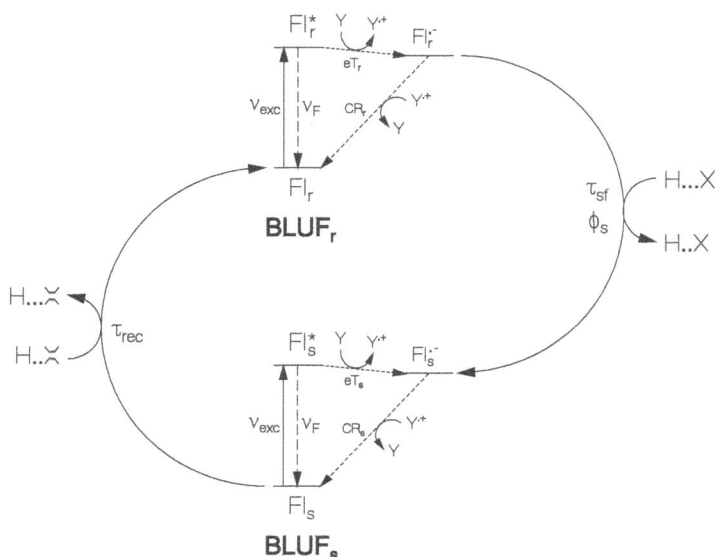

Figure 13.33b. Primary photocycle dynamics of a BLUF domain. $BLUF_r$: BLUF domain in receptor state. $BLUF_s$: BLUF domain in signaling state (from Penzkofer *et al* 2016b).

(dark-adapted state), and due to blue light absorption it changes to the red-shifted signaling state $BLUF_s$ (light-adapted state). After light-switch-off the signaling state recovers back to the receptor state on a second to minute time scale (Zirak *et al* 2005, 2006, 2007a, 2007b). As an example, in figure 13.34 the absorption cross-section spectra of the BLUF protein BlrB (= Blue-light receptor B) of the purple bacterium *Rhodobacter sphaeroides* in the receptor state ($BlrB_r$) and in the signaling state ($BlrB_s$) are shown (from Zirak *et al* 2006).

A general primary flavin photocycle scheme of BLUF domains is displayed in figure 13.33(b) taken from Penzkofer *et al* 2016b). The photoexcitation of Fl_r in the $BLUF_r$ receptor state to the first excited singlet state Fl_r^* causes electron transfer (eT_r) from Tyr to Fl_r^* ($Tyr + Fl_r^* \rightarrow Tyr^{\bullet +} + Fl_r^{\bullet -}$). During the $Tyr^{\bullet +} - Fl_r^{\bullet -}$ radical ion-pair lifetime $\tau_{Fl_r^{\bullet -}}$ there occurs a BLUF domain conformation alteration with hydrogen bond restructuring to $BLUF_s$ due to the ion-pair electrostatic force (Zirak *et al* 2007a) changing $Tyr^{\bullet +} - Fl_r^{\bullet -}$ to $Tyr^{\bullet +} - Fl_s^{\bullet -}$. The anionic flavin radical $Fl_r^{\bullet -}$ changes to $Fl_s^{\bullet -}$ with quantum yield ϕ_s of signaling state formation and to Fl_r with quantum yield $\phi_{CR,r} = 1 - \phi_s$ of charge recombination into the receptor state (CR_r). The time constant of signaling state formation is $\tau_{sf} = \tau_{Fl_r^{\bullet -}}/\phi_s$.

The anionic flavin radical $Fl_r^{\bullet -}$ recovers to neutral Fl_r with the time constant $\tau_{CR_r} = \tau_{Fl_r^{\bullet -}}/(1 - \phi_s)$ according to $Tyr^{\bullet +} + Fl_r^{\bullet -} \rightarrow Tyr + Fl_r$. The anionic flavin radical $Fl_s^{\bullet -}$ recovers to neutral Fl_s by charge recombination in the signaling state with time constant $\tau_{CR_s} = \tau_{Fl_r^{\bullet -}}$ according to $Tyr^{\bullet +} + Fl_s^{\bullet -} \rightarrow Tyr + Fl_s$. Fl_s in the $BLUF_s$ signaling state recovers back to Fl_r in the $BLUF_r$ receptor state with time constant τ_{rec} by thermal activated protein back re-conformation and hydrogen bond

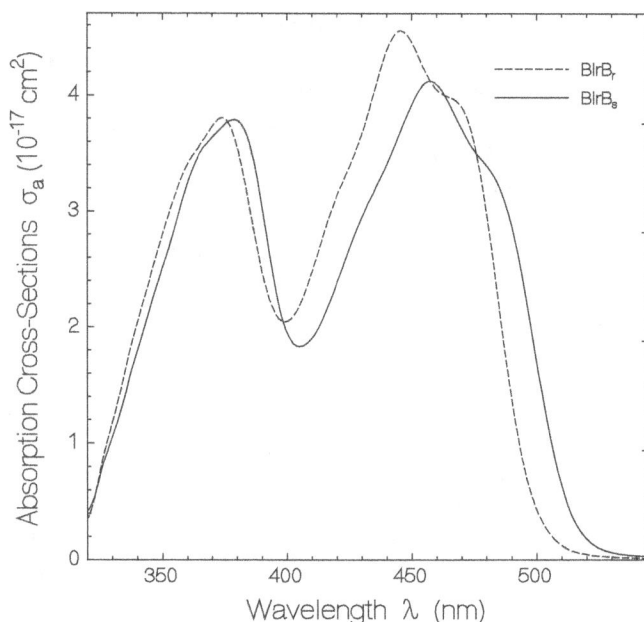

Figure 13.34. Absorption cross-section spectra of BLUF protein BlrB from *Rhodobacter sphaeroides* in the receptor state (BlrB$_r$) and in the signaling state (BlrB$_s$) (from Zirak *et al* 2006).

back restructuring. Photoexcitation of Fl$_s$ in the BLUF$_s$ signaling state to Fl$_s$* causes Tyr to Fl$_s$* electron transfer (eT$_s$) to Tyr$^{\cdot+}$– Fl$_s^{\cdot-}$ ion-pair formation and subsequent charge recombination (CR$_s$) to Tyr + Fl$_s$.

BLUF domains are found in many microorganisms either as single sensor domain proteins or as sensor part in multi-domain sensor actuator proteins. In single light sensing BLUF proteins their excitation to the signaling state causes the activation of downstream protein modules via inter-protein interaction. In BLUF domain containing multi-domain proteins the activation occurs via domain–domain interaction (Zoltowski and Gardner 2011, Masuda 2013).

Single BLUF domain proteins which have been characterized are: (i) BlrB from purple bacterium *Rhodobacter sphaeroides* of unknown function (Jung *et al* 2005, Zirak *et al* 2006, Wu *et al* 2008), (ii) Tll0078 (TePixD) from cyanobacterium *Thermosynechococcus elongates* of unknown function (Kita *et al* 2005, Okajima *et al* 2006), (iii) Slr1694 (SyPixD) from cyanobacterium *Synechocystis* sp. PCC6803 (Hasegawa *et al* 2005, Okajima *et al* 2006, Yuan *et al* 2006, Zirak *et al* 2007b) which interacts with the response regulator like protein PixE to control photo-orientation (Okajima *et al* 2006, Fujisawa and Masuda 2018), and (iv) PapB from purple bacterium *Rhodopseudomonas palustris* (Kanazawa *et al* 2010) which interacts with PapA EAL protein in the blue light-dependent degradation of the cyclic diguanylate monophosphate (c-di-GMP).

BLUF domain containing proteins combine a BLUF light sensing domain with an effector domain. Up to now studied effector domains are (i) SCHIC (= *sensor* containing *h*eme *i*nstead of *c*obalamin), (ii) EAL (= Glu(*E*)-*A*la-*L*eu), (iii) PAC (= *p*hotoactivated *a*denylyl *c*yclase), (iv) PGC (= *p*hotoactivated *g*uanylyl *c*yclase) and (v) PAE (= *p*hoto*a*ctivated *e*ndonuclease III).

(i) *BLUF coupled SCHIC proteins*: AppA (= activation of photopigment and puc expression protein) from *Rhodopacter sphaeroides* has a BLUF and a SCHIC domain (Braatsch *et al* 2002). It controls photosynthesis gene expression (Masuda and Bauer 2002, Kraft *et al* 2003, Winkler *et al* 2013).

(ii) *BLUF coupled EAL proteins*: BLUF coupled EAL proteins contain a BLUF domain and an EAL domain (EAL = Glu-Ala-Leu). EAL domains exhibit phosphodiesterase activity (Schmidt *et al* 2005, Tamayo *et al* 2005) against cyclic dimeric guanosine monophosphate (c-di-GMP), a global second messenger used by bacteria to control different cellular functions (Romling *et al* 2005, Jenal and Malone 2006, Tamayo *et al* 2007). BlrP1 (= blue light regulated phosphodiesterase 1) from *Klebsiellia pneumonia* has a BLUF domain and an EAL domain. It causes photoactivated c-di-GMP second messenger deactivation (Tyagi *et al* 2008, Barends *et al* 2009, Khrenova *et al* 2011, Winkler *et al* 2014, Shibata *et al* 2018). In YcgF from *E. coli* which controls biofilm formation (Tschowri *et al* 2009, 2012) BLUF is functionally coupled to an EAL domain via a joining helix (Rajagopal *et al* 2004, Hasegawa *et al* 2006, Nakasone *et al* 2007, Schroeder *et al* 2008).

(iii) *BLUF coupled photoactivated adenylyl cyclases*: In photoactivated adenylyl cyclases (PACs) a BLUF domain is connected to a class III adenylyl cyclase domain (see Penzkofer *et al* 2016b and references therein). The class III adenylyl cyclases convert adenosine triphosphate ATP to cyclic adenosine-monophosphate (cAMP) which acts as second messenger in cells (Lindner 2006). In the unicellular flagellate *Euglena gracilis* two euPACα and two euPACβ proteins, which form a heterotetrameric crystalline complex at the flagellar basis, are responsible for the photo-avoidance reaction of the flagellate (Iseki *et al* 2002, Schröder-Lang *et al* 2007). Genes coding BLUF-PACs have been identified in the genome of several micro-organisms through metagenome analysis. These BLUF-PACs were then synthesized, and the recombinant (= biotechnologically produced) proteins were characterized with the possibility of their application as optogenetic tools. In this way the following BLUF-PAC proteins have been characterized: bPAC from the soil bacterium *Beggiatoa* sp. (Ryu *et al* 2010, Stierl *et al* 2011, 2014, Penzkofer *et al* 2014a, Lindner *et al* 2017), NgPAC1 (nPAC) (Penzkofer *et al* 2011), NgPAC2 (Penzkofer *et al* 2013), and NgPAC3 (Penzkofer *et al* 2014b) from the unicellular eukaryotic amoeboflagellate *Naegleria gruberi*, NaPAC1 and NaPAC2 from the unicellular eukaryotic amoeboflagellate *Naegleria australiensis* (Yasukawa *et al* 2013), LiPAC from the spirochete bacterium *Leptonema illini* (Penzkofer *et al* 2014c), TpPAC from the spirochete bacterium *Turneriella parva* (Penzkofer *et al* 2015), and OaPAC from the photosynthetic cyanobacterium *Oscillatoria acuminata* (Ohki *et al* 2016). euPACα from *E. gracilis* and bPAC from *Beggiatoa* sp. have already been utilized in various optogenetics applications (see references in Penzkofer *et al* 2016b, Beck *et al* 2018). NgPACs and bPAC were applied as optogenetic tools for controlling protein kinase activity in mammalian cell lines (Tanwar *et al* 2017).

(iv) *BLUF coupled photoactivated guanylyl cyclases*: Photoactivated guanylyl cyclases (PGCs) have a BLUF domain and a guanylyl cyclase domain. The guanylyl cyclases catalyze the conversion of guanosine triphosphate GTP to cyclic guanosine-monophosphate (cGMP) which acts as second messenger in cells (Potter 2011). The BLUF coupled photoactivated guanylyl cyclase (BLUF-PGC) BlgC from *Beggiatoa* was recombinant expressed. It caused blue-light activated conversion of GTP to cGMP in *E. coli* (Ryu *et al* 2010). Photoactivated adenylyl cyclases from microbes have been converted to photoactivated guanylyl cyclases by mutagenesis of the substrate binding specific residues in the cyclase homology domain (Tanwar *et al* 2018).

(v) *BLUF coupled photoactivated endonucleases*: The photoactivated endonuclease III RmPAE from the mesophilic pink to light reddish-pigmented bacterium *Rubellimicrobium mesophilum* having a BLUF and an endonuclease III domain was engineered. It may cause light enhanced DNA digesting (Penzkofer *et al* 2016a).

13.3.3 Xanthopsins (photoactive yellow proteins)

Photoactive yellow proteins are a group of blue-light photoreceptor proteins that contain *p*-hydroxycinnamic acid (= *p*-coumaric acid) as photoactive chromophore (for reviews see Hellingwerf *et al* 2003, Van der Horst *et al* 2005a, Imamoto and Kataoka 2007, Kumauchi *et al* 2008, Yang *et al* 2017a). They are also named xanthopsins in analogy to the rhodopsins (Kort *et al* 1996, Hellingwerf *et al* 2003). The cofactor is covalently linked by a thiol-ester bond to a cysteine residue in the protein. The xanthopsins have been discovered in a range of proteobacteria (Kumauchi *et al* 2008). They act in a wide range of processes from genetic regulation of chalcone synthesis to tactic migration of bacteria, but the physiology is still quite unclear. The archetype of xanthopsins is the photoactive yellow protein from the purple-sulfur bacterium *Ectothiorhodospira halophila* (now called *Halorhodospira halophila*) which is referred to as E-PYP or simply PYP (Hellingwerf *et al* 2003). PYP is a highly soluble globular protein. The α/β fold structure of PYP is a structural prototype of the PAS domain superfamily (Imamoto and Kataoka 2007).

A simplified photocycle of E-PYP is shown in Scheme 6 (adapted from Ujj *et al* 1998, Groenhof *et al* 2002, Van der Horst *et al* 2005b, Ihee *et al* 2005, Tenboer *et al* 2014, Pande *et al* 2016). The single anionic *p*-coumaric acid chromophore in the dark adapted state of PYP (abbreviated by pG) is in the 7-*trans*,9-*s*-*cis* conformation. Its absorption maximum is at $\lambda_{a,max} = 446$ nm. Photoexcitation (pG*) causes 7-*trans*,9-*s*-*cis* to 7-*cis*,9-*s*-*trans* isomerization within 3 ps (I_0 intermediate) with $\lambda_{a,max} = 510$ nm (structural formulae are shown in figure 13.15). Some chromophore-protein interaction with H-bond restructuring changes the absorption maximum to $\lambda_{a,max} = 465$ nm (pR intermediate, also named I_1 intermediate). It follows the phenolate anion protonation to neutral *p*-coumaric acid within 200 µs (pB intermediate, also named I_2 intermediate) with $\lambda_{a,max} = 355$ nm. In this *cis*, protonated pB state the protein partially unfolds. The light adapted intermediate pB (signaling state) recovers back in the dark to the initial receptor state pG with a time constant of 140 ms by deprotonation and *cis* to *trans* back-isomerization.

$$pG(trans,\ 446\ nm) \xrightarrow{h\nu_{exc}} pG^*(trans) \xrightarrow{\leqslant\ 3ps,\ cis-isomerization}$$

$$I_0(cis,\ 510\ nm) \xrightarrow{3\ ns,\ H-bond\ restructuring} pR(I_1,\ cis,\ 465\ nm) \xrightarrow{200\ \mu s,\ protonation} \quad (Scheme\ 6)$$

$$pB(I_2,\ cis,\ 355\ nm) \xrightarrow{140\ ms,\ trans-isomerization,\ deprotonation} pG(trans,\ 446\ nm)$$

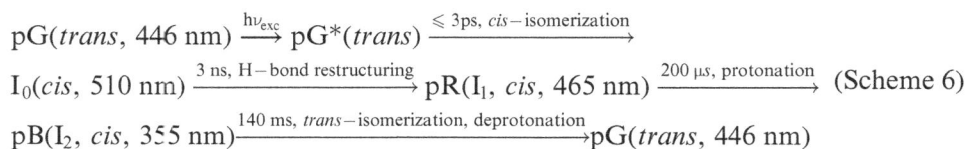

The fusion of genetically encoded PYP blue-light sensing domains to biological effector domains opens powerful optogenetic applications for cell-type-specific manipulation of living systems (Reis and Woolley 2016). For such an application one should keep in mind that in the dark state of PYP the *p*-coumaric acid chromphore is in the *trans* conformation and the PYP domain is well folded. Upon blue-light irradiation the chromophore switches to the *cis* form and the PYP domain partially unfolds involving large movements of the N-terminus and thereby activating the connected effector domain.

A genetically encoded photo-switchable DNA binding protein was designed by fusing PYP to the prototypical leucine-zipper-type DNA binding protein GCN4-bZIP (Morgan *et al* 2010). A circularly permuted photoactive yellow protein (cPYP) was created as a scaffold for photo-switch design (Kumar *et al* 2013). Optical control of protein–protein interactions was achieved via blue-light induced domain swapping in a variant of PYP with a modified surface loop (Reis *et al* 2014). A blue light controlled inhibitor of the transcription factor CREB (= cAMP response element-binding protein) was created by fusing the dominant negative inhibitor A-CREB to PYP (Ali *et al* 2015).

In live cell imaging, fluorogenic labeling systems based on PYP have been developed to probe chemical environment, movement, trafficking and interactions of proteins in live cells (Hori *et al* 2009, 2012, 2013, 2015, Kamikawa *et al* 2016, for a review see Gao *et al* 2016).

13.3.4 Phytochromes

Phytochromes are a collection of bilin-containing photoreceptors that regulate diverse processes in response to the light environment (for reviews see Sineshchekov 1995, Gärtner and Braslavsky 2003, Tu and Lagarias 2005, Vierstra and Karniol 2005, Rockwell *et al* 2006, Rockwell and Lagarias 2010, Auldridge and Forest 2011, Vierstra and Zhang 2011, Li *et al* 2011, Burgie and Vierstra 2014, Wang and Wang 2015, Xu *et al* 2015, Ziegler and Möglich 2015, Chernov *et al* 2017, Rockwell *et al* 2017, Rockwell and Lagarias 2017, Oliinyk *et al* 2017). They have an N-terminal photosensory core including a GAF domain (Aravind and Ponting 1997) where the chromophore is localized and covalently bound to a Cys residue of the GAF domain or of a PAS domain, and a C-terminal regulatory region mostly including a histidine-kinase-related domain (HK). The phytochrome super-family may be subdivided into the family of canonical phytochromes and the family of non-canonical phytochromes.

The canonical phytochromes are photo-switchable photosensors between two stable states, a red light absorbing P_r receptor form and a far-red light-absorbing P_{fr} signaling form. Their chromophores, phytochromobilin (PΦB), phycocyanobilin

(PCB), or biliverdin IXα (BV) are covalently thiol-ester bound to the GAF domain (or sometimes a PAS domain) of the phytochromes at the C3 position (see figure 13.18). Photoexcitation causes a photocycle with C15-Z,*anti* – C15-E,*anti*-photoisomerization (Rockwell *et al* 2006). Members of the canonical phytochromes are plant phytochromes (Phy) with PΦB chromophore, cyanobacteria phytochromes (Cph) with PCB chromophore, bacteria phytochromes (BphP) with BV chromophore, and fungi phytochromes (Fph) with BV chromophore (Rockwell *et al* 2006).

The non-canonical phytochromes are photo-switchable photosensors between two stable states, which include receptor states at longer wavelength with signaling states at shorter wavelength and vice versa, spanning the spectral range from the near UV to the near IR (Ulijasz *et al* 2009, Rockwell *et al* 2012, 2014, 2015, 2016, 2017). This wide diversity is reached by changing double bonds to single bonds, changing protonation stages, and isomeric forms. They may have double thio-ester cysteine linkages between chromophore and protein GAF domain. To the non-canonical phytochromes belong algal phytochromes with PCB or PΦP chromophores (Rockwell *et al* 2014) and the cyanobacteriochromes (cyanochromes) (CBCR) with the chromophores phycocyanobilin (PCB) or phytoviolobilin (PVB) (Ulijasz *et al* 2009, Rockwell *et al* 2012, 2016, 2017). In most cyanobacteriochromes PAS domains are missing and HK domains may be replaced by other functional domains (Burgie and Vierstra 2014).

The two stable states, P_r and P_{fr}, of phytochromobilin of a canonical plant phytochrome and the primary photoisomer of P_r excitation, Lumi-R, and the primary photoisomer of P_{fr} excitation, Lumi-F, are illustrated in figure 13.35(a). The structural formulae of PΦB dark-adapted inactive P_r state and the photoisomer Lumi-F are of ZZZssa form, and the structural formulae of PΦB light-adapted active P_{fr} state and the photoisomer Lumi-R are of ZZEssa form (according to Rockwell *et al* 2006).

The photo-conversion of P_r to P_{fr} is sketched in Scheme 7a, and the photo-conversion of P_{fr} to P_r is sketched in Scheme 7b (Rockwell *et al* 2006, Mroginski *et al* 2007).

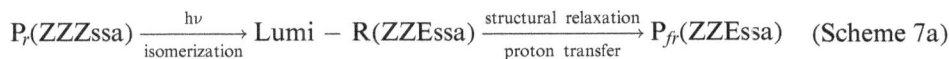

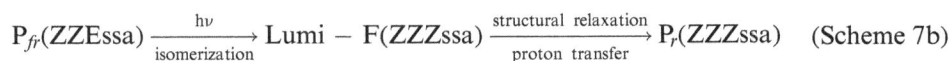

$$P_r(ZZZssa) \xrightarrow[\text{isomerization}]{h\nu} Lumi - R(ZZEssa) \xrightarrow[\text{proton transfer}]{\text{structural relaxation}} P_{fr}(ZZEssa) \quad \text{(Scheme 7a)}$$

$$P_{fr}(ZZEssa) \xrightarrow[\text{isomerization}]{h\nu} Lumi - F(ZZZssa) \xrightarrow[\text{proton transfer}]{\text{structural relaxation}} P_r(ZZZssa) \quad \text{(Scheme 7b)}$$

The photoisomerization occurs on a sub-picosecond timescale. The subsequent structural relaxation at the chromophore binding site and the chromophore–protein proton transfer occur on a millisecond timescale. At room temperature in the dark generally a very slow direct recovery from P_{fr} to P_r occurs in the minutes to hours time range (Piatkevich *et al* 2013).

In figure 13.35(b) the absorption cross-section spectra of the 124 kDa *Avena* phytochrome in TEGE buffer (0.025 M Tris-HCl, 4.0 M ethylene glycol, and 1 mM EDTA, pH 7.8) at 5 °C in the inactive P_r state and in the active P_{fr} state are shown (adpted from Kelly and Lagarias 1985 and Tu and Lagarias 2005). The

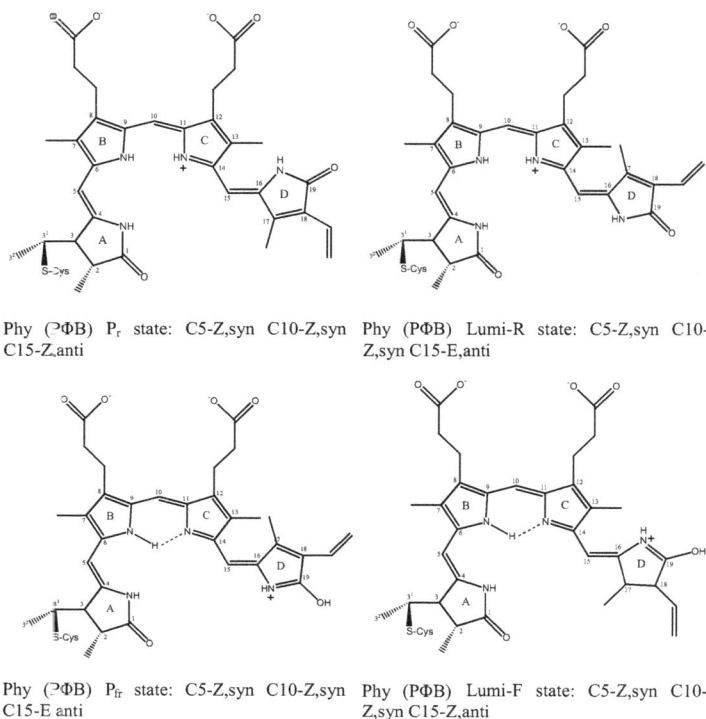

Phy (PΦB) P_r state: C5-Z,syn C10-Z,syn C15-Z,anti Phy (PΦB) Lumi-R state: C5-Z,syn C10-Z,syn C15-E,anti

Phy (PΦB) P_fr state: C5-Z,syn C10-Z,syn C15-E anti Phy (PΦB) Lumi-F state: C5-Z,syn C10-Z,syn C15-Z,anti

Figure 13.35a. Structural formulae of plant phytochrome chromophore phytochromobilin PΦB in P_r state, Lumi-R intermediate state, P_fr state, and Lumi-F intermediate state (from Rockwell *et al* 2006).

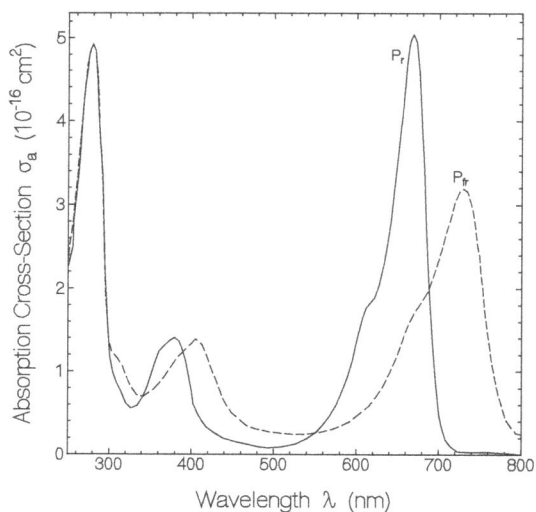

Figure 13.35b. Absorption cross-section spectra of *Avena* phytochrome (adapted from Kelly and Lagarias 1985, Tu and Lagarias 2005). Solid curve: P_r state. Dashed curve: P_fr state.

photoexcitation of P_r at a fixed wavelength λ_{exc} within the absorption spectral region of P_r the conversion of P_r to P_{fr} is incomplete, because P_r (absorption cross-section $\sigma_{a, P_r}(\lambda_{exc})$) and P_{fr} (absorption cross-section $\sigma_{a, P_{fr}}(\lambda_{exc})$) absorb simultaneously and cause excited-state inter-isomerization. At sufficiently high excitation intensity the number density ratio of $N(P_{fr})$ to $N(P_r)$ is approximately given by

$$\frac{N(P_{fr})}{N(P_r)} = \frac{\sigma_{a, P_r}(\lambda_{exc})}{\sigma_{a, P_{fr}}(\lambda_{exc})}. \qquad (13.8)$$

Photoexcitation of P_{fr} in the long-wavelength transparency region of P_r allows complete conversion of P_{fr} to P_r since in this case P_r is not absorbing and therefore no photo-induced P_r to P_{fr} back-isomerization is possible.

For optogenic applications, phytochromes were engineered to spatiotemporally control cellular processes in mammalian cells (Levskaya et al 2009, Müller et al 2013a, 2013b, Piatkevich et al 2013, Kaberniuk et al 2016, Chernov et al 2017). The engineered phytochrome photosensory core module was fused to engineered output modules of various functions. The optogenetic tools were classified on the basis of homodimerization and heterodimerization of phytochromes with binding partners as well as on conformational changes in the phytochromes. For example near-infrared light-controllable adenylate cyclases (Ryu et al 2014, Blain-Hartung et al 2018) and a red light activated phosphodiesterase were engineered (Gasser et al 2014)

Near infrared fluorescent proteins (NIR-FPs) have been engineered based on bacterial phytochromes, cyanobacterial phytochromes, cyanobacteriochromes, and allophycocyanins (Shu et al 2009, Piatkevich et al 2013, Rodriguez et al 2016, Oliinyk et al 2017, Chernov et al 2017). In these engineered linear tetrapyrrole chromophore containing proteins the non-radiative excited-state relaxation by Z/E isomerization and proton transfer had to be reduced. This reduction of non-radiative decay was approached by progressive truncation of the phytochromes to PAS-GAF and single GAF domains, and by introducing mutations in the immediate surroundings of the chromophore. Fluorescence quantum yields up to about 14% were achieved. NIR-FPs were applied in designs of reporters and biosensors to detect protein-protein interactions, proteolytic activities, and post-translational modifications (for review see Chernov et al 2017).

Intensely orange fluorescent adducts, named phytofluors, were spontaneously formed upon incubation of recombinant plant phytochrome apoproteins with the chromophore phycoerythrobilin (same structure as phytochromobilin, only lacking the C15-double bond). Phytofluors have large absorption cross-sections, fluorescence quantum yields greater than 70%, excellent photostability, stability over a wide pH range, and they can be reconstituted in living plant cells (Murphy and Lagarias 1997).

13.3.5 Cobalamin-based photoreceptors CarH and AerR

Photoreceptors with the chromophores adenosylcobalamin (AdoCbl) and/or methylcobalamin (MeCbl) are widespread in bacteria to orchestrate light perception and response. The prototypes of these cobalamin-based photoreceptors are CarH (B_{12}-dependent repressor of carotenogenesis) in *M. xanthus*, *T. thermophilus*, and *B. megaterium* and AerR (aerobic repressor of photosynthesis) in *Rhodobacter capsulatus*. CarH with AdoCbl chromophore regulates light-induced expression of carotenoid biosynthetic genes, which results in carotenoid-mediated protection against photo-oxidative damage (Ortiz-Guerrero *et al* 2011, Jost *et al* 2015, Kutta *et al* 2015a, Takano *et al* 2015, Chemaly 2016, Padmanabhan *et al* 2017). AerR with AdoCbl or MeCbl chromophore controls the DNA binding activity of the photosystem regulator CrtJ. AerR with CrtJ modulates CrtJ binding to DNA and the regulatory outcome of gene expression of photosystem promotors (Fang and Bauer 2017).

The CarH-type photoreceptors consist of an N-terminal DNA-binding domain and a C-terminal AdoCbl-binding and oligomerization domain. In the dark, AdoCbl binds to CarH and forms a AboCbl-CarH tetramer which binds to the promoter region P_{crt} of target genes to repress transcription. Light exposure disrupts the photosensitive Co–C bond in AdoCbl leading to tetramer disassembly, loss of CarH binding to the promoter region and activation of gene expression (Jost *et al* 2015). A photochemical mechanism of dark-adapted CarH conversion to light-adapted CarH is described by Kutta *et al* (2015a). In the dark state the lower axial ligand 5,6-dimethylbenzimidazole (DMB) is decoupled from Co(III) (lower base off) and instead the lower face is coupled to the CarH protein with His-177 residue (lower axial His on configuration). In the light adapted state the upper axis ligand 5′-deoxyadenosine is decoupled from Co^{3+} and instead the upper face is coupled to the CarH protein with His-132 residue. This CarH-$Cbl_{\alpha\text{-}His\text{-}177,\ \beta\text{-}His\text{-}132}$ adduct leads to monomerization of the CarH tetramer and release from the P_{crt} promoter of carotenogenic genes. So light-protection carotene is expressed. A rough scheme of the photoexcitation dynamics of CarH is shown in scheme 8 (extracted from Kutta *et al* 2015a)

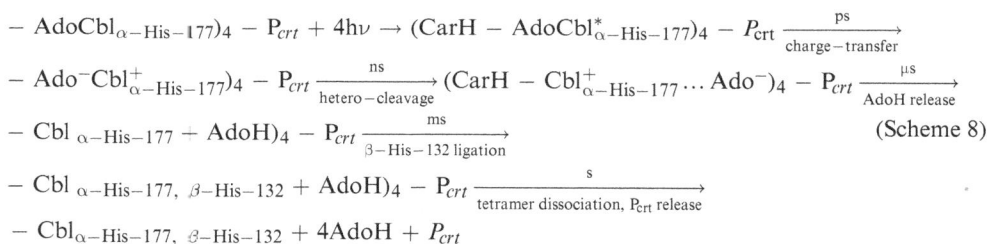

$$- \text{AdoCbl}_{\alpha-His-177})_4 - P_{crt} + 4h\nu \rightarrow (\text{CarH} - \text{AdoCbl}^*_{\alpha-His-177})_4 - P_{crt} \xrightarrow[\text{charge-transfer}]{ps}$$

$$- \text{Ado}^-\text{Cbl}^+_{\alpha-His-177})_4 - P_{crt} \xrightarrow[\text{hetero-cleavage}]{ns} (\text{CarH} - \text{Cbl}^+_{\alpha-His-177}\cdots\text{Ado}^-)_4 - P_{crt} \xrightarrow[\text{AdoH release}]{\mu s}$$

$$- \text{Cbl}_{\alpha-His-177} + \text{AdoH})_4 - P_{crt} \xrightarrow[\beta-His-132 \text{ ligation}]{ms}$$

(Scheme 8)

$$- \text{Cbl}_{\alpha-His-177,\ \beta-His-132} + \text{AdoH})_4 - P_{crt} \xrightarrow[\text{tetramer dissociation, } P_{crt} \text{ release}]{s}$$

$$- \text{Cbl}_{\alpha-His-177,\ \beta-His-132} + 4\text{AdoH} + P_{crt}$$

AerR is a small stand-alone AdoCbl or MeCbl binding domain protein. Photoexcitation of AdoCbl or MeCbl leads to photolysis of the upper axial ligand and generates OHCbl, which subsequently forms a strong upper axial covalent linkage between His-10 of AerR and Co(III) of cobalamin. Subsequent regulation of gene expression occurs through formation of a stable interaction between cobalamin-bound AerR and the CrtJ repressor of photosystem gene expression. This interaction lowers the CrtJ dimer binding to the bacteriochlorophyll promotor bchC. This de-represses the photosynthesis gene expression (Cheng *et al* 2014, 2016, Vermeulen and Bauer 2015, Fang and Bauer 2017). A rough scheme of the AerJ photoexcitation dynamics in *R. capsulatus* is presented in Scheme 9 (extracted from Fang and Bauer 2017).

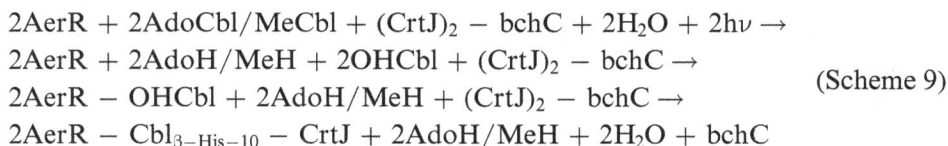

$$2AerR + 2AdoCbl/MeCbl + (CrtJ)_2 - bchC + 2H_2O + 2h\nu \rightarrow$$
$$2AerR + 2AdoH/MeH + 2OHCbl + (CrtJ)_2 - bchC \rightarrow$$
$$2AerR - OHCbl + 2AdoH/MeH + (CrtJ)_2 - bchC \rightarrow \qquad \text{(Scheme 9)}$$
$$2AerR - Cbl_{3-His-10} - CrtJ + 2AdoH/MeH + 2H_2O + bchC$$

13.3.6 UVR8 plant photoreceptors

Plants have to resist the deleterious effects of UV-B light (wavelength region 280–315 nm) in the sunlight, and for this purpose they need a UV-B sensing and activation system (Kliebenstein *et al* 2002, Huang *et al* 2014, Kaiserli and Jenkins 2007). The sensing and activation system was found and characterized in the plant *A. thaliana*. It consists of the bZIP transcription factor HY5 (ELONGATED HYPOCOTYL 5), the E3 ubiquitin ligase COP1 (CONSTUTIVELY PHOTOMORPHOGENETIC 1), and the β-propeller protein UVR8 (UV RESISTANCE LOCUS 8) (Favory *et al* 2009, Jenkins 2009, Heijde and Ulm 2012, Huang *et al* 2014, Yin *et al* 2015). The action leads to gene expression changes. It causes UV-B-induced photomorphogenesis and accumulation of UV-B-absorbing flavonols. Negative feedback regulation of the activation pathway is provided by the WD40-repeat proteins RUP1 and RUP2 (RUP = REPRESSOR OF PHOTOMORPHOGENESIS) disrupting the UVR8-COP1 interaction (Tilbrook *et al* 2013). UV-B perceived by the UVR8 photoreceptor inhibits plant thermomorphogenesis (Hayes *et al* 2017).

The photoreceptor UVR8 uses its intrinsic Trp residues as chromophores. In the dark-adapted state UVR8 dimerizes to homodimers with a Trp dominated dimer interface stitched together by a complex salt-bridge network. Salt-bridging arginines flank the excitonically coupled cross-dimer tryptophan 'pyramid' responsible for UV-B sensing. Each UVR8 monomer has 14 highly conserved tryptophans, one in the C-terminus, seven in the dimer interface and six in the β-strands. These latter tryptophans help to maintain the propeller structure as they form hydrogen bonds and hydrophobic interactions between adjacent blades. Three tryptophans at the dimer interface (W233, W285 and W337) are sufficiently close that their electronic orbitals overlap. These three tryptophans form a pyramidal arrangement with W94 on the adjacent monomer. There are two such pyramids per dimer (Jenkins 2014a, 2014b). UV-B photoreception reversibly disrupts salt bridges, triggering dimer dissociation and signal initiation (Rizzini *et al* 2011, Christie *et al* 2012b, Wu *et al* 2012, Heilmann *et al* 2015, Jenkins 2014a, 2014b, Yang *et al* 2015, Mathes *et al* 2015). UVR8 dimers are inactive. The UV-B excitation of Trp in the dimer interface causes UVR8 dimer dissociation to UVR8 monomers. These monomers are active causing binding to COP1, rapid nuclear accumulation and orchestering genes expression for UV-B light protection. A detailed UV-B photocycle scheme of Trp excitation in UVR8 dimer, ^{3}Trp formation, tryptophan radical Trp$^{\cdot}$ formation,

URV8 dimer dissociation to UVR8 monomers, and slow re-dimerization in the dark is presented by Mathes *et al* (2015).

For optogenetics applications UVR8 was used to control protein–protein interactions involved in protein secretion (Chen *et al* 2013), to UV-B-mediated induction of protein-protein interactions in mammalian cells (Crefcoeur *et al* 2013, Zhang and Cui 2015), and to multichromatic control of mammalian gene expression and signaling (Müller *et al* 2013c). A short description of UV-B-based optogenetic tools is given by Kianianmomeni (2016).

13.3.7 Upconversion nanoparticle mediated optogenetics

The organic dye chromophores in photoreceptors of optogenetic tools generally absorb in the visible spectral range. Visible light does not effectively penetrate through biological tissue. This fact makes it difficult to photo-activate non-invasively optogenetic tools inside animal beings by external visible light sources. Near-infrared light (650–1450 nm) penetrates deep into biological tissues (near infrared imaging window, Pansare *et al* 2012). The placement of upconversion nanoparticles in the vicinity of optogenetic tools, like neurons expressed with photo-reactive molecules as channelrhodopsin 2, opens the possibility of external photo-excitation using a near-infrared light source (Hososhima *et al* 2015, Huang *et al* 2016).

A near-infrared stimulable optogenetic platform (termed 'Opto-CRAC') with $NaYF_4:Yb^{3+}/Tm^{3+}$ upconversion nanoparticles was used to selectively and remotely control Ca^{2+} oscillations and Ca^{2+}-responsive gene expression in order to regulate the function of non-excitable cells, including T lymphocytes, macrophages and dendritic cells (He *et al* 2015). To the Ca^{2+} release-activated (CRAC) channel, a genetically-encoded photoswitch LOV2 domain of *Avena sativa* phototropin-1 was added to enable light control.

Blue-light-emitting upconversion nanoparticles $(NaYF_4:Yb^{3+}/Tm^{3+})$ were embedded in thin polymer films of poly(lactic-*co*-glycolic acid (PLGA) and cultured with channelrhodopsin-2 expressed hippocampal neurons for NIR-mediated opto-genetic control by balancing multiple physicochemical properties of the nano-material (Shah *et al* 2015).

Dye-sensitized core/active shell upconversion nanoparticles for optogenetics and bioimaging applications were developed by Wu *et al* 2016. The near-infrared absorbing dye IR-806 was anchored to the Yb^{3+} (sensitizer) doped shell of the $NaYF_4$ crystal with Yb^{3+} and Er^{3+} (activator) doped core. With the help of the dye-sensitized nanoparticles, red-activated channelrhodopsin ReaChR (optimal excita-tion in the range from 530 to 630 nm, absorption maximum at 532 nm, Lin *et al* 2013, Krause *et al* 2017) were photo-activated in cultured hippocampal neurons.

Molecularly tailored blue-light emitting upconversion nanoparticles $(NaYF_4:Yb^3$ $^+/Tm^{3+}$ core, $NaYF_4$ shell, SiO_2 cover) served as optogenetic actuators of trans-cranial NIR light to stimulate deep brain neurons (Chen *et al* 2018). It evoked dopamine release from genetically tagged neurons in the ventral tegmental area, induced brain oscillations through activation of inhibitory neurons in the medial septum, silenced seizure by inhibition of hippocampal excitatory cells, and triggered memory recall (Chen *et al* 2018).

Core(NaYF$_4$)–shell(NaYF$_4$:Yb^{3+}/Er^{3+})–shell(NaYF$_4$) upconversion nanoparticles with hexagonal phase were synthesized for NIR light conversion to 540–570 nm (Lin *et al* 2018) matching the excitation spectrum of enhanced *Natronomonas pharaonis* halorhodopsin eNpHR (Gradinaru *et al* 2008), a light-activated chloride pump commonly used for inhibition of neuron activity. Implantation of the upconversion nanoparticles into targeted sites deep in rat brain inhibited neuronal function and action potential firing with NIR light, which was restored after NIR light switch-off. The upconversion nanoparticles were further used to perform tetherless unilateral inhibition of the secondary motor cortex in behaving mice (Lin *et al* 2018).

13.4 Application of optogenetic tools

13.4.1 Application of optogenetic tools in neuroscience

In neuroscience, precise tools for controlling specific types of neurons are needed to understand their function in a complex network. High spatial resolution is required to localize the region of specific neuron activation, and high temporal resolution is required to follow the dynamics of neural signal response after activation. For this purpose natural and engineered photoreceptors have been employed and the new field of optogenetic brain research has evolved (Boyden *et al* 2005, Zhang *et al* 2007, Chow *et al* 2010, Boyden 2011, Yizhar *et al* 2011, Knöpfel and Boyden 2012, Packer *et al* 2013, Deisseroth 2015, Ilango and Lobo 2015, Zhou *et al* 2015b, Appasani 2017, Ordaz *et al* 2017, Rost *et al* 2017, Förster *et al* 2017, Wiegert *et al* 2017, Shemesh *et al* 2017, Galvan *et al* 2017, Wang *et al* 2017a, Lee *et al* 2017a, Coffey 2018). Recent advances in optogenetics combining optogenetic actuators and fluorescent reporters especially in all-optical electrophysiology opens new routes to drug discovery, particularly in neuroscience (Zhang and Cohen 2017, Kiskinis *et al* 2018).

Techniques of gene therapy are used to enter engineered photoreceptor and fluorescent protein genes in a viral vector. The virus is then delivered to the targeted neuronal cells in the brain of rodents or nonhuman primates typically by injection. The encoded neurons then express the photoreceptor protein domains which are generally engineered rhodopsins. A light source is used to excite the photoreceptors which activate or silence the neurons to cause organ action like muscle contraction, a blinking eye, or a stimulus in the brain that will be the basis for many actions and therapies.

Neural activation is achieved by membrane action potential depolarization (potential rise above threshold potential which is around −55 mV, neuronal firering) with encoded proton and cation conducting channelrhodopsins like channelrhodopsin 2 and several closely related variants (Fenno *et al* 2011, Yizhar *et al* 2011, Han 2012, Guru *et al* 2015, Lerner *et al* 2016). Neural inhibition has been achieved by membrane potential hyperpolarization (potential fall below resting potential which is around −70 mV) with light-driven inward chloride pumps (halorhodopsins,

Han 2012, Guru *et al* 2015, Kato *et al* 2016), light-driven outward proton pumps (archeorhodopsins like archeorhodopsin-3, Han 2012, Guru *et al* 2015, Kato *et al* 2016), light-driven outward sodium pumps (*Krokinobacter* rhodopsin 2, Kato *et al* 2016), and anion conducting channelrhodopsins (Berndt *et al* 2014, Wietek *et al* 2015, 2017, Govorunova *et al* 2015, 2016). Control of intracellular signaling in the brain was achieved with a family of opsin-receptor chimeras (G protein-coupled receptors) called optoXRs, where intracellular loops of vertebrate rhodopsins were replaced with those of adrenergic reporters (Airan *et al* 2009, Kushibiki *et al* 2014, Guru *et al* 2015, Spangler and Bruchas 2017).

For light excitation of genetic modified neurons with photoreceptors in brains of living mammals various techniques are in use: a fiberoptic cable may transmit the light from an external light source to the brain (Ung and Arenkiel 2012, Doronina-Amitonova *et al* 2015, Mohanty and Lakshminarayananan 2015, Sidor *et al* 2015, Miyamoto and Murayama 2016). Optical fibers have been combined with recording electrodes to allow simultaneous optical stimulation and electrophysiological read-out (Lee *et al* 2015 and references therein). Wireless excitation was achieved by implanting miniature light emitting diodes into the brain with wireless powering (Kim *et al* 2013, McCall *et al* 2013, Montgomery *et al* 2015, Shin *et al* 2017, Lu *et al* 2018, Gutruf and Rogers 2018). The small light penetration depth in tissue in the visible spectral region can be overcome by using short pulse intense near infrared laser external excitation which causes two-photon absorption in the photoreceptor inside the tissue (Rickgauer and Tank 2009, Prakash *et al* 2012, Bovetti *et al* 2016, Shemesh *et al* 2017, Yang *et al* 2018) or by implanting in the brain adjacent to the photoreceptor-encoded neurons fluorescence upconversion nanoparticles which are tetherless excited to transfer their emitted upconverted light to the photoreceptors absorbing in the visible range (Wu *et al* 2016, Lin *et al* 2018, Chen *et al* 2018).

13.4.2 Application of optogenetic tools in cell biology

The expression of light-activatable molecules (photoreceptors) into cells of bacteria, algae, plants, fungi, insects, and mammals has opened the field of studying the intra-cellular and inter-cellular behavior by light-triggered action (Deisseroth 2011, Toettcher *et al* 2011a, Tischer and Weiner 2014, Zhang and Cui 2015, Guglielmi *et al* 2016, Repina *et al* 2017, Mukherjee *et al* 2017, Rost *et al* 2017, Mühlhäuser *et al* 2017, O'Banion *et al* 2018). The optogenetics methods have been used besides others to direct subcellular localization of protein activity, to turn on or off protein functionality, to promote transcriptional gene expression or gene repression, and to induce protein degradation.

The strategies for optogenetic studying of protein functions and cellular functions are the photoexcitation of the light-sensitive photoreceptor domain in the non-signaling state to the signaling state and thereby activation of the photoreceptor effector domain which affects the cell biological behavior. It may cause light-induced photoreceptor uncaging, light-induced protein accociation to the photoreceptor, photoreceptor homo-dimerization and oligomerization, photoreceptor–target protein hetero-dimerization and oligomerization. Precisely modulated spatial, temporal, intensity and wavelength varied light inputs allow detailed study of the target output pathways. Localization of a signaling molecule may induce asymmetric cell division or specific migration to a tissue layer. The photoreceptor interaction with transcription factors may control gene expression or repression including transcription, translation and genome editing.

Optogenetic tools applied in cell biology include caged proteins, dimerizing and oligomerizing systems.

13.4.2.1 Caged proteins (allosteric proteins)

Optgenetic systems have been developed to retain a target protein in an inactive state via steric hindrance from a specialized light-responsive domain. Light stimulation induces a conformational change that liberates the protein (light-induced uncaging) to interact with its substrate or target protein (conformational coupling of distant functional sites with allosteric proteins). Caged optogenetic tools are predominantly based on a plant LOV domain.

The LOV2 domain from *Avena sativa* (AsLOV2) has been engineered as an optogenetic tool for light control of mammalian protein activity (Christie *et al* 2012a). The LOV based optogenetic tools consist of a LOV domain with FMN cofactor and C-terminal Jα-helix to which an effector domain is connected (caged). In the light-unexposed (receptor) state the effector domain is stacked to the LOV domain and cannot activate its target protein of interest. In the case of light exposure the Jα-helix undocks and unfolds enabling the interaction of the effector domain with the target protein or substrate. As effector domains have been used: (i) the tryptophan-activated protein (named TAP) giving the light-activable allosteric switch LovTAP (Strickland *et al* 2008, 2010, Peter *et al* 2012), (ii) the GTPase Rac1 giving the photoactivatable Rac1, named PA-Rac1 (Wu and Gardner 2009) which controls the motility of living cells, (iii) the stromal interaction molecule 1 STIM1 giving the photoactivatable protein LOVS1K which allows light activation of the plasma membrane calcium channel Orai1 in mammalian cells (Pham *et al* 2011), (iv) the cysteine-aspartic protease caspase-7 giving a photoactivatable caspase-7 fusion protein (named L57V) used for rapid induction of apoptosis (Mills *et al* 2012), (v) the cysteine-aspartic protease caspase-3 giving light-activated human caspase-3 (Caspase-LOV) used for studying neurodegeneration in larval and adult *Drosophila* (Smart *et al* 2017), and (vi) the split DnaE intein from *Nostoc punctiforme* giving the photoactivatable intein (LOVInC) for protein splicing activity in mammalian cells (Wong *et al* 2015).

    (i) In the light-activable allosteric switch LovTAP, the LOV2-Jα photoswitch of phototropin 1 from *Avena sativa* (AsLOV2-Jα) has been ligated to the tryptophan-repressor (TrpR) protein from *E. coli*. The tryptophan repressor is a transcription factor involved in controlling the amino acid metabolism. In the dark state, the AsLOV2-Jα photoswitch is inactive and exerts a repulsive electrostatic force on the DNA surface causing a disruption of the LovTAP from the DNA. In the case of photoexcitation the FMN-C4a cysteinyl adduct formation induces an undocking of the peripheral Jα-helix from the LOV core and unfolding of a hairpin-like helix–loop–helix region interlinking the AsLOV2-Jα and TrpR domains enabling the condensation of LovTAP onto the DNA surface and enabling the TrpR action (Peter *et al* 2012).

(ii) In the photoactivatable PA-Rac1, the LOV2-Jα domain is fused to Rac1, a key GTPase regulating actin cytoskeletal dynamics in metazoan cells, thereby sterically blocking Rac1 interactions in the dark. In the case of blue light exposure the Jα helix is undocked from the LOV2 domain and Rac1 is unlocked and becoming able to bind to and activate a PAK1 serine/threonine-protein kinase (belonging to the p21-activated kinases) to generate precisely localized cell protrusions and ruffling (Wu and Gardner 2009).

(iii) The photoactivatable LOVS1K protein consisting of a LOV2-Jα photoswitch and a bound stromal interaction molecule 1 (Stim1) was engineered to generate local or global Ca^{2+} signals upon blue-light exposure. The compact LOVS1K protein gets unwound under blue-light excitation and the Stim1 domain locates to the plasma membrane adjacent to the calcium release-activated calcium channel protein 1 (Orai1). Stim1 and Orai1 synergistically activate store-operated Ca^{2+} entry (SOCE), a predominant mechanism used to replenish Ca^{2+} levels when intracellular stores are depleted (Pham *et al* 2011).

(iv) The photoactivatable caspase-7 protein switch L57V is a tandem fusion of the light-sensing LOV2-Jα domain and the apoptosis-executing domain from caspase-7. Cells transfected with L57V rapidly undergo apoptosis after blue-light stimulation (Mills *et al* 2012).

(v) Cells are naturally and cleanly ablated through apoptosis due to the terminal activation of caspases. The engineered light-activated human caspase-3 (Caspase-LOV) consists of a LOV2-Jα domain and a caspase-3 domain. The spring-loaded protein expands upon blue-light exposure and gets active to kill cells with temporal and spatial selectivity (Smart *et al* 2017).

(vi) Protein splicing is mediated by inteins that auto-catalytically join two separated protein fragments with a peptide bond. The engineered photoactivatable intein LOVInC consisting of a LOV2-Jα domain and a naturally split DnaE intein from *Nostoc punctiforme* (*Npu*DnaE) was used to modulate protein splicing activity in mammalian cells (Wong *et al* 2015).

13.4.2.2 Dimerizing and oligomerizing protein systems

Light-inducible dimerizers are the most diverse class of optogenetic systems with varying properties. They allow the control of various protein–protein interactions in cells. There exist homo-dimerization systems and hetero-dimerization systems. The protein–protein interaction in dimerizing system is varied by photoexcitation (Repina *et al* 2017). Dimerizing systems are found in optogenetic tools consisting of (i) phytochromes (Phy/PIF, BphP1/PpsR2), (ii) cryptochromes (Cry2/CIB1), (iii) UVR8 plant photoreceptors (UVR8/COP1), and (iv) LOV proteins (FFK1/

GIGANTEA, EL222/EL222, VVD/VVD, nMag/pMag, TULIP (LOVpep/ePDZ), iLID (LOV2-Jα-SsrA/SspB)).

(i) Phytochrome hetero-dimerizing systems. Phytochromes regulate plant growth and development in response to red light. In *A. thaliana* the phytochrome B (PhyB) and the transcription factor PIF3 (phytochrome interaction factor 3) form a PhyB/PIF3 heterodimer upon red light exposure and direct the plant photomorphogenesis (Ni *et al* 1999, Quail 2002, 2010, Khanna *et al* 2004). PhyB undergoes conformational changes between the red-absorbing P_r state and the far-red absorbing P_{fr} state upon red and near-infrared light excitation, respectively, according to scheme 10(a).

$$P_r \text{ (inactive)} \xrightarrow[\text{750 nm or dark}]{\text{650 nm}} P_{fr} \text{ (active)} \qquad \text{(Scheme 10a)}$$

Excitation around 650 nm isomerizes P_r to P_{fr}, and exitation around 750 nm isomerizes P_{fr} to P_r. Only PhyB in the P_{fr} state associates with the PIF3 domain forming a PhyB/PIF3 heterodimer. The PhyB/PIF3 dimer formed in the cytoplasm translocates into the cell nucleus where it initiates a cascade of changes in gene expression directing the plant photomorphogenesis. In the dark or upon near infrared light application the dimer separates to its monomers and the morphogenesis action ends (Quail 2002 and 2010). PhyB phytochrome and PIF transcription factors have been engineered to form optogenetic tools for cell biological studies (Levskaya *et al* 2009, Yang *et al* 2013). PhyB with chromophore phycocyanobilin (PCB) and transcription factor PIF6 were coexpressed in mammalian cells and used for spatiotemporal control of plasma membrane recruitment (Levskaya *et al* 2009, Toettcher *et al* 2011b). A PhyB/PIF6 light-gated dimerization system according to scheme 10(b) was used to achieve rapid, reversible, and titratable control of protein localization for different organelles/positions in budding yeast (Yang *et al* 2013).

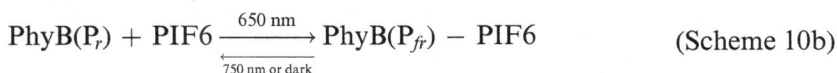

$$PhyB(P_r) + PIF6 \xrightarrow[\text{750 nm or dark}]{\text{650 nm}} PhyB(P_{fr}) - PIF6 \qquad \text{(Scheme 10b)}$$

A bacterial phytochrome-based optogenetic system controllable with near-infrared light was engineered based on the reversible light-induced binding between the bacterial phytochrome BphP1 and its natural transcriptional repressor PpsR2 from *Rhodopseudomonas palustris* (Kaberniuk *et al* 2016). RpBph1 bacterial phytochrome with biliverdin IXα is a 'bathy phytochrome' that adopts the P_{fr} state (dominant absorption at 740–780 nm) as inactive ground-state and the P_r state (dominant absorption at 660–700 nm) as active excited state. The active RpBphP1 P_r state hetero-dimerizes with the transcriptional repressor RpPpsR2, which may be localized at the plasma membrane or in the cell nucleus. The reversible photo-induced near-infrared dimerization and red or dark de-dimerization to monomers is shown in scheme 11.

$$RpBphP1(P_{fr}) + RpPpsR2 \xrightleftharpoons[\text{650 nm or dark}]{\text{740 nm}} RpBphP1(P_r) - RpPpsR2 \quad \text{(Scheme 11)}$$

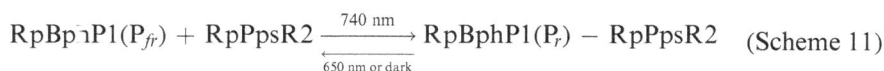

The optogenetic system was used to translocate target proteins to specific cellular compartments, such as the plasma membrane and the nucleus in mammalian cells (Kaberniuk *et al* 2016).

(ii) Cryptochrome hetero-dimerizing and homo-oligomerizing systems. The photoreceptor Cry2 and the cryptochrome-interacting basic-helix–loop–helix CIB1 from *A. thaliana* have become a powerful optogenetic tool that allows light-inducible manipulation of various signaling pathways and cellular processes in mammalian cells with high spatiotemporal resolution (Pathak *et al* 2017, Duan *et al* 2017 and references therein). This tool can undergo blue light-dependent CRY2-CRY2 homo-oligomerization and CRY2-CIB1 hetero-dimerization according to scheme 12(a) and (b).

$$nCRY2 \xrightleftharpoons[\text{dark}]{\text{blue light}} CRY2_n \quad \text{(Scheme 12a)}$$

$$CRY2 + CIB1 \xrightleftharpoons[\text{dark}]{\text{blue light}} CRY2-CIB1 \quad \text{(Scheme 12b)}$$

CRY2–CIB1 and CRY2–CRY2 interactions are governed by well separated protein interfaces at the two termini of CRY2. N-terminal charges are critical for CRY2–CIB1 interaction. Two C-terminal charges impact CRY2 homo-oligomerization, whereby positive charges facilitate oligomerization and negative charges inhibit oligomerization (Duan *et al* 2017).

(iii) UVR8/COP1 hetero-dimerization system. The UVR8 UV-B photoreceptor UV RESISTANCE LOCUS 8 exists as a homodimer that instantly monomerizes upon UV-B absorption by specific intrinsic tryptophans. The UVR8 monomer hetero-dimerizes with the E3 ubiquitin ligase COP1 CONSTITIVELY PHOTOMORPHENETIC 1. This heterodimer initiates a molecular signaling pathway that leads to gene expression changes. After UV-B light switch-off, the UVR8 in the heterodimer re-monomerizes with the help of the WD40-repeat proteins REPRESSORS OF PHOTOMORPHOGENESIS RUP1 and RUP2. (Tilbrook *et al* 2013). The UV-B light induced hetero-dimerization and dark separation is illustrated in scheme 13.

$$UVR8_2 + 2\,COP1 \xrightleftharpoons[\text{dark, RUP1, RUP2}]{\text{UV-B light}} 2\,UVR8-COP1 \quad \text{(Scheme 13)}$$

Two separate domains of UVR8 interact with COP1: the β-propeller domain of UVR8 mediates UV-B dependent interaction with the β-propeller domain of COP1, whereas the COP1 activity is regulated through the UVR8 C-terminal C27 domain (Yin *et al* 2015).

(iv) LOV protein dimerization systems. Certain photoreceptors that contain LOV domains possess the ability to homo-dimerize or hetero-dimerize after blue light exposure inducing cell biological interactions. Natural dimerizing LOV photoreceptors are FKF1 (Flavin-binding, Kelch repeat, F-box 1) from *A. thaliana* (Yazawa *et al* 2009), EL222 light-regulated DNA binding protein from *Erythrobacter litorlis* (Motta-Mena *et al* 2014), and VVD (Vivid) from fungus *Neurospora crassa* (Zoltowski and Crane 2008, Kawano *et al* 2015). Engineered dimerizing LOV based photoreceptors are TULIPs (tunable light-controlled interacting protein tags, Strickland *et al* 2012) and iLID (improved light-inducible dimer system, Guntas *et al* 2015).

(a) FKF1/GIGANTEA hetero-dimerization. The Flavin-binding, Kelch repeat, F-box 1 (FKF1) photoreceptor and the unique plant specific nuclear protein GIGANTAE (GI) from *A. thaliana* hetero-dimerize under blue light exposure, and the formed complex is involved in controlling the flowering of *A. thaliana* (Sawa *et al* 2007, Song *et al* 2014, Lee *et al* 2017b). The hetero-dimerization is illustrated in scheme 14.

$$FKF1 + GI \xrightarrow[\text{dark}]{\text{blue light}} FKF1-GI \qquad \text{(Scheme 14)}$$

FKF1 and GI were artificially expressed and localized in mammalian cells and there used for light-induced recruitment of FKF1 to GI localized at the plasma membrane (Yazawa *et al* 2009).

(b) EL222/EL222 homo-dimerization. The light-regulated DNA-binding protein (transcription factor) EL222, a 222 amino acid protein isolated from the marine bacterium *Erythrobacter litoralis* HTCC2594, consists of a LOV domain sensor, a Jα helix connector, and a HTH (helix–turn–helix) domain effector (Nash *et al* 2011, Zoltowski *et al* 2013, Takakado *et al* 2017 and 2018). The blue light-induced opening of the stacked LOV-Jα-HTH structure, which enables HTH-HTH homo-dimerization and subsequent binding to DNA is illustrated in scheme 15 (Nash *et al* 2011).

$$2 \text{ EL222(LOV}-J\alpha-\text{HTH stacked)} \xrightarrow[\text{dark}]{\text{blue light}} 2 \text{ EL222(LOV}-J\alpha-\text{HTH unstacked)} \xrightarrow{\text{DNA}} \text{EL222}_2-\text{DNA} \qquad \text{(Scheme 15)}$$

An engineered version of the EL222 system was used as optogenetic tool for light-gated transcription in several mammalian cell lines and intact zebrafish embryos (Motta-Mena *et al* 2014).

(c) Vivid (VVD) dimerization. The fungal *Neurospora crassa* photoreceptor Vivid (VVD) tunes blue-light responses and modulates gating of the circadian clock (Schwerdtfeger and Linden 2003, Zoltowski *et al* 2007). The protein consists of a LOV domain with an unusual N-terminal cap region and a loop insertion that accommodates the flavin cofactor. Light activation of VVD causes conformational changes which generate a rapidly exchanging VVD dimer (Zoltowski and Crane 2008, Vaidya *et al* 2011,

Zhou et al 2017, Hernández-Candia et al 2018). Substitution of residues critical for the switch between the monomeric and the dimeric states of the protein had profound effects on light adaption in *Neurospora crassa* (Vaidya et al 2011). A multi-directional engineering of Vivid was performed to develop pairs of distinct photoswitches named Magnets (Kawano et al 2015). In the N-terminal cap region either negative amino acids (giving Magnet nMag) or positive amino acids (giving Magnet pMag) were incorporated. Upon blue light exposure nMag and pMag preferentially hetero-dimerize because electrostatic repulsion impairs the natural homo-dimerization. The hetero-dimerization of nMag and pMag is illustrated in Scheme 16.

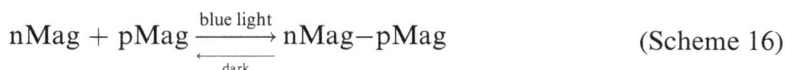

$$\text{nMag} + \text{pMag} \underset{\text{dark}}{\overset{\text{blue light}}{\rightleftharpoons}} \text{nMag}-\text{pMag} \qquad \text{(Scheme 16)}$$

Additional mutations within the Per-Arnt-Sim (PAS) core allowed one to tune the switch-off kinetics in the dark (Kawano et al 2015).

(d) TULIPs tunable light-inducible dimerization protein tags. TULIPs (tunable light-controlled interacting protein tags) have been developed based on a synthetic interaction between the LOV2 domain of *Avena sativa* phototropin 1 (AsLOV2) and an engineered PDZ affinity clamp protein (ePDZ) (Strickland et al 2012). A peptide epitope (pep) was fused to the C-terminus of the Jα helix of AsLOV2 (complex named LOVpep = LOV2-Jα-pep) to enable blue light-induced hetero-dimerization with ePDZ as illustrated in scheme 17.

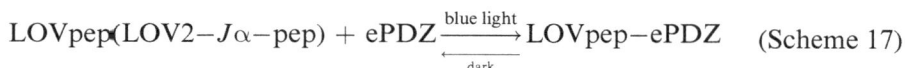

$$\text{LOVpep(LOV2}-J\alpha-\text{pep)} + \text{ePDZ} \underset{\text{dark}}{\overset{\text{blue light}}{\rightleftharpoons}} \text{LOVpep}-\text{ePDZ} \qquad \text{(Scheme 17)}$$

Upon blue light exposure the Jα helix undocks from the LOV core and unfolds enabling the fused peptide epitope to bind to ePDZ. Engineered TULIPs were applied to localize proteins to specific regions of yeast or mammalian cells, and to trigger specific cellular signaling pathways (Strickland et al 2012). TULIPs were used to recruit specific cytoskeletal motor proteins (kinesin, dynein or myosin) to selected cargoes (Van Bergeijk et al 2016, Harterink et al 2016).

(e) iLID improved light-induced dimer. An improved light-induced dimer (iLID) optogenetic tool was developed fusing the *E. coli* SsrA peptide to the C-terminal Jα helix of mutated AsLOV2 and coexpressing the *E. coli* SspB binding partner (Guntas et al 2015). Blue light exposure of this iLID system causes undocking of the Jα helix from the LOV core, and the unfolding enables the fused SsrA peptide to bind to the SspB binding partner, as is illustrated in scheme 18.

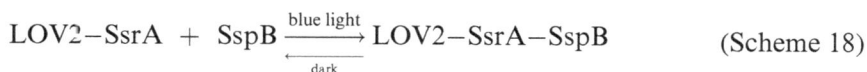

$$\text{LOV2}-\text{SsrA} + \text{SspB} \underset{\text{dark}}{\overset{\text{blue light}}{\rightleftharpoons}} \text{LOV2}-\text{SsrA}-\text{SspB} \qquad \text{(Scheme 18)}$$

The utility of the iLID photoswitch was demonstrated through light-mediated subcellular localization in mammalian cell culture and reversible control of small GTPase signaling (Guntas *et al* 2015).

13.5 Conclusions

In this chapter organic dyes have been characterized, which serve as chromophores in photoreceptors or sensors of optogenetic tools. The photodynamics of the free dyes in liquid solution was described first and then their photodynamics in the photoreceptor proteins was studied.

The field of optogenetics is rapidly growing and expanding in new fields of chemistry, biology, biotechnology, pharmacology, and medicine both in research and practical application. The discovery of new natural and engineered photobiological receptors, effectors and sensors is expected with the involvement of further organic dyes, biological dyes, and inorganic pigments.

13.6 Recent advances as to organic dyes in optogenetics

The research progress in optogenetics since the publication of the first edition of this book is presented in this section. The numbering of the subsections here follows the numbering of the original sections in order to see the progress that happened on the original topics (i.e. progress in the topic of section 13.i.j is reported in section 13.6.i.j).

Optogenetics combines optics and genetics in life science. Photoreceptor proteins (optogenetic tools) are naturally present or are genetically expressed and delivered into cells to trigger and monitor biological action by light exposure. The light sensitive cofactors (chromophores) in photoreceptor proteins are organic dyes (absorbing and/or emitting light). The research progress in optogenetics from January 2019 till September 2023 is reviewed concerning the topics published in the first edition of this chapter. New topics listed below are added.

Concerning characterization of organic dyes applied in optogenetics: The section on fluorescent probes is extended to fluorogen-activating protein/fluorogen complexes, bioluminescent probes, and chemical calcium indicators. In the section on fluorescent proteins, the development of genetically encoded calcium indicator is reviewed. The characterization of carotenoids, chlorophylls and bacteriochlorophylls is added.

Concerning the action of organic dyes in photoreceptors: The research on rhodopsins has grown enormously. Heliorhodopsins have been discovered and are described. The research on microbial rhodopsin-based genetically encoded voltage indicators is extended. A description of luminopsins is added. They are fusion proteins of luciferase and opsin and allow bioluminescent rhodopsin excitation in non-invasive deep-brain sensing. The characterization and application of orange carotenoid protein, a photoactive carotenoprotein involved in photoprotection, is added.

Concerning application of optogenetic tools: The application of optogenetic tools is continuously growing. The medical treatments (heart, eye, ear, regeneration, primary cilia function), and the studies belonging to neuronal diseases and psychiatric disorders are reviewed. Advances in cell biological applications are reviewed.

13.6.1 Introduction

New reviews on optogenetics have been published concerning optogenetic tools (Oh *et al* 2021), neuroscience (Losi *et al* 2018, Deubner *et al* 2019, Montagni *et al* 2019, Adamczyk and Zawadzki 2020, Lee *et al* 2020, Xu *et al* 2020, Yawo *et al* 2021, Rost *et al* 2022, Emiliani *et al* 2022), cardiovascular research (Ferenczi *et al* 2019, Joshi *et al* 2020, Madrid *et al* 2021), cellular physiology (Tan *et al* 2022), organelle dynamics and control (Kichuk *et al* 2021, Passmore *et al* 2021), gene expression and gene regulation (de Mena *et al* 2018, Asano *et al* 2018, Wichert *et al* 2021, Manoilov *et al* 2021, Ohlendorf and Möglich 2022), synthetic cell based optogenetics (Mansouri and Fussenegger 2021, Baumschlager and Khammash 2021, Adir *et al* 2022, Chia *et al* 2022), signal transduction (photo-switching, Ueda and Sato 2018), plant research (Banerjee and Mitra 2020, Christie and Zurbriggen 2021, Zhou *et al* 2021, Konrad *et al* 2023), developmental biology (Krueger *et al* 2019), fungi research (Yu and Fischer 2019, Losi and Gärtner 2021), bacterial biology (Losi and Gärtner 2021, Lindner and Diepold 2022), cancer drug discovery (Kielbis *et al* 2018) and others.

The field of optobiochemistry was reviewed, categorizing photosensory domains by chromophores, describing photo-regulatory systems by mechanism of action, and discussing protein classes frequently investigated using optical methods (Seong and Lin 2021).

The research field of bioluminescence-optogenetics (BL-OG) was updated which combines chemogenetic and optogenetic approaches. There the light emission of luciferins was used to photoexcite optogenetic proteins in neuroscience (Moore and Berglund 2019, Sureda-Vives and Sarkisyan 2020, Stern *et al* 2022, Porta-de-la-Riva *et al* 2023).

The proton and electron transfer in photosensory proteins and their involvement in primary photochemistry and subsequent processes of signaling state formation was reviewed (Kottke *et al* 2018). Time-resolved structural analysis of light-activated proteins was carried out (Poddar *et al* 2022). The challenges in the theoretical description of photoreceptor proteins using multiscale modeling were discussed (Mrozinski *et al* 2021).

In order to keep track on the optogenetics literature the publicly available database www.optobase.org was established (Kolar *et al* 2018).

13.6.2 Characterization of organic dyes applied in optogenetics

13.6.2.1 Retinals (rhodopsin chromophors)

The protonated retinal Schiff base photoisomerization in the light-driven proton-pump bacteriorhodopsin was studied with a femtosecond x-ray free-electron laser (Nogly *et al* 2018). Aspartic acid residues and functional water molecules in the close proximity to the retinal Schiff base responded collectively to the photoisomerization process. It was proposed that ultrafast charge transfer along the retinal was the driving force for the collective motions within the protein scaffold.

The photoisomerization dynamics of protonated retinal Schiff base in the gas phase was studied experimentally combining time-resolved action spectroscopy with femtosecond pump-probe techniques in an ion-storage ring (Kiefer *et al* 2019). The results indicate barrier-less photoisomerization of 11-*cis* retinal to all-*trans* retinal and barrier-involved photoisomerization of all-*trans* retinal to 13-*cis* retinal. The findings are discussed in terms of protonated retinal Schiff-base photoisomerization in visual rhodopsin proteins (11-*cis* → all-*trans*) and bacterial rhodopsin proteins (all-*trans* → 13-*cis*).

The retinal photoisomerization in the sodium outward pump *Krokinobacter eikastus* rhodopsin 2 (KR2) was studied with watermarked femtosecond to sub-millisecond stimulated Raman spectroscopy (Hontani *et al* 2016) and time-resolved serial femtosecond crystallography in a temporal window from 800 fs to 20 ms with the Swiss x-ray Free Electron Laser (Skopintsev *et al* 2020). A model of the KR2 photocycle was derived.

The photoisomerization of protonated retinal Schiff base in the heliorhodopsin 48C12 (HeR 48C12) was studied by low-temperature light-induced Fourier transform infrared (FTIR) spectroscopy (Tomida *et al* 2021). Light-induced difference spectra were recorded. Structural changes of the protein backbone with hydrogen-bonding changes between amino acid residues and water molecules inside the protein upon retinal isomerization were deduced.

The photocycle dynamics of retinal in channelrhodopsin 2 (ChR2) was clarified using single-turnover electrophysiology, time-resolved step-scan FTIR spectroscopy and Raman spectroscopy (Kuhne *et al* 2019). A branched photocycle was proposed explaining electrical and photochemical channel properties and establishing the structure of intermediates during channel turnover.

The retinal photocycle dynamics in proteorhodopsin was studied using dynamic nuclear polarization enhanced solid-state magnetic-angle spinning nuclear magnetic resonance spectroscopy (Maciejko *et al* 2019).

Various retinal analogs were synthesized and entered as replacement of the natural retinal chromophore in microbial rhodopsins to modify their absorption and fluorescence properties (AzimiHashemi *et al* 2014, Ganapathy *et al* 2019, Hontani *et al* 2019, Mei *et al* 2020).

The research on retinals was very active in the quantum chemical field. Theoretical insights into the isomerization mechanisms of retinal proteins were reviewed (Sen *et al* 2022). Extended multistate complete active space second-order perturbation theory (XMS-CASPT2) was applied to investigate retinal protonated Schiff base in the gas phase (Park and Shiozaki 2018). Trajectory surface hopping molecular dynamics simulations with conventional time dependent density functional theory were performed for retinal protonated Schiff-base photoisomerization (Liu and Zhu 2021). The involvement of triplet states in the isomerization of protonated retinal Schiff base, neutral retinal Schiff base, and retinal aldehyde was explored using the XMS-CASPT2 method (Filiba *et al* 2022).

The nonadiabatic dynamics of the ultrafast photoreaction in bacteriorhodopsin was studied using *ab initio* multiple spawning (AIMS) with spin-restricted ensemble Kohn–Sham (REKS) theory (Yu *et al* 2019). It was found that the photoisomerization was accompanied with weakening of the interaction between retinal protonated Schiff base and its counterion cluster.

The retinal isomerization and water-pore formation in channelrhodopsin-2 was explored by classical molecular mechanical (MM) and quantum mechanical (QM) molecular simulations. A detailed insight into the mechanism of ChR2 photo-activation and early events of pore formation was obtained (Ardevol and Hummer 2018). *Ab initio* multiple spawning (AIMS) with spin-restricted ensemble Kohn–Sham (REKS) theory were applied to study the nonadiabatic photodynamics of retinal protonated Schiff base in channelrhodopsin-2. The calculations revealed that the retinal protonated Schiff base isomerization was highly specific around the $C_{13}=C_{14}$ bond and followed the 'aborted bicycle-pedal' mechanism (Liang *et al* 2019). The effect of amino acid counterions and water molecules on the absorption spectrum of channelrhodopsin-C1C2 (a chimera composed of transmembrane helices H1 to H5 from ChR1 and H6 and H7 from ChR2) was studied theoretically using quantum mechanical/molecular mechanical dynamics (Adam *et al* 2021). The absorption maximum wavelength position was found to be dominantly influenced by bond length alternation and bond order alternation.

Molecular dynamics simulations were used to model the changes in internal hydration of the blue variant of proterohodopsin during photoactivation with the proton donor in protonated and deprotonated states (Faramarzi *et al* 2018). The origin of the absorption difference between blue-absorbing proteorhodopsin (BPR, absorption maximum at 490 nm, having amino acid Gln105) and green-absorbing proteorhodopsin (GPR, absorption maximum 525 nm, having amino acid Leu105) was deciphered using hybrid QM/MM simulations. It was found that a change in the electrostatic potential from the residue in position 105 near the retinal chromophore is the primary cause of the spectral shift (Church *et al* 2022).

13.6.2.2 Flavins (isoalloxazine derivatives)

A book on flavin-based catalysis (Cibulka and Fraaije 2021) gives a unique overview on the flavin research. There the structure and properties of flavins (Pavlovska and Cibulka 2021) and the spectral properties of flavins (Sikorski *et al* 2021) are reviewed.

Chemical modifications of the parent isoalloxazine structure were investigated towards the design of flavin catalysts with special reactivity and selectivity in organic synthesis (Rehpenn *et al* 2021). Air-stable reduced molecular flavins were synthesized based on a conformational bias strategy and used as photoreductants (Foja *et al* 2022). Mono-fluorinated flavin chromophores were synthesized and photophysically characterized (Reiffers *et al* 2018). The significance of riboflavin and its derivatives as potential photosensitizers in the photodynamic treatment of skin cancers was reviewed (Insińska-Rak *et al* 2023).

In quantum chemical investigations the impact of fluorination on the photophysics of the flavin chromophore was studied (Bracker *et al* 2019, Bracker *et al* 2022). Various fluorinated flavin derivatives were found with expected increasing triplet and fluorescence quantum yields, respectively. Vibrationally resolved absorption and fluorescence spectra of five forms of flavins in the gas phase (fully oxidized form Fl_{ox}, neutral radical form Fl_{sq}, anionic radical form Fl_{sq}^-, neutral reduced form Fl_{red}, and anionic reduced form Fl_{red}^-) were simulated by adiabatic or vertical methods combined with time-dependent or time-independent formalisms (Wang and Liu 2023a). A systematic theoretical study on the pH-dependent absorption and fluorescence spectra of flavins was carried out employing density functional theory and time-dependent density functional theory (Wang and Liu 2023b).

13.6.2.3 Folates (involved in photolyases and cryptochromes)
A new review on folates (Wusigale and Liang 2020) presents the current knowledge on the stability of folates in foods and the interaction of folates with biological molecules. The influences of light, heat, oxygen, pH, antioxidants, proteins, and nucleic acids is investigated. The protective effect and mechanisms of antioxidants against folic acid photodecomposition were investigated by using fluorescence and absorption spectroscopy, high-performance liquid chromatography, and antioxidant assay (Wusigale et al 2020).

13.6.2.4 p-Coumaric acid (chromophore of xanthopsins)
The hydrogen bond interaction between p-coumaric acid (4-hydroxycinnamic acid) and the amino acid residues of the photoactive yellow protein photocycle intermediate states was studied by x-ray crystal structure analysis (Wang 2019) and a quantum mechanical/molecular mechanical (QM/MM) approach (Tsujimura et al 2022).

Various medicinal applications of p-coumaric acid were reviewed (Abazari et al 2021). Therapeutic perspectives of p-coumaric acid were explored considering anti-necrotic, anti-cholestatic and anti-amoebic activities (Aldaba-Muruato et al 2021).

13.6.2.5 Linear tetrapyrroles (phytochrome chromophores)
The biosynthesis of modified tetrapyrroles was reviewed including linear tetrapyrroles as biliverdin IXα and cyclic tetrapyrroles as chlorophyll a_P, coenzyme F_{430}, siroheme, heme, uroporphyrinogen III, cobalamin and heme d_1 (Bryant et al 2020). Gold nanoshell-linear tetrapyrrole conjugates were fabricated and applied for near infrared-activated dual photodynamic and photothermal therapies (Wang et al 2019).

13.6.2.6 Corrinoid-based cyclic tetrapyrroles (chromophores of cobalamin-based photoreceptors)
The biosynthesis of cyclic tetrapyrroles was reviewed (Bryant et al 2020). A novel pathway for corrinoid compounds production in *Lactobacillus* was developed (Torres et al 2018). Classic highlights in porphyrin and porphyrinoid total synthesis and biosynthesis were reported (Senge et al 2021). The biosynthesis of tetrapyrrole cofactors by bacterial community inhabiting porphyrin-containing shale rock was studied (Stasiuk et al 2021). The nutrition action and importance of adenosylcobalamin coenzyme B_{12} was reviewed (Rizzo and Laganà 2020).

13.6.2.7 Tryptophan (UVR8 chromophore)
The role of tryptophan in health and disease was reviewed concentrating on anti-oxidant, anti-inflammation and nutritional aspects (Nayak et al 2019). The tryptophan analog, 7-aza-Trp, was incorporated in the BLUF domain of the

AppA (Activation of photopigment and pucA) photoreceptor in order to investigate the functional dynamics of the crucial W104 residue during photoactivation of the protein (Karadi *et al* 2020). The extreme ultraviolet-induced charge dynamics in a tryptophan cation was studied in an attosecond pump-probe experiment accompanied with elaborate theoretical calculations of the sub-4 fs dynamics (Lara-Astiaso *et al* 2018).

13.6.2.8 Fluorescent probes

13.6.2.8.1 Organic dyes
The development of new fluorophores, based on rhodamines and cyanines, enabling sophisticated imaging experiments leading to new biological insights, were overviewed (Schnermann and Lavis 2023).

13.6.2.8.2 Detection of chemical species
The development and application of fluorescent indicators for imaging of monoatomic ions in biological systems was reviewed (Carter *et al* 2014, Wu *et al* 2022). The fluorescent indicators were categorized in small molecule-based synthetic indicators and genetically encodable protein-based indicators. The considered monatomic inorganic ions in biology were Ca^{2+}, Zn^{2+}, K^+, Mg^{2+}, Na^+, H^+ (pH), Cl^-, Cu^+, and the toxic ions were Pb^{2+}, Cd^{2+}, As^{3+} and Hg^{2+}.

The progress in the development of fluorescent probes for hydrazine (N_2H_4), a highly toxic but important chemical reagent, was reviewed (Nguyen *et al* 2018).

Fluorescent probes guided by a protection-deprotection strategy using fluorescein as the signal scaffold were developed to monitor glutathione, an important antioxidant, in living systems (She *et al* 2018).

The progress in the development of fluorescent probes for thiophenol, a highly reactive and toxic aromatic thiol, was reviewed (Hao *et al* 2019).

Advances in fluorescent probes for the detection of the reactive oxygen species HOCl and the reactive nitrogen species HNO in live cells, tissues and organisms were reviewed (Ashoka *et al* 2020).

A critical review on organic small fluorescent probes for monitoring carbon monoxide in biology was published (Yan *et al* 2023a).

Newly developed fluorescent probes for imaging the process of ferroptosis, an iron-dependent form of regulated cell death, were reviewed (Li *et al* 2022).

A novel near-infrared fluorescent probe (heptamethine cyanine dye IR787) for live cell imaging was synthesized (Wan *et al* 2020).

The development of several near-infrared probes used for various protein-tag systems was reviewed (Reja *et al* 2021). The coupling of self-labeling protein tags with synthetic fluorescent probes is one of the most promising research areas in chemical biology.

13.6.2.8.3 Fluorogen-activating protein/fluorogen complexes
Fluorogens are chemical compounds which are initially not fluorescent, but become fluorescent through a chemical reaction with a target molecule. Fluorogen-

activating protein (FAP)/fluorogen complexes have become a new generation of fluorescence probes for biological research (Gallo 2019). They are fluorescent bimodular sensors composed of a nonfluorescent single-chain antibody (FAP) and a nonfluorescent small-molecule fluorogen. The steric restriction of fluorogens on binding to FAPs makes them fluorescent.

13.6.2.8.4 Bioluminescent probes

Chemiluminescence is the light emitted by a chemical reaction, and bioluminescence is a type of chemiluminescence in which the chemical reaction is catalyzed by an enzyme (Turner 1985). Bioluminescent probes are powerful tools for visualizing biology in live tissue and whole animals. The bioluminescent enzymes (luciferases) catalyze light emission via the oxidation of small molecule substrates (luciferins) (Yao *et al* 2018). The general basic reaction is: $\text{Luciferin} + O_2 \xrightarrow[\text{other cofactors}]{\text{Luciferase}} \text{Oxyluciferin}^* \rightarrow \text{Oxyluciferin} + h\nu$ and slow deoxidation of oxyluciferin to luciferin (Hastings 1983). The caging of a luciferin substrate to a biomarker-responsive trigger opens the field of activity-based sensing (ABS) for selective bioimaging (Yang *et al* 2022, Messina *et al* 2022, Yadav and Chan 2023) bioluminescence resonance energy transfer (BRET) from luciferin to light-sensitive optogenetic active molecules allows the activation of optogenetic tools without an exogenic light source (Komatsu *et al* 2018, Parag-Sharma *et al* 2020, Li *et al* 2021). This fact allows deep tissue optogenetic application without external invasion (Berglund *et al* 2021).

As an example, the bioluminescence reaction of firefly luciferin is shown in figure 13.36 (Messina *et al* 2022).

13.6.2.8.5 Chemical Ca²⁺ indicators

Calcium ions, as second messengers in cellular biology, generate versatile intracellular signals to regulate cell function (Carafoli 2002, Hofer and Lefkimmiatis 2007, Cioffi *et al* 2011, Grienberger and Konnerth 2012, Carter *et al* 2014, Bagur and Hajnóczky 2017, Wu *et al* 2022, Sanyal *et al* 2023). Molecular probes have been developed to measure the Ca^{2+} concentration in cells. These probing molecules can chelate calcium ions (form molecule-Ca^{2+} chelates). They are based on EGTA homologs (EGTA = ethylene glycol-bis(ß-aminoethyl ether)-N,N,N',N'-tetraacetic acid) with high selectivity for calcium ions versus magnesium ions (Tsien 1980, 1981 and 1989, Grynkiewicz *et al* 1985, Cobbold and Rink 1987, Paredes *et al* 2008,

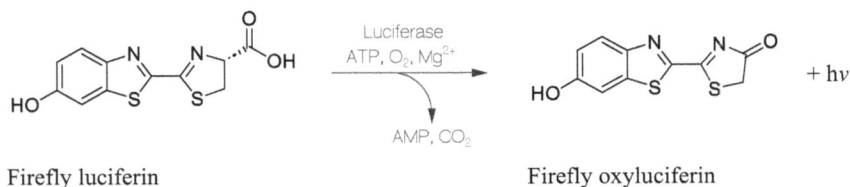

Firefly luciferin   Firefly oxyluciferin

Figure 13.36. Reaction of firefly luciferin for bioluminescence emission. ATP = adenosine triphosphate. AMP = adenosine monophosphate.

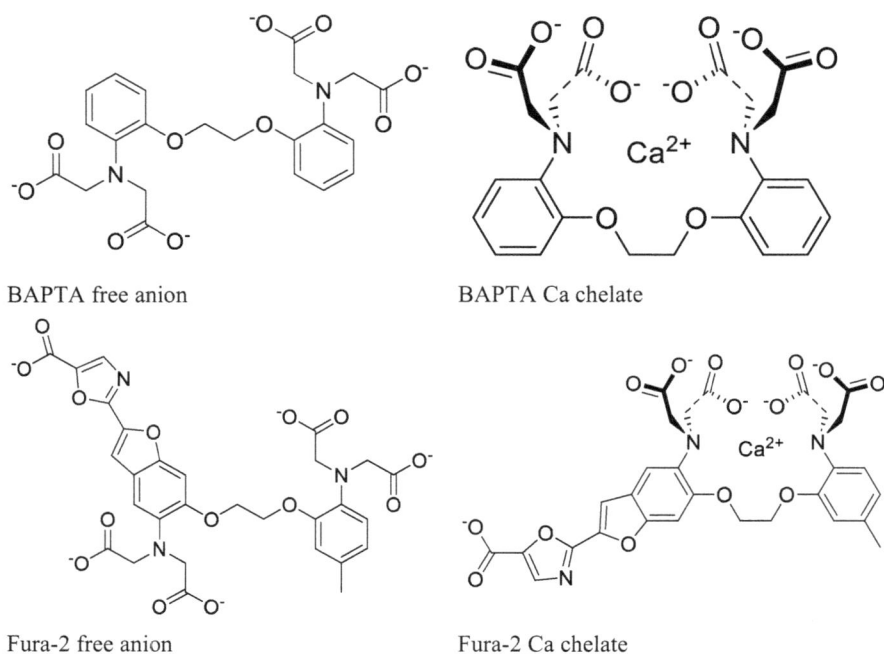

Figure 13.37. Structural formulae of BAPTA free anion and BAPTA Ca chelate (see Tsien 1980, Ducrot *et al* 2019, Wu *et al* 2022), fura-2 free anion and fura-2 Ca chelate (see Grynkiewicz *et al* 1985, Cobbold and Rink 1987, Wu *et al* 2022).

Grinberger and Konnerth 2012, Carter *et al* 2014, Ducrot *et al* 2019, Csomos *et al* 2021, Wu *et al* 2022).

As examples of chemical Ca^{2+} indicators the aminopolycarboxylic acid BAPTA (1,2-bis(o-aminophenoxy)ethane-N,N,N',N'-tetraacetic acid) and the fluorescence dye fura-2 are presented here. The structural formulae of BAPTA free anion, BAPTA Ca chelate, fura-2 free anion, and fura-2 Ca chelate are displayed in figure 13.37. The formed tetraacetic acid binding cavity has the right size for Ca^{2+} coordination complex formation (Tsien 1980).

The absorption cross-section spectra of BAPTA free anion and BAPTA Ca chelate are shown in the left part of figure 13.38 (taken from Tsien 1980, the decadic extinction coefficients, ε, there are transformed to absorption cross-sections, σ_a, here using the relation $\sigma_a = \varepsilon * \ln(10)*1000/N_A$, where $N_A = 6.022\ 214\ 199 \times 10^{23}\ \text{mol}^{-1}$ is the Avogadro constant). The dependence of the absorption spectrum of BAPTA on the concentration of free Ca^{2+} in solution allows the determination of the Ca^{2+} concentration (Tsien 1980). The fluorescence emission spectrum of calcium-free BAPTA has its maximum at $\lambda_{F,max} = 363$ nm, and its fluorescence quantum yield was found to be $\phi_F = 0.027$. Binding of Ca^{2+} did not significantly alter the shape of the fluorescence spectrum but reduced the fluorescence quantum yield by a factor of 2.8 (Tsien 1980). The use of BAPTA for the determination of free Ca^{2+} in cells is aggravated by the overlap with the absorption of amino acids in the UV spectral region.

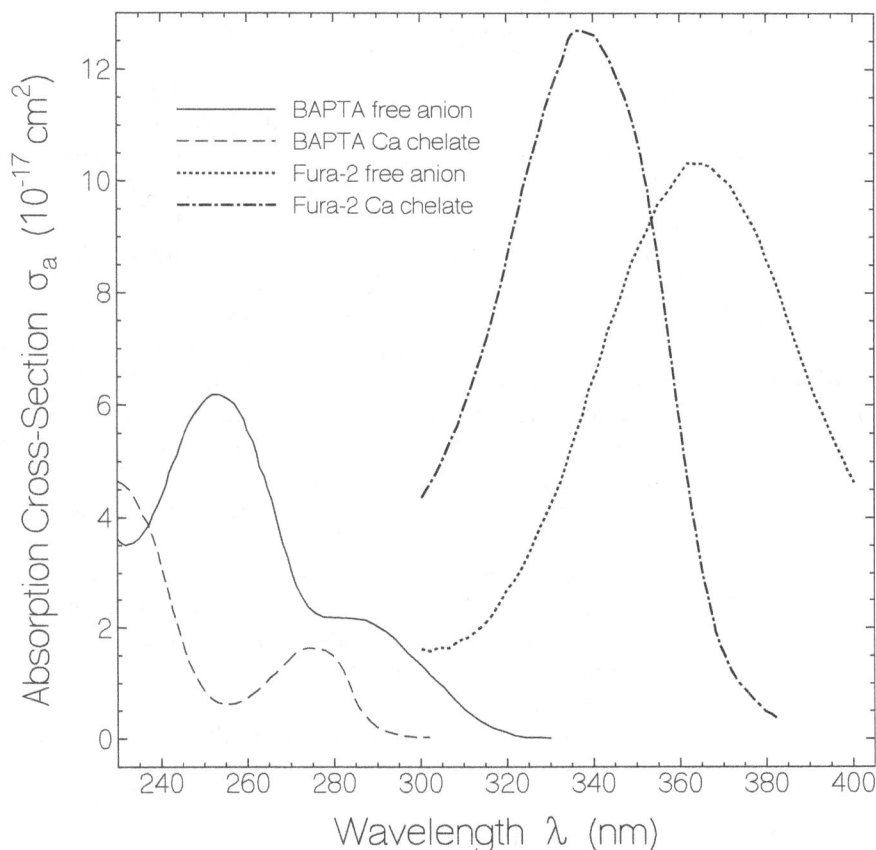

Figure 13.38. Absorption cross-section spectra of **BAPTA** free anion and **BAPTA** Ca chelate (data taken from Tsien 1980), and of fura-2 free anion and fura-2 Ca chelate (data taken from Grynkiewicz *et al* 1985 and Cobbold and Rink 1987).

The absorption cross-section spectra of fura-2 free anion and fura-2 Ca chelate are shown in right part of figure 13.38 (taken from Grynkiewicz *et al* (1985), and Cobbold and Rink (1987)). The dependence of the absorption spectrum of fura-2 on the concentration of free Ca^{2+} in solution allows the determination of the Ca^{2+} concentration (Grynkiewicz *et al* 1985). The wavelength maxima of the fluorescence emission spectra of calcium-free fura-2 and fura-2 Ca chelate were found to be at $\lambda_{F, max}$ = 518 nm and 510 nm, respectively. The fluorescence quantum yields were found to be ϕ_F(fura-2 free anion) = 0.23 and ϕ_F(fura-2 Ca complex) = 0.49 (Grynkiewicz *et al* 1985). The use of fura-2 for the determination of free Ca^{2+} in cells is convenient since there is no overlap with the absorption spectra of amino acids.

13.6.2.9 Fluorescent proteins
In a new book the recent progress in the practical application of fluorescent protein technology has been published (Sharma 2023). The development of fluorescent

proteins, the design of fluorescent protein-based biosensors, and genetically encoded fluorescent probes were reviewed (Wang *et al* 2023). A quantitative assessment of near-infrared fluorescent proteins was conducted (Zhang H *et al* 2023).

The photoswitching mechanism of the reversibly switchable fluorescent protein rsEGFP2 (chromophore: 4-hydroxybenzylidene imidazolinone HBI) was revealed by time-resolved crystallography and transient absorption spectroscopy (Coquelle *et al* 2018, Woodhouse *et al* 2020). An overview of reversibly photoswitchable fluorescent proteins (rsFPs) for reversible saturable optical linear fluorescence transition (RESOLFT) nanoscopy was given (Jensen *et al* 2020).

Bright monomeric red fluorescent proteins were generated via computational protein design to increase fluorescence quantum yield by optimizing chromophore rigidification with aliphatic amino acid residues (Legault *et al* 2022). The directed evolution resulted in the fluorescent protein mSandy2 with peak emission wavelength $\lambda_{em} = 609$ nm and fluorescence quantum yield $\phi_F = 0.35$ (Legault *et al* 2022). mScarlet3, a brilliant and fast-maturing red fluorescent protein ($\phi = 0.75$), was generated applying several rounds of targeted and random mutagenesis (Gadella *et al* 2023).

The expression of fluorescent proteins in cell-free systems was reviewed (Silverman *et al* 2020, Carenne *et al* 2021), and a so synthesized fluorescent protein (mNeonGreen) was used for low limit of detection sensing (Copeland *et al* 2022).

A monomeric green fluorescent protein sensor for chloride (GFPxm163) was discovered by structure-guided bioinformatics (Peng *et al* 2022).

Circularly permuted fluorescent protein-based indicators that led to an extensive class of biosensors have been reviewed (Kostyuk *et al* 2019).

Chemically stable yellow fluorescent proteins for advanced microscopy were engineered and characterized (Campbell *et al* 2022).

Fluorescent timers are fluorescent proteins that change their fluorescent color over time, allowing to image the protein localization changes over time (Terskikh *et al* 2000). Blue-to-red and green-to-FarRed genetically encoded true and tandem fluorescent timers were developed and characterized (Subach *et al* 2023a).

13.6.2.9.1 Genetically encoded calcium indicators
Genetically encoded calcium indicators (GECIs) are encoded by a nucleic acid sequence and are synthesized entirely by a cell. They are comprised of protein-based probes that utilize one or two fluorescent proteins (FP) as the fluorophore and a calcium-binding domain. The development of GECIs was reviewed (Mank and Griesbeck 2008, McCombs and Palmer 2008, Grienberger and Konnerth 2012, Koldenkova and Nagai 2013, Carter *et al* 2014, Piatkevich *et al* 2019, Lohr *et al* 2021, Nietz *et al* 2022, Wu *et al* 2022).

A calcium-binding domain is calmodulin (= calcium-modulated protein, CaM, for reviews see Stevens 1983, Chien and Means 2000, Andrews *et al* 2021) together with the 20-residue peptide RS20 corresponding to the calmodulin-binding domain of the smooth muscle myosin light chain kinase (Lukas *et al* 1986, Tsvetkov *et al* 1999) or the M13 synthetic peptide derived from the skeletal muscle myosin light chain kinase (Klevit *et al* 1985, Tsvetkov *et al* 1999). CaM together with a fragment of endothelial nitric oxide synthase was used for fast and sensitive calcium indicators for imaging neural populations (Zhang *et al* 2023a). Another calcium-binding domain applied in GECIs is troponin C present in skeletal and myocardial muscles of Chordata (Takeda *et al* 2003, Marston and Zamora 2020, Subach *et al* 2022). Both, in calmodulin and in troponin C, Ca^{2+} is bound in the helix-loop-helix EF-hand motifs EF1, EF2, EF3 and EF4 (Gifford *et al* 2007). Troponin C in GECIs was used in full-length form (with N-terminal EF1, EF2 and C-terminal EF3, EF4) (Heim and Griesbeck 2004, Mank *et al* 2006, Mank and Griesbeck 2008) or in truncated form (minimal motif having only EF3 and EF4) (Wilms and Häusser 2014, Thestrup *et al* 2014, Barykina *et al* 2016 and 2018, Zhang D *et al* 2021, Piatkevich *et al* 2019, Subach *et al* 2022, Subach *et al* 2023b).

For single FP-based Ca^{2+} sensors, the presence and accompanied binding of Ca^{2+} changes the chemical or electronic environment around the chromophore which causes a Ca^{2+} concentration dependent (intensometric) changing of the emission behavior (intensometric GECIs, Helassa *et al* 2016, Mohr *et al* 2020, Nasu *et al* 2021, Bi *et al* 2021). In figure 13.39(a) structural illustrations of single-fluorophore based GECIs are shown. The top part belongs to a genetically encoded calcium indicator consisting of calcium binding protein calmodulin (CaM) together with the calmodulin binding domain (M13 or RS20) connected to a fluorescent protein (FP). Without presence of Ca^{2+}, the EF-hand motives (EF1, EF2, EF3, EF4) of CaM are without Ca^{2+} and the M13 (or RS20) motive is separated from CaM (left part). In the presence of Ca^{2+}, the EF-hand motives bind Ca^{2+} and cause a clustering of CaM with M13 (or RS20) (right part). This structural change influences the fluorescence efficiency of the FP (Nakai *et al* 2001, Zhao *et al* 2011). The bottom part of

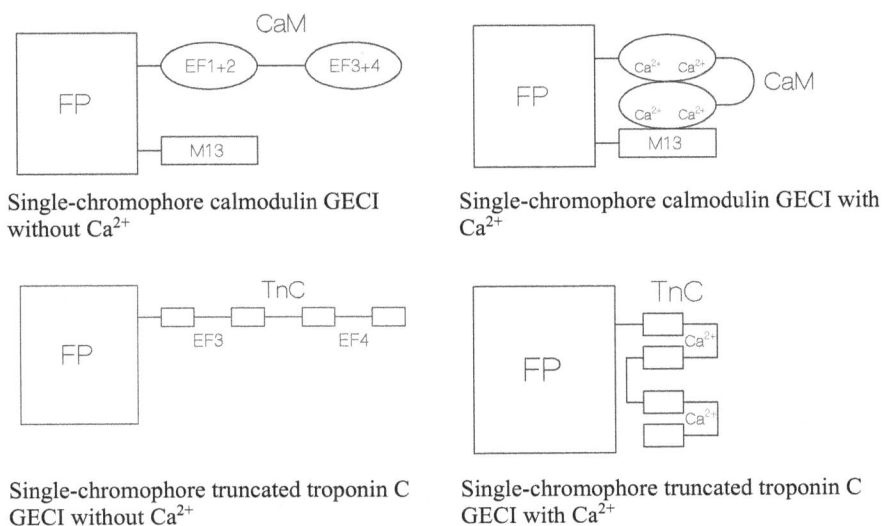

Single-chromophore calmodulin GECI without Ca^{2+}

Single-chromophore calmodulin GECI with Ca^{2+}

Single-chromophore truncated troponin C GECI without Ca^{2+}

Single-chromophore truncated troponin C GECI with Ca^{2+}

Figure 13.39a. Schematic drawings of single-fluorescent-protein-based GECIs (intensiometric GECIs). FP: fluorescent protein. CaM: calmodulin. M13 (or RS20): calmodulin-binding domain. EF1, EF2, EF3, EF4: EF-hand Ca^{2+}-binding motifs. TnC: truncated troponin C.

figure 13.39(a) belongs to a genetically encoded calcium indicator consisting of a fluorescent protein with a truncated troponin C protein. The left part shows the arrangement without presence of Ca^{2+}, and the right part shows the situation with Ca^{2+} presence.

Ca sensors containing two FPs exploit the principle of FRET (Förster resonance energy transfer) where the Ca^{2+} binding causes conformational changes which influences the FRET between the chromophores (change of energy transfer from photoexcited fluorescent protein FP1 (donor) to nearby fluorescent protein FP2 (acceptor) with decrease of fluorescence emission from FP1 and increase of fluorescence emission from FP2 (ratiometric GECIs, Miyawaki et al 1997, Wilms and Häusser 2014, Thestrup et al 2014, Shemetov et al 2021, Zhang D et al 2021, Subach et al 2023c). In figure 13.39(b) structural illustrations of double-fluorophore based GECIs are shown. The top part belongs to a genetically encoded calcium indicator consisting of calcium binding protein calmodulin (CaM) together with the calmodulin binding domain (M13 or RS20) connected to two fluorescent proteins (FP1 and FP2). Without presence of Ca^{2+}, the EF-hand motives (EF1, EF2, EF3, EF4) of CaM are without Ca^{2+} and the M13 (or RS20) motive and CaM are

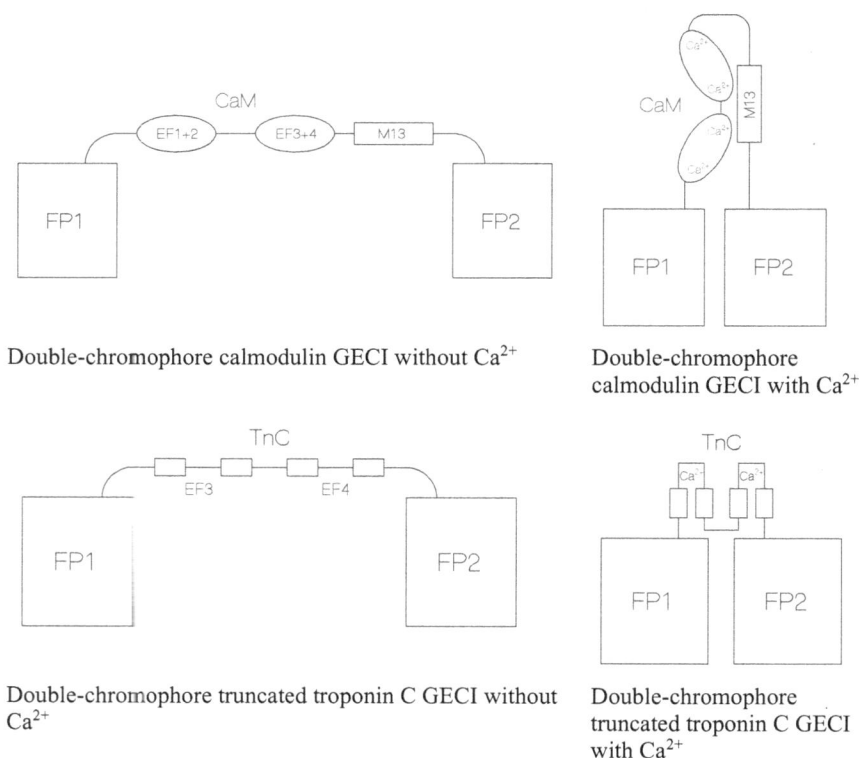

Double-chromophore calmodulin GECI without Ca^{2+}

Double-chromophore calmodulin GECI with Ca^{2+}

Double-chromophore truncated troponin C GECI without Ca^{2+}

Double-chromophore truncated troponin C GECI with Ca^{2+}

Figure 13.39b. Schematic drawings of double-fluorescent-protein-based GECIs (FRET-based GECIs, ratiometric GECIs). FP1, FP2: fluorescent proteins. CaM: calmodulin. M13 (or RS20): calmodulin-binding domain. EF1, EF2, EF3, EF4: EF-hand Ca^{2+}-binding motifs. TnC: truncated troponin C.

elongated (left part). FP1 and FP2 are widely separated hindering efficient Förster-type energy transfer from photoexcited FP1 (donor) to ground-state FP2 (acceptor). The fluorescence is emitted from FP1. In the presence of Ca^{2+}, the EF-hand motives bind Ca^{2+} and cause a clustering of CaM with M13 (or RS20) (right part). FP1 and FP2 are near together favoring efficient Förster-type energy transfer from photo-excited FP1 (donor) to ground-state FP2 (acceptor). The fluorescence is dominantly emitted from FP2 (fluorescence spectrum of FP1 is weakened and fluorescence spectrum of FP2 is enhanced). The bottom part of figure 13.39(b) belongs to a genetically encoded calcium indicator consisting of two fluorescent proteins (FP1 and FP2) with a truncated troponin C protein (Zhang *et al* 2021). The left part shows the arrangement without presence of Ca^{2+} (separation of FP1 and FP2 with weak FRET), and the right part shows the situation with Ca^{2+} presence (FP1 and FP2 near together with strong FRET).

Calcium indicators (chemical Ca^{2+} indicators and genetically encoded Ca^{2+} indicators) may be used to follow neural activity in nervous systems. When a neuron fires action potential, voltage-gated calcium channels in the plasma membrane open up and lead to a rise in cytosolic calcium concentration within a few milliseconds, which can be detected by the calcium indicators placed in the neurons (Smetters *et al* 1999, Mank and Griesbeck 2008, Hires *et al* 2008, Ross 2012, Rose *et al* 2014, Podor *et al* 2015, Peron *et al* 2015, Storace *et al* 2016, Rad *et al* 2017, Piatkevich *et al* 2019, Nasu *et al* 2021, Bi *et al* 2021, Zhu *et al* 2021, Nietz *et al* 2022, Grienberger *et al* 2012, Zhang D *et al* 2023a).

The cytoplasmic Ca^{2+} concentration, measured by cytosolic calcium indicators, is determined by extracellular Ca^{2+} influx affected by the transmembrane voltage potential and the intracellular release of Ca^{2+} from organelles like the endoplasmic reticulum (ER) in all eukaryotic cells and the sarcoplasmic reticulum (SR) in muscle cells (Bagur and Hajnóczky 2017). The intracellular Ca^{2+} dynamics was studied by a channelrhodopsin-fluorescent protein-based organelle optogenetics method (Asano *et al* 2018, light-sensitive channelrhodopsin cation channel was specifically targeted to the ER/SR).

13.6.2.10 Nanomaterials

The research on nanomaterials and their application in biology is very active. Recent reviews are: Wolfbeis 2015, Wei *et al* 2019, Wan *et al* 2020, Algar *et al* 2021, Li *et al* 2022, and Feng *et al* 2022. The optical characteristics of different fluorescent nanoprobes and the application of NIR-II nanoprobes (wavelength range 1000–1700 nm) in different biological tissues was reviewed (Zhang N-n *et al* 2021).

Multi-color fluorescent nanodiamonds containing various color centers are promising fluorescent markers for biomedical applications (Alkahtani *et al* 2018).

Organic dye guests doped into polymer hosts like rhodamine 6G in a copolymer (Holzer *et al* 2000) may form particles (Barykina *et al* 2018, Visaveliya and Köhler 2021) or microbeads (Scholtz *et al* 2022) and used as tools for imaging applications in life science and in material science. Dye-encapsulated polymer nanoparticles find applications in *in vivo* fluorescence imaging (Du *et al* 2020, Umezawa *et al* 2023).

The organic dyes may also be doped into silica dots and used in bio-applications (Gubala *et al* 2020). Recent advances in silicon quantum dot-based fluorescent biosensors have been reviewed (Zhang Y *et al* 2023b).

Carbon quantum dots are an emerging class of carbon-based nanomaterials with considerable interest in biological applications due to their photoluminescence properties (Atabaev 2018, Liu *et al* 2019, Yan *et al* 2019, Li *et al* 2019, Desmond *et al* 2021, Antar *et al* 2021, Zhang L *et al* 2022, Kumar *et al* 2022, Soumya *et al* 2023). Graphene quantum dots usually derived from graphene and/or graphene oxide are also candidates for applications in sensing, bioimaging, photocatalysis, energy storage and flexible electronics (Li *et al* 2019, Sharma *et al* 2021). Fluorescent carbon nanotubes have been developed for biosensing (Hendler-Neumark and Bisker 2019, Ackermann *et al* 2022).

Metallic nanoparticles are used in various biological application. They are synthesized by chemical, physical or biological methods (Jamkhande *et al* 2019, Zhang D *et al* 2020, Shi *et al* 2021, Sowmya *et al* 2023).

Metallic nanoparticles of the size from sub-nm to around 2 nm are called metallic nanoclusters (nanodots). They consist of few to tens of atoms. Due to enhanced quantum confinement they possess molecule-like optical transitions with molecule-similar absorption and fluorescence features (Yu *et al* 2015). The application of fluorescent gold nanoclusters in sensing and imaging was reviewed (Chen *et al* 2015). The progress in copper nanocluster-based fluorescent probing was collected (Qing *et al* 2019). The detection of biomarkers related to diverse diseases with fluorescence nanoclusters was reviewed (Romeo *et al* 2021). The challenges and opportunities of metal nanocluster-based devices was discussed (Chen L *et al* 2021). The functioning and application of ligand-protected noble metal nanoclusters was worked out (Matus and Häkkinen 2023).

Metal nanoparticles of 2–10 nm size, referred to as quantum-sized metal nano-particles, have molecule-like properties of metal nanoclusters and localized surface plasmon resonance properties of plasmonic nanoparticles. They are applied in photocatalysis surface chemistry of adsorbates (Stewart *et al* 2021).

Metal nanoparticles of the size of 10–100 nm are called plasmonic metal nanoparticles. In them localized surface plasmon resonances are generated by light exposure. These are coherent and collective electron oscillations confined at the dielectric-metal interface (Petryayeva and Krull 2011, Amendola *et al* 2017, Yu *et al* 2019). They are applied as photosensitizers to produce reactive chemical species (like $^1O_2$, O_2^-, OH^-, H_2O_2) in plasmonic nanoparticle-mediated photodynamic therapy of bacterial infections and cancers (Vankayala *et al* 2011, Singh *et al* 2023). They are used in plasmonic photothermal therapy (energy of photons absorbed by plasmonic nanoparticles is transferred to heat, Indhu *et al* 2023).

The intensity of a fluorescent biosensor may be considerably enhanced by placing a metallic nanostructure and a fluorophore in close proximity due to the effect of metal enhanced fluorescence (Fothergill *et al* 2018, Sabzalipoor *et al* 2022) also called the effect of plasmon-enhanced fluorescence (Bauch *et al* 2014).

13.6.2.10.1 Fluorescent semiconductor nanocrystals (quantum dots)

Conventional semiconductor quantum dots like CdSe, CdTe etc were compared with metal halide perovskite semiconductor quantum dots like $CH_3NH_3PbBr_3$, $CsPbI_3$ etc (Aldakov and Reiss 2019). Polystyrene microbeads encoded with CdSe/CdS quantum dots were prepared for application in life and material sciences (Scholtz *et al* 2022).

Semiconducting polymer dots consisting of luminescent polymers like MEH-PPV (Holzer *et al* 2004) in their core gain importance in biology and medicine (Wu and Chiu 2013, Yu *et al* 2017, Li *et al* 2018, Wang *et al* 2019). Semiconducting polymer dots were developed with reasonable fluorescence emission in the NIR-IIa spectral region (1300–1400 nm) for through-skull mouse-brain imaging (Zhang Z *et al* 2020), where the effects of aggregation-induced emission by restriction of intermolecular motion were exploited (Guo *et al* 2020, Suzuki *et al* 2020, Cai and Liu 2020). The complex photobleaching behavior of semiconducting polymer dots was studied in detail (Gupta *et al* 2022). Photoswichable semiconducting polymer dots were synthesized for effective pattern encoding and superresolution imaging (Yao *et al* 2023).

13.6.2.10.2 Lanthanide upconversion nanoparticles

Lanthanide-doped upconversion nanoparticles (UCNPs) have the ability to convert multiple near-infrared photons into higher energy ultraviolet-visible photons. Concerning biological applications, they have the advantage of high tissue penetration depths. The progress in the development and application of lanthanide-doped UCNPs has been reviewed in several papers (Duan *et al* 2018, Wen *et al* 2018, Li *et al* 2019, Himmelstoß and Hirsch 2019, Li and Chen 2020, Zhang Z *et al* 2020, Das *et al* 2020, Liang *et al* 2020, Samhadaneh *et al* 2020, Zhang H *et al* 2021, Du *et al* 2022, Lei *et al* 2022, Xu *et al* 2022, Malhotra *et al* 2023). Biological (Li *et al* 2019), biomedical (Duan *et al* 2018, Liang *et al* 2020), bioimaging (Liang *et al* 2020), bioanalytical (Malhotra *et al* 2023), theragnostic (Malhotra *et al* 2023), and biological super-resolution fluorescence imaging (Dong *et al* 2021, Zhang H *et al* 2021, Xu *et al* 2022) applications were discussed besides others.

13.6.2.11 Carotenoids

Carotenoids are yellow, orange and red organic pigments (water insoluble dyes, liposoluble dyes) that are produced in plants, algae, several bacteria, and fungi (Alcaíno *et al* 2016). They are categorized into carotenes, which are purely hydrocarbons, and xanthophylls, which are hydrocarbons containing oxygen (Cazzanuiga *et al* 2016). The carotenoids have important function in photosynthesis (Hashimoto *et al* 2016, Zulfiqar *et al* 2021).

The xanthophylls echinenone (ECN), 3′-hydroxyechinenone (3-hECN) and canthaxanthin (CAN) are noncovalently bound chromophores of the orange carotenoid protein (OCP), a blue–green light photoactive protein (Kirilovsky and Kerfeld 2013). The structural formulae of β-carotene, ECN, 3-hECN and CAN are displayed in figure 13.40.

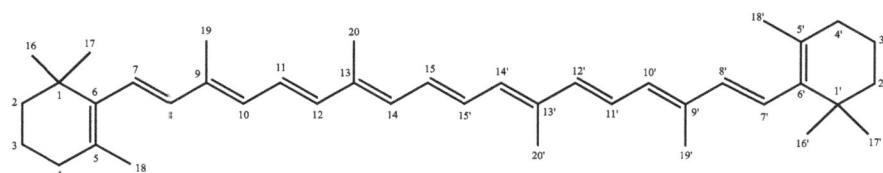

β-Carotene

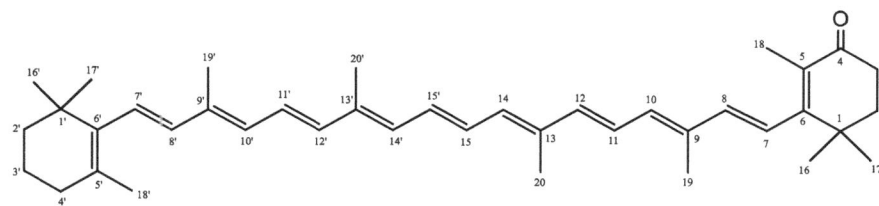

Echinenone

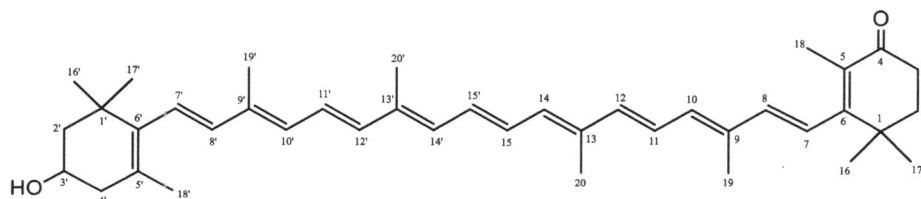

3'-Hydroxyechinenone

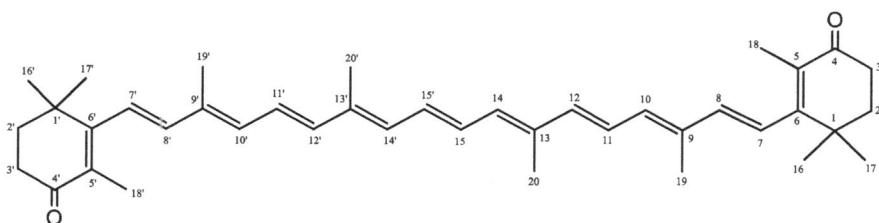

Canthaxanthin

Figure 13.40. Structural formulae of some carotenoids: β-carotene (provitamin A), echinenone, 3'-hydroxyechinenone, and canthaxanthin.

The conjugated double bonds of the polyene chain, the double bond extension to the β-ionone rings, and the carbonyl groups are responsible for the absorption in the blue and green spectral range. The absorption cross-section spectrum, $\sigma_a(\lambda)$, of β-carotene in n-hexane at room temperature is shown in the top part of figure 13.41 (from PhotochemCAD ID: D10). Normalized absorption coefficient spectra, $\alpha_a(\lambda)/\alpha_a(\lambda_{max})$, are shown in the bottom part of figure 13.41 for β-carotene in n-hexane (from PhotochemCAD ID: D10) and in benzene (from Chábera et al 2009), 3'-hydroxyechinenone in n-hexane (Polívka et al 2005), and echinenone and canthaxanthin both in benzene (from Chábera et al 2009).

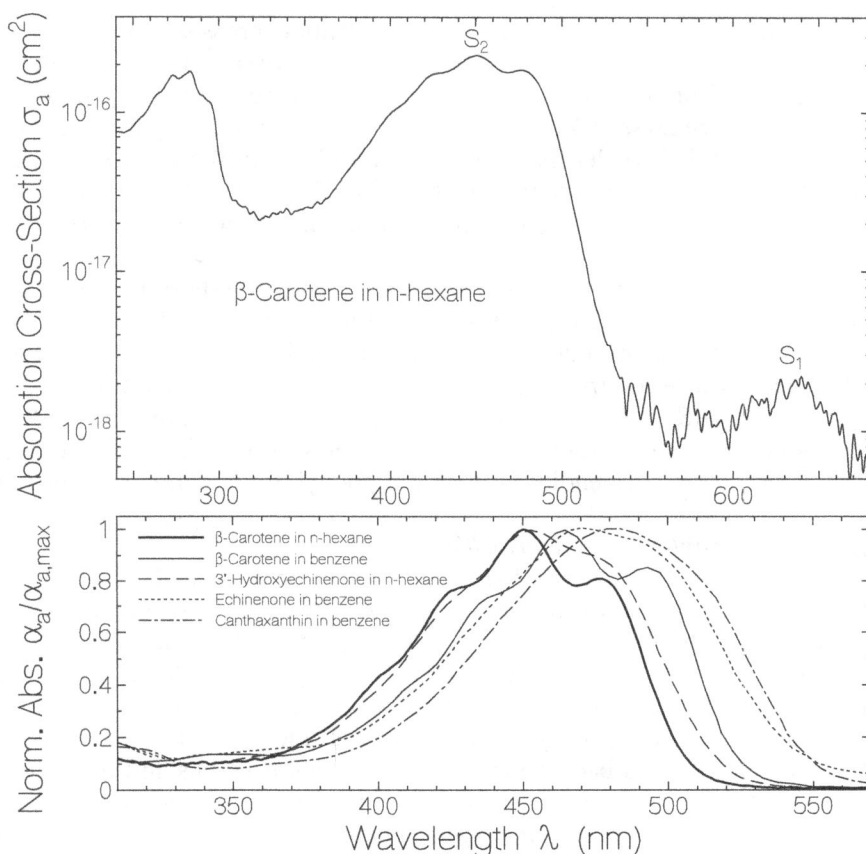

Figure 13.41. Top part: Absorption cross-section spectrum of β-carotene in n-hexane (from PhotochemCAD ID: D10). Bottom part: Normalized absorption coefficient spectra of β-carotene in n-hexane (from PhotochemCAD ID: D10) and in benzene (from Chábera *et al* 2009), 3′-hydroxyechinenone in n-hexane (from Polívka *et al* 2005), echinenone and canthaxanthin in benzene (from Chábera *et al* 2009).

The absorption behavior of the presented carotenoids (carotenes and xantho-phylls) is determined by their C_{2h} symmetry of the polyene chain and the end rings (Polívka and Sundström 2004). For them (polyene chains with length of more than 3 conjugated double bonds (Tavan and Schulten 1987)), the one-photon transition from the ground-state S_0 ($1\ ^1A_g^-$) to the first excited singlet state S_1 ($2\ ^1A_g^-$) is symmetry forbidden and to the second excited singlet state S_2 ($1\ ^1B_u^+$) is allowed. A detailed analysis of electronic properties and spectroscopy of polyenes is found in Hudson and Kohler 1974, Hudson *et al* 1982, Tavan and Schulten 1987, Orlandi *et al* 1991, Fuß *et al* 2000.

The absorption cross-section spectrum of β-carotene in n-hexane, shown in the top part of figure 13.41, resolves the weak S_0–S_1 absorption band ($\lambda_{a,max} \approx 640$ nm, $\sigma_{a,max} \approx 2 \times 10^{-18}$ cm^2) and shows the strong S_0-S_2 absorption band ($\lambda_{a,max} \approx 445$ nm, $\sigma_{a,max} \approx 2.25 \times 10^{-16}$ cm^2). In the bottom part of figure 13.41 normalized S_0–S_2

absorption bands are shown (no absolute absorption cross-sections were found besides β-carotene in n-hexane). The spectra positions are slightly solvent dependent as seen by comparison of β-carotene in n-hexane and in benzene ($\lambda_{a,max}$(n-hexane) ≈ 450 nm, $\lambda_{a,max}$(benzene) ≈ 464 nm). The absorption band of 3′-hydroxyechinenone in n-hexane is slightly broader than the absorption band of β-carotene in n-hexane. The absorption bands of echinenone in benzene ($\lambda_{a,max}$ ≈ 470 nm) and canthaxanthin in benzene ($\lambda_{a,max}$ ≈ 482 nm) are spectrally more smeared out (vibrational structure less resolved).

The fluorescence behavior of all-*trans* β-carotene in n-hexane, toluene, and carbon disulfide was studied by Bondarev and Knyukshto 1994. The $S_2 \rightarrow S_0$ ($\lambda_{F,max}$(n-hexane) ≈ 522 nm) fluorescence spectrum and the $S_1 \rightarrow S_0$ ($\lambda_{F,max}$(n-hexane) ≈ 830 nm) fluorescence spectrum was resolved. The fluorescence quantum yields were determined to be $\phi_F(S_2$, n-hexane$)$ ≈ 1.3×10^{-4} and $\phi_F(S_1$, n-hexane$)$ ≈ 5×10^{-6}. For carotenoids containing a carbonyl group the fluorescence yield from the S_1 state increased compared to carotenoids without a carbonyl group (Mimuro *et al* 1993).

13.6.2.12 Chlorophylls and bacteriochlorophylls

The cyclic tetrapyrroles chlorophylls and bacteriochlorophylls, noncovalently bound to specific apoproteins, serve as principal light-harvesting and energy-transforming pigments in photosynthetic organisms (Scheer 2006, Blankenship 2021, Lokstein *et al* 2021, Simkin *et al* 2022). Protochlorophyllides (having a porphyrin core) and chlorophyllides (having a chlorin core) are penultimate and last biosynthetic precursors, respectively, of chlorophylls and bacteriochlorophylls. The free-base pheophytins and bacteriopheophytins are chlorophylls and bacteriochlorophylls without the magnesium atom in the center (Milenković *et al* 2012).

Chlorophylls (Chls) are present in oxygenic photosynthetic systems (there take-up of CO_2 from air and H_2O from soil to synthesize under the action of light starch, cellulose, sugar and release O_2; basic reaction: $6CO_2 + 6H_2O + h\nu \rightarrow C_6H_{12}O_6 + 6O_2$, see for example Stirbet *et al* 2020). To the oxygenic photosynthetic systems belong green plants, algae, and cyanobacteria. Their main members are Chl a, Chl b, Chl d, and Chl f having a chlorin core and phytyl tail and protochlorophyll a having a poryhyrin core and a phytyl tail (for details see Taniguchi *et al* 2021). Chl a is present in all oxygenic photosynthetic systems (Katz *et al* 1978). Pheophytin is the primary electron acceptor in the photosystem PSII of the oxygenic photosynthetic reaction centers of cyanobacteria, algae and plants (Klimov 2003).

Bacteriochlorophylls (Bchl) are present in anoxygenic photosynthetic systems (basic reaction: $6CO_2 + 12H_2S + h\nu \rightarrow C_6H_{12}O_6 + 12S + 6H_2O$, no oxidation of H_2O with release of O_2) to which belong phototrophic bacteria. Their main members are Bchl a, Bchl b, Bchl g having a bacteriochlorin core and phytyl tail and Bchl c, Bchl d, Bchl e, Bchl f having a chlorin core and farnesyl tail (for details see Taniguchi *et al* 2021). Bchl a is the principal bacteriochlorophyll in purple photosynthetic bacteria (Katz *et al* 1978). Bacteriopheophytin is the intermediate electron acceptor in the type II anoxygenic photosynthetic reaction center of purple bacteria (George *et al* 2020).

The chlorophyll and bacteriochlorophyll biosynthesis were reviewed (Eckhardt *et al* 2004, Bollivar 2006, Tanaka R and Tanaka A 2007, Heyes *et al* 2021). In the penultimate step of the biosynthesis of chlorophyll and bacteriochlorophyll, the nonhomologous enzymes light-independent protochlorophyllide reductase (DPOR) and light-dependent protochlorophyllide reductase (LPOR) catalyze the reduction of protochlorophyllide (Pchlide) to chlorophyllide (Chlide) (Masuda and Takamiya 2004, Vedalankar and Tripathy 2019, Heyes *et al* 2021, Silva and Cheng 2022). DPOR utilizes ATP and reduced ferredoxin to reduce Pchlide to Chlide (Vedalankar and Tripathy 2019). In LPOR photon absorption by the pigment protochlorophyllide bound in the ternary LPOR-Pchlide-NADPH complex causes the reduction of Pchlide to Chlide by accompanying oxidation of nicotinamide adenine dinucleotide phosphate (NADPH) to $NADP^+$ (Dietzek *et al* 2004, Colindres-Rojas *et al* 2011, Heyes *et al* 2015, Gabruk and Mysliwa-Kurdziel 2015, Schneidewind *et al* 2019, Zhang S *et al* 2019, Heyes *et al* 2021, Johannissen *et al* 2022). Further enzymatic reactions convert chlorophyllides to chlorophylls and bacteriochlorophylls (Vedalankar and Tripathy 2019, Heyes *et al* 2021).

The central moieties of protochlorophylls (porphyrin), chlorophylls (chlorin), and bacteriochlorophyll (bacteriochlorin) are displayed in the top part of figure 13.42. Sites of substitution with different reactivity are: (i) the meso positions 5, 10,15,20, (ii) the β-pyrrole positions 2, 3, 7, 8, 12, 13, 17, 18, and (iii) the core positions 21, 22, 23, 24. The structural formulae of the phytyl group and the farnesyl group are shown in the bottom part of figure 13.42.

The structural formulae of pheophytin a (equal to chlorophyll a without Mg), protochlorophyll a, chlorophyll a, chlorophyll b, and bacteriochlorophyll a are displayed in figure 13.43.

Absorption cross-section spectra of protochloropyll a in diethyl ether (top part), chlorophyll a and chlorophyll b in diethyl ether as well as pheophytin a in acetone (middle part), and bacteriochlorophyll a in toluene (bottom part) are shown in figure 13.44. The spectra are understood in terms of a four-orbital model for porphyrins involving the two highest occupied π orbitals and the two lowest unoccupied π* orbitals with vibronic bands first applied by Gouterman

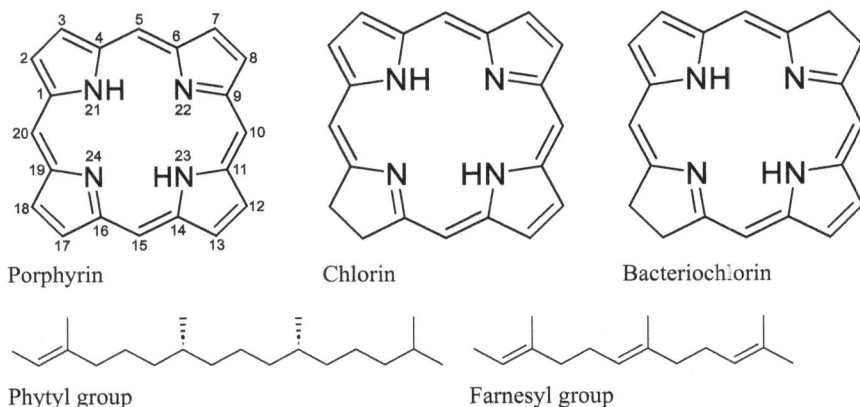

Porphyrin   Chlorin   Bacteriochlorin

Phytyl group   Farnesyl group

Figure 13.42. Structural formulae of porphyrin (having 22 π electrons), chlorin (having 20 π electrons), bacteriochlorin (having 18 π electrons), the phytyl group and the farnesyl group.

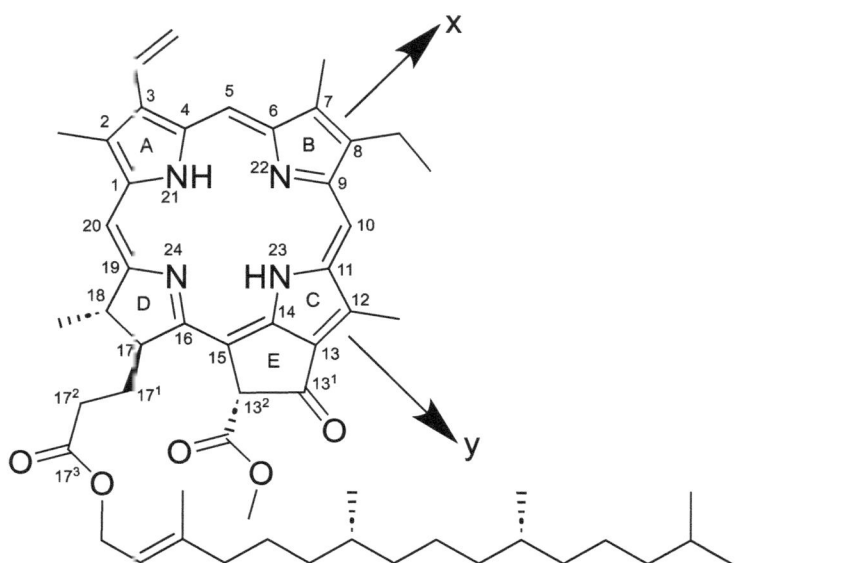

Pheophytin a

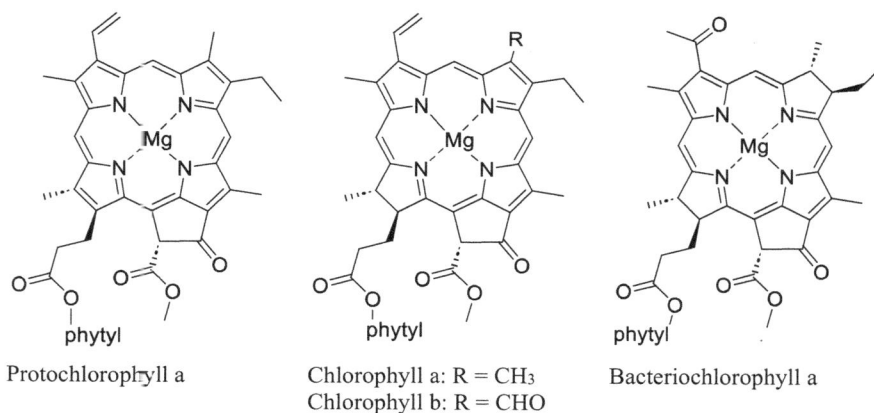

Protochlorophyll a   Chlorophyll a: R = CH₃   Bacteriochlorophyll a
Chlorophyll b: R = CHO

Figure 13.43. Structural formulae of pheophytin a (numbering according to IUPAC-IUB nomenclature is included according to Scheer 2006), protochlorophyll a, chlorophyll a and b, and bacteriochlorophyll a (see Taniguchi *et al* 2021).

(Gouterman 1959, 1961, Gouterman *et al* 1963, Weiss 1978, Hanson 1991, Scheer 2006, Giovannetti 2012, de Souza *et al* 2020, Büchner *et al* 2021, Reiter *et al* 2022). The four-orbital model level scheme with the Q-band transitions Q_y (HOMO → LUMO $a_{1u} \rightarrow e_{gx}$) and Q_x (HOMO-1 → LUMO $a_{2u} \rightarrow e_{gx}$) and the B-band (Soret-band) transitions B_x (HOMO → LUMO+1 $a_{1u} \rightarrow e_{gy}$) and B_y (HOMO-1→ LUMO +1 $a_{2u} \rightarrow e_{gy}$) is included in figure 13.44.

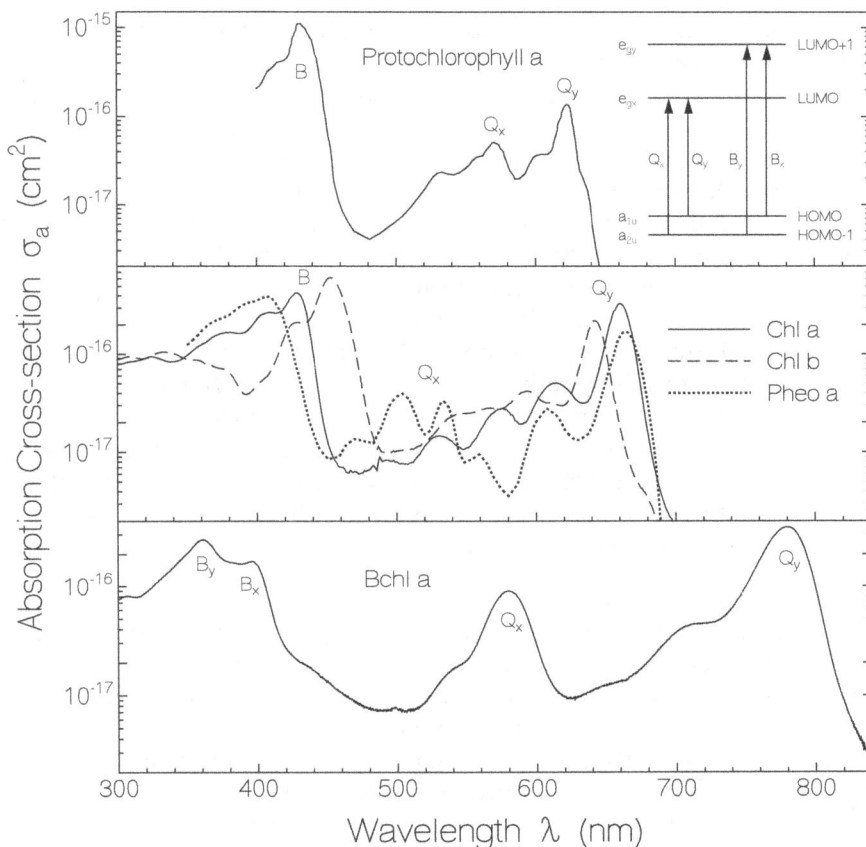

Figure 13.44. Absorption cross-section spectra of protochlorophyll a in diethyl ether (data from Boardman 1966), chlorophyll a in diethyl ether (from PhotochemCAD ID T01), chlorophyll b in diethyl ether (from PhotochemCAD ID T02), pheophytin a (Pheo a) in acetone (from Milenković *et al* 2012 with calibration from Lorenzen and Jeffrey 1980), and bacteriochlorophyll a in toluene (from PhotochemCAD ID T16).

13.6.3 Action of organic dyes in photoreceptors

13.6.3.1 Rhodopsins

Rhodopsins are light-sensitive, photoreceptive transmembrane proteins consisting of opsin protein and a bound retinal cofactor. The convergent evolution of animal and microbial rhodopsins was worked out (Kojima and Sudo 2023).

A new book on rhodopsins was published recently concentrating on microbial rhodopsins (Gordeliy 2022). The paleo biological history of ancestral microbial rhodopsins was reviewed (Sephus *et al* 2022). The evolutionary rhodopsin-based retinalophototrophy was examined through synthetic construction (Peterson *et al* 2023). The importance of microbial rhodopsins concerning phototropic metabolism in the sea has been worked out (Gómez-Consarnau *et al* 2019, Hassanzadeh *et al* 2021). A short review on rhodopsins is given by Nagata and Inoue (2021).

The importance of microbial rhodopsins as multi-functional photoreactive membrane proteins for optogenetics was worked out (Kojima *et al* 2020a, Nakao *et al* 2021, Deisseroth 2021). Novel modular rhodopsins with variant domains were identified (Awasthi *et al* 2020). The biophysics of rhodopsins and their relation to optogenetics was reviewed (Kandori 2020a, 2020b). The progress on microbial rhodopsin research (type-1 rhodopsins and recently discovered heliorhodopsins) was discussed (Rozenberg *et al* 2021). Advances of rhodopsin-based optogenetics in plant research was reported (Zhou *et al* 2021). The discovery, heterologous expression and purification, photochemical properties, bioengineering and prospects of type-1 (microbial) and type-2 (animal) rhodopsins was reviewed (de Grip and Ganapathy 2022).

The biological function of microbial rhodopsins is due to all-*trans* retinal photoisomerization to 13-*cis* retinal (Chang *et al* 2022). It may induce: (i) ion transport (ion-transporting rhodopsins, Kandori (2021), Furutani and Yang (2023)) as active unidirectional ion transport in ion pumps (ion-pumping rhodopsins) and passive bidirectional transport in ion channels (channelrhodopsins); (ii) sensory action in generating phototaxis signals (sensory rhodopsins); and (iii) enzyme activation (enzymerhodopsins).

13.6.3.1.1 Animal rhodopsins

Monostable animal rhodopsins, found in the retina rod photoreceptors of the eyes of vertebrates, are G-protein-coupled receptors with 11-*cis* retinal bound chromophore for low-light vision and motion detection (Hussey *et al* 2022, Hofmann and Lamb 2023). Light absorption triggers photoisomerization from 11-*cis* retinal to all-*trans* retinal, leading to movements in the rhodopsin helix to form the active state of G-protein activation of vision. All-*trans* retinal is released from the transmembrane to the retinal pigment epithelium where it is enzymatically converted in the dark to 11-*cis* retinal and reentered into the opsin protein closing the cycle and enabling new vision in light absorption. The first direct visualization of the retinal isomerization in a mammalian rhodopsin was achieved recently by time-resolved serial femtosecond crystallography with an x-ray free-electron laser (Gruhl *et al* 2023).

Bistable animal rhodopsins are photopigments expressed in both invertebrates and vertebrates. They undergo a reversible reaction upon illumination. A first photon initiates the 11-*cis* to all-*trans* photoisomerization to the thermally stable and active Meta state, and a second photon reverts this process to recovery to the original ground state (Ehrenberg *et al* 2019). The two-photon reversible reaction of the bistable rhodopsin-1 from the jumping spider *Hasarius adansoni* was studied (Ehrenberg *et al* 2019). Lamprey parapinopsin ('Lamplight'), a Gi/o-coupled bistable animal opsin, was applied for switchable and scalable optogenetic inhibition of neurons (Rodgers *et al* 2021). The optogenetic manipulation of Gq- and Gi/o coupled bistable animal rhodopsin signaling in neurons and heart muscle cells was studied (Hagio *et al* 2023).

13.6.3.1.2 Microbial rhodopsin-based ion pumps

The research on rhodopsin-based ion pumps was reviewed by Inoue 2021. They consist of outward H^+ pumps, inward H^+ pumps, outward Na^+ pumps, and inward Cl^- pumps.

An overview of outward H$^+$ pumps in archaebacteria, eubacteria and lower eukaryotes was given by Tamogami and Kikukawa 2021. New reviews were published on the outward H$^+$ pumps bacteriorhodopsin (Li *et al* 2018, Wickstrand *et al* 2019, Perrino *et al* 2021), proteorhodopsin (Olson *et al* 2018), and cyanobacterial N2098R from *Calothrix* sp. NIES-2098 (Hasegawa *et al* 2020).

Inward H$^+$ pumps have been discovered and were described. They were found in xenorhodopsins (Ugalde *et al* 2011) from the Nanohaloarchaea family (Inoue *et al* 2016, Shevchenko *et al* 2017, Inoue K *et al* 2018, Inoue S *et al* 2018) and in schizorhodopsins (Bulzu *et al* 2019) from the Asgard archaea family (Inoue *et al* 2020, Kawasaki *et al* 2021, Shionoya *et al* 2021).

The mechanism of outward Na$^+$ pumping in KR2 rhodopsin from *Krokinobacter eikastus* was deciphered (Kandori *et al* 2018, Kovalev *et al* 2019, Kovalev *et al* 2020a, Eberhardt *et al* 2021). A Li$^+$/Na$^+$ pumping rhodopsin, MrP from *Methylobacterium populi* was discovered and analyzed (Cho *et al* 2021). A light-driven coexisting Na$^+$/H$^+$ pumping was found in *Nonlabens (Donghaeana) dokdonensis* rhodopsin 2 (DDR2) (Zhao *et al* 2017). Retinal-carotenoid interactions in DDR2 were studied (Ghosh *et al* 2023).

Inward Cl$^-$ pumping rhodopsins find application as neural silencer in optogenetics. Functional mechanisms of Cl$^-$ pumping rhodopsin groups were characterized by Kikukawa 2021. The functional and structural characteristics of light-driven chloride pumping halorhodopsins was reviewed by Engelhard *et al* 2018. The early-stage dynamics of a Cl$^-$ pumping rhodopsin from the marine bacterium *Nocardioides marinus* was revealed by pump-probe experiments with a femtosecond x-ray free electron laser (Yun *et al* 2021). The pumping mechanism of a newly identified Cl$^-$ pumping rhodopsin, NM-R3, from the marine flavobacterium *Nonlabens marinus* was investigated (Yun *et al* 2020a). A novel microbial chloride pumping rhodopsin (MrHR) from the cyanobacterium *Mastigocladopsin repens* was discovered and characterized (Hasemi *et al* 2016). On MrHR, photochemical studies were carried out (Hasemi *et al* 2019), molecular details of the chloride transport mechanism were worked out (Harris *et al* 2018), and the crystal structure was determined (Besaw *et al* 2020).

Light-driven inward SO$_4^{2-}$ ion pumping was found in the cyanobacterial *Synechocystis* halorhodopsin SyHR (Niho *et al* 2017). The functional mechanism of this divalent anion transport was analyzed (Yun *et al* 2020b). Mutations enabled SO$_4^{2-}$ ion pumping also in the cyanobacterial *Mastigocladopsin repens* rhodopsin MrHR (Doi *et al* 2022).

13.6.3.1.3 Microbial channelrhodopsins

The research on channelrhodopsin (ChR) development is very active (Wietek and Prigge 2016, Kato 2021, Govorunova *et al* 2021, Govorunova *et al* 2022, Tucker *et al* 2022). The channelrhodopsins may be divided in cation channelrhodopsins (CCRs) and anion channelrhodopsins (ACRs).

Cation channelrhodopsins are primarily proton channels, but some types of them also conduct other metal cations (Na$^+$, Ca^{2+}). In neurons, photoactivation of CCRs depolarizes the neuronal membrane and stimulates spiking (neuron activation).

A red-light absorbing proton conduction channelrhodopsin, Chrimson from *Chlamydomonas noctigama*, was discovered (Klapoetke *et al* 2014) and its reaction dynamics was spectroscopically resolved (van Stokkum *et al* 2023).

Potassium channelrhodopsins were found and investigated recently (Govorunova *et al* 2022a, 2022b, 2023, Vierock *et al* 2022, Morizumi *et al* 2023). They are hyperpolarizing tools to suppress excitable cell firing upon illumination.

Recently calcium-permeable channelrhodopsins have been developed that allow a light-driven increase of cytosolic calcium with temporal resolution in the sub-second range (Lahore *et al* 2022).

Anion channelrhodopsins (ACRs) conduct Cl⁻. They hyperpolarize the neuronal membrane and thereby suppress neuronal activity (neuron inhibition) (Wietek *et al* 2017, Govorunova *et al* 2018, Rozenberg *et al* 2020, Zhou *et al* 2021, Tsujimura M *et al* 2021, Rodriguez-Rozada *et al* 2022, Schleissner *et al* 2023).

The combined expression of blue-light absorbing ACRs and red-light absorbing CCRs in neurons allows the optogenetic manipulation of neuron inhibition by blue-light exposure and neuron excitation by red-light exposure (Klapoetke *et al* 2014, Vierock *et al* 2021).

A step-function opsin (SFO) family of high light sensitivity was engineered by directed molecular engineering of channelrhodopsin-2 to modify protein residues in order to achieve substantially delayed activation state recovery after photoexcitation in closing the photocycle (Berndt *et al* 2009, Diester *et al* 2011, Yizhar *et al* 2011, Gong *et al* 2020, Rodriguez-Rozada *et al* 2022, Bansal *et al* 2022).

Bi-stable neural state switches use step-function opsins (bi-stable channelrhodopsins) whose photocurrents can be precisely initiated and terminated on a millisecond timescale with different colors of light (Berndt *et al* 2009, Diester *et al* 2011, Yizhar *et al* 2011, Rodriguez-Rozada *et al* 2022, Bansal *et al* 2022).

New discoveries extend beyond classical channelrhodopsins. A group of pump-like channelrhodopsins from cryptophyte algae was discovered functioning as light-gated cation channels (Govorunova *et al* 2016, Yamauchi *et al* 2017). The structure of ChRmine belonging to this group was resolved by cryogenic electron microscopy (Tucker *et al* 2022, Kishi *et al* 2022). A microbial rhodopsin subfamily from marine unicellular algae named bestrhodoposins was discovered and characterized (Rozenberg *et al* 2022). In them one or two rhodopsin domains of eight-trans-membrane helices are C-terminally fused to a bestrophin channel. The rhodopsin-rhodopsin -bestrophin fusion forms a pentameric megacomplex with five rhodopsin pseudodimers surrounding the bestrophin channel in the center. Heterologous expressed bestrhodopsin behaves as a light-modulated anion channel.

Red-shifted channelrhodopsins have been engineered (absorption maximum in the orange to red spectral region) which enable deep transcranial optogenetic neuron excitation (Chen R *et al* 2021) and deep cardiac excitation (O'Shea *et al* 2019, Pyatri *et al* 2023). Red-light activatable cation channelrhodopsins are ReaChR (Mairon *et al* 2018), Chrimson (Klapoetke *et al* 2014, Oda *et al* 2018), and ChRmine (Chen R *et al* 2021, Kishi *et al* 2022). Red-shifted anion channelr-hodopsins RubyACRs from the heterotrophic protists labyrinthulea have been identified and characterized which allow deep brain neural inhibition (Govorunova *et al* 2020). The RubyACR photocycle, involving the sequential absorption of two photons which creates a bistable form, was studied (Sineshchekov *et al* 2023).

13.6.3.1.4 Microbial sensory rhodopsins

Sensory rhodopsins have a cognate transducer protein that mediates the photo-isomerization response of the rhodopsin transmembrane protein to an actuator. The signal propagation by the sensory rhodopsin II/transducer complex in *Natronomonas pharaonic* was studied (Ishchenko *et al* 2017). The blue–green light-sensing sensory rhodopsin SRM from *Haloarcula marismortui* attenuated both positive and negative phototaxis responses mediated by sensory rhodopsin I (SRI) and II (SRII) in *Halobacterium salinarum*, respectively (Chen J-L *et al* 2019). The dynamics of *Anabaena* sensory rhodopsin action was studied (Roy *et al* 2019, Lee *et al* 2021). Proteorhodopsin-related sensory rhodopsins have been identified and characterized (Salminasab *et al* 2023).

13.6.3.1.5 Microbial enzymerhodopsins

In enzymerhodopsins the enzyme domain and the rhodopsin domain are fused into a single amino acid sequence. They are a class of natural rhodopsin-based photo-receptors with light regulated enzyme activity (Mukherjee *et al* 2019, Tsunoda *et al* 2021). Enzymerhodopsins have eight trans-membrane helices (8-TM) and both their N and C terminal reaching into the cytoplasm (Zhou *et al* 2021, Tian *et al* 2022a). They form dimers (Tian *et al* 2018). These 8-TM microbial rhodopsins may be classified as type-Ib rhodopsins and the 7-TM microbial rhodopsins may be classified as type-1a rhodopsins (Tian *et al* 2018 and 2022a).

Three different types of enzymerhodopsins have been identified: histidine kinase rhodopsins (HKRs), rhodopsin phosphodiesterases (RhoPDEs), and rhodopsin cyclases (RhACs and RhGCs) (kinases are enzymes that transfer phosphate residues; phospho-diesterases are enzymes that convert the second messenger cyclic nucleotides cyclic adenosine monophosphate (cAMP) and cyclic guanosine monophosphate (cGMP) to linear nucleotides adenosine monophosphate (AMP) and guanosine monophosphate (GMP); cyclases are enzymes that catalyze a chemical reaction to form a cyclic compound).

Histidine kinase rhodopsins trigger a phosphorylation cascade (Tian *et al* 2018, Mukherjee *et al* 2019, Luck *et al* 2019, Möglich 2019).

Rhodopsin phosphodiesterases regulate second messenger levels (cGMP and cAMP) in eukaryotic microbes (Yoshida *et al* 2017, Sugiura *et al* 2020, Ikuta *et al* 2020, Tian *et al* 2022a). They degrade cAMP to AMP and cGMP to GMP (Tian *et al* 2022b).

Rhodopsin guanylyl cyclases (RhGCs) photocontrol cGMP levels and rhodopsin adenyl cyclases (RhACs) photocontrol cAMP levels (conversion of GTP to cGMP, conversion of ATP to cAMP) (Scheib *et al* 2018, Broser *et al* 2020, Fischer *et al* 2021, Tian *et al* 2021, Palombo *et al* 2022, Broser 2022, Tian *et al* 2022b, Zhang Y *et al* 2022).

13.6.3.1.6 Microbial heliorhodopsins

A new group of microbial rhodopsins, named heliorhodopsins, was discovered using functional metagenetics (Pushkarev *et al* 2018). They were detected in archaea, bacteria, eukarya and their viruses. Different from microbial type-1 rhodopsins and animal type-2 rhodopsins, for these 7-transmembrane domain proteins the N-terminus is inside and the C-terminus is outside. The photocycle accompanies retinal isomerization and proton transfer, but the proton is not released from the protein. The proton transfer and induced structrural changes along the H-bond network was investigated using a quantum mechanical/molecular mechanical approach and molecular dynamics simulations (Tsujimura *et al* 2022). Light-sensory activity is expected. The crystal structure of heliorhodopsin has been resolved (Lu *et al* 2019, Shihoya *et al* 2019, Kovalev *et al* 2020b). A photochemical character-ization was given for the Gram-positive eubacterium HeR-48C12 discovered in an actinobacterial fosmid from freshwater lake Kinneret (Pushkarev *et al* 2018) and for the Gram-negative eubacterium *Bellilinea caldifistulae* (BcHeR) (Shibukawa *et al* 2019) (Gram-positive (monoderm) bacteria have only an inner membrane, Gram-negative (diderm) bacteria have an outer and an inner membrane). The helio-rhodpsin evolution was described (Flores-Uribe *et al* 2019, Bulzu *et al* 2021). Heliorhodopsin from *Thermoplasmatales archaeon* (TaHeR) shows Zn^{2+} binding (Hashimoto *et al* 2020, 2023). The absorption spectra of natural variants of heliorhodopsins were studied and a color-tuning was achieved (Kim *et al* 2021). A novel function of heliorhodopsin as a regulatory rhodopsin with the capacity to bind and regulate enzyme activity required for nitrogen assimilation has been discovered (Cho *et al* 2022). A viral heliorhodopsin from *Emiliania huxleyi* virus 202 (V2HeR3) was found to function as a light-activated proton channel (Hososhima *et al* 2022). Diverse heliorhodopsins, recently detected via functional metagenomics in freshwater *Atinobacteria, Chloroflexi* and *Archaea* give indication of their involvement in light-induced membrane lipid modifications (Chazan *et al* 2022).

13.6.3.1.7 Microbial rhodopsin-based genetically encoded voltage indicators
Microbial rhodopsin-based genetically encoded voltage indicators (GEVIs) applied for plasma membrane voltage detection have been further developed. Several reviews on GEVIs have been published (Platisa and Pieribone 2018, Bando *et al* 2019a and 2019b, Knöpfel and Song 2019, Kannan *et al* 2019, Lazzari-Dean *et al* 2019, Mollinedo *et al* 2021, Zhang X M *et al* 2021, Chien *et al* 2021, Andreoni and Tian 2023). The essential aspects of microbial rhodopsin photocycles that are critical to understand the mechanisms of voltage sensing in engineered proton-pumping rhodopsins have been reviewed (Meng *et al* 2023).

In hyperpolarized membranes (extracellular positive and intracellular negative charged) the retinal Schiff base in rhodopsin is dominantly present in neutral form (RSB) absorbing and emitting in the violet spectral region. In depolarized mem-branes (extracellular negative and intracellular positive charged) the retinal Schiff base in rhodopsin is dominantly present in protonated form (PRSB) absorbing in the green to red spectral region and emitting in the red spectral region (Zou *et al* 2014, Xu *et al* 2018, Abdelfattah *et al* 2019 Zhang X M *et al* 2021). The change of the red fluorescence emission signal with membrane voltage is used to determine the membrane voltage in stand-alone genetically encoded microbial rhodopsin voltage indicators (Adam *et al* 2019, Milosevic *et al* 2020). In rhodopsin–fluorescent protein couples with green to red emitting fluorescent proteins the fluorescent protein emission is quenched by Förster-type resonant energy transfer from the fluorescent protein to PRSB in depolarized membranes because the fluorescence quantum yield of PRSB is considerably weaker than the fluorescence efficiency of fluorescent proteins (Zou *et al* 2014). So, the fluorescent protein emission efficiency indicates the membrane potential situation (Kannan *et al* 2018).

The absorption and emission spectroscopic behavior of Archaerhodopsin 3 based fluorescent voltage sensor QuasAr1 (Penzkofer *et al* 2019) and Archon2 (Penzkofer *et al* 2020a) were studied. The photocycle dynamics of QuasAr1 (Penzkofer *et al* 2020b) and Archon2 (Penzkofer *et al* 2021, Penzkofer *et al* 2023) were investigated. The origin of fluorescence and its voltage sensitivity was revealed (Silapetere *et al* 2022). For stand-alone rhodopsin based GEVIs possibly high fluorescence efficiency is of advantage. Therefore, studies were carried out to find genetically encoded rhodopsin mutations with enhanced fluorescence efficiency (Martin *et al* 2019, Kojima *et al* 2020b, Barneschi *et al* 2022, Palombo *et al* 2022, Pedraza-González *et al* 2022 and 2023).

The combined encoding of a voltage indicator and an optogenetic neuron activator or inhibitor allows the observation and steering of light-induced neuron manipulation (Fan *et al* 2020, Bergs *et al* 2023). An all-optical closed-loop voltage clamp was established for the precise control of muscles and neurons in live animals (Bergs *et al* 2023). It consists of the voltage-indicator QuasAr2 and the bidirectional optogenetic actuator BiPOLES (comprising the H^+ cation channelrhodopsin depolarizer Chrimson and the Cl^- anion channelrhodopsin hyperpolarizer *Gt*ACR2 from *Guillardia theta*).

13.6.3.1.8 Microbial luminopsins

Luminopsins are fusion proteins of luciferase and opsin allowing the bioluminescent photoexcitation of the fused rhodopsin thereby integrating chemogenetics and optogenetics (Berglund *et al* 2016a, 2016b and 2021, Park *et al* 2020, Sureda-Vives and Sarkisyan 2020, Stern *et al* 2022a and 2022b, Jiang *et al* 2023). The catalytic oxidation of injected luciferin by the luciferase enzyme (Syed and Anderson 2021) provides the bioluminescence photon for retinal photoisomerization causing optogenetic rhodopsin action. Bioluminescence resonance energy transfer (BRET) occurs from luciferin to rhodopsin (Li *et al* 2021). Luminopsins are applied in neuroscience for non-invasive deep-brain sensing and controlling biological processes (Jiang *et al* 2023). Inhibitory luminopsins are composed of a luciferase and a light-sensitive outward chloride ion pump or anion channel rhodopsin that causes cell membrane hyperpolarization leading to silencing neural spiking. Activatable luminopsins are composed of a luciferase and an inward proton ion pump or proton channel rhodopsin that causes cell membrane depolarization leading to activated neural spiking (Sureda-Vives and Sarkisyan 2020). Step-function luminopsins use lumiferase and a channelrhodopsin mutation with slow light-activated state deactivation (Berglund *et al* 2020). They were applied for bimodal prolonged neuromodulation (Berglund *et al* 2020).

13.6.3.2 Flavoproteins

Flavoprotein photoreceptors are blue-light receptors and applied in optogenetics (Losi *et al* 2018). Photoreaction mechanisms of flavoprotein photoreceptors and their applications has been reviewed (Iwata and Masuda 2021). Recent progress in understanding the fundamental photochemical reactions in flavoproteins was outlined and novel photocatalytic approaches were explored (Zhuang *et al* 2022). Plant flavoprotein photoreceptors that mediate the effects of blue light in the dicotyledonous genetic model *Arabidopsis thaliana* were described (Christie *et al* 2015). Physical methods for studying flavoprotein photoreceptors were collected (Yee *et al* 2019).

13.6.3.2.1 LOV proteins

The progress in LOV (Light-Oxygen-Voltage) domain-based optogenetic tool development and its applications for controlling and interrogating specific cellular events was reviewed (Hoffmann *et al* 2018). The signal transduction in photoreceptor histidine kinases was reviewed (Möglich 2019). A network analysis of the chromophore binding site in LOV domains was presented by Panda *et al* (2022). The photobiophysics of LOV proteins was described by Losi (2020). A class of fungal LOV proteins dynamically associate with anionic plasma membrane phospholipids by a blue light-switched electrostatic interaction. They regulate G-protein signaling by controlled recruitment of fused regulator of G-protein signaling (RGS) domains (Glantz *et al* 2018). A short LOV protein (DsLOV) form the marine phototrophic bacterium *Dinoroseobacter shibae* was characterized, for which the dark state represents the physiological relevant signaling state of regulation of photopigment synthesis (Endres *et al* 2015, Fettweiss *et al* 2018). The LOV photoreceptor activation dynamics of the LOV-STAS protein YtvA from *Bacillus subtilis*, the LOV-HTH transcription factor EL222 from *Erythrobacter litoralis*, and the LOV-histidine kinase LovK from *Caulobacter crescentus* were studied by time resolved infrared spectroscopy (Iuliano *et al* 2018). A LOV-activated diguanylate cyclase (LadC) with high dynamic range of enzymatic activity was characterized (Vide *et al* 2023).

Phototropins. The research on plant phototropins, consisting of two N-terminal LOV domains (LOV1 and LOV2) and a C-terminal Ser/Tyr kinase domain, has been continued (Hart and Gardner 2021). The two phototropins, *phot1* and *phot2*, in *A. thaliana* influence photosynthesis, UV-C induced photooxidative stress responses, and cell death (Rusaczonek *et al* 2021). The signaling mechanism of phototropin-mediated chloroplast movement in *Arabidopsis* was reviewed (Suetsugu and Wada 2020). A chemical genetic approach was applied to engineer phototropin kinases for substrate labeling (Schnabel *et al* 2018).

ZEITLUPE. The LOV domain containing blue-light photoreceptor ZEITLUPE (ZTL) directs circadian timing by degrading clock proteins in plants (Pudasaini *et al* 2017 and 2020). The dimeric allostery mechanism of ZEITLUPE in *A. thaliana* was analyzed recently via an integrated computational approach (Trozzi *et al* 2021).

YtvA. YtvA is a blue-light photoreceptor from the common soil bacterium *B. subtilis*. YtvA has two domains: A N-terminal LOV (light oxygen voltage) domain and a C-terminal STAS (sulfate transporter and anti-sigma factor antagonist) domain (Möglich and Moffat 2007). The functioning of YtvA in the general stress response (GSR) of *B. subtilis* by activating a large protein complex (the stressosome) was discussed by van der Steen *et al* 2013. The influence of cellular environment on the photocycle dynamics of YtvA was studied (Pennacchietti *et al* 2014). Thermodynamic and structural aspects of the flavin-LOV interaction in YtvA were analyzed (Dorn *et al* 2013). The application of YtvA as reversibly switchable fluorescent protein in super-resolution microscopy was reported (Losi *et al* 2013, Gregor *et al* 2018).

VIVID and white collar complex. VIVID (VVD) and White Collar Complex (WCC), photoreceptors incorporating LOV domains, act together in fungi as blue light regulatory proteins for photoadaptation (Scherdtfeger and Linden 2003, Zoltowski *et al* 2007, Zoltowski and Crane 2008, Hunt *et al* 2010, Gin *et al* 2013, Dasgupta *et al* 2015, Yu and Fischer 2019, Corrochano 2019, Losi and Gärtner 2021). FUN_LOV, an optogenetic switch based on the fungal light-oxygen-voltage domains WC-1 and VVD from the fungus *Neurospora crassa*, was applied for optogenetic control of gene expression and flocculation in yeast (Salinas *et al* 2018).

LOV-based fluorescent proteins. LOV-based fluorescent proteins (also called flavin-based fluorescent proteins FbFP), where flavin-C4a cysteinyl adduct photocycling is hindered by cysteine replacement with alanine, are used as oxygen-independent fluorescent reporters and applied in qualitative and quantitative studies of biological processes (Wingen *et al* 2014, Nazarenko *et al* 2019, Remeeva *et al* 2020, Yudenko *et al* 2021, Wehler *et al* 2022, Bitzenhofer *et al* 2023).

LOV-based fluorescent proteins have been used as photosensitizers generating reactive oxygen species ROS ($^1O_2$ type-II mechanism; $O_2^{\cdot-}$, H_2O_2, and $HO^{\cdot}$ type-I mechanism) causing chemical reactions on the environing matter and inactivation of enzyme actions (Shu *et al* 2011, Endres *et al* 2018, Westberg *et al* 2019, Torra *et al* 2019, Hilgers *et al* 2019, Raber *et al* 2020, Hally *et al* 2020, Mogensen *et al* 2021, Takemoto 2021, Gerlach *et al* 2022).

13.6.3.2.2 Photolyases and cryptochromes

Photolyases act as DNA-repair enzymes (chromophore is FAD_{red}^-, antenna chromophore is MTHF or 8-HDF). Cryptochromes are UV-A/blue light photoreceptors (chromophore is FAD_{ox}, antenna chromophore is MTHF) consisting of a N-terminal photolyase homologous region (PHR) and a divergent cryptochrome C-terminal extension (CCE). Cryptochrome/photolyase family (CPF) members are present in all major taxa from archaea to mammals (Mei and Dvornyk 2015, Deppisch *et al* 2022). They are activated by UV-A/blue light (chromophore FAD).

Photolyases. New reviews on photolyases have been published (Kavakli *et al* 2019, Vechtomova *et al* 2021, Ramírez-Gamboa *et al* 2022).

Structure-functions and reaction mechanisms of different classes of photolyases were discussed (Kavakli *et al* 2019). The damaging action of UVB (wavelength region 280 nm—315 nm) and UVA (315 nm—400 nm), and the mechanisms of

DNA repair with the participation of the DNA-photolyase enzyme or of the nucleotide excision repair (NER) system was discussed (Vechtomova *et al* 2021). The functions, characteristics, and types of photolyases, their therapeutic and cosmetic applications as well as some photolyase-producing microorganisms and delivery systems were described (Ramírez-Gamboa *et al* 2022). The DNA repair efficiency of photolyase was enhanced by ligand-directed labeling of the photolyase protein with an artificial coumarin chromophore whose fluorescence spectrum overlapped with the $FADH^-$ absorption spectrum (optimum use of photoexcitation energy by application of optimum Förster resonance energy transfer, Terai *et al* 2020).

Theoretical studies including molecular dynamics simulations were carried out to investigate the mechanism of repair of cyclobutane pyrimidine dimer lesions in DNA by electron transfer from the flavin adenine dinucleotide cofactor (Roussseau *et al* 2018, 2023). It was found that adenine mediates the electron transfer in mesophile and extremophile DNA photolyases through a similar mechanism.

DASH-type cryptochromes Cry-DASH. DASH is the abbreviation of *Drosophila, Arabidopsis, Synechocystis, Homo.* In *A. thaliana* cry-DASH was named cry3 by Kleine *et al* (2003). The research on DASH-type cryptochromes, Cry-DASH, has been continued (Kiontke *et al* 2020, Navarro *et al* 2020, Rredhi *et al* 2021). A short review was given by Kiontke *et al* (2020). Cry-DASH from the fungus *Mucor circinelloides* was found to act like CPD-photolyase in dsDNA repair (Navarro *et al* 2020). The DASH cryptochrome CRY-DASH1 form the green alga *Chlamydomonas reinhardtii*, which binds the chromophore FAD_{red} ($FADH_2$) and the antenna MTHF, was found to have a balancing function in the photosynthetic machinery (Rredhi *et al* 2021, 2024).

Plant cryptochromes pCRYs. The progress in plant cryptochrome (pCRY) research has been reviewed (D'Amico-Damião and Carvalho 2018, Wang *et al* 2018, Fantini and Facella 2020, Wang and Lin 2020, Goett-Zink and Kottke 2021, Ponnu and Hoecker 2022, Patnaik *et al* 2022). Plant cryptochromes are central blue light receptors for the control of land plant and algal development including the circadian clock and the cell cycle (Goett-Zink and Kottke 2021). In the dark form the chromophore is FAD_{ox}. Aspartic acid (Asp, D) and adenosine triphosphate (ATP) are bound near to FAD_{ox} and are involved in the photocycle (Iwata *et al* 2020, Goett-Zink *et al* 2021, Goett-Zink and Kottke 2021). Upon blue light exposure the monomeric inactive plant CRYs undergo phosphorylation and oligomerization which are crucial to CRY function (Shao *et al* 2020, Wang and Lin 2020, Palayam *et al* 2021, Goett-Zink *et al* 2021, Ponnu and Hoecker 2022). Two phylogenetically distinguishable clads of plant cryptochromes are CRY1 and CRY2. They are nucleocytoplasmic proteins (relating to nucleus and cytoplasm). Many studies were performed on cryptochromes from *A. thaliana* (Palayam *et al* 2021, Pooam *et al* 2020, Cho *et al* 2021) and *C. reinhardtii* (Müller *et al* 2017, Goett-Zink *et al* 2021). The pivotal role of cryptochromes in controlling the development and physiology of tomato was found out (Liu *et al* 2018, Fantini *et al* 2019).

Magnetic field effects on *A. thaliana* have been reported (Pooam *et al* 2019, Pooam *et al* 2020, Hammad *et al* 2020, Dhiman *et al* 2023).

Animal cryptochromes aCRYs. The research on animal cryptochromes (aCRYs) is very active (Sancar 2004, Foley and Emery 2020, Karki *et al* 2021, Deppisch *et al* 2023). According to sequence homology, animal CRYs are divided in three lineages: Type I (insects), Type II (mammals) and Type IV (birds, fish, turtles). Type I CRYs are regulators of the circadian clock in light-dependent mechanisms (Sancar 2004, Öztürk *et al* 2008, Öztürk *et al* 2014, Lin *et al* 2018, Parico and Partch 2020, Rasmussen *et al* 2022). Type II CRYs are regulators of the circadian clock in light-independent mechanisms (Wang *et al* 2018, Einwich *et al* 2022).

Type IV CRYs demonstrate expression patterns and photochemistry ideal for magnetoreception (Wang *et al* 2018, Zoltowski *et al* 2019, Günther *et al* 2017, Einwich *et al* 2020, Karki *et al* 2021). The photocycle mechanism of an animal-like cryptochrome starting from FAD_{ox} was studied (Lacombat *et al* 2019). The light-dependent and light-independent magnetoreception in animals involving cryptochromes is intensely studied (Wiltschko R and Wiltschko W 2019, Dempsey 2021, Karki *et al* 2021, Kavet and Brain 2021, Kyriacou and Rosato 2022). A recent experimental study found no evidence for magnetic field effects on the behavior of *Drosophila* (Bassetto *et al* 2023). The cryptochrome involved magnetoreception is explained in terms of photo-excitation-caused radical pair formation ($FAD_{ox}^{\cdot-} + Trp^{\cdot+}$) or chemical radical pair formation ($FADH + O_2^{\cdot-}$) (for discussion see Karki *et al* 2021).

13.6.3.2.3 BLUF proteins

BLUF (Blue Light sensor Using Flavin) proteins are blue light receptors having a well-preserved domain with noncovalently bound flavin. Photoexcitation changes the hydrogen bond network around the flavin chromophore causing a spectral redshift of the absorption band. The BLUF domain excitation activates intra-molecularly or intermolecularly functional regions (Nakasone and Terazima 2022).

The research on BLUF proteins continued (Iwata *et al* 2018, Kaushik *et al* 2019, Goings and Hammes-Schriffer 2019, Golic *et al* 2019, Goings *et al* 2020, Chitrakar *et al* 2020, Karadi *et al* 2020, Lukacs *et al* 2022, Tokonami *et al* 2022, Nakasone and Terazima 2022, Hontani *et al* 2023, Chretien *et al* 2023).

Research concerning single BLUF domain proteins (short BLUF proteins). The mechanism of the photocycle process for the Slr1694 BLUF domain involving forward and reverse proton-coupled electron transfer (PCET) between Gln, Tyr and flavin was unraveled by hybrid quantum mechanical/molecular mechanical dynamics simulations (Goings and Hammes-Schriffer 2019, Goings *et al* 2020) and further femtosecond stimulated Raman spectroscopy and site-selective isotope labeling Fourier-transform infrared spectroscopy (Hontani *et al* 2023).

The photoreaction of the PapB BLUF domain with the PapA EAL protein was studied (Tokonami *et al* 2022, Nakasone and Terazima 2022). Slow conformational changes in the millisecond time range were revealed for PapB from

Rhodopseudomonas palusiris, TePixD from *Thermosynechococcus elongatus* BP-1, and SyPixD from *Synechocystis* sp. PCC6803 (Tokonami *et al* 2022).

The BLUF protein BlsA from the human pathogen *Acinetobacter baumannii* was found to be a low to moderate temperature acting photo-regulator of motility (Golic *et al* 2019). The regulation of biofilm formation and iron uptake in *A. baumannii* was studied by coimmunoprecipitation and x-ray crystal structure determination (Chitrakar *et al* 2020).

Research on BLUF domain containing proteins (multidomain BLUF proteins). BLUF domain containing proteins with a diversity of effector domains of various functions were identified from available bioinformatics databases (Kaushik *et al* 2019).

A light-gated adenylate cyclase (LgAC) from *Beggiatoa* sp. was utilized as an optogenetic tool to decipher the cAMP-modulated bacterial processes in *Escherichia coli* like growth, biofilm formation and virulence (Kaushik *et al* 2020).

The involvement of conserved Gln and Tyr residues near FAD in the photocycle process of AppA-BLUF was clarified by Fourier transform infrared spectroscopy and quantum mechanics/molecular mechanics calculations (Iwata *et al* 2018). The functional dynamics of a Trp residue in the AppA-BLUF protein was investigated by unnatural amino acid 7-aza-Trp mutagenesis applying fluorescence anisotropy decay and fluorescence lifetime measurement (Karadi *et al* 2020).

The photoreaction of the BLUF proteins BlrP1 and YcgF with EAL functional groups was studied (Nakasone and Terazima 2022). Slow conformational changes in the millisecond time range were revealed for AppA from *Rhodobacter sphaeroides*, OaPAC from *Oscillatoria accuminata*, BlrP1 from *Klebsiella pneumoniae*, and YcgF from *E. coli* (Tokonami *et al* 2022).

Photoactivated adenylyl cyclases (PACs) consisting of a sensing BLUF domain and a reacting adenylyl cyclase domain were further investigated (Hirano *et al* 2019, Iseki and Park 2021, Xia *et al* 2022, Henss *et al* 2022, Zhou *et al* 2022, Collado *et al* 2022, Chretien *et al* 2023). Upon blue light exposure they catalyze the production of cyclic adenosine monophosphate (cAMP) from adenosine triphosphate (ATP) as reviewed by Iseki and Park 2021. The crystal structures of the two PACs, OaPAC from photosynthetic cyanobacterium *Oscillatoria acuminata* and bPAC from sulfur bacterium *Beggiatoa* sp. were determined and their photoactivity compared (Hirano *et al* 2019). The photoactivation mechanism of OaPAC was investigated (Zhou *et al* 2022, Collado *et al* 2022, Chretien *et al* 2023, Nakasone *et al* 2023). Membrane-bound PACs were generated in *Caenorhabditis elegans* and used as optogenetic modulators of neuronal activity (Henss *et al* 2022). In mice immunology experiments it was found that bPAC from *Beggiatoa* sp. and biPAC from *Beggiatoa* sp. IS2 attenuated sepsis-induced cardiomyopathy by suppressing macrophage-mediated inflammation (Xia *et al* 2022).

13.6.3.3 Xanthopsins (photoactive yellow proteins)
Photoactive yellow protein (PYP) binds *p*-coumaric acid as chromophore and acts as blue light sensor mediating negative phototaxis as a photoprotective effect (Sprenger *et al* 1993). A comprehensive comparative genomic analysis of photoactive yellow

proteins (PYPs) was carried out whereby phylogenetic analyses revealed that PYP is an evolutionary novel family of the Per-Arnt-Sim (PAS) superfamily (Xing *et al* 2022). It was shown that PYP is associated with signal transduction involved in gene expression, motility, and biofilm formation (Xing *et al* 2022). A heterogeneity in PYP of *Halorhodospira halophila* was resolved by excitation wavelength dependent photocycle dynamics studies (Mix *et al* 2018). The intermolecular interaction dynamics of photo-illuminated PYP from *Rhodobacter capsulatus* with the downstream PYP-binding protein PBP was investigated (Kim *et al* 2021). The photodynamics of the PYP isolated from *Salinibacter ruber* was characterized with a comprehensive range of spectroscopic techniques revealing differences to the photocycles dynamics of the PYP from *H. halophila* (Mix *et al* 2021).

In live-cell imaging of proteins of interest, their specific labeling with chemical tags that interact with synthetic fluorescent probes has become an attractive molecular approach (Marks and Nolan 2006, Jing and Cornish 2011). Photoactive yellow protein-based protein labeling systems for fluorescence imaging were developed (Reja *et al* 2021) and are in use for living cell imaging studies (Kumar *et al* 2019 and 2020).

13.6.3.4 Phytochromes

Phytochromes are linear tetrapyrrole-binding photoreceptors. They are present in photosynthetic organisms (algae, plants, cyanobacteria) and non-photosynthetic organisms (bacteria and fungi). Depending on the number of domains in their photosensory core module they are divided into Group I, Group II and Group III phytochromes (Anders and Essen 2015). Group I phytochromes are composed of the domains PAS, GAF and PHY (PAS = Per-Arnt-Sim; GAF = domain found in c-*G*MP-specific phosphodiesterases, *A*denylate cyclases, and formate hydrogenlyase transcriptional activator *F*hlA; PHY = phytochrome-explicit area). Group II phytochromes are composed of the domains GAF and PHY. Group III phytochromes, the cyanobacteriochromes, contain only the domain GAF in their core module (Anders and Essen, 2015). The phytochrome evolution, concerning deletion, duplication, and diversification, was reviewed by Rockwell and Lagarias 2020.

The research on phytochromes continued and several reviews have been published (Wang *et al* 2018, Lymperopoulos *et al* 2018, Kreslavski *et al* 2018, Pham *et al* 2018, Fushimi and Narikawa 2019, Legris *et al* 2019, Rockwell and Lagarias 2020, Klose *et al* 2020, Balcerowicz 2020, Takala *et al* 2020, Cheng *et al* 2021, Lamparter *et al* 2021, Church *et al* 2021, Fushimi and Narikawa 2021, Tang *et al* 2021, Sivaprakasam *et al* 2022).

The molecular mechanisms underlying phytochrome-controlled morphogenetics in plants were summarized (Legris *et al* 2019, Fushimi and Narikawa 2021). The interplay of phytochromes with phytochrome interacting factors (PIFs) in plant growth and development was studied (Pham *et al* 2018, Balcerowicz 2020). The interaction of plant phytochromes with phytohormones (the chemical messengers that include auxins, cytokinins, abscisic acid, and gibberellins) in the control of plant growth and development was reviewed (Lymperopoulos *et al* 2018). The phytochrome-mediated signaling pathways with the downstream signaling partners (including transcription factors and enzymes such as kinases, phosphatases, and E3-ligases) were described in detail in the review paper of Cheng *et al* 2021. The photoreceptor coaction of phytochrome and cryptochrome to regulate genome expression and plant development was studied (Wang *et al* 2018). The role of phytochromes in higher plants and cyanobacteria in the regulation of the expression of genes encoding key photosynthetic proteins, antioxidant enzymes and other components involved in stress protection of the photosynthetic apparatus were elucidated (Kreslavski *et al* 2018). The light-independent but temperature-regulated deactivation of the physiologically active far-red-absorbing P_{fr} form of phytochrome occurring in darkness and in light was studied considering light and temperature sensing in plants (Klose *et al* 2020).

Properties of bacteriophytochromes have been summarized (Gourinchas *et al* 2019, Takala *et al* 2020). The phytochromes Agp1 and Agp2 from *Agrobacterium fabrum* were used as model phytochromes (Lamparter *et al* 2021).

The diverse light responses of cyanobacteria mediated by phytochromes were reviewed (Wiltbank and Kehoe 2019). Cyanobacteriochromes covering the entire UV-to-visible spectrum were reviewed (Fushimi and Narikawa 2019, 2021). The nanosecond protein dynamics in a red/green cyanobacteriochrome was revealed by transient IR spectroscopy (Buhrke *et al* 2020). The application of cyanophytochromes and cyanobacteriochromes in optogenetics was addressed (Sivaprakasam *et al* 2022, Hoshino *et al* 2022).

Fungal phytochrome biosynthesis at mitochondria in *Aspergillus nidulans* was studied (Streng *et al* 2021)

Computational studies of the photochemistry in phytochrome proteins were carried out (Church *et al* 2021).

13.6.3.5 Cobalamin-based photoreceptors CarH and AerR

The research on cobalamin-based photoreceptors continued. CarH has the chromophore 5'-deoxy-5'-adenosylcobalamin (AdoCbl, coenzyme B_{12}) and is involved in the light regulation of carotenoid biosynthesis. AerR has the chromphore AdoCbl or methylcobalamin (MeCbl) and is involved in the light regulation of bacteriochlorophyll biosynthesis (Cooper *et al* 2021).

Purple photosynthetic bacteria synthesize the photoreceptor AerR to control photosystem synthesis. AerR directly interacts with the redox responding transcription factor CrtJ, affecting CrtJ's interaction with photosystem promoters. It was found that differing isoforms of AerR oppositely regulate photosystem expression in the purple photosynthetic bacterium *R. capsulatus* (Yamamoto *et al* 2018). Two size variants of AerR, short form SAerR and long form LAerR, were found in *R. capsulatus*, whereby the short form promotes CrtJ repression while the long form converts CrtJ into an activator (Dragnea *et al* 2022).

The vitamin B_{12}-based photoreceptor CarH, controlling light-dependent gene expression in photoprotective cellular responses, was studied concerning structure, function and application in optogenetics and synthetic biology (Padmanabhan *et al* 2019, 2022). The role of the CarH photoreceptor protein environment on the cobalamin photochemistry was revealed by molecular dynamics and hybrid quantum mechanical/molecular mechanical calculations (Cooper *et al* 2021). A mechanistic understanding of the coenzyme B_{12} in the CarH protein from *Thermus thermophilus* is presented in Poddar *et al* 2022. Quantum chemical calculations were employed to clarify photolytical properties of B_{12} in CarH (Toda *et al* 2020 and 2022, Ghosh *et al* 2022). Transient absorption spectroscopy in the UV-visible and x-ray regions were carried out to characterize the excited state of CarH (Miller *et al* 2020).

13.6.3.6 UVR8 plant photoreceptors

The photoreceptor UVR8 (UV RESISTANCE LOCUS 8) mediates photomorphogenic responses to UV-B light (280–315 nm) by regulating transcription of a set of target genes. It uses specific Trp amino acids for light absorption during UV-B photoreception, whereby the dark-adapted UVR8 dimer dissociates into monomers during UV-B exposure (light-adapted state), initiating signal transduction through interaction with COP1 (CONSTITUTIVELY PHOTOMORPHOGENETIC 1) (Jenkins 2014a, 2014b).

The continued research on the UV-B photoreceptor UVR8 was summarized in several reviews (Jenkins 2014a, 2014b, Liang *et al* 2019, Tossi *et al* 2019, Yadav *et al* 2020, Podolec *et al* 2021, Rai *et al* 2021, Takeda 2021). The dimeric and monomeric structure of UVR8 and its *in vivo* function was discussed in the reviews of Jenkins 2014a, 2014b. The UVR8 mediated UV-B signal transduction pathways including various interacting proteins was summarized (Liang *et al* 2019). The UV-B-mediated plant growth regulation in interaction with environmental factors and phytohormones was discussed (Yadav *et al* 2020). The perception of solar UV radiation by plants with interaction of URV8 and CRY photoreceptors was analyzed (Rai *et al* 2021). The molecular mechanism of UVR8-mediated photomorphogenesis was derived from revaluation of action spectra (Takeda 2021). A detailed review on the perception and signaling of UV-B radiation in plants, including molecular mechanisms of UVR8 signaling and interaction with other plant signaling pathways and environmental responses, was given by Podolec *et al* 2021. Differential UV-B response mediated by UVR8 in diverse species beyond the plant model *Arabidopsis* was reviewed by Tossi *et al* 2019.

13.6.3.7 Upconversion nanoparticle mediated optogenetics

Upconversion nanoparticle-mediated optogenetics enables remote delivery of upconverted visible light from a near-infrared light source to targeted neurons for *in vivo* effective deep-tissue neuromodulation (Yi *et al* 2021).

The progress in upconversion nanoparticle-mediated optogenetics was reviewed (Wang *et al* 2018, Yu *et al* 2019, All *et al* 2019, Yi *et al* 2021, Patel *et al* 2022). The NIR regulation of membrane ion channels via upconversion optogenetics in biomedical research was reviewed (Wang *et al* 2018). The engineering and application of upconversion optogenetic systems by the incorporation multiple emissive UCNPs (*upconversion nanoparticles*) into various light-gated ChRs/ligands (channelrhodopsins) were elaborated (Wang *et al* 2018). Advances in design strategies and synthetic methods of NIR-activatable nanomaterials for deep-tissue-penetrating wireless optogenetic applications were summarized (Yu *et al* 2019). Advances in developing new modalities for neural circuitry modulation utilizing upconversion-nanoparticle-mediated optogenetics were presented (All *et al* 2019). Recent advances of lanthanide-doped UCNPs design strategies and their mechanisms were reviewed and their application in neural circuitry modulation was discussed (Patel *et al* 2022). Mammalian near-infrared image vision was achieved through ocular injectable retinal photoreceptor-binding upconversion nanoparticles (Ma *et al* 2019).

13.6.3.8 Orange carotenoid protein

Orange carotenoid protein (OCP) is a photoactive carotenoprotein involved in photoprotection of cyanobacteria, which uses a ketocarotenoid (echinenone, 3′-hydroxyecheinone, or canthaxanthin) as a chromophore. When it absorbs blue–green light, it converts from an inactive OCP^O orange form to an activated OCP^R red form. OCP^R is able to bind to light-harvesting complexes (phycobilisomes) facilitating thermal dissipation of the excess absorbed light energy (Leccese *et al* 2023).

The normalized absorption coefficient spectra of orange carotenoid protein (OCP) isolated from *Synechocystis* sp. PCC6803 in 0.1 M Tris–HCl pH 8 buffer containing the ketocarotenoid 3′-hydroxyechinenone (hECN) in the dark state (OCP^0) and after photoconversion (illumination with blue–green light in the range 400–550 nm at 10 °C) to the active light-adapted state (OCP^R) is shown in figure 13.45 (from Wilson *et al* 2008). The spectral red-shift of OCP^R compared to OCP^0 indicates a slightly increased double-bond conjugation length due to hECN isomerization (Gurchiek *et al* 2018, Chukhutsina *et al* 2022) and hECN-protein interaction charge (Leverenz *et al* 2015). The loss of vibronic structure of OCP^R indicates a faster S_2 excited state relaxation (line broadening caused by sub-100 fs relaxation).

A slightly simplified photocycle scheme of OCP in solution is depicted in figure 13.46. It is drawn taking information from Konold *et al* 2019, Pigni *et al* 2020, Muzzopappa and Kirilovsky 2020, Yaroshevich *et al* 2021, and Niziński *et al* 2022. In the dark-adapted state (OCP^0) the N-terminal domain NTD and the C-terminal domain CTD of OCP^0 are close together and the noncovalently bound ketocarotenoid chromophore (abbreviated kCar) extends over both domains. Photoexcitation of kCar (S_0-S_2 transition) initiates the photocycle by breaking the hydrogen bonds between the carbonyl group of the ionone ring and the amino acids Y201 and W288 in the CTD domain (notation belongs to *Synechocystis* sp. PCC6803) forming the first product state P_1. Translocation of kCar away from CTP into NTP occurs via the product states P_2 and P_3. It follows a spatial separation of the NTD domain and the CTD domain forming the light-adapted active state OCP^R with red-shifted and spectrally smoothed absorption spectrum. After excitation light switch-off OCP^R recovers thermally activated back to the dark-adapted form OCP^0 on a timescale of seconds.

In native form in cyanobacteria, OCP acts in photoprotection of phycobilisome, the light harvesting antenna of their photosystem II. In the dark and under low irradiation, OCP is in its inactive form OCP^0 and not bound to the phycobilisome (PBS) which transfers absorbed energy to chlorophyll a of the photosystem II in the

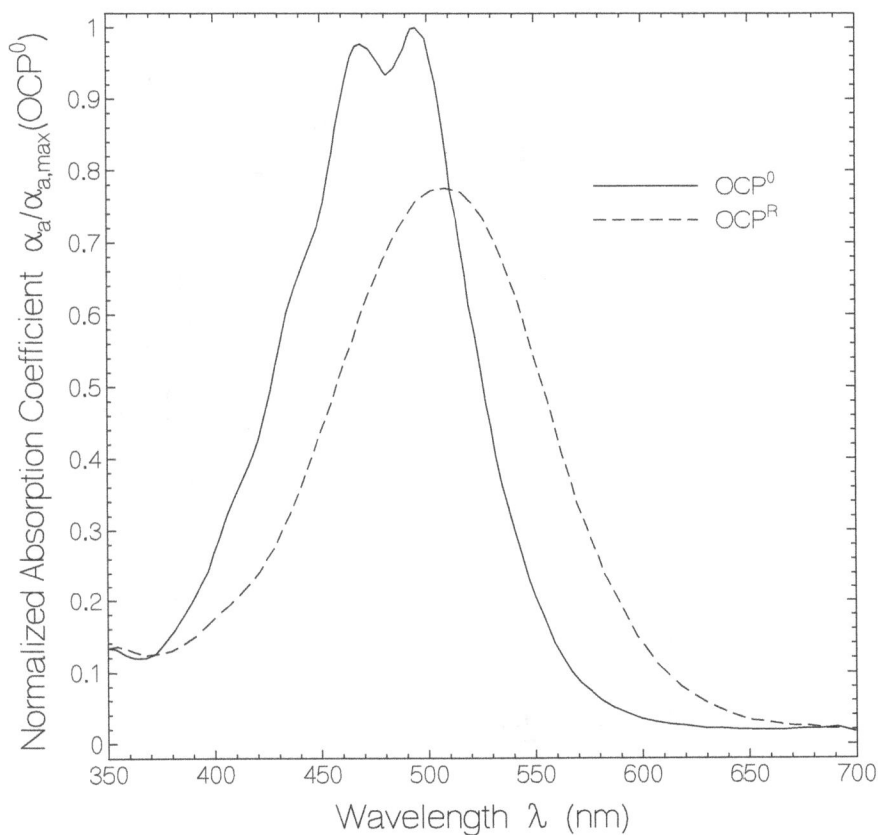

Figure 13.45. Normalized absorption coefficient spectra of OCP^0 and OPC^R isolated from *Synechocystis* sp. PCC6803 in 0.1 M Tris–HCl pH 8 buffer containing the ketocarotenoid 3′-hydroxyechinenone (data from Wilson *et al* 2008).

cyanobacteria reaction center. Strong light exposure causes photoactivation of OCP^0 to the open active red form OCP^R according to the scheme shown in figure 13.46. OCP^R binds to the PBS core and induces the transfer of excess absorbed light in PBS to OCP^R where it dissipates as heat, thereby minimizing the excitation energy transfer to photosystem II for photosynthesis (mechanism of non-photochemical quenching). The dimeric fluorescence recovery protein (FRP) present in cyanobacteria helps to detach OCP^R from PBS under all light conditions (Moldenhauer *et al* 2018). After OCP^R is detached from PBS, it converts back to OCP^0 where FRP accelerates the back-conversion of OCP^R to OCP^0. This process of OCP-related photoprotection in cyanobacteria was reviewed by Muzzopappa and Kirilovsky 2020 and by Slonimskiy *et al* 2020.

The orange carotenoid protein (OCP) family of cyanobacterial photoreceptors has been engineered for applications in synthetic biology optogenetics (Dominguez-Martin and Kerfeld 2019). An OCP-based photoswitch was synthesized to control blue–green light responses in chloroplasts (Piccinini *et al* 2022).

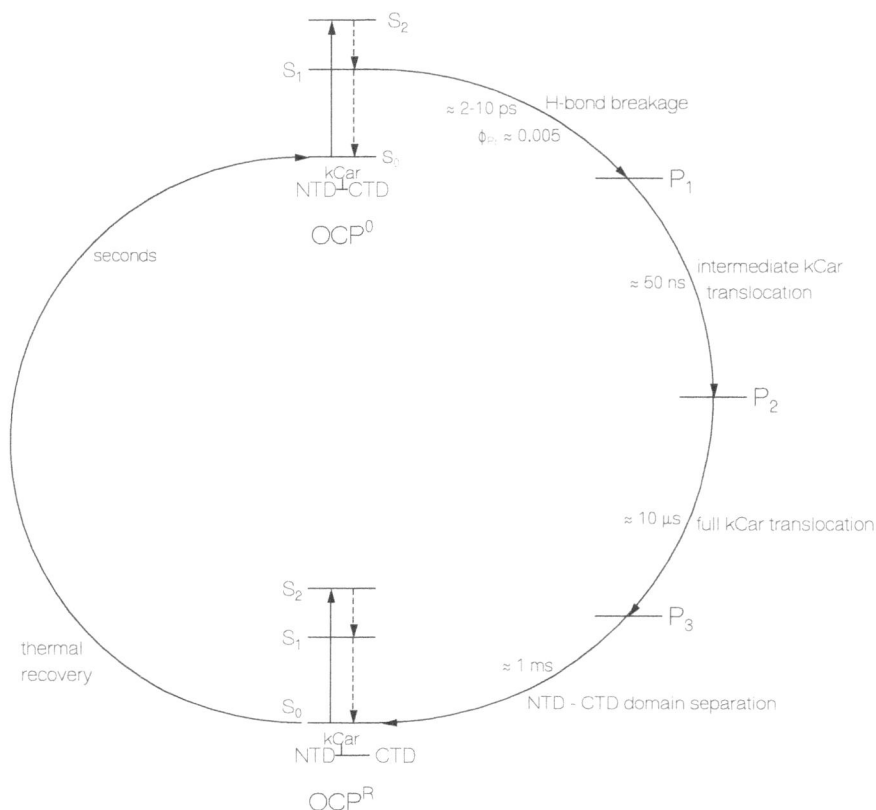

Figure 13.46. Simplified photocycle scheme of OCP in solution (information taken from Konold *et al* 2019, Pigni *et al* 2020, Muzzopappa and Kirilovsky 2020, Yaroshevich *et al* 2021, Niziński *et al* 2022). NTD: N-terminal domain of OCP. CTD: C-terminal domain of OCP. kCar: ketocarotenoid chromophore in OCP. OCP^0: OCP in inactive dark-adapted state. OCP^R: OCP in active light-adapted state. P_1: state with broken hydrogen bonds to kCar (Tyr-201 and Trp-288 in CTD). P_2: intermediate state with intermediate translocation of kCar into NTD. P_3: state with complete (12 Å) translocation of kCar into NTD. OCP^R: active state with maximum separation of NTD and CTD domains. Approximate time constants of product state formations are indicated. $\phi_{Pr} \approx 0.5\%$ is approximate quantum yield of P_1 formation (OCP^R formation).

13.6.4 Application of optogenetic tools

Optogenetic tools are applied in medicine (medical optogenetics, Häusser 2021), neuroscience (neurobiological optogenetics, Chen W *et al* 2022), cell biology (molecular optogenetics, Kolar *et al* 2018), cellular physiology (optophysiology, Tan *et al* 2022), developmental biology (Krueger *et al* 2019), material science (material science optogenetics, Månsson *et al* 2022) and other fields.

The application of optogenetics technology in biologic research follows the procedure: (i) find a suitable light-sensitive protein, (ii) bring it to target cell through methods like transfection, viral transduction, establishment of transgenic animal lines, (iii) use light source for spatial and temporal controlled sample excitation, (iv) use appropriate signal detection system, and (v) evaluate the impact of caused cell activities (Wang 2020).

Optogenetics is opening new opportunities for precision-guided medicine by achieving spatiotemporal control of cellular activities. The research in medical optogenetics (Ye and Fussenegger 2019, Kushibiki 2021, Häusser 2021) includes cardiac diagnostics and therapy (Joshi *et al* 2020, Entcheva and Kay 2021, Madrid *et al* 2021, Sung *et al* 2022), vision restoration (Sahel *et al* 2021, Ratner 2021, Tomita and Sugano 2021, Lu and Pan 2021, Gauvain *et al* 2021, Too *et al* 2022, Lindner *et al* 2022, van Gelder *et al* 2022), inner ear reconstruction (Dombrowski *et al* 2019), regenerative medicine (Hu *et al* 2020) and others.

13.6.4.1 Medical optogenetic applications

13.6.4.1.1 Cardiovascular research
The application of optogenetics in cardiovascular research in *in vitro* and *in vivo* animal studies was reviewed by Joshi *et al* 2020. Important developments and applications of optogenetics to the heart were collected (Entcheva and Kay 2021). Recent advances in novel implantable optogenetic devices and their feasibility in cardiac research and medicine were reviewed by Madrid *et al* 2021. The cardiac electrophysiology and the technical progress about experimental and clinical cardiac optogenetics were discussed (Sung *et al* 2022).

13.6.4.1.2 Vision research
The optogenetic strategies for vision restoration (Ostrovsky and Kirpichnikov 2019, Simon *et al* 2020, Farnum and Pelled 2020, Bansal *et al* 2021, Lu and Pan 2021, Prosseda *et al* 2022, Harris and Gilbert 2022) and optogenetics-mediated gene therapy of retinal diseases (Tomita and Sugano 2021) were reviewed.

Restoration of visual function by transplantation of optogenetically engineered photoreceptors was achieved in blind mice (Garita-Hernandez *et al* 2019). Visual function restoration in retinal ganglion cells of non-human primates (Gauvian *et al* 2021) and blind mice (Chen F *et al* 2022) was achieved with special engineered channelrhodopsins.

Partial recovery of visual function in a blind patient because of retinitis pigmentosa was achieved by intraocular injection of an adeno-associated viral vector encoding the channelrhodopsin protein ChrimsonR with light stimulation via engineered googles (Sahel *et al* 2021).

Optogenetic restoration of high sensitivity vision with red-shifted channelrhodopsins was approached by optogenetic gene therapy on normally light-insensitive bipolar cells or retinal ganglion cells (Too *et al* 2022).

13.6.4.1.3 Inner ear reconstruction
Sensorineural hearing impairment results from cochlear dysfunction. Cochlear implants (CIs) are considered the most successful neuro-prosthesis as they enable speech comprehension for people suffering from sensorineural hearing loss. Electrical cochlear implants (eCIs) remain the key options for hearing rehabilitation.

They electrically stimulate the auditory nerve. Optical cochlear implants (oCIs) are under development. They promise increased spectral selectivity of artificial sound encoding. The research on oCIs is reviewed by Dombrowski *et al* 2019, Moser and Dieter 2020, Dieter *et al* 2020, Khurana *et al* 2023.

13.6.4.1.4 Regenerative medicine

Regenerative medicine deals with the process of replacing, engineering or regenerating human or animal cells, tissues or organs to restore or establish normal function (Mason and Dunnill 2008). Optogenetic tools for tissue engineering and regenerative medicine were reviewed (Hu *et al* 2020). The impact of optogenetics on regenerative medicine was reviewed (Spagnuolo *et al* 2020). The perspectives of optogenetic neuroregeneration were discussed (Janovjak and Kleinlogel 2022).

13.6.4.1.5 Primary cilia research

Primary cilia (sensory cilia) are antenna-like sensory organelles protruding from the surface of most vertebrate cell types. They are essential for regulating signaling pathways development and adult homeostasis (Mill *et al* 2023). Optogenetic methods were applied to study their function as antennae that translate sensory information into cellular response (Hansen *et al* 2020, Prosseda *et al* 2020, Wachten and Mick 2021) Recent discoveries regarding primary cilia and filopodia and their role in pancreatic islet homeostasis and intercellular islet communication were reviewed (Moruzzi *et al* 2023).

Ciliopathies (diseases caused by primary cilia disorders, Shamseldin *et al* 2020, Chen H Y *et al* 2021) were studied by optogenetic approaches (Guo *et al* 2019, Hansen *et al* 2020).

13.6.4.2 Application of optogenetic tools in neuroscience

The goal of most neuroscientific studies, carried out in living animals *in vivo*, was to find out the role of a certain structure or cell type in perception (Marshel *et al* 2019), cognition and behavior (Chen W *et al* 2022). Optogenetic techniques usually use adenoviruses that encode photosensitive proteins in specific neural regions of interest which can be controlled with spatiotemporal resolution by light exposure (Lehtinen *et al* 2022). Optogenetic tools combined with transgenic mouse lines enabled unprecedented spatiotemporal precision in neural circuit analysis (Lee *et al* 2020). The site-specific Cre-Lox recombination technology (McLellan *et al* 2017) was used to enter opsin photoreceptors at specific sites in the DNA of cells and allowed light triggered neuron activation or deactivation. For neuron excitation cation conducting channelrhodopsins (mainly ChR2) were used which depolarize the membrane potential. For neuron inhibition anion conducting channelrhodopsins, inward rhodopsin Cl^- ion pumps (like halorhodopsin), or outward proton pumps (like Arch) were used which hyperpolarize the membrane potential (Lee *et al* 2020).

Progress in neurobiological optogenetics has been documented in several reviews (Mahmoudi *et al* 2017, Lee *et al* 2020, Wang 2020, Shen *et al* 2020, Zhang H *et al* 2022, Lehtinen *et al* 2022, Chen W *et al* 2022). Neuron functions and neural circuits were explored (Shao *et al* 2018, Luo *et al* 2018, Ehmann and Pauls 2020, Lee *et al* 2020).

13.6.4.2.1 Research on neuronal diseases

Research on optogenetic investigation and treatment of neuronal diseases is very active (Camporeze *et al* 2018, Wang 2020, Xu *et al* 2020, Kushibiki 2021, Chen *et al* 2022, Geng *et al* 2023).

The research on neuronal diseases includes:

Alzheimer's disease (Etter *et al* 2019, Mirzayi *et al* 2022, Tiwari and Tolwinski 2023),

Parkinson's disease (Kravitz *et al* 2010, Chen Y *et al* 2015, Magno *et al* 2019, Valverde *et al* 2020, Fougère *et al* 2021, Ingles-Prieto *et al* 2021),

Epilepsy (Bentley *et al* 2013, Wykes *et al* 2016, Choy *et al* 2017, Tønnesen and Kokaia 2017, Camporeze *et al* 2018, Smirnova and Zaitsev 2019, Cela and Sjöström 2019, Osawa and Tominaga 2021, Shimoda *et al* 2022, Bauer *et al* 2023),

Stroke rehabilitation (Cheng *et al* 2014, Pendharkar *et al* 2016, Shah *et al* 2017, Lu *et al* 2019, Prestori *et al* 2020, Geng *et al* 2023),

Tumors of the central nervous system (Camporeze *et al* 2018).

13.6.4.2.2 Research on psychiatric disorders

Optogenetic investigations on psychiatric disorders were reviewed (Tye and Deisseroth 2012, Touriño *et al* 2013, Shirai and Hayashi-Takagi 2017, Cheng *et al* 2020, Liu *et al* 2021, Fakhoury 2021, Nakai *et al* 2021, Vickstrom *et al* 2022, Arslan *et al* 2022, Duncan and Deisseroth 2023).

They include:

Anxiety (Allsop *et al* 2014, Jarrin and Finn 2019, Hsueh *et al* 2023),

Addiction (Vickstrom *et al* 2022),

Autism (Nakai *et al* 2021),

Depression (Camporeze *et al* 2018, Cheng *et al* 2020, Fakhoury 2021, Arslan *et al* 2022),

Fear (Belzung *et al* 2014, Klavir *et al* 2017, Luchkina and Bolshakov 2018, Yan *et al* 2023b),

Mood disorders (Liu *et al* 2021),

Schizophrenia (Peled 2011, Wolff *et al* 2018, Obi-Nagata *et al* 2019, Patrono *et al* 2021).

13.6.4.3 Application of optogenetic tools in cell biology

The application of optogenetic tools in cell biology has become a huge expanding field (Seong and Lin 2021, Emiliani *et al* 2022, Tan *et al* 2022). The light dependent regulation of gene expression (transcription of DNA template of a gene to mRNA and translation of mRNA to protein) has become a major technique to control biological systems (Hartmann *et al* 2020, Yamada *et al* 2020, Isomura 2021, Manoilov *et al* 2021, Ohlendorf and Möglich 2022, Ranzani *et al* 2022).

13.6.4.3.1 Caged proteins (allosteric proteins)

Optogenetic tools that function via inducible allostery are usually single-component systems consisting of a photoreceptor domain and an effector protein whereby their light-switching behavior is mediated by steric or allosteric interaction between the photoreceptor domain and the effector protein (Möglich 2019, Mathony and Niopek 2021). They are mainly based on light-voltage-oxygen (LOV) domains and phytochromes.

Optogenetic tools applied in cell biology based on allosteric proteins have been reviewed by Hoffmann *et al* 2018, Mathony and Niopek 2021, Ohlendorf and Möglich 2022, McCue and Kuhlman 2022, Fauser *et al* 2022, and Fischer *et al* 2022. Spatiotemporal control of proximity labels in living cells was achieved by installing a LOV domain into the proximity labeling enzyme TurboID (Cho *et al* 2020) to rapidly and reversibly control its labeling activity with low-power blue light (Lee *et al* 2023).

13.6.4.3.2 *Dimerizing and oligomerizing protein systems*
Proximity is an essential regulatory factor in biological processes which can be controlled by light induced and chemically induced dimerization and oligomerization approaches. Applications range from manipulation of protein folding, activation, localization, and degradation to controlling gene transcription or cell therapy. Optogenetic dimerization systems employ photosensitive proteins that undergo a conformational change upon illumination and thereby induce protein interaction (Klewer and Wu 2019), organelle tethering (Shi *et al* 2018), recruiting of proteins to subcellular organelles (Benedetti *et al* 2020), and controlled intracellular organelle transport (Nijenhuis *et al* 2020). Chemo-optogenetic dimerization systems use photoactivatable (photocleavable) small-molecule dimerizers which induce (disrupt) proximity by photoexcitation (Chen X *et al* 2018, Klewer and Wu 2019).

The light-regulated dimerization/oligomerization reactions have been reviewed (Klewer and Wu 2019, Terazima 2021, Pearce and Tucker 2021, Ohlendorf and Möglich 2022, Takao *et al* 2022). The red/far-red light-responsive PhyB/PIF (PhyB = phytochrome with chromophore phytochromobilin PΦB, PIF = phytochrome interacting factor) or the blue light-responsive CRY2/CIB1 (CRY2 = cryptochrome 2, CIB1 = cryptochrome-interacting basic-helix-loop-helix 1 transcription factor) hetero-dimerization systems are examples for association-dependent optogenetic tools.

13.6.5 Conclusions

The field of optogenetics expanded enormously in the last years. The application of optogenetic tools to the research on animals approaches towards the beneficial research on humans in health and disease monitoring as well as in medical and clinical treatment.

The combination of optogenetics with new imaging techniques like multiphoton microscopy opens new ways of deep-brain neuroscience (Adesnik and Abdeladim 2021, Xiao *et al* 2023, Klioutchnikov *et al* 2023, Zhao *et al* 2023). Fluorescence nanoscopy may find combination with optogenetics in biological research (Wang *et al* 2023, Ostersehlt *et al* 2022, Weber *et al* 2023, Ortkrass *et al* 2023, Wolff *et al* 2023, Deguchi *et al* 2023). The *anti-*Brownian *el*ectrokinetic (ABEL) trap single-molecule fluorescence microscopy allows real-time observation of dynamic photophysical changes in biological complexes (Squires *et al* 2022).

List of Abbreviations

Abbreviation	Full name
ABS	Activity-based sensing
aCRY	Animal cryptochrome
ACR	Anion-conducting channelrhodopsin
AdoCbl, AdoB$_{12}$	Adenosylcobalamin, coenzyme B$_{12}$
AerR	Aerobic repressor of photosynthesis
Agp1	A phytochrome from *Agrobacterium fabrum*
Agp2	A phytochrome from *Agrobacterium fabrum*
AIMS	*ab initio* multiple spawning
AMP	Adenosine monophosphate
AppA	Activation of photopigment and puc expression protein
Arch	Archaerhodopsin 3
Arg	Arginine, R
ATP	Adenosine triphosphate
BAPTA	1,2-*bis*(*o*-aminophenoxy)ethane-*N*,*N*,*N'*,*N'*-*t*etraacetic acid
BcHeR	Heliorhodopsin from eubacterium *Bellilinea caldifistulea*
Bchl	Bacteriochlorophyll
bchC	Bacteriochlorophyll promoter
BlrB	Blue-light receptor B
BL-OG	Bioluminescence-optogenetics
BlrP1	Blue-light-regulated phosphodiesterase 1 from *Klebsiella pneumoniae*
BlsA	BLUF protein from *Acinetobacter baumannii*
BLUF	Sensor for blue light using FAD (or flavin)
BLUF-PAC	BLUF domain coupled photoactivated adenylyl cyclase
BLUF-PGC	BLUF domain coupled photoactivated guanylyl cyclase
BphP	Bacteria phytochrome photoreceptor
biPAC	Photoactivated adenylyl cyclase from sulfur bacterium *Beggiatoa* sp. IS2
bPAC	Photoactivated adenylyl cyclase from sulfur bacterium *Beggiatoa* sp.
BRET	Bioluminescence resonance energy transfer
BPR	Blue-absorbing proteorhodopsin
BR	Bacteriorhodopsin
BV	Biliverdin
BVE	Biliverdin dimethyl ester
bZIP	Basic leucine zipper domain
CA	Coumaric acid
cAMP	Cyclic adenosine monophosphate
CarH	B$_{12}$-dependent repressor of carotenogenesis
Caspases	Cysteine-aspartic proteases
CBCR	Cyanobacteriochrome
CaM	Calmodulin, calcium-modulated protein
CAN	Canthaxanthin
CCR	Cation channelrhodopsin
CCE	Cryptochrome C-terminal extension
c-di-GMP	Cyclic dimeric guanosine monophosphate
cDNA	Complementary DNA
cGMP	Cyclic guanosine monophosphate
ChR	Channelrhodopsin

Chlide	Chlorophyllide
Chl a	Chlorophyll a
Chl b	Chlorophyll b
Chl d	Chlorophyll d
Chl f	Chlorophyll f
ChR2	Channelrhodopsin 2
CIB1	Cryptochrome-interacting basic-helix–loop–helix 1
CNCbl, CNB$_{12}$	Cyanocobalamin, vitamin B$_{12}$
COP1	CONSTUTIVELY PHOTOMORPHENETIC 1
CPF	Cryptochrome/Photolyase family
CPD	Cyclobutane pyrimidine dimer
Cph	Cyanobacteria phytochrome
CR	Charge recombination
Cre	cAMP responsive element
CREB	cAMP response element-binding protein
CrtJ	Aerobic repressor of carotenoid (crt) gene expression
Cry	Cryptochrome
CRY-DASH	(*Drosophila*, *Arabidopsis*, *Synechocystis*, Human)—type cryptochrome
Cry2PHR	Photolyase homology region (PHR) of cryptochrome 2
CT	Charge transfer, charge transfer state
CTD	C-terminal domain
Cys	Cysteine, C
dcry	Cryptochrome of *Drosophila melanogaster*
DDR2	*Donghaeana dokdonensis* rhodopsin 2
DMB	5,6-Dimethylbenzimidazole
DNA	Deoxyribonucleic acid
DPOR	Light-independent protochlorophyllide reductase
DsRed	*Discosoma* sp. red fluorescent protein
dsDNA	Double-stranded DNA
DsLOV	LOV domain from marine phototrophic bacterium *Dinoroseobacter shibae*
EAL	Glu-Ala-Leu
EBFP	Enhanced blue fluorescent protein
ECFP	Enhanced cyan fluorescent protein
ECN	Echinenone
EF hand	A helix–loop–helix structural domain found in a large family of calcium-binding proteins
EGFP	Enhanced green fluorescent protein
EGTA	Ethylene glycol-bis(ß-aminoethyl ether)-*N,N,N',N'*-tetraacetic acid
EL222	Light-activated DNA-binding protein from *Erythrobacter litorlis*
eNpHR	Enhanced *Natronomonas pharaonis* halorhodopsin
ePDZ	Engineered PDZ affinity clamp protein
ER	Endoplasmic reticulum
eR	Electron return
ET	Energy transfer
eT	Electron transfer
EYFP	Enhanced yellow fluorescent protein
FA	Folic acid
FAD	Flavin adenine dinucleotide
FAP	Fluorogen-activating protein

FbFP	Flavin-based fluorescent protein
FhlA	Formate hydrogenlyase transcriptional activator
FKF1	Flavin-binding, Kelch repeat, F-box 1
Fl_{ox}	Fully oxidized form of flavin
Fl_{red}	Neutral reduced form of flavin
Fl_{red}^-	Anionic reduced form of flavin
Fl_{sq}	Neutral radical form of flavin
Fl_{sq}^-	Anionic radial form of flavin
Fl	Flavin
FMN	Flavin mononucleotide
FP	Fluorescent protein
Fph	Fungi phytochrome
FRP	Fluorescence recovery protein
FTIR	Fourier transform infrared
GAF	Protein domain named after some of the proteins it is found in: cGMP-specific phosphodiesterases, **a**denylyl cyclases and **FhlA**
GECI	Genetically encoded calcium indicator
GEVI	Genetically encoded voltage indicator
GFP	Green fluorescent protein
GIGANTEA	Unique plant specific nuclear protein from *Arabidopsis thaliana*
Glu	Glutamic acid, E
GMP	Guanosine monophosphate
GPR	Green-absorbing proteorhodopsin
GSR	General stress response
GTP	Guanosine-5′-triphosphate
GTPase	Hydrolase enzyme that can bind and hydrolyze guanosine triphosphate (GTP)
8-HDF	8-Hydroxydeazaflavin
H_2OCbl, H_2OB_{12}	Aquocobalamin, vitamin B_{12a}
HBI	4-*h*ydroxy*b*enzylidene *i*midazolinone
hECN, 3-hECN	3′-hydroxyechinenone
HeR	Heliorhodopsin
HK	Histidine kinase
HKR	Histidine kinase rhodopsin
His	Histidine, H
HOMO	Highest occupied molecular orbital
HR	Halorhodopsin
HTH	Helix–turn–helix
HY5	ELONGATED HYPOCOTYL 5
IC	Internal conversion
ICT	Intra-molecular charge transfer
iLID	Improved light-induced dimer
iLOV	Photoreversible LOV protein with improved fluorescent properties
IPT	Intra-molecular proton transfer
ISC	Singlet-triplet intersystem crossing
kCar	Ketocarotenoid
KR2	*Krokinobacter eikastus* rhodopsin 2
LC	Lumichrome

LE	Locally excited state
LgAC	Light-gated adenylate cyclase
LMCT	Ligand-to-metal charge transfer
LOV	Light-Oxygen-Voltage
LOVpep	AsLOV2-Jα with fused peptide epitope
LovK	LOV-histidine kinase from *Caulobacter crescentus*
LPOR	Light-dependent protochlorophyllide reductase
LUMO	Lowest unoccupied molecular orbital
MeCbl	Methylcobalamin
MLCT	Metal-to-ligand charge transfer
mCherry	Monomeric red fluorescent protein
mOrange	Monomeric orange fluorescent protein
MM	Molecular mechanical
MrHR	Cl^- pumping rhodopsin from *Mastigocladopsin repens*
MrP	Li^+/Na^+ pumping rhodopsin from *Methylobacterium populi*
M13	Synthetic peptide derived from the skeletal muscle myosin light chain kinase
MTHF	5,10-Methenyltetrahydrofolate
NADPH	Nicotinamide adenine dinucleotide phosphate
NER	Nucleotide excision repair
NIR-II	Near infrared II (spectral range 1000–1700 nm)
NM-R3	Cl^- pumping rhodopsin from *Nonlabens marinus*
NIR FP	Near infrared fluorescent protein
nMag	Negative magnet = with negative amino acids modified Vivid fungal photoreceptor
NTD	N-terminal domain
OaPAC	Photoactivated adenylate cyclase from the photosynthetic cyanobacterium *Oscillatoria acuminata*
OCP	Orange carotenoid protein
OHCbl, OHB$_{12}$	Hydroxocobalamin, vitamin B_{12b}
Orai1	Calcium release-activated calcium channel protein 1
PAC	Photoactivated adenylyl cyclase
PAE	Photoactivated endonuclease III
PAS domain	Per-Arnt-Sim domain. It is named after the three proteins in which it was first discovered: Per = period circadian protein, Arnt = aryl hydrocarbon receptor nuclear translocator protein, and Sim = single-minded protein
PapA	Multi-domain protein with GGDEF sequence and EAL domain from *Rhodopseudomonas palustris*
PapB	Short BLUF protein from *Rhodopseudomonas palustris*
PBP	Photoactive yellow protein binding protein
PBS	Phycobilisome
PCET	Proton-coupled electron transfer
Pchlide	Protochlorophyllide
pCRY	Plant cryptochrome
PCB	Phycocyanobilin
pcFPs	Photoconversion fluorescent proteins
P$_{crt}$	Promoter of carotenogenic gene(s)
PΦB	Phytochromobilin
PGC	Photoactivated guanylyl cyclase
Pheo a	Pheophytin a

Phot	Phototropin
PHR	Photolyase homology region
Phy	Phytochrome
PHY	Phytochrome-explicit area
PI	Photo-ionization
PICT	Planar intra-molecular charge transfer
PIF	Phytochrome-interacting factor
pMag	Positive Magnet = with positive amino acids modified Vivid fungal photoreceptor
POR	Protochlorophyllide reductase
PpsR	Photosynthesis gene transcription repressors
PR	Proteorhodopsin
pR	Proton return
PRSB	Protonated retinal Schiff base
PSII	Photosystem II
pT	Proton transfer
PVB	Phycoviolobilin
PYP	Photoactive yellow protein
QM	Quantum mechanical
Rac1	Ras-related C3 botulinum toxin substrate 1
ReaChR	Red-activatable channelrhodopsin
REKS	Restricted ensemble Kohn–Sham
RESOLFT	Reversible saturable optical linear fluorescence transition
Ret	Retinal
RGS	Regulator of G protein signaling
RhAC	Rhodopsin adenyl-cyclase
RhGC	Rhodopsin guanylyl-cyclase
RhoPDE	Rhodopsin phosphodiesterase
Rib	Riboflavin
ROS	Reactive oxygen species
rsFPs	Reversibly switchable fluorescent proteins
RS20	20-residue peptide of the calmodulin-binding domain of the smooth muscle myosin light chain kinase
RSB	Retinal Schiff base
rsFP	Reversibly photoswitchable fluorescent protein
RUP	REPRESSOR OF PHOTOMORPHOGENESIS
SCHIC	Sensor containing heme instead of cobalamin
SFO	Step-function opsin
SOMO	Single occupied molecular orbital (half-filled HOMO)
SpoIIE	Sporulation stage II protein E
SR	Sensory rhodopsin
SR	Sarcoplasmic reticulum
ssDNA	Single-stranded DNA
SsrA	Peptide tag that marks a protein for degradation by the proteins recycling machinery
SspB	Adapter protein involved in stimulating degradation of SsrA tagged proteins
STAS	Sulphate transporter and anti-sigma factor antagonist
STIM1	Stromal interaction molecule 1
SyHR	*Synechocystis* halorhodopsin

SyPixD	BLUF protein from *Synechocystis* sp. PCC6803
TaHeR	Heliorhodopsin from *Thermoplasmatales archaeon*
TAP	Tryptophan-activated protein
TePixD	BLUF protein from *Thermosynechococcus elongatus* BP-1
Thr	Threonine, T
TICT	Twisted intra-molecular charge transfer
TnC	Troponin C
Trp	Tryptophan, W
TULIP	Tunable light-controlled interacting protein tag
Tyr	Tyrosine, Y
UCNP	Upconversion nanoparticle
UVR8	UV RESISTANCE LOCUS 8
V2HeR3	Heliorhodopsin from *Emiliania huxleyi* virus 202
VR	Vibrational relaxation
VVD	Fungal photoreceptor Vivid from *Neurospora crassa*
WCC	White Collar Complex
WC-1	White Collar-1
XMS-CASPT2	Extended multistate complete active space second-order perturbation theory
YcgF	Multidomain BLUF protein from *Escherichia coli* having the EAL domain at the C-terminal of the BLUF domain
YtvA	LOV-STAS protein from *Bacillus subtilis*
ZTL	ZEITLUPE

Symbols

Symbol	Meaning
ϑ	Temperature
ϕ_F	Fluorescence quantum yield
$\phi_T = \phi_{ISC}$	Quantum yield of triplet formation = quantum yield of intersystem crossing
ϕ_P	Phosphorescence quantum yield
τ_F	Fluorescence lifetime
τ_{rad}	Radiative lifetime
k_B	Boltzmann constant
k_{rad}	Radiative rate constant

Acknowledgements

A P thanks Professor F J Gießibl, University of Regensburg, for his continued kind hospitality.

References

Abazari M F *et al* 2021 An updated review of various medicinal applications of *p-coumaric* acid: from antioxidative and anti-inflammatory properties to effects on cell cycle and proliferation *Mini-Reviews in Medicinal Chemistry* **21** 2187–201

Abdelfattah A S *et al* 2019 Bright and photostable chemokinetic indicator for extended *in vivo* voltage imaging *Science* **364** 699–704

Acharya A, Bogdanov A M, Grigorenko B L, Bravaya K B, Nemukhin A V, Lukyanov K A and Krylov A I 2017 Photoinduced chemistry in fluorescent proteins: curse or blessing? *Chem. Rev.* **117** 758–95

Ackermann J, Metternich J T, Herbertz S and Kruss S 2022 Biosensing with fluorescent carbon nanotubes *Angew Chem. Int. Ed* **61** e202112372

Adam S, Wiebeler C and Schapiro I 2021 Structural factors determining the absorption spectrum of channelrhodopsins: a case study of the chimera C1C2 *J. Chem. Theory Comput.* **17** 6302–13

Adam Y *et al* 2019 Voltage imaging and optogenetics reveal behavior-dependent changes in hippocampal dynamics *Nature* **569** 413–7

Adamczyk A K and Zawadzki P 2020 The memory-modifying potential of optogenetics and the need for neuroethics *Nanoethics* **14** 207–25

Adesnik H and Abdelamin L 2021 Probing neural codes with two-photon holographic opto-genetics *Nature Nanoscience* **24** 1356–66

Adir O *et al* 2022 Synthetic cells with self-activating optogenetic proteins communicate with natural cells *Nat. Commun.* **13** 2328

Agliassa C, Narayana R, Christie J M and Maffei M E 2018 Geomagnetic field impacts on cryptochrome and phytochrome signaling *J. Photochem. Phtobiol. B: Biol.* **185** 32–40

Ahmad M 2016 Photocycle and signaling mechanisms of plant cryptochromes *Curr. Opin. Plant Biol.* **33** 108–15

Airan R D, Thompson K R, Fenno L F, Bernstein H and Deisseroth K 2009 Temporally precise *in vivo* control of intracellular signalling *Nature* **458** 1025–9

Alcaíno J, Baeza M and Cifuentes V 2016 Carotenoid distribution in nature *Carotenoids in Nature* ed C Stange (Switzerland: Springer International Publishing) pp 3–33

Aldaba-Muruato L R, Ventura-Juárez J, Perez-Hernandez A M, Hernández-Morales A, Muñoz-Ortega M H, Martínez-Hernández S L, Alvarado -Sánchez B and Macías-Pérez J R 2021 Therapeutic perspectives of *p*-coumaric acid: anti-necrotic, anti-cholestatic and anti-amoebic activities *World Acad. Sci. J.* **3** 47

Aldakov D and Reiss P 2019 Safer-by-design fluorescent nanocrystals: metal halide perovskites vs semiconductor quantum dots *J. Phys. Chem. C* **123** 12527–41

Alexandre M T, Domratcheva T, Bonetti C, van Wilderen L J G W, van Grondelle R, Groot M L, Hellingwerf K J and Kennis J T M 2009 Primary reactions of the LOV2 domain of phototropin studied with ultrafast mid-infrared spectroscopy and quantum chemistry *Biophys. J.* **97** 227–37

Alexandre M T, Purcell E B, van Grondelle R, Robert B, Kennis J T M and Crosson S 2010 Electronic and protein structural dynamics of a photosensory LOV histidine kinase *Biochem.* **49** 4752–9

Algar W R *et al* 2021 Photoluminescent nanoparticles for chemical and biological analysis and imaging *Chem. Rev.* **121** 9243–358

Ali A M *et al* 2015 Optogenetic inhibitor of the transcription factor CREB *Chem. Biol.* **22** 1531–9

Alkahtani M H, Alghannam F, Jiang L, Almethen A, Rampersaud A A, Brick R, Gomes C L, Scully M O and Hemmer P R 2018 Fluorescent nanodiamonds: past, present, and future *Nanophotonics* **7** 1423–53

All A H, Zeng X, Teh D B L, Yi Z, Prasad A, Ishizuka T, Thakor N, Hiromu Y and Liu X 2019 Expanding the toolbox of upconversion nanoparticles for *in vivo* optogenetics and neuro-modulation *Adv. Mater.* **31** 1803474

Allsop S A, van der Weele C M, Wichmann R and Tye K M 2014 Optogenetic insights on the relationship between anxiety-related behaviors and social deficits *Front. Behavioral Neurosci.* **8** 241

Amendola V, Pilot R, Frasconi M, Maragò O M and Iati M A 2017 Surface plasmon resonance in gold nanoparticles: a review *J. Phys. Condens. Matter* **29** 203002

Anders K and Essen L-O 2015 The family of phytochrome-like photoreceptors: diverse, complex and multicolored, but very useful *Curr. Opin. Struct. Biol.* **35** 7–16

Andrews C, Xu Y, Kirberger M and Yang J J 2021 Structural aspects and prediction of calmodulin-binding proteins *Int. J. Mol. Sci.* **22** 308

Andrasfalvy B K, Zemelman B V, Tang J and Vaziri A 2010 Two-photon single-cell optogenetic control of neuronal activity by sculpted light *PNAS* **107** 11981–6

Andreoni A and Tian L 2023 Lighting up action potentials with fast and bright voltage sensors *Nat. Methods* **20** 990–2

Anuar N K K, Tan H L, Lim Y P, So'aib M S and Bakar N F A 2021 A review on multifunctional carbon-dots synthesized from biomass waste: design/fabrication, characterization and applications *Front. Energy Res.* **9** 626549

Appasani K (ed) 2017 *Optogenetics: From Neuronal Function to Mapping and Disease Biology* (Cambridge: Cambridge University Press)

Aravind L and Ponting C P 1997 The GAF domain: an evolutionary link between diverse phototransducing proteins *Trends Biochem. Sci.* **22** 458–9

Ardevol A and Hummer G 2018 Retinal isomerization and water-pore formation in channelrhodopsin-2 *PNAS* **115** 3557–62

Arlt T, Schmidt S, Zinth W, Haupts U and Oesterhelt D 1995 The initial reaction dynamics of the light-driven chloride pump halorhodopsin *Chem. Phys. Lett.* **241** 559–65

Arslan A, Unal-Aydin P, Dogan T and Aydin O 2022 Optogenetic animal models of depression: from mice to men *Translational Research Methods for Major Depressive Disorder* ed Yong-Ku Kim and Meysam Amidfar (New York: Humana Press) pp 167–91

Asano T, Igarashi H, Ishizuka T and Yawo H 2018 Organelle optogenetics: direct manipulation of intracellular Ca^{2+} dynamics by light *Front. Neurosci.* **12** 561

Ashoka A H, Ali F, Tiwari R, Kumari R and Pramanik S K 2020 Recent advances in fluorescent probes for detection of HOCl and HNO *ACS Omega* **5** 1730–42

Atabaev T S 2018 Doped carbon dots for sensing and bioimaging applications: a minireview *Nanomaterials* **8** 342

Auldridge M E and Forest K T 2011 Bacterial phytochromes: more than meets the light *Crit. Rev. Biochem. Mol. Biol.* **46** 67–88

Auzel F 2004 Upconversion and anti-Stokes processes with f and d ions in solids *Chem. Rev.* **104** 139–73

Avelar G M, Schumacher R I, Zaini P A, Leonard G, Richards T A and Gomes S L 2014 A rhodopsin-guanylyl cyclase gene fusion functions in visual perception in a fungus *Curr. Biol.* **24** 1234–40

Avila-Pérez M, Vreede J, Tang Y, Bende O, Losi A and Gärtner W 2009 *In vivo* mutational analysis of YtvA from *Bacillus subtilis*: mechanism of light activation of the general stress response *J. Biol. Chem.* **284** 24958–64

Awasthi M, Sushmita K, Kaushik M S, Ranjan P and Kateriya S 2020 Novel modular rhodopsins from green algae hold great potential for cellular optogenetic modulation across the biological model systems *Life* **10** 259

AzimiHashemi N *et al* 2014 Synthetic retinal analogues modify the spectral and kinetic characteristics of microbial rhodopsin optogenetic tools *Nat. Commun.* **5** 5810

Bagur R and Hajnóczky G 2017 Intracellular Ca^{2+} sensing: its role in calcium homeostasis and signaling *Mol. Cell* **66** 780–8

Balashov S P, Imasheva E S, Govindjee R and Ebrey T S 1996 Titration of aspargate-85 in bacteriorhodopsin: What it says about chromophore isomerization and proton release *Biophys. J.* **70** 473–81

Balcerowicz M 2020 Phytochrome-interacting factors at the interface of light and temperature signaling *Physiologia Planatarum* **169** 347–56

Balzani V 2001 *Electron Transfer in Chemistry* (Weinheim: Wiley)

Bamann C, Bamberg E, Wachtveitl J and Glaubitz C 2014 Proteorhodopsin *Biochim. Biophys. Acta Bioenerg.* **1837** 614–25

Bando Y, Grimm C, Cornejo V H and Yuste R 2019a Genetic voltage indicators *BCM Biology* **17** 71

Bando Y, Sakamoto M, Kim S, Ayzenshtat I and Yuste R 2019b Comparative evaluation of genetically encoded voltage indicators *Cell Reports* **26** 802–13

Banerjee S and Mitra D 2020 Structural basis of design and engineering for advanced plant optogenetics *Trends Plant Sci.* **25** 35–65

Bansal H, Gupta N and Roy S 2021 Theoretical analysis of optogenetic spiking with ChRmine, bReaChES and CsChrimson-expressing neurons for retinal prostheses *J. Neural Eng.* **18** 0460b8

Bansal H, Pyari G and Roy S 2022 Co-expressing fast channelrhodopsin with step-function opsin overcomes spike failure due to photocurrent desensitization in optogenetics: a theoretical study *J. Neural Eng.* **19** 026032

Bao H, Melnicki M R and Kerfeld C A 2017 Structure and functions of Orange Carotenoid Protein homologs in cyanobacteria *Curr. Opin. Plant Biol.* **37** 1–9

Barends T R M, Hartmann E, Griese J J, Beitlich T, Kirienko N V, Ryjenkov D A, Reinstein J, Shoeman R L, Gomelsky M and Schlichting I 2009 Structure and mechanism of a bacterial light-regulated cyclic nucleotide phosphodiesterase *Nature* **459** 1015–8

Barykina N V *et al* 2016 A new design for a gree3n calcium indicator with smaller size and a reduced number of calcium-binding sites *Sci. Rep.* **6** 34447

Barykina N V *et al* 2018 NTnC-like genetically encoded calcium indicator with a positive and enhanced response and fast kinetics *Sci. Rep.* **8** 15233

Barneschi L *et al* 2022 On the fluorescence enhancement of arch neuronal optogenetic reporters *Nat. Commun.* **13** 6432

Bassetto M, Reichl T, Kobylkov D, Kattnig D R, Winklhofer M, Hore P J and Mouritsen H 2023 No evidence for magnetic field effects on the behaviour of *Drosophila Nature* **620** 595–9

Bassolino G, Sovdat T, Liebel M, Schnedermann C, Odell B, Claridge T D W, Kukura P and Fletcher S P 2014 Synthetic control of retinal photochemistry and photophysics in solution *J. Am. Chem. Soc.* **136** 2650–8

Bassolino G, Sovdat T, Duarte A S, Lim J M, Schnedermann C, Liebel M, Odell B, Claridge T D W, Fletcher S P and Kukura P 2015 Barrierless photoisomerization of 11-*cis* retinal protonated Schiff base in solution *J. Am. Chem. Soc.* **137** 12434–7

Batschauer A (ed) 2003 *Photoreceptors and Light Signalling* (Cambridge: The Royal Society of Chemistry)

Bauch M, Toma K, Toma M, Zhang Q and Dostalek J 2014 Plasmon-enhanced fluorescence biosensors: a review *Plasmonics* **9** 781–99

Bauer J, Devinsky O, Rothermel M and Koch H 2023 Autonomic dysfunction in epilepsy mouse models with implication for SUDEP research *Front. Neurol.* **13** 1040648

Baumschlager A and Khammash M 2021 Synthetic biological approaches for optogenetics and tools for transcriptional light-control in bacteria *Advanced Biology* **5** 2000256

Beck S, Yu-Strzelczyk J, Pauls D, Constantin O M, Gee C E, Ehmann N, Kittel R J, Nagel G and Gao S 2018 Synthetic light-gated ion channels for optogenetic activation and inhibition *Front. Neurosci.* **12** 643

Becker R S, Freedman K and Causey G 1982 Schiff base of 11-*cis*-retinal: photoisomerization as a function of solvent and protonation *J. Am. Chem. Soc.* **104** 5797–8

Becker R S and Freedman K 1985 A comprehensive investigation of the mechanism and photophysics of isomerization of a protonated and unprotonated Schiff base of 11-*cis*-retinal *J. Am. Chem. Phys.* **105** 1477–85

Becker-Baldus J *et al* 2015 Enlightening the photactive site of channelrhodopsin-2 by DNP-enhanced solid-state NMR spectroscopy *PNAS* **112** 9896–901

Béja O *et al* 2000 Bacterial rhodopsin: evidence for a new type of phototrophy in the sea *Science* **289** 1902–6

Bent D V and Hayon E 1975 Excited state chemistry of aromatic amino acids and related peptides. III. Tryptophan *J. Am. Chem. Soc.* **97** 2612–9

Bentley J N, Chstek C, Stacey C and Patil P G 2013 Optogenetics in epilepsy *Neurosurgical Focus* **34** E4

Benedetti L, Marvin J S, Falahati H, Guillén-Samander A, Looger L L and De Camilli P 2020 Optimized Vivid-derived Magnets photodimerizers for subcellular optogenetics in mammalian cells *eLife* **9** e63230

Belzung C, Turiault M and Griebel G 2014 Optogenetics to study circuits of fear- and depression-like behaviors: a critical analysis *Pharmacol. Biochem. Behav.* **122** 144–57

Berglund K *et al* 2016a Luminopsins integrate opto- and chemogenetics by using physical and biological light sources for opsin activation *PNAS* **113** E358–67

Berglund K, Tung J K, Higashikubo B, Gross R E, Moore C I and Hochgeschwender U 2016b Combined optogenetic and chemogenetic control of neurons *Optogenetics: Methods and Protocols* ed A Kianianmomeni (New York: Humana Press) pp 207–25

Berglund K, Fernandez A M, Gutekunst C-A N, Hochgeaschwender U and Gross R E 2020 Step-function luminopsins for bimodal prolonged neuromodulation *J. Neuro Res.* **98** 422–36

Berglund K, Stern M A and Gross R E 2021 Bioluminescence-Optogenetics *Optogenetics: Light-sensing Proteins and their Applications in Neuroscience and beyond* 2nd edn ed H Yawo, H Kandori, A Koizumi and R Kageyama (Singapore: Springer) pp 281–93

Bergs A C F *et al* 2023 All-optical closed-loop voltage clamp for precise control of muscles and neurons in live animals *Nat. Commun.* **14** 1939

Berndt A, Kottke T, Breitkreuz H, Dvorsky R, Hennig S, Alexander M and Wolf E 2007 A novel photoreaction mechanism for the circadian blue light photoreceptor Drosophila cryptochrome *J. Biol. Chem.* **282** 13011–21

Berndt A, Lee S Y, Ramakrishnan C and Deisseroth K 2014 Structure-guided transformation of channelrhodopsin into a light-activated chloride channel *Science* **344** 420–4

Berndt A *et al* 2016 Structural foundations of optogenetics: determinants of channelrhodopsin ion selectivity *PNAS* **113** 822–9

Berndt A, Yizhar O, Gunaydin L A, Hegemann P and Deisseroth K 2009 Bi-stable neural state switches *Nat. Neurosci.* **12** 229–34

Besaw J E, Ou W-L, Morizumi T, Eger B T, Vasquez J D S, Chu J H Y, Harris A, Brown L S, Miller R J D and Ernst O P 2020 The crystal structures of a chloride-pumping microbial rhodopsin and its proton-pumping mutant illuminate proton transfer determinants *J. Biol. Chem.* **295** 14793–804

Bi X, Beck C and Gong Y 2021 Genetically encoded fluorescent indicators for imaging brain chemistry *Biosensors* **11** 116

Birge R R, Schulten K and Karplus M 1975 Possible influence of low-lying "covalent" excited state on the absorption spectrum and photoisomerization of 11-*cis* retinal *Chem. Phys. Lett.* **31** 451–4

Birks J B 1970 *Photophysics of Aromatic Molecules* (New York: Wiley)

Bismuth O, Friedman N, Sheves M and Ruhman S 2007 Photochemical dynamics of all-*trans* retinal protonated Schiff-base in solution: excitation wavelength dependence *Chem. Phys.* **341** 267–75

Bitzenhofer N L *et al* 2023 Development and characterization of flavin-binding fluorescent proteins, Part II: advanced characterization *Fluorescent Proteins: Methods and Protocols, Methods in Molecular Biology* vol. 2564 ed M Sharma (New York: Humana) pp 143–83

Blain-Hartung M, Rockwell N C, Moreno M V, Martin S S, Gan F, Bryant D A and Lagarias J C 2018 Cyanobacteriochrome-based photoswitchable adenylyl cyclases (cPACs) for broad spectrum light regulation of cAMP levels in cells *J. Biol. Chem.* **293** 8473–83

Blakley R L 1969 *The Biochemistry of Folic Acid and Related Pteridines* (Amsterdam: North-Holland)

Blakley R L and Benkovic S J 1984 *Folates and Pterins, Chemistry and Biochemistry of Folates* vol 1 (New York: Wiley)

Blancafort L, González D, Olivucci M and Robb M A 2002 Quenching of tryptophan $^1(\pi, \pi^*)$ fluorescence induced by intramolecular hydogen abstraction via an aborted decarboxylation mechanism *J. Am. Chem. Soc.* **124** 6398–406

Blankenstein R E 2021 *Molecular Mechanisms of Photosynthesis* 3rd edn (Hoboken, New Jersey: Wiley)

Boardman B F 1966 Protochlorophyll *The Chlorophylls* ed L P Vernon and G R Seely (New York: Academic) pp 437–79

Bollivar D W 2006 Recent advances in chlorophyll biosynthesis *Photosynth. Res.* **89** 1–22

Bondarev S L and Knyukshto V N 1994 Fluorescence from the S_1 (2 ^1A_g) state of all-trans-β-carotene *Chem. Phys. Lett.* **225** 346–50

Bourgeois D and Adam V 2012 Reversible photoswitching in fluorescent proteins: a mechanistic view *IUBMB Life* **64** 482–91

Bovetti S, Moretti C, Zucca S, Maschio M D, Bonifazi P and Fellin T 2017 Simultaneous high-speed imaging and optogenetic inhibition in the intact mouse brain *Sci. Rep.* **7** 40041

Boyden E S, Zhang F, Bamberg E, Nagel G and Deisseroth K 2005 Millisecond-timescale, genetically targeted optical control of neural activity *Nat. Neurosci.* **8** 1263–8

Boyden E S 2011 A history of optogenetics: the development of tools for controlling brain circuits with light *F1000 Rep. Biol.* **3** 11

Braatsch S, Gomelsky M, Kuphal S and Klug G 2002 A single flavoprotein, AppA, integrates both redox and light signals in *Rhodobacter sphaeroides Mol. Microbiol.* **45** 827–36

Bracker M, Dinkelbach F, Weingart O and Kleinschmidt M 2019 Impact of fluorination on the photophysics of the flavin chromophore: a quantum chemical perspective *Phys. Chem. Chem. Phys.* **21** 9912–23

Bracker M, Kubitz M K, Czekelius C, Marian C M and Kleinschmidt M 2022 Computer-aided design of fluorinated flavin derivatives by modulation of intersystem crossing and fluorescence *ChemPhotoChem.* **6** e202200040

Braslavsky S E, Holzwarth A R, Lehner H and Schaffner K 1978 The fluorescence of biliverdin dimethyl ester *Helv. Chim. Acta* **61** 2219–22

Braslavsky S E, Holzwarth A R and Schaffner K 1983 Solution conformations, photophysics, and photochemistry of bile pigments; bilirubin and biliverdin dimethyl esters and related linear tetrapyrroles *Angew. Chem. Int. Ed.* **22** 656–74

Briggs W R and Spudich J L (ed) 2005 *Handbook of Photosensory Receptors* (Weinheim: Wiley)

Broser M *et al* 2020 NeoR, a near-infrared absorbing rhodopsin *Nat. Commun.* **11** 5682

Broser M 2022 Far-red absorbing rhodopsins, insights from heterodimeric rhodopsin-cyclases *Front. Mol. Biosci.* **8** 806922

Brudler R *et al* 2003 Identification of a new cryptochrome class: structure, function, and evolution *Mol. Cell* **11** 59–67

Bruun S *et al* 2015 Light-dark adaption of channelrhodopsin involves photoconversion between the all-*trans* and 13-*cis* retinal isomers *Biochem.* **54** 5389–400

Bryant D A, Hunter N and Warren M J 2020 Biosynthesis of the modified tetrapyrroles – the pigments of life *J. Biol. Chem.* **295** 6888–925

Buckley A M, Petersen J, Roe A J, Douce G R and Christie J M 2015 LOV-based reporters for fluorescence imaging *Curr. Opin. Chem. Biol.* **27** 39–45

Bugaj L, Choksi A T, Mesuda C K, Kane R S and Schaffer D V 2013 Optogenetic protein clustering and signaling in mammalian cells *Nat. Meth.* **10** 249–52

Büchner R, Fondell M, Haverkamp R, Pietzsch A, da Cruz V V and Föhlisch A 2021 The porphyrin center as a regulator for metal-ligand covalency and π hybridization in the entire molecule *Phys. Chem. Chem. Phys.* **23** 24765–72

Buhrke D, Oppelt K T, Heckmeier P J, Fernández-Terán R and Hamm P 2020 Nanosecond protein dynamics in a red/green cyanobacteriochrome revealed by transient IR spectroscopy *J. Chem. Phys.* **153** 245101

Bulzu P-A, Andrei A-Ş, Salcher M M, Mehrshad M, Inoue K, Kandori H, Béjà O, Ghai R and Banciu H L 2019 Casting light on Asgardarchaeota metabolism in a sunlit microoxic niche *Nat. Microbiol.* **4** 1129–37

Bulzu P-A, Kavagutti V S, Chiriac M-C, Vavourakis C D, Inoue K, Kandori H, Andrei A-S and Ghai R 2021 Heliorhodopsin evolution is driven by photosensory promiscuity in monoderms *mSphere* **6** e00661–21

Burgie E S and Vierstra R D 2014 Phytochrome signaling: an atomic perspective on photo-activation and signaling *Plant Cell* **26** 4568–83

Cai X and Liu B 2020 Aggregation-induced emission: recent advances in material and biomedical applications *Angew. Chem. Int. Ed.* **59** 9868–86

Campbell R E 2008 Fluorescent proteins *Scholarpedia* **3** 5410

Campbell B C, Paez-Segala M G, Looger L L, Petsko G A and Liu C F 2022 Chemically stable fluorescent proteins for advanced microscopy *Nat. Meth.* **19** 1612–21

Camporeze B, Manica B A, Bonafé G A, Ferreira J J C, Diniz A L, de Oliveira C T P, Junior L R M, de Aguiar P H P and Ortega M M 2018 Optogenetics: the new molecular approach to control functions of neural cells in epilepsy, depression and tumors of the central nervous system *Am. J. Cancer Res.* **8** 1900–18

Cao Z, Livoti E, Losi A and Gärtner W 2010 A blue light-inducible phosphodiesterase activity in the cyanobacterium *Synechococcus elongatus Photochem. Photobiol.* **86** 606–11

Carafoli E 2002 Calcium signaling: A tale for all seasons *PNAS* **99** 1115–22

Carenne D, Haines M C, Romantseva E F, Freemont P, Strychalski E A and Noireaux V 2021 Cell-free gene expression *Nat. Rev. Meth. Prim.* **1** 49

Carniol K, Ben-Yehuda S, King N and Losick R 2005 Genetic dissection of the sporulation protein SpoIIE and its role in asymmetric division in *Bacillus subtilis J. Bacteriol.* **187** 3511–20

Carter K P, Young A M and Palmer A E 2014 Fluorescent sensors for measuring metal ions in living systems *Chem. Rev.* **114** 4564–601

Cazzaniga S, Bressan M, Carbonera D, Agostini A and Dall'Osto L 2016 Differential roles of carotenes and xanthophylls in photosystem I photoprotection *Biochemistry* **55** 36–36-49

Cela E and Sjöström J 2019 Novel optogenetic approaches in epilepsy research *Front. Neurosci.* **13** 947

Chábera P, Fuciman M, Hříbek P and Polívka T 2009 Effect of carotenoid structure on excited-state dynamics of carbonyl carotenoids *Phys. Chem. Chem. Phys.* **11** 8795–803

Chalfie M, Tu Y, Euskirchen G, Ward W W and Prasher D C 1994 Green fluorescent protein as a marker for gene expression *Science* **263** 802–5

Chang C-F, Kuramochi H, Singh M, Abe-Yoshizumi R, Tsukuda T, Kandori H and Tahara T 2022 A unified view on varied ultrafast dynamics of the primary process in microbial rhodopsins *Angew. Chem. Int. Ed.* **61** e202111930

Changenet-Baret P, Plaza P and Martin M M 2001 Primary events in the photoactive yellow protein chromophore in solution *Chem. Phys. Lett.* **336** 439–44

Changenet-Baret P, Espagne A, Katsonis N, Charier S, Baudin J-B, Jullien L, Plaza P and Martin M M 2002 Excited-state relaxation dynamics of a PYP chromophore model in solution: influence of the thioester group *Chem. Phys. Lett.* **365** 285–91

Chapman S, Faulkner C, Kaiserli E, Garcia-Mata C, Savenkov E I and Roberts A G 2008 The photoreversible fluorescent protein iLOV outperforms GFP as a reporter of plant virus infection *PNAS* **105** 20038–43

Chaves I, Pokorny R, Byrdin M, Hoang N, Ritz T, Brettel K, Essen L-O, van der Horst G T J, Batschauer A and Ahmad M 2011 The cryptochromes: blue light photoreceptors in plants and animals *Annu. Rev. Plant Biol.* **62** 335–64

Chazan A, Rozenberg A, Mannen K, Nagata T, Tahan R, Yaish S, Larom S, Inoue K, Béjà O and Pushkarev A 2022 Diverse heliorhodopsins detected via functional metagenetics in freshwater *Actinobacteria, Chloroflexi and Archaea Environ. Microbiol.* **24** 110–21

Che D L, Duan L, Zhang K and Cui B 2015 The dual characteristics of light-induced cryptochrome 2, homo-oligomerization and heterodimerization, for optogenetic manipulation in mammalian cells *ACS Synth. Biol.* **4** 1124–35

Chemaly S M, Jack L A, Yellowlees L J, Harper P L S, Heeg B and Pratt J M 2004 Vitamin B_{12} as an allosteric cofactor; dual fluorescence, hysteresis, oscillations and the selection of corrin over porphyrin *Dalton Trans.* **2004** 2125–34

Chemaly S M 2016 New light on vitamin B_{12}: the adenosylcobalamin-dependent photoreceptor protein CarH *S. Afr. J. Sci.* **112** 0106

Chen R F 1990 Fluorescence of proteins and peptides *Practical Fluorescence* 2nd edn ed G G Guilbault (New York: Dekker) pp 575–682

Chen K-Y, Hsieh C-C, Cheng Y-M, Lai C-H, Chou P-T and Chow T J 2006 Tuning excited-state electron transfer from adiabatic to nonadiabatic type in donor-bridge-acceptor systems and the associated energy transfer process *J. Phys. Chem.* A **110** 12136–44

Chen D, Gibson E S and Kennedy M J 2013 A light-triggered protein secretion system *J. Cell Biol.* **201** 631–40

Chen G, Qiu H, Prasad P N and Chen X 2014 Upconversion nanoparticles: design, nano-chemistry, and applications in theranostics *Chem. Rev.* **114** 5161–214

Chen G *et al* 2015 Energy-cascaded upconversion in an organic dye-sensitized core/shell fluoride nanocrystal *Nano Lett.* **15** 7400–7

Chen S *et al* 2018 Near-infrared deep brain stimulation via upconversion nanoparticle-mediated optogenetics *Science* **359** 679–84

Chen Z-H, Raffelberg S, Losi A, Schaap P and Gärtner W 2014 A cyanobacterial light activated adenylyl cyclase partly restores development of a *Dictyostelium discoideum*, adenylyl cyclase a null mutant *J. Biotechnol.* **191** 246–9

Chen F, Duan X, Yu Y, Yang S, Chen Y, Gee C E, Nagel G and Shen Y 2022 Visual function restoration with a highly sensitive and fast channelrhodopsin in blind mice *Signal Transduct. Target. Ther.* **7** 104

Chen H Y, Kelley R A, Li T and Swaroop A 2021 Primary cilia biogenesis and associated ciliopathies *Semin. Cell Dev. Biol.* **110** 70–88

Chen J-L, Lin Y-C, Fu H-Y and Yang C-S 2019 The blue-green sensory rhodopsin SRM from *Haloarcula marismortui* attenuates both phototactic responses mediated by sensory rhodopsin I and II in *Halobacterium salinarum Sci. Rep.* **9** 5672

Chen L, Black A, Parak W J, Klinke C and Chakraborty I 2021 Metal nanocluster-based devices: challenges and opportunities *Aggregate* **3** e132

Chen L-Y, Wang C-W, Yuan Z and Chang H-T 2015 Fluorescent gold nanoclusters: recent advances in sensing and imaging *Anal. Chem.* **87** 216–29

Chen R *et al* 2021 Deep brain optogenetics without intracranial surgery *Nat. Biotechnol.* **39** 161–4

Chen X, Ventatachalapathy M, Dehmelt L and Wu Y-W 2018 Multidirectional activity control of cellular processes by a versatile chemo-optogenetic approach *Angew. Chem.* **130** 12169–73

Chen W *et al* 2022 The roles of optogenetics and technology in neurobiology: a review *Front. Aging Neurosci.* **14** 867863

Chen Y, Xiong M and Zhang S-C 2015 Illuminating Parkinson's therapy with optogenetics *Nat. Biotechnol.* **33** 149–50

Cheng M-C, Kathare P K, Paik I and Huq E 2021 Phytochrome signaling networks *Annu. Rev. Plant Biol.* **72** 217–44

Cheng M Y, Wang E H and Steinberg G K 2014 Optogenetic approaches to study stroke recovery *ACS Chem. Neurosci.* **5** 1144–5

Cheng Z, Cui R, Ge T, Yang W and Li B 2020 Optogenetics: What it has uncovered in potential pathways of depression *Pharmacol. Res.* **152** 104596

Cheng Z, Li K, Hammad L A, Karty J A and Bauer C E 2014 Vitamin B$_{12}$ regulates photosystem gene expression via the CrtJ antirepressor AerR in *Rhodobacter capsulatus Mol. Biol.* **91** 649–64

Cheng Z, Yamamoto H and Bauer C E 2016 Cobalamin's (vitamin B$_{12}$) surprising function as a photoreceptor *Trends Biochem. Sci.* **41** 647–50

Chernov K G, Redchuk T A, Omelina E S and Verkhusha V V 2017 Near-infrared fluorescent proteins, biosensors, and optogenetic tools engineered from phytochromes *Chem. Rev.* **117** 6423–46

Chia N, Lee S Y and Tong Y 2022 Optogenetic tools for microbial synthetic biology *Biotechnol. Adv.* **59** 107953

Chien M-P, Brinks D, Testa-Silva G, Tian H, Brooks F P, Adam Y, Bloxham B, Gmeiner B, Keifets S and Cohen A E 2021 Photoactivated voltage imaging in tissue with an archaerhodopsin-derived reporter *Sci. Adv.* **7** eabe3216

Chin D and Means A R 2000 Calmodulin: a prototypical calcium sensor *Trends Cell Biol.* **10** 322–8

Chitrakar I, Iuliano J N, He Y, Woroniecka H A, Collado J T, Wint J M, Walker S G, Tonge P J and French J B 2020 Structural basis for the regulation of biofilm formation and iron uptake in *A. baumannii* by the blue-light-using photoreceptor, BlsA *ACS Infect. Dis* **6** 2592–603

Cho K F, Branon T C, Udeshi N D, Myers S A, Carr S A and Ting A Y 2020 Proximity labeling in mammalian cells with TurboID and split TurboID *Nat. Protoc.* **15** 3971–99

Cho S-G, Shim J-g, Choun K, Meas S, Kang K-W, Kim J-h, Cho H-S and Jung K-H 2021 Discovery of a new light-driven Li$^+$/Na$^+$-pumping rhodopsin with DTG motif *J. Photochem. Photobiol. B: Biol.* **223** 112285

Cho S-G, Song M, Chuon K, Shim J-g, Meas S and Jung K-H 2022 Heliorhodopsin binds and regulates glutamine synthetase activity *PLoS Biol.* **20** e3001817

Chow B Y et al 2010 High-performance genetically targetable optical neural silencing by light-driven proton pumps *Nature* **463** 98–102

Choy M K, Duffy B A and Lee J H 2017 Optogenetic study of networks in epilepsy *J. Neurosci. Res.* **95** 2325–35

Chretien A et al 2024 Light-induced Trp$_{in}$/Met$_{out}$ switching during BLUF domain activation in ATP-bound photoactivatable adenylate cyclase OaPAC *J. Mol. Biol.* **436** 168439

Christie J M, Reymond P, Powell G K, Bernaconi P, Raibekas A A, Liscum E and Briggs W R 1998 *Arabidopsis* NPH1: a flavoprotein with the properties of a photoreceptor for phototropism *Science* **282** 1698–701

Christie J M, Gawthorne J, Young G, Fraser N J and Roe A J 2012a LOV to BLUF: flavoprotein contributions to the optogenetic toolkit *Mol. Plant* **5** 533–44

Christie J M et al 2012b Plant UVR8 photoreceptor senses UV-B by tryptophan-mediated disruption of cross-dimer salt bridges *Science* **335** 1492–6

Christie J M, Hitomi K, Arvai A S, Hartfield K A, Mettlen M, Pratt A J, Tainer J A and Getzoff E D 2012c Structural tuning of the fluorescent protein iLOV for improved photostability *J. Biol. Chem.* **287** 22295–304

Christie J M, Blackwood L, Petersen J and Sullivan S 2015 Plant flavoprotein photoreceptors *Plant Cell Physiol.* **56** 401–13

Christie J M and Zurbriggen M D 2021 Optogenetics in plants *New Phytol.* **229** 3108–15

Chudakov D M, Matz M V, Lukyanov S and Lukyanov K A 2010 Fluorescent proteins and their applications in imaging living cells and tissues *Physiol. Rev.* **90** 1103–63

Chukhutsina V U, Baxter J M, Fadini A, Morgan R M, Pope M A, Maghlaoui K, Orr C M, Wagner A and van Thor J J 2022 Light activation of orange carotenoid protein reveals bicycle-pedal single-bond isomerization *Nature Commun.* **13** 6420

Church J R, Rao A G, Barnoy A, Wiebeler C and Schapiro I 2021 Computational studies of photochemistry in phytochrome proteins *Challenges Adv. Comput. Chem. Phys.* **31** 197–226

Church J R, Amoyal G S, Borin V A, Adam S, Olsen J M H and Schapiro I 2022 Deciphering the spectral tuning mechanism in proteorhdopsin: the dominant role of electrostatics instead of chromophore geometry *Chem. Eur. J.* **28** e202200139

Cibulka R and Fraaije M W (ed) 2021 *Flavin-Based Catalysis: Principles and Applications* (Weinheim: Wiley-VCH)

Cioffi D L, Barry C J and Stevens T 2011 Role of calcium as a second messenger in signaling: a focus on endothelium *Textbook of Pulmonary Vascular Disease* ed J X-J Yuan, J G N Garcia, C A Hales, S Rich, S L Archer and J B West (Berlin: Springer) pp 261–72

Cobbold P H and Rink T J 1987 Fluorescence and bioluminescence measurement of cytoplasmic free calcium *Biochem. J* **248** 313–28

Coffey V C 2018 Optogenetics controlling neurons with photons *Optics and Photonic News* **April** 24–31

Colindres-Rojas M, Wolf M M N, Groß R, Seidel S, Dietzek B, Schmitt M, Popp J, Hermann G and Diller R 2011 Excited-state dynamics of protochlorophyllide revealed by subpicosecond infrared spectroscopy *Biophys. J.* **100** 260–7

Collado J T *et al* 2022 Unraveling the photoactivation mechanism of a light-activated adenylyl cyclase using ultrafast spectroscopy coupled with unnatural amino acid mutagenesis *ACS Chem. Biol.* **17** 2643–54

Conrad K S, Manahan C C and Crane B R 2014 Photochemistry of flavoprotein light sensors *Nat. Chem. Biol.* **10** 801–9

Cooper C L, Panitz N, Edwards T A and Goyal P 2021 Role of the CarH photoreceptor protein environment in the modulation of cobalamin photochemistry *Biophys. J.* **120** 3688–96

Copeland C E, Kim J, Copeland P L, Heitmeier C J and Kwon Y-C 2022 Characterizing a new fluorescent protein for low limit of detection sensing in the cell-free system *ACS Synth. Biol.* **11** 2800–10

Coquelle N *et al* 2018 Chromophore twisting in the excited state of a photoswitchable fluorescent protein captured by time-resolved serial femtosecond crystallography *Nat. Chem.* **10** 31–7

Corrochano L M 2019 Light in the fungal world: from photoreception to gene transcription and beyond *Annu. Rev. Genet.* **53** 149–70

Correa F, Ko W-H, Ocasio V, Bogomolni R A and Gardner K H 2013 Blue light regulated two-component systems: enzymatic and functional analyses of light-oxygen-voltage (LOV)-histidine kinases and downstream response regulators *Biochemistry* **52** 4656–66

Creed D 1984 The photophysics and photochemistry of the near-UV absorbing amino acids – I. Tryptophan and its simple derivatives *Photochem. Photobiol.* **39** 537–62

Crefcoeur R P, Yin R, Ulm R and Halazonetis T D 2013 Ultraviolet-B-mediated induction of protein–protein interactions in mammalian cells *Nat. Commun.* **4** 1779

Crosson S and Moffat K 2001 Structure of a flavin-binding plant photoreceptor domain: insights into light mediated signal transduction *PNAS* **98** 2995–3000

Crosson S, Rajagopol S and Moffat K 2003 The LOV domain family: photoresponsive signaling modules coupled to diverse output domains *Biochemistry* **42** 2–10

Csomos A, Kontra B, Jancsó A, Galbács G, Deme R, Kele Z, Rózsa B J, Kovács E and Mucsi Z 2021 A comprehensive study of the Ca^{2+} ion binding of fluorescently labelled BAPTA analogues *Eur. J. Org. Chem.* **2021** 5248–61

D'Amico-Damião V and Carvalho R F 2018 Cryptochrome-related abiotic stress responses in plants *Front. Plant Sci.* **9** 1897

Das A, Bae K and Park W 2020 Enhancement of upconversion luminescence using photonic nanostructures *Nanophotonics* **9** 1359–71

Dasgupta A, Chen C-H, Lee C H, Gladfelter A S, Dunlap J C and Loros J J 2015 Biological significance of photoreceptor photocycle length: VIVID photocycle governs the dynamic

VIVID-White Collar Complex pool mediating photo-adaptation and response to changes in light intensity *PLoS Genet.* **11** e1005215

Day R N and Davidson M W 2009 The fluorescent protein palette: tools for cellular imaging *Chem. Soc. Rev.* **38** 2887–921

De Grip W J and Ganapathy S 2022 Rhodopsins: an excitingly versatile protein species for research, development and creative engineering *Front. Chem.* **10** 879609

Deguchi T *et al* 2023 Direct observation of motor protein stepping in living cells using MINFLUX *Science* **379** 1010–15

Deisseroth K 2011 Optogenetics *Nat. Meth.* **8** 26–9

Deisseroth K 2015 Optogenetics: 10 years of microbial opsins in neuroscience *Nat. Neurosci.* **18** 1213–25

Deisseroth K 2021 From microbial membrane proteins to the mysteries of emotion *Cell* **184** 5279–85

Deisseroth K, Feng G, Majewska A K, Miesenböck G, Ting A and Schnitzer M J 2006 Next-generation optical technologies for illuminating genetically targeted brain circuits *J. Neurosci.* **26** 10380–6

Deisseroth K and Hegemann P 2017 The form and function of channelrhodopsin *Science* **357** eaan5544

De Mello Donegá C (ed) 2014 *Nanoparticles* (Berlin: Springer)

de Mena L, Rizk P and Rincon-Limas D E 2018 Bringing light to transcription: the optogenetic repertoire *Front. Genet.* **9** 518

Dempsey N M 2021 Magnetism in biology *Handbook of Magnetism and Magnetic Materials* ed J M D Coey and S S P Parkin (Cham: Springer) pp 1633–77

Deppisch P, Helferich-Förster C and Senthilan P R 2022 The gain and loss of cryptochrome/photolyase family members during evolution *Genes* **13** 1613

Deppisch P, Kirsch V, Helfrich-Förster C and Senthilan P R 2023 Contribution of cryptochromes and photolyases for insect life under sunlight *J. Comp. Physiol.* A **209** 373–89

Desmond L J, Phan A N and Gentile P 2021 Critical overview on the green synthesis of carbon quantum dots and their application to cancer therapy *Environ. Sci. Nano* **8** 848–62

De Souza J R, de Moraes M M F, Aoto Y A and Homem-de-Mello P 2020 Can one use the electronic absorption of metalloporphyrins to benchmark electronic structure methods? A case study on the cobalt porphyrin *Phys. Chem. Chem. Phys.* **22** 23886–98

Deubner J, Coulon P and Diester I 2019 Optogenetic approaches to study the mammalian brain *Curr. Opin. Struc. Biol.* **57** 157–63

Dexter D L 1953 A theory of sensitized luminescence in solids *J. Chem. Phys.* **21** 836–50

Dhiman S K, Wu F and Galland P 2023 Effects of weak static magnetic fields on the development of seedlings of *Arabidopsis thaliana Protoplasma* **260** 767–86

Dieke G H 1968 *Spectra and Energy Levels of Rare Earth Ions in Crystals* (New York: Wiley)

Diester I, Kaufman M T, Mogri M, Pashale R, Goo W, Yizhar O, Ramakrishnan C, Deisseroth K and Shenoy K V 2011 An optogenetic toolbox designed for primates *Nat. Neurosci.* **14** 387–97

Dieter A, Keppeler D and Moser T 2020 Towards the optical cochlear implant: optogenetic approaches for hearing restoration *EMBO Mol. Med.* **12** e11618

Dietzek B, Maksimenka R, Siebert T, Birckner E, Kiefer E, Popp J, Hermann G and Schmitt M 2004 Excited-state processes in protochlorophyllide *a*—a femtosecond time-resolved absorption study *Chem. Phys. Lett.* **397** 110–5

Dittrich M, Freddolino P L and Schulten K 2005 When light falls in LOV: a quantum mechanical/molecular mechanical study of photoexcitation in Phot-LOV1 of *Chlamydomonas reinhardtii* *J. Phys. Chem.* B **109** 13006–13

Dobler J, Zinth W, Kaiser W and Oesterhelt D 1988 Excited-state reaction dynamics of bacteriorhodopsin studied by femtosecond spectroscopy *Chem. Phys. Lett.* **144** 215–20

Doi Y, Watanabe J, Nii R, Tsukamoto T, Demura M, Sudo Y and Kikukawa T 2022 Mutations conferring SO_4^{2-} pumping ability on the cyanobacterial anion pump rhodopsin and the resultant unique features of the mutant *Sci. Rep.* **12** 16422

Dombrowski T, Rankovic V and Moser T 2019 Toward the optical cochlear implant *Cold Spring Harb. Perspect. Med.* **9** a033225

Dominguez-Martin M A and Kerfeld C A 2019 Engineering the orange carotenoid protein for applications in synthetic biology *Curr. Opin. Struct. Biol.* **57** 110–7

Dong H, Sun L-D and Yan C-H 2021 Lanthanide-doped upconversion nanoparticles for super-resolution microscopy *Front. Chem.* **8** 619377

Dorn M, Jurk M, Wartenberg A, Hahn A and Schmieder P 2013 LOV takes a pick: thermodynamic and structural aspects of the Flavin-LOV-interaction of the blue-light sensitive photoreceptor YtvA from *Bacillus subtilis* *PLoS One* **8** e81268

Doronina L V, Fedotov I V, Fedotov A B, Anokhin K V and Zheltikov A M 2015 Neurophotonics: optical methods to study and control the brain *Physics-Uspekhi* **58** 345–64

Dragnea V, Gonzalez-Gutierrez G and Bauer C E 2022 Structural analyses of CrtJ and its B_{12}-binding co-regulators SAerR and LAerR from the purple photosynthetic bacterium *Rhodobacter capsulatus Microorganisms* **10** 912

Drepper T, Eggert T, Circolone F, Heck A, Krauß U, Guterl J-K, Wendorff M, Losi A, Gärtner W and Jaeger K-E 2007 Reporter proteins for *in vivo* fluorescence without oxygen *Nat. Biotechnol.* **25** 443–5

Drössler P, Holzer W, Penzkofer A and Hegemann P 2002 pH dependence of the absorption and emission behaviour of riboflavin in aqueous solution *Chem. Phys.* **282** 429–39

Du K, Feng J, Gao X and Zhang H 2022 Nanocomposites based on lanthanide-doped upconversion nanoparticles: diverse designs and applications *Light: Sci. Appl.* **11** 222

Du Y, Alifu N, Wu Z, Chen R, Wang X, Ji G, Li Q, Qian J, Xu B and Song D 2020 Encapsulation-dependent enhanced emission of near-infrared nanoparticles using *in vivo* three-photon fluorescence imaging *Front. Bioeng. Biotechnol.* **8** 1029

Duan L, Hope J, Ong Q, Lou H-Y, Kim N, McCarthy C, Acero V, Lin M Z and Cui B 2017 Understanding CRY2 interactions for optical control of intracellular signaling *Nat. Commun.* **8** 547

Duan C, Liang L, Li L, Zhang R and Xu Z P 2018 Recent progress in upconversion luminescence nanomaterials for biomedical applications *J. Mater. Chem.* B **6** 192–209

Ducrot A, Tron A, Bofinger R, Beguer I S, Pozzo J-L and McClenaghan 2019 Photoreversible stretching of a BAPTA chelator marshalling Ca^{2+}-binding in aqueous media *Beilstein J. Org. Chem.* **15** 2801–11

Dugave C (ed) 2006 *cis-trans Isomerization in Biochemistry* (Weinheim: Wiley)

Duncan L and Deisseroth K 2023 Are novel treatments for brain disorder in plain sight? *Neuropsychopharmacology* **49** 276–81

Eberhardt P, Slavov C, Sörmann J, Bamann C, Braun M and Wachtveitl J 2021 Temperature dependence of the *Krokinobacter* rhodopsin 2 kinetics *Biophys. J.* **120** 568–75

Eckhardt U, Grimm B and Hörtensteiner S 2004 Recent advances in chlorophyll biosynthesis and breakdown in higher plants *Plant Mol. Biol.* **56** 1–14

Ehmann N and Pauls D 2020 Optogenetics: illuminating neuronal circuits of memory formation *Neurogenet* **34** 47–54

Ehrenberg D, Varma N, Deupu X, Koyanagi M, Terakita A, Schertler G F X, Heberle J and Lesca E 2019 The two-photon reversible reaction of the bistable jumping spider rhodopsin-1 *Biophys. J.* **116** 1248–58

Einwich A, Dedek K, Seth P K, Laubinger S and Mouritsen H 2020 A novel isoform of cryptochrome 4 (Cry4b) is expressed in the retina of a night-migratory songbird *Sci. Rep.* **10** 15794

Einwich A, Seth P K, Bartölke R, Bolte P, Feederle R, Dedek K and Mouritsen H 2022 Localisation of cryptochrome 2 in the avian retina *J. Comp. Physiol.* A **208** 69–81

Emiliani V *et al* 2022 Optogenetics for light control of biological systems *Nat. Rev. Methods Primers* **2** 55

Endres S *et al* 2015 Structure and function of a short LOV protein from the marine phototrophic bacterium *Dinoroseobacter shibae BMC Microbiol.* **15** 30

Endres S *et al* 2018 An optogenetic toolbox of LOV-based photosensitizers for light driven killing of bacteria *Sci. Rep.* **8** 15021

Engelhard C, Chizhov I, Siebert F and Engelhard M 2018 Microbial halorhodopsins: light-driven chloride pumps *Chem. Rev.* **118** 10629–45

Entcheva E and Kay M W 2021 Cardiac optogenetics: a decade of enlightenment *Nat. Rev. Cardiol.* **18** 349–67

Ernst O P, Lodowski D T, Elstner M, Hegemann P, Brown L S and Kandori H 2014 Microbial and animal rhodopsins: structures, functions, and molecular mechanisms *Chem. Rev.* **114** 126–63

Eschenmoser A 1970 Roads to corrins *Quart. Rev. Chem. Soc.* **24** 366–415

Espagne A, Paik D H, Changenet-Barret P, Martin M M and Zewail A H 2006 Ultrafast photoisomerization of photoactive yellow protein chromophore analogues in solution: influence of the protonation state *ChemPhysChem.* **7** 1717–26

Espagne A, Changenet-Barret P, Baudin J-B, Plaza P and Martin M M 2007a Photoinduced charge shift as the driving force for the excited-state relaxation of analoges of the photoactive yellow protein chromophore in solution *J. Photochem. Photobiol. A: Chem.* **185** 245–52

Espagne A, Paik D H, Changenet-Barret P, Plaza P, Martin M M and Zewail A H 2007b Ultrafast light-induced response of photoactive yellow protein chromophore analogues *Photochem. Photobiol. Sci.* **6** 780–7

Etter G, van der Veldt S, Manseau F, Zarrinkoub I, Trillaud-Doppia E and Williams S 2019 Optogenetic gamma stimulation rescues memory impairments in an Alzheimer's disease mouse model *Nat. Commun.* **10** 5322

Fang M and Bauer C E 2017 The vitamin B_{12}-dependent photoreceptor AerR relieves photo-system gene repression by extending the interaction of CrtJ with photosystem promoters *mBio* **8** e00261–e17

Favory J-J *et al* 2009 Interaction of COP1 and UVR8 regulates UV-B-induced photomorpho-genesis and stress acclimation in *Arabidopsis EMBO* J **28** 591–601

Fakhoury M 2021 Optogenetics: a revolutionary approach for the study of depression *Prog. Neuro-Psychopharmacol. Biol. Psych.* **106** 110094

Fan L Z *et al* 2020 All-optical electrophysiology reveals the role of lateral inhibition in sensory processing on cortial layer 1 *Cell* **180** 521–35

Fantini E, Sulli M, Zhang L, Aprea G, Jiménez-Gómez J M, Bendahmane A, Perrotta G, Guiliano G and Facella P 2019 Pivotal roles of cryptochromes 1a and 2 in tomato development and physiology *Plant Physiol.* **179** 732–48

Fantini E and Facella P 2020 *Cryptochromes: How Blue Light Perception Influences Plant Physiology eLS* (Chichester: Wiley) pp 1–10

Faramarzi S, Feng J and Mertz B 2018 Allosteric effects of the proton donor on the microbial proton pump proteorhodopsin *Biophys. J.* **115** 1240–50

Farnum A and Pelled G 2020 New vision for visual protheses *Front. Neurosci.* **14** 36

Fauser J, Leschinsky N, Szynal B N and Kaginov A V 2022 Engineered allosteric regulation of protein function *J. Mol. Biol.* **434** 167620

Feng R, Li G, Ko C-N, Zhang Z and Wan J-B 2023 Long-lived second near-infrared luminescent probes: an emerging role in time-resolved luminescence bioimaging and biosensing *Small Struct.* **4** 2200131

Ferenczi E A, Tan X and Huang C L-H 2019 Principles of optogenetic methods and their application to cardiac experimental systems *Front. Physiol.* **10** 1096

Fettweiss T, Röllen K, Granzin J, Reiners O, Endres S, Drepper T, Willbold D, Jaeger K-E, Batra-Safferling R and Krauss U 2018 Mechanistic basis of the fast dark recovery of the short LOV protein DsLOV from *Dinoroseobacter shibae Biochemistry* **57** 4833–47

Fenno L, Yizhar O and Deisseroth K 2011 The development and application of optogenetics *Annu. Rev. Neurosci.* **34** 389–412

Fife D J and Moore W M 1979 The reduction and quenching of photo-excited flavins by EDTA *Photochem. Photobiol.* **29** 43–7

Firmino T, Mangaud E, Cailliez F, Devolder A, Mendive-Tapia D, Gatti F, Meier C, Desouter-Lecomte M and de la Lande A 2016 Quantum effects in ultrafast electron transfers within cryptochromes *Phys. Chem. Chem. Phys.* **18** 21442–57

Filiba O, Borin V A and Schapiro I 2022 The involvement of triplet states in the isomerization of retinaloids *Phys. Chem. Chem, Phys.* **24** 26223–31

Fischer A A M, Kramer M M, Radziwill G and Weber W 2022 Shedding light on current trends in molecular optogenetics *Curr. Opin. Chem. Biol.* **70** 102196

Fischer P, Mukherjee S, Peter E, Broser M, Bartl F and Hegemann P 2021 The inner mechanism of rhodopsin guanylyl cyclase during cGMP-formation revealed by real-time FTIR spectroscopy *eLife* **10** e71384

Flores-Uribe J, Hevroni G, Ghai R, Pushkarev A, Inoue K, Kandori H and Béjà O 2019 Heliorhodopsins are absent in diderm (Gram-negative) bacteria: some thoughts and possible implications for activity *Environ. Microbiol. Rep.* **11** 419–24

Foja R, Walter A, Jandl C, Thyrhaug E, Hauer J and Storch G 2022 Reduced molecular flavins as single-electron reductants after photoexcitation *J. Am. Chem. Soc.* **144** 4721–26

Foley L E and Emery P 2020 *Drosophilia* cryptochrome: variations in blue *J. Biol. Rhythms* **35** 16–27

Fothergill S M, Joyce C and Xie F 2018 Metal enhanced fluorescence bioimaging: from ultraviolet towards second near-infrared window *Nanoscale* **10** 20914–29

Fougère M, van der Zouwen C I, Boutin J, Neszvecsko K, Sarret P and Ryczko D 2021 Optogenetic stimulation of glutamatergic neurons in the cuneiform nucleus controls locomotion in a mouse model of Parkinson's disease *Proc. Natl Acad. Sci.* **118** e2110934118

Förster T 1951 *Fluoreszenz organischer Verbindungen* (Göttingen: Vandenhoeck und Ruprecht)

Förster D, Maschio M D, Laurell E and Baier H 2017 An optogenetic toolbox for unbiased discovery of functionally connected cells in neural circuits *Nat. Commun.* **8** 116

Fraikin G Y, Strakhovskaya M G, Belenikina N S and Rubin A B 2016 LOV and BLUF flavoproteins of microorganisms and photosensory actuators in optogenetic systems *Moscow Univ. Biol. Sci. Bull.* **71** 50–7

Fredj A *et al* 2012 The single T65S mutation generates brighter cyan fluorescent proteins with increased photostability and pH insensitivity *PLoS One* **7** e49149

Freedman K A and Becker R S 1986 Comparitive investigation of the photoisomerization of the protonated and unprotonated *n*-butylamine Schiff bases of 9-*cis*-, 11-*cis*-, 13-*cis*-, and all-*trans*-retinals *J. Am. Chem. Soc.* **108** 1245–51

Fritz B J, Kasai S and Martsui K 1987 Photochemical properties of flavin derivatives *Photochem. Photobiol.* **45** 113–7

Fugate R D, Chin C-A and Song P-S 1976 A spectroscopic analysis of vitamin B12 derivatives *Biochim. Biophys. Acta* **421** 1–11

Fujisawa T and Masuda S 2018 Light-induced chromophore and protein responses and mechanical signal transduction of BLUF proteins *Biophys. Rev.* **10** 327–37

Furutani Y and Yang C-S 2023 Ion-transporting mechanism in microbial rhodopsins: mini-review relating to the session 5 at the 19th international conference on retinal proteins *Biophys. Physicobiol.* **20** e201005

Fushimi K and Narikawa R 2019 Cyanobacteriochromes: photoreceptors covering the entire UV-to-visible spectrum *Curr. Opin. Struct. Biol.* **57** 39–46

Fushimi K and Narikawa R 2021 Phytochromes and cyanobacteriochromes: photoreceptor molecules incorporating a linear tetrapyrrole chromophore *Optogenetics: Light-Sensing Proteins and Their Applications in Neuroscience and Beyond* 2nd edn ed H Yawo, H Kandori, A Koizumi and R Kageyama (Singapore: Springer) pp 167–87

Fuß W, Haas Y and Zilberg S 2000 Twin states and conical intersections in linear polyenes *Chem. Phys.* **259** 273–95

Gabruk M and Mysliwa-Kurdziel B 2015 Light-dependent protochlorophyllide oxidoreductase: phylogeny, regulation, and catalytic properties *Biochemistry* **54** 5255–62

Gadella T W J, van Weeren L, Stouthamer J, Hink M A, Wolters A H G, Giepmans B N G, Aumonier S, Dupuy J and Royant A 2023 mScarlet3: a brilliant and fast-maturing red fluorescent protein *Nat. Methods* **20** 541–5

Gallo E 2019 Fluorogen-activating proteins: next-generation fluorescence probes for biological research *Bioconjug. Chem.* **31** 16–27

Ganapathy S, Kratz S, Chen Q, Hellingwerf K J, de Groot H J M, Rothschild K J and de Grip W J 2019 Redshifted and near-infrared active analog pigments based upon archaerhodopsin-3 *Photochem. Photobiol.* **95** 959–68

Garenne D, Haines M C, Romantseva E F, Freemont P, Strychalski E A and Noireaux V 2021 Cell-free gene expression *Nat. Rev. Method Primers* **1** 49

Garita-Hernandez *et al* 2019 Restoration of visual function by transplantation of optogenetically engineered photoreceptors *Nat. Commun.* **10** 4524

Gauvain G *et al* 2021 Optogenetic therapy: high spatiotemporal resolution and pattern discrimination compatible with vision restoration in non-human primates *Commun. Biol.* **4** 125

Gärtner W and Braslavsky S E 2003 The phytochromes: spectroscopy and function *Photoreceptors and Light Signalling* ed A Batschauer (Cambridge: The Royal Society of Chemistry) pp 136–80

Galvan A, Stauffer W R, Acker L, El-Shamayleh Y and Inoue K-I 2017 Nonhuman primate optogenetics: recent advances and future directions *J. Neurosci.* **37** 10894–903

Gao S, Nagpal J, Schneider M W, Kozjak-Pavlovic V, Nagel G and Gottschalk A 2015 Optogenetic manipulation of cGMP in cells and animals by the tightly light-regulated guanylyl-cyclase opsin CyclOp *Nat. Commun.* **6** 8046

Gao F, Gao T, Zhou K and Zeng W 2016 Small molecule-photoactive yellow protein labeling technology in live cell imaging *Molecules* **21** 1163

Gaponenko S V 2010 *Introduction to Nanophotonics* (Cambridge: Cambridge University)

Gardiner M and Thomson A J 1974 Luminescence properties of some synthetic metallocorrins *J. Chem. Soc. Dalton Trans.* **8** 820–8

Gasser C, Taiber S, Yeh C-M, Wittig C H, Hegemann P and Ryu S 2014 Engineering of a red-light-activated human cAMP/cGMP-specific phosphodiesterase *PNAS* **111** 8803–8

Gegear R J, Foley L E, Casselman A and Reppert S M 2010 Animal cryptochromes mediate magnetoreception by an unconventional photochemical mechanism *Nature* **463** 804–7

Geng Y *et al* 2023 Advances in optogenetics applications for central nervous system injuries *J. Neurotrauma* **40** 1297–316

George D M, Vincent A S and Mackey H R 2020 An overview of anoxygenic phototrophic bacteria and their applications in environmental biotechnology for sustainable resource recovery *Biotechnol. Rep.* **28** e00563

Gerlach T, Schain J, Söltl S, van Schie M M C H, Hilgers F, Bitzenhofer N L, Drepper T and Rother D 2022 Photoregulation of enzyme activity: the inactivation of carboligase with genetically encoded photosensitizer fusion tags *Front. Catal.* **2** 835919

Ghisla S, Massey V, Lhoste J-M and Meyhew S G 1974 Fluorescence and optical characteristics of reduced flavines and flavoproteins *Biochemistry* **13** 589–97

Ghisla S 1980 Fluorescence and optical characteristics of reduced flavins and flavoproteins *Methods Enzymol.* **66** 360–73

Ghosh A P, Toda M J and Kozlowski P M 2022 Photolytic properties of B_{12}-dependent enzymes: a theoretical perspective *Vitam. Horm.* **119** 185–220

Ghosh M, Misra R, Bhattacharya S, Majhi K, Jung K-H and Sheves M 2023 Retinal-carotenoid interactions in a sodium-ion-pumping rhodopsin: implications on oligomerization and thermal stability *J. Phys. Chem.* B **127** 2128–37

Gifford J L, Walsh M P and Vogel H J 2007 Structures and metal-ion-binding properties of the Ca^{2+}-binding helix-loop-helix EF-hand motifs *Biochem. J.* **405** 199–221

Gin E, Diernfellner A C R, Brunner M and Höfer T 2013 The *Neurospora* photoreceptor VIVID exerts negative and positive control on light sensing to achieve adaptation *Mol. Syst. Biol.* **9** 667

Giovannetti R 2012 The use of spectrophotometry UV–Vis for the study of porphyrins *Macro to Nano Spectroscopy* ed J Uddin (Rijeka: InTech) pp 87–108

Glantz S T, Berlew E E, Jaber Z, Schuster B S, Gargner K H and Chow B Y 2018 Directly light-regulated binding of RGS-LOV photoreceptors to anionic membrane phospholipids *Proc. Natl Acad. Sci.* **115** E7720–7

Glantz S T, Carpenter E J, Melkonian M, Gardner K H, Boyden E S, Wong G K-S and Chow B Y 2016 Functional and topological diversity of LOV domain photoreceptors *PNAS* **113** E1442–51

Göbel T, Reisbacher S, Batschauer A and Pokorny R 2017 Flavin adenine dinucleotide and N^5, N^{10}-methenyltetrahydrofolate are the *in planta* cofactors of *Arabidopsis thaliana* crypto-chrome 3 *Photochem. Photobiol.* **93** 355–62

Goett-Zink L and Kottke T 2021 Plant cryptochromes illuminated: a spectroscopic perspective on the mechanism *Front. Chem.* **9** 780199

Goett-Zink L, Toschke A L, Petersen J, Mittag M and Kottke T 2021 C-terminal extension of a plant cryptochrome dissociates from the β-sheet of the flavin-binding domain *J. Phys. Chem. Lett.* **12** 5558–63

Goings J J and Hammes-Schiffer 2019 Early photocycle of Slr1694 blue-light using flavin photoreceptor unraveled through adiabatic excited-state quantum mechanical/molecular mechanical dynamics *J. Am. Chem. Soc.* **141** 20470–9

Goings J J, Li P, Zhu Q and Hammes-Schiffer S 2020 Formation of an unusual glutamine tautomer in a blue light using flavin photocycle characterizes the light-adapted state *Proc. Natl Acad. Sci.* **117** 26626–32

Golic A E, Valle L, Jaime P C, Álvarez C E, Parodi C, Borsarelli C D, Abatedaga I and Mussi M A 2019 BlsA is a low to moderate temperature blue light photoreceptor in the human pathogen *Acinetobacter baummannii Front. Microbiol.* **10** 1925

Gomelsky M and Klug G 2002 A novel FAD-binding domain involved in sensory transduction in microorganisms *Trends Biochem. Sci.* **27** 497–500

Gómez-Consarnau L, Raven J A, Levine N M, Cutter L S, Wang D, Seegers B, Aristegul J, Fuhrman J A, Gasol J M and Sañudo-Wilhelmy S A 2019 Microbial rhodopsins are major contributors to the solar energy captured in the sea *Sci. Adv.* **5** eaaw8855

Gong Y, Huang C, Li J Z, Grewe B F, Zhang Y, Eismann S and Schnitzer M J 2015 High-speed recording of neural spikes in awake mice and flies with a fluorescent voltage sensor *Science* **350** 1361–6

Gong X *et al* 2020 An ultra-sensitive step-function opsin for minimally invasive optogenetic stimulation in mice and macaques *Neuron* **107** 38–51

Gordeliy V (ed) 2022 *Rhodopsin: Methods and Protocols* (New York: Human Press)

Gossauer A and Hirsch W 1974 Totalsynthese des racemischen Phycocyanobilins (Phycobiliverdins) sowie eines "Homophycobiliverdins" *Liebigs Ann. Chem.* **1974** 1496–513

Gourinchas G, Etzl S and Winkler A 2019 Bacteriophytochromes—from informative model systems of phytochrome to powerful tools in cell biology *Curr. Opin. Struct. Biol.* **57** 72–83

Gouterman M 1959 Study of the effects of substitution on the absorption spectra of porphin *J. Chem. Phys.* **30** 1139–61

Gouterman M 1961 Spectra of porphyrins *J. Mol. Spectroscopy* **6** 138–63

Gouterman M and Wagnière G H 1963 Spectra of porphyrins Part II. Four orbital model *J. Mol. Spectroscopy* **11** 108–27

Govorunova E G *et al* 2021 Cation and anion channelrhodopsins: sequence motifs and taxonomic distribution *mBio* **12** e01656

Govorunova E G, Cunha S R, Sineshchekov O A and Spudich J L 2016 Anion channelrhodopsins for inhibitory cardiac optogenetics *Sci. Rep.* **6** 33530

Govorunova E G, Sinesshchekov O A, Janz R, Liu X and Spudich J L 2015 Natural light-gated anion channels: a family of microbial rhodopsins for advanced optpgenetics *Science* **349** 647–50

Govorunova E G, Sineshchekov O A and Spudich J L 2016 Structurally distinct cation channelrhodopsins from cryptophyte algae *Biophys. J.* **110** 2302–4

Govorunova E G, Sineschekov O A, Hemmati R, Janz R, Morelle O, Melkonian M, Wong G K-S and Spudich J L 2018 Extending the time domain of neuronal silencing with cryptophyte anion channelrhodopsins *eNeuro* **5** e0174

Govorunova E G, Sineshchekov O A, Li H, Wang Y, Brown L S and Spudich J L 2020 RubyACRs, nonalgal anion channelrhodopsins with highly red-shifted absorption *Proc. Natl Acad. Sci.* **117** 22833–40

Govorunova E G, Sineshchekov O A and Spudich J L 2022 Emerging diversity of channelrhodopsins and their structure-function relationships *Front. Cell. Neurosci.* **15** 800313

Govorunova E G, Gou Y, Sineshchekov O A, Li H, Lu X, Wang Y, Brown L S, St-Pierre F, Xue M and Spudich J L 2022a Kalium channelrhodopsins are natural light-gated potassium channels that mediate optogenetic inhibition *Nat. Neurosci.* **2022** 1–8

Govorunova E G, Sineshchekov O A, Brown S L, Bondar A-N and Spudich J L 2022b Structural foundations of potassium selectivity in channelrhodopsins *mBio* **13** e0303922

Govoruova E G, Sineshchekov O A and Spudich J L 2023 Potassium-selective channelrhodopsins *Biophys. Physicobiol.* **20** e201011

Gregor C, Sidenstein S C, Andresen M, Sahl S J, Danzl J G and Hell S W 2018 Novel reversibly switchable fluorescent proteins for RESOLFT and STED nanoscopy engineered from the bacterial photoreceptor YtvA *Sci. Rep.* **8** 2724

Grienberger C and Konnerth 2012 Imaging calcium in neurons *Neuron* **73** 862–85

Gruhl T *et al* 2023 Ultrafast structural changes direct the first molecular events of vision *Nature* **615** 939–44

Grynkiewicz G, Poenie M and Tsien R Y 1985 A new generation of Ca^{2+} indicators with greatly improved fluorescence properties *J. Biol. Chem.* **260** 3440–50

Grabolle M, Ziegler J, Merkulov A, Nann T and Resch-Genger U 2008 Stability and fluorescence quantum yield of CdSe-ZnS quantum dots—influence of the thickness of the ZnS shell *Ann. N. Y. Aacd. Sci.* **1130** 235–41

Grabolle M, Spieles M, Lesnyak V, Gaponik N, Eychmüller A and Resch-Genger U 2009 Determination of the fluorescence quantum yield of quantum dots: suitable procedures and achievable uncertainties *Anal. Chem.* **81** 6285–94

Grabowski Z R, Rotkiewicz K and Rettig W 2003 Structural changes accompanying intramolecular electron transfer: focus on twisted intramolecular charge-transfer states and structures *Chem. Rev.* **103** 3899–4032

Gradinaru V, Thompson K R and Deisseroth K 2008 eNpHR: a *Natronomonas* halorhodopsin enhanced for optogenetic applications *Brain Cell Biol.* **36** 129–39

Grodowski M S, Veyret B and Weiss K 1977 Photochemistry of flavins. II. Photophysical properties of alloxazines and isoalloxazines *Photochem. Photobiol.* **26** 341–52

Groenhof G, Lensink M F, Berendsen H J C, Snijders J G and Mark A E 2002 Signal transduction in the photoactive yellow protein. I. Photon absorption and the isomerization of the chromophore *Proteins* **48** 202–11

Gross L A, Baird G S, Hoffman R C, Baldridge K K and Tsien R Y 2000 The structure of the chromophore within DsRed, a red fluorescent protein from coral *PNAS* **97** 11990–5

Grossweiner L I 1984 Photochemistry of proteins: a review *Curr. Eye Res.* **3** 137–44

Grote M, Engelhard M and Hegemann P 2014 Of ion pumps, sensors and channels – perspectives on microbial rhodopsins between science and history *Biochim. Biophys. Acta: Bioenerg.* **1837** 533–45

Grotjohann T, Testa I, Leutenegger M, Bock H, Urban N T, Lavoie-Cardinal F, Willig K I, Eggeling C, Jakobs S and Hell S W 2011 Diffraction-unlimited all-optical imaging and writing with a photochromic GFP *Nature* **478** 204–8

Gubala V, Giovannini G, Kunc F, Monopoli M P and Moore C J 2020 Dye-doped silica nanoparticles: synthesis, surface chemistry and bioapplications *Cancer Nanotechnol.* **11** 1

Guglielmi G, Falk H J and De Renzis S 2016 Optogenetic control of protein function: from intracellular processes to tissue morphogenesis *Trends Cell Biol.* **26** 864–74

Guntas G, Hallett R A, Zimmerman S P, Williams T, Yumerefendi H, Bear J E and Kuhlman B 2015 Engineering an improved light-induced dimer (iLID) for controlling the loalization and activity of signaling proteins *PNAS* **112** 112–7

Günther A, Einwich A, Sjulstok E, Feederle R, Bolte P, Koch K-W, Solov'yov I A and Mouritsen H 2017 Double-cone localization and seasonal expression pattern suggest a role in magneto-reception for European robin cryptochrome 4 *Curr. Biol.* **28** 211–23

Guo J *et al* 2019 Primary cilia signaling promotes axonal tract development and is disrupted in Joubert Syndrome-related disorder modes *Develop. Cell* **51** 759–74

Guo J, Fan J, Liu X, Zhao Z and Tang B Z 2020 Photomechanical luminescence from through-space conjugated AIEgens *Angew. Chem. Int. Ed.* **59** 8828–32

Gupta R, Darwish G H and Algar W R 2022 Complex photobleaching behavior of semi-conducting polymer dots *J. Phys. Chem.* C **126** 20960–74

Gurchiek J K, Bao H, Domínguez-Martin A, McGovern S E, Marquardt C E, Roscioli J D, Ghosh S, Kerfeld C A and Beck W F 2018 Fluorescence and excited-state conformational dynamics of the orange carotenoid protein *J. Phys. Chem.* B **122** 1792–800

Guru A, Post R J, Ho Y-Y and Warden M R 2015 Making sense of optogenetics *Int. J. Neuropsychopharmacol.* **18** pyv079

Gushchin I and Gordeliy V 2018 Microbial rhodopsins *Membrane Protein Complexes: Structure and Function* ed J R Harris and E J Boekema (Singapore: Springer Nature) pp 19–56

Gutruf P and Rogers J A 2018 Implantable, wireless device platforms for neuroscience research *Curr. Opin. Neurobiol.* **50** 1–8

Haake S, Schenkl S, Vinzani S and Chergui M 2002 Femtosecond and picosecond fluorescence of native bacteriorhodopsin and a nonisomerizing analog *Biopolymers* **67** 306–9

Haase M and Schäfer H 2011 Upconverting nanoparticles *Angew. Chem. Int. Ed.* **50** 5808–29

Hagio H *et al* 2023 Optogenetic manipulation of Gq- and Gi/o-coupled receptor signaling in neurons and heart muscle cells *eLife* **12** e83974

Hally C, Delcanale P, Nonell S, Viappiani C and Appruzzetti S 2020 Photosensitizing proteins for antibacterial photodynamic inactivation *Translational Biophotonics* **2** e201900031

Hamm P, Zurek M, Röschinger T, Patzelt H, Oesterhelt D and Zinth W 1996 Femtosecond spectroscopy of the photoisomerisation of the protonated Schiff base of all-*trans* retinal *Chem. Phys. Lett.* **263** 613–21

Hammad M, Albaqami M, Pooam M, Kernevez E, Witczak J, Ritz T, Martino C and Ahmad M 2020 Cryptochrome mediated magnetic sensitivity in *Arabidopsis* occurs independently of light-induced electron transfer to the flavin *Photochem. Photobiol. Sci.* **19** 341–52

Han X 2012 *In vivo* application of optogenetics for neural circuit analysis *ACS Chem. Neurosci.* **3** 577–84

Hansen J N *et al* 2020 Nanobody-directed targeting of optogenetic tools to study signaling in the primary cilium *eLife* **9** e57907

Hanson L K 1991 Molecular orbital theory of monomer pigments *Chlorophylls* ed H Scheer (Boca Raton: CRC Press) pp 993–1014

Hao Y, Yin Q, Zhang Y, Xu M and Chen S 2019 Recent progress in the development of fluorescent probes for thiophenol *Molecules* **24** 3716

Harada K, Ito M, Wang X, Tanaka M, Wongso D, Konno A, Hirai H, Hirase H, Tsuboi T and Kitaguchi T 2017 Red fluorescent protein-based cAMP indicator applicable to optogenetics and *in vivo* imaging *Sci. Rep.* **7** 7351

Harbison G S, Smith S O, Pardoen J A, Winkel C, Lugtenburg J, Herzfeld J, Mathies R and Griffin R G 1984 Dark-adapted bacteriorhodopsin contains 13-*cis*,15-*syn* and all-*trans*,15-*anti* retinal Schiff bases *PNAS* **81** 1706–9

Harris A, Saita M, Resler T, Hughes-Visentin A, Maia R, Pranga-Sellnau F, Bondar A-N, Heberle J and Brown L S 2018 Molecular details of the unique mechanism of chloride transport by a cyanobacterial rhodopsin *Phys. Chem. Chem. Phys.* **20** 3184–99

Harris A R and Gilbert F 2022 Restoring vision using optogenetics without being blind to the risks *Graefe's Arch. Clin. Exp. Ophthalmol.* **260** 41–5

Hart J E and Gardner K H 2021 Lighting the way: recent insights into the structure and regulation of phototropin blue light receptors *J. Biol. Chem.* **296** 100594

Harterink M, van Bergeijk P, Allier C, de Haan B, van den Heuvel S, Hoogenraad C C and Kapitein L C 2016 Light-controlled intracellular transport in *Caenorhabditis elegans Curr. Biol.* **26** R153–4

Hartmann D, Smith J M, Mazzotti G, Chowdhry R and Booth M J 2020 Controlling gene expression with light: a multidisciplinary endeavour *Biochem. Soc. Trans.* **48** 1645–59

Hasegawa K, Masuda S and Ono T-a 2005 Spectroscopic analysis of the dark relaxation process of a photocycle in a sensor of blue light using FAD (BLUF) protein Slr1694 of the cyanobacterium *Synechocystis* sp. PCC6803 *Plant Cell Physiol.* **46** 136–46

Hasegawa K, Masuda S and Ono T-a 2006 Light induced structural changes of a full-length protein and its BLUF domain in YcgF(Blrp), a blue-light sensing protein that uses FAD (BLUF) *Biochemistry* **45** 3785–93

Hasegawa M, Hosaka T, Kojima K, Nishimura Y, Nakajima Y, Kimura-Someya T, Shirouzu M, Sudo Y and Yoshizawa S 2020 A unique clade of light-driven proton-pumping rhodopsins evolved in the cyanobacterial lineage *Sci. Rep.* **10** 16752

Hasemi T, Kikukawa T, Kamo N and Demura M 2016 Characterization of a cyanobacterial chloride-pumping rhodopsin and its conversion into a proton pump *J. Biol. Chem.* **291** 355–62

Hasemi T, Kikukawa T, Watanabe Y, Aizawa T, Miyauchi S, Kamo N and Demura M 2019 Photochemical study of a cyanobacterial chloride-ion pumping rhodopsin *Biochim. Biophys. Acta, Bioenerg.* **1860** 136–46

Hashimoto H, Uragami C and Cogdell R J 2016 Carotenoids and photosynthesis *Carotenoids in Nature* ed C Stange (Switzerland: Springer International Publishing) pp 111–39

Hashimoto M, Katayama K, Furutani Y and Kandori H 2020 Zinc binding to heliorhodopsin *J. Phys. Chem. Lett.* **11** 8604–9

Hashimoto M, Miyagawa K, Singh M, Katayama K, Shoji M, Furutani Y, Shigeta Y and Kndori H 2023 Specific zinc binding to heliorhodopsin *Phys. Chem. Chem. Phys.* **25** 3535–43

Hassanzadeh B *et al* 2021 Microbial rhodopsins are increasingly favored over chlorophyll in high nutrient low chlorophyll waters *Environ. Microbiol. Rep.* **13** 401–6

Hastings J W 1983 Biological diversity, chemical mechanisms, and the evolutionary origins of bioluminescent systems *J. Mol. Evol.* **19** 309–21

Häuser M 2021 Optogenetics – the might of light *N. Engl. J. Med.* **385** 1623–6

Haug M F, Gesemann M, Lazović V and Neuhauss C F 2015 Eumetazoan cryptochrome phylogeny and evolution *Genome Biol. Evol.* **7** 601–19

Haugland R P 2002 *Handbook of Fluorescent Probes and Research Products* 9th edn (Eugene: Molecular Probes, Inc.)

Hayes S, Sharma A, Fraser D P, Trevisan M, Cragg-Barber C K, Tavridou E, Fankhauser C, Jenkins G I and Franklin K A 2017 UV-B perceived by the UVR8 photoreceptor inhibits plant thermomorphogenesis *Curr. Biol.* **27** 120–7

He L *et al* 2015 Near-infrared photoactivatable control of Ca^{2+} signaling and optogenetic immunomodulation *eLIFE* **4** e10024

Hegemann P and Sigrist S (ed) 2013 *Optogenetics* (Berlin: Walter de Gruyter)

Heijde M and Ulm R 2012 UV-B photoreceptor-mediated signaling in plants *Trends Plant Sci.* **17** 230–7

Heilmann M, Christie J M, Kennis J T M, Jenkins G I and Mathes T 2015 Photoinduced transformation of UVR8 monitored by vibrational and fluorescence spectroscopy *Photochem. Photobiol. Sci.* **14** 252–7

Heim N and Griesbeck O 2004 Genetically encoded indicators of cellular calcium dynamics based on troponin C and green fluorescent protein *J. Biol. Chem.* **279** 14280–6

Heizmann C, Hemmerich P, Mengel R and Pfleiderer W 1973 Studien in der Flavin-Reihe. XIX. Mitteilung. Anomale Reduktion des Flavinkerns: Pteridine aus (Iso)alloxazinen *Helv. Chim. Acta* **56** 1908–20

Helassa N, Podor B, Fine A and Török K 2016 Design and mechanistic insight into ultrafast calcium indicators for monitoring intracellular calcium dynamics *Sci. Rep.* **6** 38276

Hellingwerf K J, Hendriks J, van der Horst M, Haker A, Crielaard W and Gensch T 2003 The family of photoactive yellow proteins, the xanthopsins: from structure and mechanism of photoactivation to biological function *Photoreceptors and Light Signaling* ed A Batschauer (Cambridge: The Royal Society of Chemistry) pp 228–71

Henderson J N, Gepshtein R, Heenan J R, Kallio K, Huppert D and Remington S J 2009 Structure and mechanism of the photoactivatable green fluorescent protein *J. Am. Chem. Soc.* **131** 4176–7

Hendler-Neumark A and Bisker G 2019 Fluorescent single-walled carbon nanotubes for protein detection *Sensors* **19** 5403

Henry J T and Crosson S 2011 Ligand-binding PAS domains in a genomic, cellular, and structural context *Annu. Rev. Miocrobiol.* **65** 261–86

Henss T, Schneider M, Vettkotter D, Costa W S, Liewald J F and Gottschalk A 2022 Photoactivated adenylyl cyclases as optogenetic modulators of neuronal activity *cAMP Signaling: Methods and Protocols* ed M Zaccolo (Springer Nature) pp 61–76

Hernández-Candia C N, Casas-Flores S and Gutiérrez-Medina B 2018 Light induces oxidative damage and protein stability in the fungal photoreceptor Vivid *PLOS ONE* **13** (7) e0201028

Herrou J and Crosson S 2011 Function, structure and mechanism of bacterial photosensory LOV proteins *Nat. Rev. Microbiol.* **9** 713–23

Heyes D J, Hardman S J O, Hedison T M, Hoeven R, Greetham Q M, Towrie M and Scrutton N S 2015 Excited-state charge separation in the photochemical mechanism of the light-driven enzyme protochlorophyllide oxidoreductase *Angew. Chem.* **127** 1532–5

Heyes D J, Zhang S, Taylor A, Johannissen L O, Hardman S J O, Hay S and Scrutton N S 2021 Photocatalysis as the 'master switch' of photomorphogenesis in early plant development *Nat. Plants* **7** 268–76

Hildebrandt N, Spillmann C M, Algar R, Pons T, Stewart M H, Oh E, Susumu K, Díaz S A, Delehanty J B and Medintz I L 2017 Energy transfer with semiconductor quantum dot bioconjugates: a versatile platform for biosensing, energy harvesting, and other developing applications *Chem. Rev.* **117** 536–711

Hilgers F, Bitzenhofer N L, Ackermann Y, Burmeister A, Grünberger A, Jaeger K-H and Drepper T 2019 Genetically encoded photosensitizers as light-triggered antimicrobial agents *Int. J. Mol. Sci.* **20** 4608

Himmelstoß S F and Hirsch T 2019 A critical comparison of lanthanide based upconversion nanoparticles to fluorescent proteins, semiconductor quantum dots, and carbon dots for use in optical sensing and imaging *Methods. Appl. Fluoresc.* **7** 022002

Hirano M, Takebe M, Ishido T, Ide T and Matsunaga S 2019 The C-terminal region affects the activity of photoactivated adenylyl cyclase from *Oscillatoria acuminata Sci. Rep.* **9** 20262

Hirai T and Subramaniam S 2009 Protein conformational changes in the bacteriorhodopsin photocycle: comparison of findings from electron and X-ray crystallographic analyses *PLoS One* **4** e5769

Hires S A, Tian L and Looger L 2008 Reporting neural activity with genetically encoded calcium indicators *Brain Cell Biol.* **36** 69–86

Hitomi K, Okamoto K, Daiyasu H, Miyashita H, Iwai S, Toh H, Ishiura M and Todo T 2000 Bacterial cryptochrome and photolyase: characterization of two photolyase-like genes of *Synechocystis* sp *Nucl. Acids Res.* **12** 2353–62

Hochbaum D R *et al* 2014 All-optical electrophysiology in mammalian neurons using engineered microbial rhodopsins *Nat. Meth* **11** 825–33

Hoff W D, Jung K-H and Suduch J L 1997 Molecular mechanism of photosignaling by archaeal sensory rhodopsins *Annu. Rev. Biophys. Biomol. Struct.* **26** 223–58

Hofer A M and Lefkimmiatis K 2007 Extracellular calcium and cAMP: second messengers as 'third messengers'? *Physiology* **22** 320–7

Hoffmann M D, Bubeck F, Eils R and Niopek D 2018 Controlling cells with light and LOV *Adv. Biosys.* **2** 1800098

Hofmann K P and Lamb T D 2023 Rhodopsin, light sensor of vison *Prog. Retinal Eye Res.* **93** 101116

Holzer W, Penzkofer A, Fuhrmann M and Hegemann P 2002 Spectroscopic characterization of flavin mononucleotide bound to the LOV1 domain of phot1 from *Chlamydomonas reinhardtii Photochem. Photobiol.* **75** 479–87

Holzer W, Penzkofer A, Susdorf T, Álvarez M, Islam S D M and Hegemann P 2004 Absorption and emission spectroscopic characterization of the LOV2-domain of phot from *Chlamydomonas reinhardtii* fused to a maltose binding protein *Chem. Phys.* **302** 105–18

Holzer W, Penzkofer A and Hegemann P 2005a Photophysical and photochemical excitation and relaxation dynamics of LOV domains of phot from *Chlamydomonas reinhardtii J. Lumin.* **112** 444–8

Holzer W, Shirdel J, Zirak P, Penzkofer A, Hegemann P, Deutzmann R and Hochmuth E 2005b Photo-induced degradation of some flavins in aqueous solution *Chem. Phys.* **308** 69–78

Holzer W, Penzkofer A, Gong S-H, Bradley D D C, Long X, Blau W J and Davey A P 1998 Excitation intensity dependent fluorescence behaviour of some luminescent polymers *Polymer* **39** 3651–6

Holzer W, Gratz H, Schmitt T, Penzkofer A, Costela A, Garcia-Moreno I, Sastre R and Duarte F J 2000 Photo-physical characterization of rhodamine 6G in a 2-hydroxyethyl-methacrylate methyl-methacrylate copolymer *Chem. Phys.* **256** 125–36

Holzer W, Penzkofer A, Tillmann H, Bader C and Hörhold H-H 2004 Spectroscopic and travelling-wave lasing characterisation of Gilch-type and Horner-Type MEH-PPV *Synth. Met.* **140** 155–70

Holzwarth A R, Lehner H, Braslavsky S E and Schaffner K 1978 The fluorescence of biliverdin dimethyl ester *Liebigs Ann. Chem.* **1978** 2002–17

Hontani Y, Marazzi M, Stehfest K, Mathes T, van Stokkum I H M, Elstner M, Hegemann P and Kennis J T M 2017 Reaction dynamics of the chimeric channelrhodopsin C1C2 *Sci. Rep.* **7** 7217

Hontani Y, Inoue K, Kloz M, Kato Y, Kandori H and Kennis J T M 2016 The photochemistry of sodium ion pump rhodopsin observed by watermarked femto- to submillisecond stimulated Raman spectroscopy *Phys. Chem. Chem. Phys.* **18** 24729–36

Hontani Y, Ganapathy S, Frehan S, Kloz M, de Grip W J and Kennis J T M 2019 Photoreaction dynamics of red-shifting retinal analogues reconstituted in proteorhodopsin *J. Phys. Chem.* B **123** 4242–50

Hontani Y, Mehlhorn J, Domratcheva T, Beck S, Kloz M, Hegemann P, Mathes T and Kennis J T M 2023 Spectroscopic and computational observation of glutamine tautomerization in the blue light sensing using flavin domain photoreaction *J. Am. Chem. Soc.* **145** 1040–52

Hopfield J 1974 Electron transfer between biological molecules by thermally activated tunneling *PNAS* **71** 3640–44

Hore P J and Mouritsen H 2016 The radical-pair mechanism of magnetoreception *Annu. Rev. Biophys.* **45** 299–34

Hori Y, Ueno H, Mizukami S and Kikuchi K 2009 Photoactive yellow protein labeling system with turn-on fluorescence intensity *J. Am. Chem. Soc.* **131** 16610–1

Hori Y, Nakaki K, Sato M, Mizukami S and Kikuchi K 2012 Development of protein-labeling probes with a redesigned fluorogenic switch based on intramolecular association for no-wash live-cell imaging *Angew. Chem. Int. Ed.* **51** 5611–4

Hori Y, Norinobu T, Sato M, Arita K, Shiirakawa M and Kikuchi K 2013 Development of fluorogenic probes for quick no-wash live-cell imaging of intracellular proteins *J. Am. Chem. Soc.* **135** 12360–5

Hori Y, Hirayama S, Sato M and Kikuchi K 2015 Redesign of a fluorogenic labeling system to improve surface charge, brightness, and binding kinetics for imaging the functional localization of bromodomains *Angew. Chem. Int. Ed.* **54** 14368–71

Hoshino H, Miyake K and Narikawa R 2022 Cyanobacteria photoreceptors and their applications *Cyanobacterial Physiology. From Fundamentals to Biotechnology* ed H Kageyama and W Sirisattha (New York: Academic) pp 201–10

Hososhima S *et al* 2022 Proton-transporting heliorhodopsins from marine giant viruses *eLife* **11** e78416

Hososhima S, Yuasa H, Ishizuka T, Hoque M R, Yamashita T, Yamanaka A, Sugano E, Tomita H and Yawo H 2015 Near-infrared (NIR) up-conversion optogenetics *Sci. Rep.* **5** 16533

Hsueh B *et al* 2023 Cardiogenic control of affective behavioural state *Nature* **615** 292–9

Hu W, Li Q, Li B, Ma K, Zhang C and Fu X 2020 Optogenetics sheds new light on tissue engineering and regenerative medicine *Biomaterials* **227** 119546

Huang X, Yang P, Ouyang X, Chen L and Deng X W 2014 Photoactivated UVR8-COP1 module determines photomorphogenic UV-B signaling output in *Arabidopsis PLOS Genetics* **10** e1004218

Huang K, Dou Q and Loh X J 2016 Nanomaterial mediated optogenetics: opportunities and challenges *RSC Adv.* **6** 60896–906

Hubbard R and Wald G 1952 *Cis-trans* isomers of vitamin A and retinene in the rhodopsin system *J. Gen. Physiol.* **36** 269–300

Hudson B and Kohler B 1974 Linear polyene electronic structure and spectroscopy *Annu. Rev. Phys. Chem.* **25** 437–60

Hudson B, Kohler B and Schulten K 1982 Linear polyene electronic structure and potential surfaces *Excited States* ed E C Lim (New York: Academic Press) Vol. 6 pp 1–95

Hunt S M, Tompson S, Elvin M and Heintzen C 2010 VIVID interacts with the WHITE COLLAR complex and FREQUENCY-interacting RNA helicase to alter light and clock responses in Neurospora *PNAS* **107** 16709–714

Hussey K A, Hadyniak S E and Johnston R J 2022 Patterning and development of photoreceptors in the human retina *Front. Cell Dev. Biol.* **10** 878350

Igarashi N, Onoue S and Tsuda Y 2007 Photoreactivity of amino acids: tryptophan-induced photochemical events via reactive oxygen species generation *Anal. Sci.* **23** 943–8

Ihee H, Rajagopal S, Srajer V, Pahl R, Anderson S, Schmidt M, Schotte F, Anfinrud P A, Wulff M and Moffat K 2005 Visualizing reaction pathways in photoactive yellow protein from nanoseconds to seconds *PNAS* **102** 7145–50

Ikeuchi M and Ishizuka T 2008 Cyanobacteriochromes: a new superfamily of tetrapyrrole-binding photoreceptors in cyanobacteria *Photochem. Photobiol. Sci.* **7** 1159–67

Ikuta T *et al* 2020 Structural insights into the mechanism of rhodopsin phosphodiesterase *Nat. Commun.* **11** 5605

Ilango A and Lobo M K (ed) 2015 *Neural Circuits Underlying Emotion and Motivation: Insights from Optogenetics and Pharmacogenetics* (Lausanne: Frontiers in Behavioral Neuroscience)

Imamoto Y and Kataoka M 2007 Structure and photoreaction of photoactive yellow protein, a structural prototype of the PAS domain superfamily *Photochem. Photobiol.* **83** 40–9

Imasheva E S, Shimono K, Balashov S P, Wang J M, Zadok U, Sheves M, Kamo N and Lanyi J K 2005 Formation of a long-lived photoproduct with a deprotonated Schiff base in proteorhodopsin, and its enhancement by mutation of Asp227 *Biochemistry* **44** 10828–38

Indhu A R, Keerthana L and Dharmalingam G 2023 Plasmonic nanotechnology for photothermal applications – an evaluation *Beilstein J. Nanotechnol.* **14** 380–419

Ingles-Prieto A *et al* 2021 Optogenetic delivery of trophic signals in a genetic model of Parkinson's disease *PLoS Genet.* **17** e1009479

Inoue K, Ito S, Kato Y, Nomura Y, Shibata M, Uchihashi T, Satoshi S and Kandori H 2016 A natural light-driven inward proton pump *Nat. Commun.* **7** 13415

Inoue K, Tahara S, Kato Y, Takeuchi S, Tahara T and Kandori H 2018 Spectroscopic study of proton-transfer mechanism of inward proton-pump rhodopsin, *Parvularcula oceani* xenorhodopsin *J. Phys. Chem. B* **122** 6453–61

Inoue S, Yoshizawa S, Nakajiama Y, Kojima K, Tsukamoto T, Kikukawa T and Sudoh Y 2018 Spectroscopic characteristics of *Rubricoccus marinus* xenorhodopsin (*Rm*XeR) and a putative model for its inward H$^+$ transport mechanism *Phys. Chem. Chem. Phys.* **20** 3172–83

Inoue K *et al* 2020 Schizorhodopsins: A family of rhodopsins from Asgard archaea that function as light-driven inward H$^+$ pumps *Sci. Adv.* **6** eaaz2441

Inoue K 2021 Diversity, mechanism and optogenetic application of light-driven ion pump rhodopsins *Optogenetics: Light-Sensing Proteins and Their Applications in Neuroscience and Beyond* 2nd edn ed H Yawo, H Kandori, A Koizumi and R Kageyama (Singapore: Springer) pp 89–126

Inouye S and Tsuji F I 1994 *Aequorea* green fluorescent protein. Expression of the gene and fluorescence characteristics of the recombinant protein *FEBS Lett.* **341** 277–80

Insińska-Rak M, Sikorski M and Wolnicka-Glubisz A 2023 Riboflavin and its derivatives as potential photosensitizers in the treatment of skin cancers *Cells* **12** 2304

Iseki M, Matsunaga A, Ohno K, Shiga K and Yoshida K 2002 A blue-light-activated adenylyl cyclase mediates photoavoidance in *Euglena gracilis Nature* **415** 1047–51

Iseki M and Park S-Y 2021 Photoactivated adenylyl cyclases: fundamental properties and applications *Optogenetics: Light-Sensing Proteins and Their Applications in Neuroscience and Beyond* 2nd edn ed H Yawo, H Kandori, A Koizumi and R Kageyama (Singapore: Springer) pp 129–39

Ishchenko A S *et al* 2017 New insights on signal propagation by sensory rhodopsin II/transducer complex *Sci. Rep.* **7** 41811

Ishizuka T, Kaduda M, Araki R and Yawo H 2006 Kinetic evaluation of photosensitivity in genetically engineered neurons expressing green algae light-gated channels *Neurosci. Res.* **54** 85–94

Ishizuka T, Shimada T, Okajiama K, Yoshihara S, Ochiai Y, Katayama M and Ikeuchi M 2006 Characterization of cyanobacteriochrome TePixJ from a thermophilic cyanobacterium *Thermosynechococcus elongatus* strain BP-1 *Plant Cell Physiol.* **49** 1251–61

Islam S D M, Penzkofer A and Hegemann P 2003a Quantum yield of triplet formation of riboflavin in aqueous solution and of flavin mononucleotide bound to the LOV1 domain of phot1 from *Chlamydomonas reinhardtii Chem. Phys.* **291** 97–114

Islam S D M, Susdorf T, Penzkofer A and Hegemann P 2003b Fluorescence quenching of flavin adenine dinucleotide in aqueous solution by pH dependent isomerisation and photo-induced electron transfer *Chem. Phys.* **295** 137–49

Isomura A 2021 Light control of gene expression dynamics *Optogenetics: Light-Sensing Proteins and Their Applications in Neuroscience and Beyond* 2nd edn ed H Yawo, H Kandori, A Koizumi and R Kageyama (Singapore: Springer) pp 235–46

Iuliano J N *et al* 2018 Variation in LOV photoreceptor activation dynamics probed by time resolved infrared spectroscopy *Biochemistry* **57** 620–30

Iwata T, Nagai T, Ito S, Osoegawa S, Iseki M, Watanabe M, Unno M, Kitagawa S and Kandori H 2018 Hydrogen bonding environments in the photocycle process around the flavin chromophore of the AppA-BLUF domain *J. Am. Chem. Soc.* **140** 11982–91

Iwata T, Yamada D, Mikuni K, Agata K, Hitomi K, Getzoff E D and Kandori H 2020 ATP binding promotes light-induced structural changes to the protein moiety of *Arabidopsis* cryptochrome 1 *Photochem. Photobiol. Sci.* **19** 1326–31

Iwata T and Masuda S 2021 Photoreaction mechanisms of flavoprotein photoreceptors and their applications *Optogenetics: Light-Sensing Proteins and Their Applications in Neuroscience and Beyond* 2nd edn ed H Yawo, H Kandori, A Koizumi and R Kageyama (Singapore: Springer) pp 189–206

Jamieson T, Bakhshi R, Petrova D, Pocock R, Imani M and Seifalian A M 2007 Biological applications of quantum dots *Biomaterials* **28** 4717–32

Jamkhande P G, Ghule N W, Bamer A H and Kalaskar M G 2019 Metal nanoparticles synthesis: an overview on methods of preparation, advantages and disadvantages, and applications *J. Drug Delivery Sci. Technol.* **53** 101174

Janovjak H and Kleinlogel S 2022 Optogenetic neuroregeneration *Neural Regen. Res.* **17** 1468–70

Jarring S and Finn D P 2019 Optogenetics and its application in pain and anxiety research *Neurosci. Biobehav. Rev.* **105** 200–11

Jen M, Lee S and Pang Y 2015 Excited-state dynamics of all-*trans*-retinal investigated by time-resolved electronic and vibrational spectroscopy *Bull. Korean Chem. Soc.* **36** 900–5

Jenal U and Malone J 2006 Mechanisms of cyclic-di-GMP signaling in bacteria *Annu. Rev. Genet.* **40** 385–407

Jenkins G I 2009 Signal transduction in responses to UV-B radiation *Annu. Rev. Plant Biol.* **60** 407–31

Jenkins G I 2014a Structure and function of the UV-B photoreceptor UVR8 *Curr. Opin. Struct. Biol.* **29** 52–7

Jenkins G I 2014b The UV-B photoreceptor UVB8: from structure to physiology *Plant Cell* **26** 21–37

Jensen N A, Jensen I, Kamper M and Jakobs S 2020 Reversibly switchable fluorescent proteins for RESOLFT nanoscopy *Nanoscale Photonic Imaging* ed T Salditt, A Egner and D R Luke (Cham: Springer Nature Switzerland AG) pp 241–60

Jiang T, Song J and Zhang Y 2023 Coelenterazine-type bioluminescence-induced optical probes for sensing and controlling biological processes *Int. J. Mol. Sci.* **24** 5074

Jing C and Cornish V W 2011 Chemical tags for labeling proteins inside living cells *Acc. Chem. Res.* **44** 784–92

Johannissen L O, Taylor A, Hardman S J O, Heyes D J, Scrutton N S and Hay S 2022 How photoactivation triggers protochlorophyllide reduction: computational evidence of a stepwise hydride transfer during chlorophyll biosynthesis *ACS Catal.* **12** 4141–8

Johnson A W, Merwyn L, Shaw N and Smith E L 1963 A partial synthesis of the vitamin B_{12} coenzyme and some of its analogues *J. Chem. Soc.* **0** 4146–56

Jortner J 1976 Temperature dependent activation energy for electron transfer between biological molecules *J. Chem. Phys.* **64** 6840–7

Joshi J, Rubart M and Zhu W 2020 Optogenetics: background, methological advances and potential applications for cardiovascular research and medicine *Front. Bioeng. Biotechnol.* **7** 466

Jost M, Fernández-Zapata J, Polanco M C, Ortiz-Guerrero J M, Chen P Y-T, Kang G, Padmanabhan S, Elías-Arnanz M and Drennan C L 2015 Structural basis for gene regulation by a B_{12}-dependent photoreceptor *Nature* **526** 536–41

Jung G (ed) 2012a *Fluorescent Proteins I. From Understanding to Design* (Berlin: Springer)

Jung G (ed) 2012b *Fluorescent Proteins II. Application of Fluorescent Protein Technology* (Berlin: Springer)

Jung A, Domratcheva T, Tarutina M, Wu Q, Ko W, Shoeman R L, Gomelsky M, Gardner K H and Schlichting I 2005 Structure of a bacterial BLUF photoreceptor: insights into blue light-mediated signal transduction *PNAS* **102** 12350–5

Kaberniuk A A, Shemetov A A and Verkhusha V V 2016 An optogenetic system based on bacterial phytochrome controllable with near-infrared light *Nat. Meth.* **13** 591–7

Kaiserli E and Jenkins G I 2007 UV-B promotes rapid nuclear translocation of the *Arabidopsis* UV-B-specific signaling component UVR8 and activates its function in the nucleus *Plant Cell* **19** 2662–73

Kamikawa Y, Hori Y, Yamashita K, Lin J, Hirayama S, Standley D M and Kikuchi K 2016 Design of a protein tag and fluorogenic probe with modular structure for live-cell imaging of intracellular proteins *Chem. Sci.* **7** 308–14

Kamo N, Hazemoto N, Kobatake Y and Mukohata Y 1985 Light and dark adaption of halorhodopsin *Arch. Biochem. Biophys.* **238** 90–6

Kanazawa T, Ren S, Maekawa M, Hasegawa K, Arisaka F, Hyodo M, Hayakawa Y, Ohta H and Masuda S 2010 Biochemical and physiological characterization of a BLUF protein-EAL protein complex involved in blue light-dependent degradation of cyclic diguanylate in the purple bacterium *Rhodopseudomonas palustris Biochemistry* **49** 10647–55

Kandori H, Shichida Y and Yoshizawa T 2001 Photoisomerization in rhodopsin *Biochem. (Moscow)* **66** 1197–209

Kandori H 2015 Ion-pumping microbial rhodopsins *Front. Mol. Biosci.* **2** 52

Kannan M, Vasan G, Huang C, Haziza S, Li J Z, Inan H, Schnitzer M and Pierebone V A 2018 Fast, *in vivo* voltage imaging using a red fluorescent indicator *Nat. Methods* **15** 1108–16

Kannan M, Vasan G and Pieribone V A 2019 Optimizing strategies for developing genetically encoded voltage indicators *Front. Cell. Neurosci.* **13** 53

Kandori H, Inoue K and Tsunoda S P 2018 Light-driven sodium pumping rhodopsin: a new concept of active transport *Chem. Rev.* **118** 10646–58

Kandori H 2020a Biophysics of rhodopsins and optogenetics *Biophys. Rev.* **12** 355–61

Kandori H 2020b Retinal proteins: Photochemistry and optogenetics *Bull. Chem. Soc. Jpn.* **93** 76–85

Kandori H 2021 History and perspectives of ion-transporting rhodopsins *Optogenetics: Light-Sensing Proteins and Their Applications in Neuroscience and Beyond* 2nd edn ed H Yawo, H Kandori, A Koizumi and R Kageyama (Singapore: Springer) pp 3–19

Kao Y-T, Saxena C, He T-F, Guo L, Wang L, Sancar A and Zhong D 2008 Ultrafast dynamics of flavins in five redox states *J. Am. Chem. Soc.* **130** 13132–9

Karadi K *et al* 2020 Functional dynamics of a single tryptophan residue in a BLUF protein revealed by fluorescence spectroscopy *Sci. Rep.* **10** 2061

Karki N, Vergish S and Zoltowski B D 2021 Cryptochromes: photochemical and structural insight into magnetoreception *Protein Sci.* **30** 1521–34

Kato H E 2021 Structure-function relationship of channelrhodopsins *Optogenetics: Light-Sensing Proteins and Their Applications in Neuroscience and Beyond* 2nd edn ed H Yawo, H Kandori, A Koizumi and R Kageyama (Singapore: Springer) pp 45–63

Kato H E, Inoue K, Kandori H and Nureki O 2016 The light-driven sodium ion pump: a new player in rhodopsin research *Bioessays* **38** 1274–82

Katz J J, Norris J R, Shipman L L, Thurnauer M C and Wasielewski M R 1978 Chlorophyll function in the photosynthetic reaction center *Ann. Rev. Biophys. Bioeng.* **7** 393–434

Kaushik M S, Sharma R, Veetil S K, Srivastava S K and Kateriya S 2019 Modular diversity of the BLUF proteins and their potential for the development of diverse optogenetic tools *Appl. Sci.* **9** 3924

Kaushik M S, Pati S R, Soni S, Mishra A, Sushmita K and Kateriya S 2020 Establishment of optogenetic modulation of cAMP for analyzing growth, biofilm formation, and virulence pathways of bacteria using a light-gated cyclase *Appl. Sci.* **10** 5535

Kavet R and Brain J 2021 Cryptochromes in mammals and birds: clock or magnetic compass? *Physiology* **36** 183–94

Kavkli I H, Ozturk N and Gul S 2019 Chapter One – DNA repair by photolyases *Adv. Protein Chem. Struct. Biol.* **115** 1–19

Kavarnos G J 1993 *Fundamentals of Photoinduced Electron Transfer* (Weinheim: VHC)

Kawanabe A, Furutani Y, Jung K-H and Kandori H 2007 Photochromism of *Anabaena* sensory rhodopsin *J. Am. Chem. Soc.* **129** 8644–9

Kawano F, Suzuki H, Furuya A and Sato M 2015 Engineered pairs of distinct photoswitches for optogenetic control of cellular proteins *Nat. Commun.* **6** 6256

Kawasaki Y, Konno M and Inoue K 2021 Thermostable light-driven inward proton pump rhodopsins *Chem. Phys. Lett.* **779** 138868

Kelly J M and Lazarias J C 1985 Photochemistry of 14-kilo Dalton *Avena* phytochrome under constant illumination *in vitro Biochemistry* **24** 6003–10

Kennedy M J, Hughes R M, Peteya L A, Schwartz J W, Ehlers M D and Tucker C L 2010 Rapid blue-light-mediated induction of protein interactions in living cells *Nat. Meth.* **7** 973–5

Kennis J T M, Crosson S, Gauden M, van Stokkum I H M, Moffat K and van Grondelle R 2003 Primary reactions of the LOV2 domain of phototropin, a plant blue-light photoreceptor *Biochemistry* **42** 3385–92

Kennis J T M, van Stokkum I H M, Crosson S, Gauden M, Moffat K and van Grondelle R 2004 The LOV2 domain of phototropin: A reversible photochromic switch *J. Am. Chem. Soc.* **126** 4512–3

Kerfeld C A, Melnicki M R, Sutter M and Dominguez-Martin M A 2017 Structure, function and evolution of the cyanobacterial orange carotenoid protein and its homologs *New Phytolog.* **215** 937–51

Khanna R, Huq E, Kikis E A, Al-Sady B, Lanzatella C and Quail P H 2004 A novel molecular recognition motif necessary for targeting photoactivated phytochrome signaling to specific basic helix–loop–helix transcription factors *Plant Cell* **16** 3033–44

Khrenova M, Domratcheva T, Grigorenko B and Nemukhin A 2011 Coupling between the BLUF and EAL domains in the blue light-regulated phosphodiesterase BlrP1 *J. Mol. Model* **17** 1579–86

Khurana L, Harczos T, Moser T and Jablonski L 2023 En route to sound coding strategies for optical cochlear implants *iScience* **26** 107725

Kianianmomeni A (ed) 2016 *Optogenetics: Methods and Protocols* (Berlin: Springer)

Kichuk T, Carrasco-López C and Avalos J L 2021 Lights up on organelles: optogenetic tools to control subcellular structure and organization *WIREs Mech. Dis.* **13** e1500

Kiefer H V, Gruber E, Langeland J, Kusochek P A, Bochenkova A V and Andersen L H 2019 Intrinsic photoisomerization dynamics of protonated Schiff-base retinal *Nat. Commun.* **10** 1210

Kiełbus M, Czapiński J, Odrzywolski A, Stasiak G, Szymańska K, Kałafut J, Kos M, Giannopoulos K, Stepulak A and Rivero-Müller A 2018 Optogenetics in cancer drug discovery *Expert Opin. Drug Discovery* **13** 459–72

Kikukawa T 2021 Functional mechanism of Cl⁻-pump rhodopsin and its conversion into H⁺ pump *Optogenetics: Light-Sensing Proteins and Their Applications in Neuroscience and Beyond* 2nd edn ed H Yawo, H Kandori, A Koizumi and R Kageyama (Singapore: Springer) pp 55–71

Kikukawa T, Tamogami J, Shimono K, Demura M, Nara T and Kamo N 2012 Photo-induced proton transfers in microbial rhodopsins *Molecular Photochemistry – Various Aspects* ed S Saha (London: IntechOpen) pp 89–108

Kim T-I *et al* 2013 Injectable, cellular-scale optoelectronics with applications for wireless optogenetics *Science* **340** 211–6

Kim H, Zou T, Modi C, Dörner K, Grunkemeyer T J, Chen L, Fromme R, Matz M V, Ozkan B and Wachter R M 2015 A hinge migration mechanism unlocks the evolution of green-to-red photoconversion in GFP-like proteins *Structure* **23** 34–43

Kim C K, Adhikari A and Deisseroth K 2017 Integration of optogenetics with complementary methodologies in systems neuroscience *Nat. Rev. Neurosci.* **18** 222–35

Kim S, Nakasone Y, Takakado A, Yamazaki Y, Kamikubo H and Terazima M 2021 A unique photochromic UV-A sensor protein, Rc-PYP, interacting with the PYP-binding protein *Phys. Chem. Chem. Phys.* **23** 17813–25

Kim S-H, Chuon K, Cho S-G, Choi A, Meas S, Cho H-S and Jung K-H 2021 Color-tuning of natural variants of heliorhodopsin *Sci. Rep.* **11** 854

Kiontke S, Göbel T, Brych A and Batschauer A 2020 DASH-type cryptochromes – solved and open questions *Biol. Chem.* **401** 1487–93

Kirilovsky D and Kerfeld C A 2013 The orange carotenoid protein: a blue-green light photoactive protein *Photochem. Photobiol. Sci.* **12** 1135–43

Kishi K E *et al* 2022 Structural basis for channel conduction in the pump-like channelrhodopsin ChRmine *Cell* **185** 672–89

Kiskinis E, Kralj J M, Zou P, Weinstein E N, Zhang H, Tsioras K, Wiskow O, Ortega J A, Eggan K and Cohen A E 2018 All-optical electrophysiology for high-throughput functional character-ization of a human iPSC-derived motor neuron model of ALS *Stem Cell Rep.* **10** 1–14

Kita A, Okajima K, Morimoto Y, Ikeuchi M and Miki K 2005 Structure of a cyanobacterial BLUF protein, Tll0078, containing a novel FAD-binding blue light sensor domain *J. Mol. Biol.* **349** 1–9

Klapoetke N C *et al* 2014 Independent optical excitation of distinct neural populations *Nat. Meth.* **11** 338–46

Klavir O, Prigge M, Sarel A, Paz R and Yizhar O 2017 Manipulating fear associations via optogenetic modulation of amygdala inputs to prefrontal cortex *Nat. Neurosci.* **20** 836–44

Kleine T, Lockhart P and Batschauer A 2003 An *Arabidopsis* protein closely related to *Synechocystis* cryptochrome is targeted to organelles *Plant J.* **35** 93–103

Klevit R E, Blumenthal D K, Wemmer D E and Krebs E G 1985 Interaction of calmodulin and a calmodulin-binding peptide from myosin light chain kinase: major spectral changes in both occur as the result of complex formation *Biochemistry* **24** 8152–7

Klewer L and Wu Y-W 2019 Light-induced dimerization approaches to control cellular processes *Chem. Eur. J.* **25** 12452–63

Kliebenstein D J, Lim J E, Landry L G and Last R L 2002 *Arabidopsis* UVR8 regulates ultraviolet-B signal transduction and tolerance and contains sequence similarity to human regulator of chromatin condensation 1 *Plant Physiol.* **130** 234–43

Klimov V V 2003 Discovery of pheophytin function in the photosynthetic energy conversion as the primary electron acceptor of photosystem II *Photosynth. Res.* **76** 247–53

Klioutchnikov A, Wallace D J, Sawinski J, Voit K-M, Groemping Y and Kerr J N D 2023 A three-photon head-mounted microscope for imaging all layers of visual cortex in freely moving mice *Nat. Methods* **20** 610–6

Klose C, Nagy F and Schäfer E 2020 Thermal reversion of plant phytochromes *Mol. Plant* **13** 386–97

Knöpfel T and Boyden E (ed) 2012 *Optogenetics: Tools for Controlling and Monitoring Neural Activity* (Amsterdam: Elsevier)

Knöpfel T and Song C 2019 Optical voltage imaging in neurons: moving from technology development to practical tool *Nat. Rev.* **20** 719–27

Kojima K, Shibukawa A and Sudo Y 2020a The unlimited potential of microbial rhodopsins as optical tools *Biochemistry* **59** 218–29

Kojima K, Kurihara R, Sakamoto M, Takanashi T, Kuramochi H, Zhang X M, Bito H, Tahara T and Sudo Y 2020b Comparative studies of the fluorescence properties of microbial rhodopsins: spontaneous emission versus photointermediate fluorescence *J. Phys. Chem.* B **124** 7361–7

Kojima K and Sudo Y 2023 Convergent evolution of animal and microbial rhodopsins *RSC Adv.* **13** 5367–81

Kolar K, Knobloch C, Stork H, Žnidarič M and Weber W 2018 OptoBase: A web platform for molecular optogenetics *ACS Synth. Biol.* **7** 1825–8

Koldenkova V P and Nagai T 2013 Genetically encoded Ca^{2+} indicators: properties and evaluation *Biochim. Biophys. Acta* **1833** 1787–97

Komatsu N, Terai K, Imanishi A, Kamioka Y, Sumiyama K, Jin T, Okada Y, Nagai T and Matsuda M 2018 A platform of BRET-FRET hybrid biosensors for optogenetics, chemical screening, and *in vivo* imaging *Sci. Rep.* **8** 8984

Konold P E, van Stokkum I H M, Muzzopappa F, Wilson A, Groot M-L, Kirilovsky D and Kennis J T M 2019 Photoactivation mechanism, timing of protein secondary structure dynamics and carotenoid translocation in the orange carotenoid protein *J. Am. Chem. Soc.* **141** 520–30

Konrad K R, Gao S, Zurbrüggen M D and Nagel G 2023 Optogenetic methods in plant biology *Annu. Rev. Plant Biol.* **74** 313–39

Kopka B, Magerl K, Savitsy A, Davaru M D, Röllen K, Bocola M, Dick B, Schwaneberg U, Jaeger K-E and Krauss U 2017 Electron transfer pathways in a light, oxygen, voltage (LOV) protein devoid of the photoactive cysteine *Sci. Rep.* **7** 13346

Kort R, Hoff W D, van West M, Kroon A R, Hoffer S M, Vlieg K H, Crielaard W, van Beeumen J J and Hellingwerf K J 1996 The xanthopsins: a new family of eubacterial blue-light photoreceptors *EMBO J.* **15** 3209–18

Kostyiuk A I, Demidovich A D, Kotova D A, Beiousov V V and Bilan D S 2019 Circularly permuted fluorescent protein-based indicators: history, principals, and classification *Int. J. Mol. Sci.* **10** 4200

Kottke T, Xie A, Larsen D S and Hoff W D 2018 Photoreceptors take charge: emerging principles for light sensing *Annu. Rev. Biophys.* **47** 291–313

Kottke T, Heberle J, Hehn D, Dick B and Hegemann P 2003 Phot-LOV1: photocycle of a blue-light receptor domain from the green alga *Chlamydomonas reindardtii Biophys. J.* **84** 1192–201

Kovalev K *et al* 2019 Structure and mechanisms of sodium-pumping KR2 rhodopsin *Sci. Adv.* **5** eaav2671

Kovalev K *et al* 2020a Molecular mechanism of light-driven sodium pumping *Nat. Commun.* **11** 2137

Kovalev K *et al* 2020b High-resolution structural insights into the heliorhodopsin family *PNAS* **117** 4131–41

Kraft B J, Masuda S, Kikuchi J, Dragnea V, Tollin G, Zaleski J M and Bauer C E 2003 Spectroscopic and mutational analysis of the blue-light photoreceptor AppA: a novel photocycle involving flavin stacking with an aromatic amino acid *Biochemistry* **42** 6726–34

Kralj J M, Hochbaum D R, Douglass A D and Cohen A E 2011a Electrical spiking in *Escherichia coli* probed with a fluorescent voltage-indicating protein *Science* **333** 345–8

Kralj J M, Douglass A D, Hochbaum D R, Maclaurin D and Cohen A E 2011b Optical recording of action potentials in mammalian neurons using a microbial rhodopsin *Nat. Meth.* **9** 90–5

Krause B S, Grimm C, Kaufmann J C D, Schneider F, Sakmar T P, Bartl F J and Hegemann P 2017 Complex photochemistry within the green-absorbing channelrhodopsin ReaChR *Biophys. J.* **112** 1166–75

Kravitz A V, Freeze B S, Parker P R L, Kay K, Thwin M T and Deisseroth K 2010 Regulation of parkinsonian motor behaviours by optogenetic control of basal ganglia circuitry *Nature* **466** 622–6

Kreslavski V D, Los D A, Schmitt F-J, Zharmukhamedov S K, Kuznetsov V V and Allakhverdiev S I 2018 The impact of phytochromes on photosynthetic processes *BBA Bioenergetics* **1859** 400–8

Kritsky M S, Telgina T A, Vechtomova Y L, Kolesnikov M P, Lyudnikova T A and Golup O A 2010 Excited flavin and pterin coenzyme molecules in evolution *Biochemistry (Moscow)* **75** 1200–16

Kropf A and Hubbard B 1970 The photoisomerization of retinal *Photochem. Photobiol.* **12** 249–60

Krueger D, Izquierdo E, Viswanathan R, Hartmann J, Cartes C P and De Renzis S 2019 Principles and applications of optogenetics in developmental biology *Development* **146** dev175067

Kuhne J, Vierock J, Tennigkeit S A, Dreier M-A, Wietek J, Petersen D, Gavriljuk K, El-Mashtoly S F, Hegemann P and Gerwert K 2019 Unifying photocycle model for light adaption and temporal evolution of cation conductance in channelrhodopsin-2 *PNAS* **116** 9380–9

Kumar N, Hori Y and Kikuchi K 2019 Photoactive yellow protein and its chemical probes: an approach to protein labelling in living cells *J. Biochem.* **166** 121–7

Kumar N, Hori Y, Nishiura M and Kikuchi K 2020 Rapid no-wash labeling of PYP-tag proteins with reactive fluorogenic ligands affords stable fluorescent protein conjugates for long-term cell imaging studies *Chem. Sci.* **11** 3694–701

Kumar P, Dua S, Kaur R, Kumar M and Bhatt G 2022 A review on advancements in carbon quantum dots and their application in photovoltaics *RSC Adv.* **12** 4714–59

Kumar A, Burns D C, Al-Abdul-Wahid M S and Woolley G A 2013 A circularly permuted photoactive yellow protein as a scaffold for photoswitch design *Biochem.* **52** 3320–31

Kumauchi M, Hara M T, Stalcup P, Xie A and Hoff W D 2008 Identification of six new photoactive proteins – diversity and structure–function relationships in a bacterial blue-light photoreceptor *Photochem. Photobiol.* **84** 956–69

Kushibiki T, Okawa S, Hirasawa T and Ishihara M 2014 Optogenetics: novel tools for controlling mammalian cell functions with light *Int. J. Photoenergy* **2014** 895039

Kushibiki T 2021 Current topics of optogenetics for medical applications towards therapy *Adv. Exp. Med. Biol.* **1293** 513–21

Kutta R J *et al* 2015a The photochemical mechanism of a B_{12}-dependent photoreceptor protein *Nat. Commun.* **6** 7907

Kutta R J, Magerl K, Kensy U and Dick B 2015b A search for radical intermediates in the photocycle of LOV domains *Photochem. Photobiol. Sci.* **14** 288–99

Kutta R J, Archipowa N, Johannissen L O, Jones A R and Scrutton N S 2017 Vertebrate cryptochromes are vestigial flavoproteins *Sci. Rep.* **7** 44906

Kuznetsov A M and Ulstrup J 1999 *Electron Transfer in Chemistry and Biology: An Introduction to the Theory* (New York: Wiley)

Kyriacou C P and Rosato E 2022 Genetic analysis of cryptochrome in insect magnetosensitivity *Front. Physio.* **13** 928416

Laan W, Bednarz T, Heberle J and Hellingwerf K J 2004 Chromophore composition of a heterologously expressed BLUF domain *Photochem. Photobiol. Sci.* **3** 1011–6

Lacombat F, Espagne A, Dozova N, Plaza P, Müller P, Brettel K, Franz-Badur S and Essen L-O 2019 Ultrafast oxidation of a tyrosine by proton-coupled electron transfer promotes light activation of an animal-like cryptochrome *J. Am. Chem. Soc.* **141** 13394–409

Lahore R G F *et al* 2022 Calcium-permeable channelrhodopsins for the photocontrol of calcium signaling *Nat. Commun.* **13** 7844

Lambeth D O and Palmer G 1973 The kinetics and mechanism of reduction of electron transfer proteins and other compounds of biological interest by dithionite *J. Biol. Chem.* **248** 6095–103

Lamparter T, Xue P, Elkurdi A, Kaeser G, Sauthof L, Scheerer P and Krauß N 2021 Phytochromes in *Agrobacterium fabrum Front. Plant Sci.* **12** 642801

Lanyi J K 1986 Photochromism of halorhodopsin. cis/trans isomerization of the retinal around the 13-14 double bond *J. Biol. Chem.* **261** 14025–30

Lanyi J K 2004 Bacteriorhodopsin *Annu. Rev. Physiol.* **66** 665–88

Lanyi J K 2006 Proton transfers in the bacteriorhodopsin photocycle *Biochim. Biophys. Acta* **1757** 1012–18

Lara-Astiaso M *et al* 2018 Attosecond pump-probe spectroscopy of charge dynamics in tryptophan *J. Phys. Chem. Lett.* **9** 4570–7

Larsen D S, Vengris M, van Stokkum I H M, van der Horst M A, Cordfunke R A, Hellingwerf K J and van Grondelle R 2003 Initial photo-induced dynamics of the photoactive yellow protein chromophore in solution *Chem. Phys. Lett.* **369** 563–9

Larson E J, Friesen L A and Johnson C K 1997 An ultrafast one-photon and two-photon transient absorption study of the solvent-dependent photophysics in all-*trans* retinal *Chem. Phys. Lett.* **265** 161–8

Law Y and Wood J M 1973 The photolysis of 5'-deoxy-adenosylcobalamin under anaerobic conditions *Biochim. Biophys. Acta* **331** 451–54

Lazzari-Dean J R, Gest A M and Miller E W 2019 Optical estimation of absolute membrane potential using fluorescence lifetime imaging *eLife* **8** e44522

Leccese S, Calcinori A, Wilson A, Kirilovsky D, Carbonera D, Onfroy T, Jolivalt C and Mezzetti A 2023 Orange carotenoid protein in mesoporous silica: a new system towards the development of colorimetric and fluorescent sensors for pH and temperature *Micromachines* **14** 1871

Lee C, Lavoie A, Liu J, Chen S X and Liz B-h 2020 Light up the brain: the application of optogenetics in cell-type specific dissection of mouse brain circuits *Front. Neural Circuits* **14** 18

Lee J J, Kim S H, Lee K A, Chuon K, Jung K-H and Kim D 2021 Kinetics of DNA looping by *Anabaena* sensory rhodopsin transducer (ASRT) by using DNA cyclization assay *Sci. Rep.* **11** 23721

Lee S-Y, Cheah J S, Zhao B, Xu C, Roh H, Kim C K, Cho K F, Udeshi N D, Carr S A and Ting A Y 2023 Engineered allostery in light-regulated LOV-Turbo enables precise spatiotemporal control of proximity labeling in living cells *Nat. Meth.* **20** 908–17

Lee J, Ozden I, Song Y-K and Nurmikko A V 2015 Transparent intracortical microprobe array for simultaneous spatiotemporal optical stimulation and multichannel electrical recording *Nat. Meth.* **12** 1157–62

Lee D, Hyun J H, Jung K, Hannan P and Kwon H-B 2017a A calcium- and light-gated switch to induce gene expression in activated neurons *Nat. Biotechnol.* **35** 858–63

Lee B-D *et al* 2017b The F-box protein FKF1 inhibits dimerization of COP1 in the control of photoperiodic flowering *Nat. Commun.* **8** 2259

Legault S, Fraser-Halberg D P, McAnelly R L, Eason M G, Thompson M C and Chica R A 2022 Generation of bright monomeric red fluorescent proteins *via* computational design of enhanced chromophore packing *Chem. Sci.* **13** 1408–18

Legris M, Ince Y Ç and Frankhauser C 2019 Molecular mechanisms underlying phytochrome-controlled morphogenesis in plants *Nat. Commun.* **10** 5219

Lehtinen K, Nokia M S and Takala H 2022 Red light optogenetics in neuroscience *Front. Cell. Neuroscience* **15** 778900

Lei L, Wang Y, Kuzmin A, Hua Y, Zhao J, Xu S and Prasad P N 2022 Next generation lanthanide doped nanoscintillators and photon converters *eLight* **2** 17

Lerner T N, Ye L and Deisseroth K 2016 Communication in neural circuits: tools, opportunities, and challenges *Cell* **164** 1136–50

Leverenz R L *et al* 2015 A 12 Å carotenoid translocation in a photoswitch associated with cyanobacterial photoprotection *Science* **348** 1463–6

Levskaya A, Weiner O D, Lim W A and Voigt C A 2009 Spatiotemporal control of cell signalling using a light-switchable protein interaction *Nature* **461** 997–1001

Li X, Gutierrez D V, Hanson M G, Han J, Mark M D, Chiel H, Hegemann P, Landmesser L T and Herlitze S 2005 Fast noninvasive activation and inhibition of neural and network activity by vertebrate rhodopsin and green algae channelrhodopsin *PNAS* **102** 17816–21

Li J, Li G, Wang H and Deng X W 2011 Phytochrome signaling mechanisms *Arabidopsis Book* **9** e0148

Li A-H and Chen G 2020 Controlling lanthanide-doped upconversion nanoparticles for brighter luminescence *J. Phys. D: Appl. Phys.* **53** 043001

Li H, Wang X, Huang D and Chen G 2019 Recent advances of lanthanide-doped upconversion nanoparticles for biological applications *Nanotechnology* **31** 072001

Li H, An Y, Gao J, Yang M, Luo J, Li X, Lv J, Li X, Yuan Z and Ma H 2022 Recent advances of fluorescent probes for imaging of ferroptosis process *Chemosensors* **10** 233

Li J, Rao J and Pu K 2018 Recent progress on semiconducting polymer nanoparticles for molecular imaging and cancer phototherapy *Biomaterials* **155** 217–35

Li M, Chen T, Gooding J J and Liu J 2019 Review of carbon and graphene quantum dots for sensing *ACS Sens.* **4** 1732–48

Li T, Chen X, Qian Y, Shao J, Li X, Liu S, Zhu L, Zhao Y, Ye H and Yang Y 2021 A synthetic BRET-based optogenetic device for pulsatile transgene expression enabling glucose homeostasis in mice *Nat. Commun.* **12** 615

Li W, Kaminski Schierle G S, Lei B, Liu Y and Kaminski C F 2022 Fluorescent nanoparticles for super-resolution imaging *Chem. Rev.* **122** 12495–543

Li Y-T, Tian Y, Tian H, Tu T, Gou G-Y, Wang Q, Qiao Y-C, Yang Y and Ren T-L 2018 A review on bacteriorhodopsin-based bioelectronic devices *Sensors* **18** 1368

Liang G, Wang H, Shi H, Wang H, Zhu M, Jing A, Li J and Li G 2020 Recent progress in the development of upconversion nanomaterials in bioimaging and disease treatment *J. Nanobiotechnol.* **18** 154

Liang R, Liu F and Martínez T J 2019 Nonadiabatic photodynamics of retinal protonated Schiff base in channelrhodopsin 2 *J. Phys. Chem. Lett.* **10** 2862–8

Liang T, Yang Y and Liu H 2019 Signal transduction mediated by the plant UV-B photoreceptor UVR8 *New Phytol.* **221** 1247–52

Lin C, Top D, Manahan C C, Young M W and Crane B R 2018 Circadian clock activity of cryptochrome relies on tryptophan-mediated photoreduction *PNAS* **115** 3822–7

Lin J Y, Knutsen P M, Muller A, Kleinfeld D and Tsien R Y 2013 ReaChR: a red-shifted variant of channelrhodopsin enables deep transcranial optogenetic excitation *Nat. Neurosci.* **16** 1499–508

Lin X *et al* 2018 Core–shell–shell upconversion nanoparticles with enhanced emission for wireless optogenetic inhibition *Nano Lett.* **18** 948–56

Lindner J U 2006 Class III adenylyl cyclases: molecular mechanisms of catalysis and regulation *Cell. Mol. Life Sci.* **63** 1736–51

Lindner R, Hartmann E, Tarnawski M, Winkler A, Frey D, Reinstein J, Meinhart A and Schlichting I 2017 Photoactivation mechanism of a bacterial light-regulated adenylate cyclase *J. Mol. Biol.* **429** 1336–51

Lindner F and Diepold A 2022 Optogenetics in bacteria – applications and opportunities *FEMS Microbiol. Rev.* **46** 1–17

Lindner M, Gilhooley M J, Hughes S and Hankins M W 2022 Optogenetics for visual restoration: from proof of principle to translational challenges *Prog. Retinal Eye Res.* **91** 101089

Lindsey J 2005 PhotochemCAD spectra by category. Available online: http://omlc.ogi.edu/spectra/PhotochemCAD/html/

Liu X, Yan C-H and Capobianco J A 2015 Photon upconversion nanomaterials *Chem. Soc. Rev.* **44** 1299–301

Liu Z, Wang L and Zhong D 2015 Dynamics and mechanism of DNA repair by photolyase *Phys. Chem. Chem. Phys.* **17** 11933–49

Liu C C, Ahammed G J, Wang G T, Xu C J, Chen K S, Zhou Y H and Yu J Q 2018 Tomato CRY1a plays a critical role in the regulation of phytohormone homeostasis, plant development, and carotenoid metabolism in fruits *Plant Cell Environ.* **41** 354–66

Liu M L, Chen B B, Li C M and Huang C Z 2019 Carbon dots: synthesis, formation mechanism, fluorescence origin and sensing applications *Green Chem.* **21** 449–71

Liu Q, Zhang Z and Zhang W 2021 Optogenetic dissection of neural circuits underlying stress-induced mood disorders *Front. Psychology* **12** 600999

Liu Y and Zhu C 2021 Trajectory surface hopping molecular dynamics simulations for retinal Schiff-base photoisomerization *Phys. Chem. Chem. Phys.* **23** 23861–74

Lodowski P, Jaworska M, Garabato B D and Kozlowski P M 2015 Mechanism of Co-C bond photolysis in methylcobalamin: influence of axial base *J. Phys. Chem.* A **119** 3913–28

Lodowski P, Jaworska M, Kornobis K, Andruniów T and Kozlowski P M 2011 Electronic and structural properties of low-lying excited states of vitamin B_{12} *J. Phys. Chem.* B **115** 13304–19

Logunov S L, Song L and El-Sayed M A 1996 Excited-state dynamics of a protonated retinal Schiff base in solution *J. Phys. Chem.* **100** 18586–91

Lohr C, Beiersdorfer A, Fischer T, Hirnet D, Rotermund N, Sauer J, Schulz K and Gee C E 2021 Using genetically encoded calcium indicators to study astrocyte physiology: a field guide *Front. Cell. Neurosci.* **15** 690147

Lokstein H, Renger G and Götze J P 2021 Photosynthetic light-harvesting (antenna) complexes-structures and functions *Molecules* **26** 3378

Lorenzen C J and Jeffrey S W 1980 Determination of chlorophyll in seawater *UNESCO Technical Papers in Marine Science* **No. 35** 1–20

Losi A, Gärtner W, Raffelberg S, Zanacchi F C, Bianchini P, Diaspro A, Mandalari C, Abbruzzetti S and Viappiani C 2013 A photochromic bacterial photoreceptor with potential for super-resolution microscopy *Photochem. Photobiol. Sci.* **12** 231–5

Losi A, Gardner K H and Möglich A 2018 Blue-light receptors for optogenetics *Chem. Rev.* **118** 10659–709

Losi A 2020 LOV proteins: photobiophysics *Encyclopedia of Biophysics* ed G C K Roberts and A Watts (Berlin: Springer) pp 1312–6

Losi A and Gärtner W 2021 A light life together: photosensing in the plant microbiota *Photochem. Photobiol. Sci.* **20** 451–73

Losi A, Polverini E, Quest B and Gärtner W 2002 First evidence for phototropin-related blue-light receptors in prokaryotes *Biophys. J.* **82** 2627

Losi A 2007 Flavin-based blue-light photosensors: a photobiophysics update *Photochem. Photobiol.* **83** 1283–300

Losi A and Gärtner W 2011 Old chromophores, new photoactivation paradigms, trendy applications: flavins in blue light-sensing photoreceptors *Photochem. Photobiol.* **87** 491–510

Losi A and Gärtner W 2012 The evolution of flavin-binding photoreceptors: An ancient chromophore serving trendy blue-light sensors *Annu. Rev. Plant Biol.* **63** 49–72

Losi A, Gärtner W, Raffelberg S, Zanacchi F C, Bianchini P, Diaspro A, Mandalari C, Abbruzzetti S and Viappiani C 2013 A photochromic bacterial photoreceptor with potential for super-resolution microscopy *Photochem. Photobiol. Sci.* **12** 231–5

Losi A and Gärtner W 2017 Solving blue light riddles: new lessons from flavin-binding LOV photoreceptors *Photochem. Photobiol.* **93** 141–58

Losi A, Gardner K H and Möglich A 2018 Blue-light receptors for optogenetics *Chem. Rev.* **118** 10659–709

Lu L *et al* 2018 Wireless optoelectronic photometers for monitoring neuronal dynamics in the deep brain *PNAS* E1374–83

Lu C, Wu X, Ma H, Wang Q, Wang Y, Luo Y, Li C and Xu H 2019 Optogenetic stimulation enhanced neuronal plasticities in motor recovery after ischemic stroke *Neural Plast.* **2019** 5271573

Lu Q and Pan Z-H 2021 Optogenetic strategies for vision restoration *Optogenetics: Light-Sensing Proteins and Their Applications in Neuroscience and Beyond* 2nd edn ed H Yawo, H Kandori, A Koizumi and R Kageyama (Singapore: Springer) pp 545–55

Lu Y *et al* 2019 Crystal structure of heliorhodopsin 48C12 *Cell Res.* **30** 88–90

Luchkina N V and Bolshakov V Y 2018 Diminishing fear: optogenetic approach toward understanding neural circuits of fear control *Pharmacol. Biochem. Behav.* **174** 64–79

Luck M, Escobar F V, Glass K, Sabotke M-I, Hagedorn R, Corellou F, Siebert F, Hildebrandt P and Hegemann P 2019 Photoreactions of the Histidine kinase rhodopsin Ot-HKR from the marine picoalga *Ostreococcus tauri Biochemistry* **58** 1878–91

Luck M, Mathes T, Bruun S, Fudim R, Hagedorn R, Nguyen T M T, Kateriya S, Kennis J T M, Hildebrandt P and Hegemann P 2012 A photochromic histidine kinase rhodopsin (HKR1) that is bimodally switched by ultraviolet and blue light *J. Biol. Chem.* **287** 40083–90

Lugo K, Miao X, Rieke F and Lin L Y 2012 Remote switching of cellular activity and cell signaling using light in conjunction with quantum dots *Biomed. Opt. Express* **3** 447–54

Lukacs A, Tonge P J and Meech S R 2022 Photophysics of the blue light using flavin domain *Acc. Chem. Res.* **55** 402–14

Lukas T J, Burgess W H, Prendergast F G, Lau W and Watterson D M 1986 Calmodulin binding domains: characterization of a phosphorylation and calmodulin binding site from myosin light chain kinase *Biochemistry* **25** 1458–64

Luo L, Callaway E M and Svoboda K 2018 Genetic dissection of neural circuits: a decade of progress *Neuron* **98** 256–81

Lymperopoulos P, Msanne J and Rabara R 2018 Phytochrome and phytohormones: working in tandem for plant growth and development *Front. Plant Sci.* **9** 1037

Ma Y, Bao J, Zhang Y, Li Z, Zhou X, Wan C, Huang C, Zhao Y, Han G and Xue T 2019 Mammalian near-infrared image vision through injectable and self-powered retinal nano-antennae *Cell* **177** 243–55

Maciejko J, Kaur J, Becker-Baldus J and Glaubitz C 2019 Photocycle-dependent conformational changes in the proteorhodopsin cross-protomer Asp-His-Trp triad revealed by DPN-enhanced MAS-NMR *PNAS* **116** 8342–9

Madelung O 2004 *Semiconductor Data Handbook* (Berlin: Springer)

Madrid M K, Brennan J A, Yin R T, Knight H S and Efimov I R 2021 Advances in implantable optogenetic technology for cardiovascular research and medicine *Front. Physiol.* **12** 720190

Magno L A V *et al* 2019 Optogenetic stimulation of the M2 cortex reverts motor dysfunction in a mouse model of Parkinson's disease *J. Neurosci.* **39** 3234–48

Mahmoudi P, Veladi H and Pakdel F G 2017 Optogenetics, tools and applications in neuro-biology *J. Med. Signals Sens.* **7** 71–9

Maimon B E, Sparks K, Srinivasan S, Zorzos A N and Herr H M 2018 Spectrally distinct channelrhodopsins for two-colour optogenetic peripheral nerve simulation *Nat. Biomed. Eng.* **2** 485–96

Malhotra K, Hrovat D, Kumar B, Qu G, Van Houten J, Ahmed R, Piunno P A E, Gunning P T and Krull U J 2023 Lanthanide-doped upconversion nanoparticles: exploring a treasure trove of NIR-mediated emerging applications *ACS Appl. Mater. Interfaces* **15** 2499–528

Mališ M, Novak J, Zgrablić G, Parmigiani F and Došlić N 2017 Mechanism of ultrafast non-reactive deactivation of the retinal chromophore in non-polar solvents *Phys. Chem. Chem. Phys.* **19** 25970–8

Mank M, Reiff D F, Heim N, Friedrich M W, Borst A and Griesbeck O 2006 A FRET-based calcium biosensor with fast signal kinetics and high fluorescence change *Biophysical J.* **90** 1790–6

Mank M and Griesbeck O 2008 Genetically encoded calcium indicators *Chem. Rev.* **108** 1550–64

Manoilov K Y, Verkhusha V V and Shcherbakova D M 2021 A guide to the optogenetic regulation of endogenous molecules *Nat. Methods* **18** 1027–37

Mansouti M and Fussenegger M 2021 Synthetic biology-based optogenetic approaches to control therapeutic designer cells *Curr. Opin. Syst. Biol.* **28** 100396

Månsson L K, Pitenis A A and Wilson M Z 2022 Extracellular optogenetics at the interface of synthetic biology and materials science *Front. Bioeng. Biotechnol.* **10** 903982

Marchena M, Gil M, Martín C, Organero J A, Sanchez F and Douhal A 2011 Stability and photodynamics of lumichrome structures in water at different pHs and in chemical and biological caging media *J. Phys. Chem.* B **2011** 2424–35

Marcus R A 1993 Elektronentransferreaktionen in der Chemie—Theorie und Experiment (Nobel-Vortrag) *Angew. Chem.* **105** 1163–72

Marks K M and Nolan G P 2006 Chemical labeling strategies for cell biology *Nat. Methods* **3** 591–6

Marshel J H *et al* 2019 Cortical layer-specific critical dynamics triggering perception *Science* **365** 558

Marston S and Zamora J E 2020 Troponin structure and function: a view of recent progress *J. Muscle Res. Cell Motil.* **41** 71–89

Martin M del C *et al* 2019 Fluorescence enhancement of a microbial rhodopsin via electronic reprogramming *J. Am. Chem. Soc.* **141** 262–71

Martin R, Lacombat F, Espagne A, Dozova N, Plaza P, Yamamoto J, Müller P, Brettel K and de la Lande A 2017 Ultrafast flavin photoreduction in an oxidized animal (6-4) photolyase through an unconventional tryptophan tetrad *Phys. Chem. Chem. Phys.* **19** 24493–504

Mason C and Dunnill P 2008 A brief definition of regenerative medicine *Regen. Med.* **3** 1–5

Massey V and Palmer G 1966 On the existence of spectrally distinct classes of flavoprotein semiquinones. A new method for the quantitative production of flavoprotein semiquinones *Biochemistry (Moscow)* **5** 3181–9

Masuda S and Bauer C E 2002 AppA is a blue light photoreceptor that antirepresses photosynthesis gene expression in *Rhodobacter sphaeroides Cell* **137** 613–23

Masuda S 2013 Light detection and signal transduction in the BLUF photoreceptors *Plant Cell Physiol.* **54** 171–9

Masuda T and Takamiya K-i 2004 Novel insights into the enzymology, regulation and physiological functions of light-dependent protochlorophyllide oxidoreductase in angiosperms *Photosynth. Res.* **81** 1–29

Mathes T *et al* 2015 Proton-coupled electron transfer constitutes the photoactivation mechanism of the plant photoreceptor UVR8 *J. Am. Chem. Soc.* **137** 8113–20

Mathes T and Kennis J T M (ed) 2016 *Optogenetic Tools in the Molecular Spotlight* (Lausanne: Frontiers in Molecular Biosciences)

Mathes T and Götze J P 2015 A proposal for a dipole-generated BLUF domain mechanism *Front. Mol. Biosci.* **2** 62

Mathony J and Niopek D 2021 Enlightening allostery: designing switchable proteins by photoreceptor fusion *Adv. Biosys.* **5** 2000181

Matus M F and Häkkinen 2023 Understanding ligand-protected noble metal nanoclusters at work *Nat. Rev. Mater.* **8** 372–89

Matz M V, Fradkov A F, Labas Y A, Savitsky A P, Zaraisky A G, Markelov M L and Lukyanov S A 1999 Fluorescent proteins from nonbioluminescent Anthozoa species *Nat. Biotechnol.* **17** 969–73

Matz M V, Lukyanov K A and Lukyanov S A 2002 Family of the green fluorescent protein: journey to the end of the rainbow *BioEssays* **24** 953–9

Mayhew S G and Massey V 1973 Studies on the kinetics and mechanism of reduction of flavodoxin from *Peptostreptococcus elsdenii* by sodium dithionite *Biochim. Biophys. Acta – Enzymology* **315** 181–90

McCall J G, Kim T-i, Shin G, Huang X, Jung Y H, Al-Hasani R, Omenetto F G, Bruchas M R and Rogers J A 2013 Fabrication and application of flexible, multimodal light-emitting devices for wireless optogenetics *Nat. Protoc.* **8** 2413–28

McCombs J E and Palmer A E 2008 Measuring calcium dynamics in living cells with genetically encodable calcium indicators *Methods* **46** 152–9

McCue A C and Kuhlman B 2022 Design and engineering of light-sensitive protein switches *Cur. Opin. Struct. Biol.* **74** 102377

McGlynn S P, Azumi T and Kinoshita M 1969 *Molecular Spectroscopy of the Triplet State* (Englewood Cliffs, NJ: Prentice-Hall)

McIsaac R S *et al* 2014 Directed evolution of a far-red fluorescent rhodopsin *PNAS* **111** 13034–9

McLellan M A, Rosenthal N A and Pinto A R 2017 Cre-*lox*P mediated recombination: general principles and experimental considerations *Curr. Protoc. Mouse Biol.* **7** 1–12

Mei G, Mamaeva N, Ganapathy S, Wang P, DeGrip W J and Rothschild K J 2020 Analog retinal redshifts visible absorption of QuasAr transmembrane voltage sensor into near-infrared *Photochem. Photobiol.* **96** 55–66

Mei Q and Dvornyk V 2015 Evolutionary history of the photolyase/cryprochrome superfamily in eukaryotes *PLoS One* **10** e0135940

Meng X, Ganapathy S, van Roemburg L, Post M and Brinks D 2023 Voltage imaging with engineered proton-pumping rhodopsins: insights from the proton transfer pathway *ACS Phys. Chem. Au.* **3** 320–33

Merz T, Sadeghian K and Schütz M 2011 Why BLUF photoreceptors with roseoflavin lose their biological functionality *Phys. Chem. Chem. Phys.* **13** 14775–83

Messina M S, Quargnali G and Chang C J 2022 Activity-based sensing for chemistry-enabled biology: illuminating principles, probes, and prospects for boronate reagents for studying hydrogen peroxide *ACS Bio. Med. Chem. Au.* **2** 548–64

Michael A K, Fribourgh J L, van Gelder R N and Partch C L 2017 Animal cryptochromes: divergent roles in light perception, circadian timekeeping and beyond *Photochem. Photobiol.* **93** 128–40

Milenkov S M, Zvezdanović J B and Anđelković T D 2012 The identification of chlorophyll and its derivatives in the pigment mixtures: HPLC-chromatography, visible and mass spectroscopy studies *Adv. Technol.* **1** 16–24

Mill P, Christensen S T and Pederesen L B 2023 Primary cilia as dynamic and diverse signalling hubs in development and disease *Nat. Rev. Genet.* **24** 421–41

Miller N A *et al* 2020 The photoactive excited state of the B_{12}-based photoreceptor CarH *J. Phys. Chem. B* **124** 10732–8

Mills E, Chen X, Pham E, Wong S and Truong K 2012 Engineering a photoactivated caspase-7 for rapid induction of apoptosis *ACS Synth. Biol.* **1** 75–82

Milosevic M M, Jang J, McKimm E J, Zhu M H and Antic S D 2020 *In vitro* testing of voltage indicators: Archon1, ArcLightD, ASAP1, ASAP2s, ASAP3b, Bongwoori-Pos6, eRSZ1, FlicR1, and Chi-VSFP-Butterfly *eNeuro* **7** 1–19

Mimuro M, Nishimura Y, Takaichi S, Yamano Y, Ito M, Nagaoka S-I, Yamazaki I, Katoh T and Nagashima U 1993 The effect of molecular structure on the relaxation processes of carotenoids containing a carbonyl group *Chem. Phys. Lett.* **213** 576–80

Mirzayi P, Shobeiri P, Kalantari A, Perry G and Rezaei N 2022 Optogenetics: implications for Alzheimer' disease research and therapy *Mol. Brain* **15** 20

Mix L T *et al* 2018 Excitation-wavelength-dependent photocycle initiation dynamics resolve heterogeneity in the photoactive yellow protein from *Halorhodospira halophila Biochemistry* **57** 1733–47

Mix L T, Hara M, Fuzell J, Kumauchi M, Kaledhonkar M, Xie A, Hoff W D and Larsen D S 2021 Not all photoactive yellow proteins are built alike: surprises and insights into chromophore photoisomerization, protonation, and thermal reisomerization of the photoactive yellow protein isolated from *Salinibacter ruber J. Am. Chem. Soc.* **143** 19614–28

Miyawaki A, Liopis J, Heim R, McCaffery J M, Adams J A, Ikura M and Tsien R Y 1997 Fluorescent indicators for Ca^{2+} based on green fluorescent proteins and calmodulin *Nature* **388** 882–7

Miyamoto D and Murayama M 2016 The fiber-optic imaging and manipulation of neural activity during animal behavior *Neurosci. Res.* **103** 1–9

Mizuno H, Mal T K, Tong K I, Ando R, Furuta T, Ikura M and Miyawaki A 2003 Photo-induced peptide cleavage in the green to red conversion of a fluorescent protein *Mol. Cell* **12** 1051–8

Mogensen D J, Westberg M, Breitenbach T, Etzerodt M and Ogilby P R 2021 Stable transfection of the singlet oxygen photosensitizing protein SOPP3: examining aspects of intracellular behavior *Photochem. Photobiol.* **97** 1417–30

Möglich A and Moffat K 2007 Structural basis for light-dependent signaling in the dimeric LOV domain of the photosensor YtvA *J. Mol. Biol.* **373** 112–26

Möglich A 2019 Signal transduction in photoreceptor histidine kinases *Protein Sci.* **28** 1923–46

Mohanty S K and Lakshminarayananan V 2015 Optical techniques in optogenetics *J. Mod. Opt.* **62** 949–70

Mohr M A *et al* 2020 jYCaMP: an optimized calcium indicator for two-photon imaging at fiber laser wavelengths *Nat. Methods* **17** 694–7

Moldenhauer M *et al* 2018 Interaction of the signaling state analog and the apoprotein form of the orange carotenoid protein with the fluorescence recovery protein *Photosynth. Res.* **135** 125–39

Mollinedo-Gajate I, Song C and Knöpfel T 2021 Genetically encoded voltage indicators *Optogenetics: Light-Sensing Proteins and Their Applications in Neuroscience and Beyond* 2nd edn ed H Yawo, H Kandori, A Koizumi and R Kageyama (Singapore: Springer) pp 209–24

Montagni E, Resta F, Letizia A, Mascaro A and Pavone F S 2019 Optogenetics in brain research: from a strategy to investigate physiological function to a therapeutic tool *Photonics* **6** 92

Montgomery K L *et al* 2015 Wireless powered, fully internal optogenetics for brain, spinal and peripheral circuts in mice *Nat. Meth.* **12** 969–74

Moore T C, Mera P E and Escalante-Semerena J C 2014 The EutT enzyme of *Salmonella enterica* is a unique ATP:Cob(I)alamin adenosyltransferase metalloprotein that requires ferrous ions for maximal activity *J. Bacteriol.* **196** 903–10

Moore C I and Berglund K 2019 BL-OG: BioLuminescent-Optogenetics *J. Neurosci. Res.* **98** 469–70

Morgan S-A, Al-Abdul-Wahid S and Woolley G A 2010 Structure-based design of a photo-controlled DNA binding protein *J. Mol. Biol.* **399** 94–112

Morizumi T *et al* 2023 Structures of channelrhodopsin paralogs in peptidiscs explain their contrasting K^+ and Na^+ selectivities *Nat. Commun.* **14** 4365

Moruzzi N, Leibiger B, Barker C J, Leibiger I B and Berggren P-O 2023 Novel aspects of intra-islet communication: Primary cilia and filopodia *Adv. Biol. Reg.* **87** 100919

Moser T and Dieter A 2020 Towards optogenetic approaches for hearing restoration *Biochem. Biophys. Res. Commun.* **527** 337–42

Motta-Mena L B, Reade A, Mallory M J, Glantz S and Weiner O D 2014 An optogenetic gene expression system with rapid activation and deactivation kinetics *Nat. Chem. Biol.* **10** 196–202

Moulton P F 1982 Paramagnetic ion lasers *Lasers and Masers* ed M J Weber (Boca Raton, FL: CRC) (CRC Handbook of Laser Science and Technology) vol 1 pp 21–146

Mouritsen H 2018 Long-distance navigation and magnetoreception in migratory animals *Nature* **558** 50–9

Mroginski M-A *et al* 2021 Frontiers in multiscale modeling of photoreceptor proteins *Photochem. Photobiol.* **97** 243–69

Mroginski M A, Murgida D H and Hildebrandt P 2007 The structural changes during the photocycle of phytochrome: a combined resonance Raman and quantum chemical approach *Acc. Chem. Res.* **40** 258–66

Mühlhäuser W W D, Fischer A, Weber W and Radziwill G 2017 Optogenetics – bringing light into the darkness of mammalian signal transduction *Biochim. Biophys. Acta Mol. Cell Res.* **1864** 280–92

Mukherjee S, Hegemann P and Broser M 2019 Enzymerhodopsins: novel photoregulated catalysts for optogenetics *Curr. Opin. Struct. Biol.* **57** 118–26

Mukherjee A and Schroeder C M 2015 Flavin-based fluorescent proteins: emerging paradigms in biological imaging *Curr. Opin. Biotechnol.* **31** 16–23

Mukherjee A, Repina N A, Schaffer D V and Kane R S 2017 Optogenetic tools for cell biological applications *J. Thorac. Dis.* **9** 4867–70

Müller F and Massey V 1971 Sulfite interaction with free and protein bound flavin *Methods Enzymol.* **18B** 468–73

Müller F (ed) 1991a *Chemistry and Biochemistry of Flavoenzymes* **vols 1–3** (Boca Raton, NJ: CRC)

Müller F 1991b Free flavins: synthesis, chemical and physical properties *Chemistry and Biochemistry of Flavoenzymes* ed F Müller (Boca Raton, FL: CRC) vol 1 pp 1–71

Müller K, Engesser R, Timmer J, Nagy F, Zurbriggen M D and Weber W 2013a Synthesis of phycocyanobilin in mammalian cells *Chem. Commun.* **49** 8970–2

Müller K *et al* 2013b A red/far-red light-responsive bi-stable toggle switch to control gene expression in mammalian cells *Nucleic Acids Res.* **41** e77

Müller K, Engesser R, Schulz S, Steinberg T, Tomakidi P, Weber C C, Ulm R, Timmer J, Zurbriggen M D and Weber W 2013c Multi-chromatic control of mammalian gene expression and signaling *Nucleic Acids Res.* **41** e124

Müller N, Wenzel S, Zou Y, Künzel S, Sasso S, Weiß D, Prager K, Grossman A, Kottke T and Mittag M 2017 A plant cryptochrome controls key features of the *Chlamydomonas* circadian clock and its life cycle *Plant Physiol.* **174** 185–201

Murphy J T and Lagarias J C 1997 The phytofluors: a new class of fluorescent protein probes *Curr. Biol.* **7** 870–6

Muzzopappa F and Kirilovsky D 2020 Changing color for photoprotection: the orange carotenoid protein *Trends Plant Sci.* **25** 92–104

Nagata T and Inoue K 2021 Rhodopins at a glance *J. Cell Sci.* **134** jscc258989

Nagel G, Szellas T, Huhn W, Kateriya S, Adeishvili N, Berthold P, Ollig D, Hegemann P and Bamberg E 2003 Channelrhodopsin-2, a directly light-gated cation-selective membrane channel *Proc. Natl. Sci. Am.* **100** 13940–5

Nagel G, Brauner M, Liewald J F, Adeishvili N, Bamberg E and Gottschalk A 2005 Light activation of channelrhodopsin-2 in excitable cells of *Caenorhabditis elegans* triggers rapid behavioral responses *Curr. Biol.* **15** 2279–84

Nakai J, Ohkura M and Imoto K 2001 A high signal-to-noise Ca^{2+} probe composed of a single green fluorescent protein *Nat. Biotechnol.* **19** 137–41

Nakai N, Overton E T N and Takumi T 2021 Optogenetic approaches to understand the neural circuit mechanism of social deficits seen in autism spectrum disorders *Optogenetics: Light-Sensing Proteins and Their Applications in Neuroscience and Beyond* 2nd edn ed H Yawo, H Kandori, A Koizumi and R Kageyam (Singapore: Springer) pp 557–62

Nakao S, Kojima K and Sudo Y 2021 Microbial rhodopsins as multi-functional photoreactive membrane proteins for optogenetics *Biol. Pharm. Bull.* **44** 1357–63

Nakasone Y and Terazima M 2022 Time-resolved diffusion reveals photoreactions of BLUF proteins with similar functional domains *Photochem. Photobiol. Sci.* **21** 493–507

Nakasone Y, Murakami H, Tokonami S, Oda T and Terazima M 2023 Time-resolved study on signaling pathway of photoactivated adenylate cyclase and its nonlinear response *J. Biol. Chem.* **299** 105285

Nakasone Y, Ono T A, Ishii A, Masuda S and Terazima M 2007 Transient dimerization and conformational change of a BLUF protein: YcgF *J. Am. Chem. Soc.* **129** 7028–35

Nash A I, McNulty R, Shillito M E, Swartz T E, Bogomolni R A, Luecke H and Gardner K H 2011 Structural basis of photosensitivity in a bacterial light-oxygen-voltage/helix–turn–helix (LOV-HTH) DNA binding protein *PNAS* **108** 9449–54

Nasu Y, Shen Y, Kramer L and Campbell R E 2021 Structure- and mechanism-guided design of single fluorescent protein-based biosensors *Nat. Chem. Biol.* **17** 509–18

Navarro E, Niemann N, Kock D, Dadaeva T, Gutiérrez G, Engelsdorf T, Kiontke S, Corrochano L M, Batschauer A and Garre V 2020 The DASH-type cryptochrome from the fungus *Mucur circinelloides* is a canonical CPD-photolyase *Curr. Biol.* **30** 4483–90

Nayak B N, Singh R B and Buttar H S 2019 Role of tryptophan in health and disease: Systematic review of the anti-oxidant, anti-inflammation, and nutritional aspects of tryptophan and its metabolites *World Heart J.* **11** 161–78

Nazarenko V V *et al* 2019 A thermostable flavin-based fluorescent protein from *Chloroflexus aggregans*: a framework for ultra-high resolution structural studies *Photochem. Photobiol. Sci.* **18** 1793–805

Neumann K, Verhoefen M-K, Weber I, Glaubitz C and Wachtveitl J 2008 Initial reaction dynamics of proteorhodopsin observed by femtosecond infrared and visible spectroscopy *Biophys. J.* **94** 4796–807

Nguyen K H, Hao Y, Chen W, Zhang Y, Xu M, Yang M and Liu Y-N 2018 Recent progress in the development of fluorescent probes for hydrazine *J. Biol. Chem. Luminesc.* **33** 816–36

Ni M, Tepperman J M and Quail P H 1999 Binding the phytochrome B to its nuclear signalling partner PIF3 is reversibly induced by light *Nature* **400** 781–4

Nienhaus G U and Wiedenmann J 2009 Structure, dynamics and optical properties of fluorescent proteins: perspectives for marker development *ChemPhysChem* **10** 1369–79

Nießner C, Denzau S, Stapput K, Ahmad M, Peichl L, Wiltschko W and Wilschko R 2013 Magnetoreception: activated cryptochrome 1a concurs with magnetic orientation in birds *J. R. Soc. Interface* **10** 20130638

Nietz A K, Popa L S, Streng M L, Carter R E, Kodandaramaiah S B and Ebner T J 2022 Wide-field calcium imaging of neuronal network dynamics *in vivo Biology* **11** 1601

Niho A *et al* 2017 Demonstration of a light-driven SO_4^{2-} transporter and its spectroscopic characteristics *J. Am. Chem. Soc.* **139** 4376–89

Nijenhuis W, van Grinsven M M P and Kapitein L C 2020 An optimized toolbox for the optogenetic control of intracellular transport *J. Cell Biol.* **219** e201907149

Niziński S, Wilson A, Uriarte L M, Ruckebusch C, Andreeva E A, Schlichting I, Colletier J-P, Kirilovsky D, Burdzinski G and Sliwa M 2022 Unifying perspective of the ultrafast photodynamics of orange carotenoid proteins from *Synechocystis*: peril of high-power excitation, existence of different S* states, and influence of tagging *JACS Au.* **2** 1084–95

Nogly P *et al* 2018 Retinal isomerization in acteriorhodopsin captured by a femtosecond x-ray laser *Science* **361** eaat0094

Obi-Nagata K, Temma Y and Hayashi-Takagi A 2019 Synaptic functions and their disruption in schizophrenia: from clinical evidence to synaptic optogenetics in an animal model *Proc. Jpn. Acad. Ser.* B **95** 179–97

O'Banion C P and Lawrence D S 2018 Optogenetics: a primer for chemists *ChemBioChem* **19** 1201–16

O'Banion C P, Pristman M A, Hughes R M, Herring L E and Capuzzi S J 2018 Design and profiling of a subcellular targeted optogenetic cAMP-dependent protein kinase *Cell Chem. Biol.* **25** 100–9

Oda K, Vierock J, Oishi S, Rodriguez-Rozada S, Taniguchi R, Yamashita K, Wiegert J S, Nishizawa T and Hegemann P 2018 Crystal structure of the red light-activated channelrho-dopsin Chrimson *Nat. Commun.* **9** 3949

Oesterhelt D, Hegemann P and Tittor J 1985 The photocycle of the chloride pump halorhodopsin. II: quantum yields and a kinetic model *EMBO J.* **4** 2351–6

Oh T-J, Fan H, Skeeters S S and Zhang K 2021 Steering molecular activity with optogenetics: recent advances and perspectives *Adv. Biol.* **5** 2000180

Ohki M *et al* 2016 Structural insight into photoactivation of an adenylate cyclase from a photosynthetic cyanobacterium *PNAS* **113** 6659–64

Ohlendorf R and Möglich A 2022 Light-regulated gene expression in bacteria: fundamentals, advances, and perspectives *Front. Bioeng. Biotechnol.* **10** 1029403

Okajima K *et al* 2006 Fate determination of the flavin photoreceptions in the cyanobacterial blue light receptor TePixD (Tll0078) *J. Mol. Biol.* **363** 10–8

Okajima K *et al* 2006 Biochemical and functional characterization of BLUF-type flavin-binding proteins of two species of cyanobacteria *J. Biochem.* **137** 741–50

Oliinyk O S, Chernov K G and Verkhusha V V 2017 Bacteriophytochromes, cyanobacterio-chromes and allophycocyanines as a source of near-infrared fluorescent probes *Int. J. Mol. Sci.* **18** 1691

Olson D K, Yoshizawa S, Boeuf D, Iwasaki W and DeLong E F 2018 Proteorhodopsin variability and distribution in the North Pacific Subtropical Gyre *ISME J.* **12** 1047–60

Ordaz J D, Wu W and Xu X-M 2017 Optogenetics and its application in neural degeneration and regulation *Neural Regen. Res.* **12** 1197–209

Orlandi G, Zerbetti F and Zgierski M Z 1991 Theoretical analysis of spectra of short polyenes *Chem. Rev.* **91** 867–91

Ortiz-Guerrero J M, Polanco M C, Murillo F J and Padmanabhan S 2011 Light-dependent gene regulation by coenzyme B_{12}-based photoreceptor *PNAS* **108** 7565–70

Ortkrass H, Schürstedt J, Wiebusch G, Szafranska K, McCourt P and Huser T 2023 High-speed TIRF and 2D super-resolution structured illumination microscopy with a large field of view based on fiber optic components *Opt. Express* **31** 29156–65

Osawa S-I and Tominaga T 2021 Application of optogenetics in epilepsy *Optogenetics: Light-Sensing Proteins and Their Applications in Neuroscience and Beyond* 2nd edn ed H Yawo, H Kandori, A Koizumi and R Kageyama (Singapore: Springer) pp 557–62

O'Shea C *et al* 2019 Cardiac optogenetics and optical mapping - overcoming spectral congestion in all-optical cardiac electrophysiology *Front. Physiol.* **10** 182

Ostersehlt L M, Jans D C, Wittek A, Keller-Findeisen J, Inamdar K, Sahl S J, Hell S W and Jaökobs S 2022 DNA-PAINT MINFLUX nanoscopy *Nat. Methods* **19** 1072–5

Ostrovsky M A and Kirpichnikov M P 2019 Prospects of optogenetic prosthesis of the degenerate retina of the eye *Biochemistry (Moscow)* **84** 479–90

Öztürk N, Song S-H, Özgur S, Selby C P, Morrison L, Partch C, Zhong D and Sancar A 2007 Structure and function of animal cryptochromes *Cold Spring Hab. Symp. Quant. Biol.* **72** 119–31

Öztürk N, Song S-H, Selby C P and Sancar A 2008 Animal type 1 cryptochromes. Analysis of the redox state of the flavin cofactor by site-directed mutagenesis *J. Biol. Chem.* **283** 3256–63

Öztürk N, Selby C P, Zhong D and Sancar A 2014 Mechanism of photosignaling by *Drosophila* cryptochrome. Role of redox status of the flavin chromophore *J. Biol. Chem.* **289** 4634–42

Ozturk N 2017 Phylogenetic and functional classification of the photolyase/cryptochrome family *Photochem. Photobiol.* **93** 104–11

Packer A M, Roska B and Häusser M 2013 Targeting neurons and photons for optogenetics *Nat. Neurosci.* **16** 805–15

Padmanabhan S, Jost M, Drennan C L and Elías-Arnanz M 2017 A new facet of vitamin B_{12}: gene regulation by cobalamin-based photoreceptors *Annu. Rev. Biochem.* **86** 485–514

Padmanabhan S, Pérez-Castaño R and Elías-Arnanz E 2019 B_{12}-based photoreceptors: from structure and function to applications in optogenetics and synthetic biology *Curr. Opin. Struct. Biol.* **57** 47–55

Padmanabhan S, Pérez-Castaño R, Osete-Alcaraz L, Polanco M C and Elías-Arnanz M 2022 Vitamin B_{12} photoreceptors *Vitam. Horm.* **119** 149–84

Pal R, Sekharan S and Batista V S 2013 Spectral tuning in halorhodopsin: the chloride pump photoreceptor *J. Am. Chem. Soc.* **135** 9624–7

Palayam M, Ganapathy J, Guercio A M, Tal L, Deck S L and Shabek N 2021 Structural insights into photoactivation of plant cryptochrome-2 *Commun. Biol.* **4** 28

Palombo R, Barnweschi L, Pedraza-González L, Padula D, Schapiro I and Olivucci M 2022 Retinal chromophore charge delocalization and confinement explain the extreme photophysics of Neorhodopsin *Nat. Commun.* **13** 6652

Panda R, Panda P K, Krishnamoorthy J and Kar R K 2022 Network analysis of chromophore binding site in LOV domain *Comput. Biol. Med.* **161** 106996

Pande K *et al* 2016 Femtosecond structural dynamics drives the *trans/cis* isomerization in photoactive yellow protein *Science* **352** 725–9

Pansare V, Hejazi S, Faenza W and Prud'homme R K 2012 Review of long-wavelength optical and NIR imaging materials: contrast agents, fluorophores and multifunctional nano carriers *Chem. Mater.* **24** 812–27

Papagiakoumou E, Anselmi F, Bègue A, de Sars V and Glückstad J 2010 Scanless two-photon excitation of channelrhodopsin-2 *Nat. Methods* **7** 848–54

Parag-Sharma K, O'Banion C P, Henry E C, Musicant A M, Cleveland J L, Lawrence D S and Amelio A L 2020 Engineered BRET-based biologic light sources enable spatiotemporal control over diverse optogenetic systems *ACS Synth. Biol.* **9** 1–9

Paredes R M, Etzler J C, Watts L T and Lechleiter J D 2008 Chemical calcium indicators *Methods* **46** 143–51

Parico G C G and Partch C L 2020 The tail of cryptochromes: an intrinsically disordered cog within the mammalian circadian clock *Cell Commun. Signal.* **18** 182

Park J W and Shiozaki T 2018 On the accuracy of retinal protonated Schiff base models *Mol. Phys.* **116** 2583–90

Park S Y *et al* 2020 Novel luciferase – opsin combinations for improved luminopsins *J. Neurosci. Res.* **98** 410–21

Park H, Kim N Y, Lee S, Kim N, Kim J and Heo W D 2017 Optogenetic protein clustering through fluorescent protein tagging and extension of CRY2 *Nat. Commun.* **8** 30

Park S-Y and Tame J R H 2017 Seeing the light with BLUF proteins *Biophys. Rev.* **9** 169–76

Parker C A 1968 *Photoluminescence of Solutions* (Amsterdam: Elsevier)

Passmore J B, Nijenhuis W and Kapitein L C 2021 From observing to controlling: inducible control of organelle dynamics and interactions *Curr. Opin. Cell Biol.* **71** 69–76

Patel M, Meenu M, Pandey J K, Kumar P and Patel R 2022 Recent development in upconversion nanoparticles and their application in optogenetics: a review *J. Rare Earths* **40** 847–61

Pathak G P, Spiltoir J I, Höglund C, Polstein L R, Heine-Kosinen S, Gersbach C A, Rossi J and Tucker C L 2017 Bidirectional approaches for optogenetic regulation of gene expression in mammalian cells using *Arabidopsis* cryptochrome 2 *Nucl. Acids Res.* **45** e167

Patnaik A, Alavilli H, Rath J, Panigrahi K C S and Panigraphy M 2022 Variations in circadian clock organization & function: a journey from ancient to recent *Planta* **256** 91

Patrono E, Svoboda J and Stuchlík A 2021 Schizophrenia, the gut microbiota, and new opportunities from optogenetic manipulations of the gut-brain axis *Behav. Brain Funct.* **17** 7

Patterson G H, Knobel S M, Sharif W D, Kain S R and Piston D W 1997 Use of green fluorescent protein and its mutants in quantitative fluorescence microscopy *Biophys. J.* **73** 2782–90

Patterson G H, Piston D W and Barisas B G 2000 Förster distances between green fluorescent protein pairs *Analyt. Biochem.* **284** 438–40

Pavlovska T and Cibulka R 2021 Structure and properties of flavins *Flavin-Based Catalysis: Principles and Applications* ed R Cibulka and M W Fraaije (Weinheim: Wiley-VCH) pp 1–27

Pearce S and Tucher C L 2021 Dual systems for enhancing control of protein activity through induced dimerization approaches *Adv. Biol.* **5** 2000234

Pedraza-González L, Barneschi L, Padula D, De Vico L and Olivucci M 2022 Evolution of the automatic rhodopsin modeling (ARM) model *Top. Curr. Chem.* **380** 21

Pedraza-González L, Barneschi L, Marszałek M, Padula D, De Vico L and Olivucci M 2023 Automated QM/MM screening of rhodopsin variants with enhanced fluorescence *J. Chem. Theory Comput.* **19** 293–310

Peled A 2011 Optogenetic neuronal control in schizophrenia *Med. Hypotheses* **76** 914–21

Pendharkar A V, Levy S L, Ho A L, Sussman E S, Cheng M Y and Steinberg G K 2016 Optogenetic modulation in stroke recovery *Neurosurg. Focus* **40** E6

Peng W, Maydew C, Kam H, Lynd J, Tutol J N, Phelps S M, Abeyrathna S, Meloni G and Dodani S C 2022 Discovery of a monomeric green fluorescent protein sensor for chlorine by structure-guided bioinformatics *Chem. Sci.* **13** 12659

Pennacchietti F, Abbruzzetti S, Losi A, Mandalari C, Bedotti R, Viappiani C, Zanacchi F C, Diaspro A and Gärtner W 2014 The dark recovery rate in the photocycle of the bacterial photoreceptor YtvA is affected by the cellular environment and by hydration *PLoS* **9** e107489

Pennacchietti F *et al* 2018 Fast reversibly photoswitching red fluorescent proteins for live-cell RESOLFT nanoscopy *Nat. Meth.* **15** 601–4

Penzkofer A 1988 Solid state lasers *Prog. Quant. Electr.* **12** 291–427

Penzkofer A, Endres L, Schiereis T and Hegemann P 2005 Yield of photo-adduct formation of LOV domains from *Chlamydomonas reinhardtii* by picosecond laser excitation *Chem. Phys.* **316** 185–94

Penzkofer A, Stierl M, Hegemann P and Kateriya S 2011 Photo-dynamics of the BLUF domain containing soluble adenylate cyclase (nPAC) from the amoeboflagellate *Naegleria gruberi* NEG-M strain *Chem. Phys.* **387** 25–38

Penzkofer A 2012 Reduction-oxidation photocycle dynamics of flavins in starch films *Int. J. Mol. Sci.* **13** 9157–83

Penzkofer A, Simmel M and Riedl D 2012 Room temperature phosphorescence lifetime and quantum yield of erythrosine B and rose bengal in aerobic alkaline aqueous solution *J. Lumin.* **132** 1055–62

Penzkofer A, Tanwar M, Veetil S K, Kateriya S and Hegemann P 2013 Photo-dynamics and thermal behavior of the BLUF domain containing adenylate cyclase NgPAC2 from the amoeboflagellate *Naegleria gruberi* NEG-M strain *Chem. Phys.* **412** 96–108

Penzkofer A, Stierl M, Mathes T and Hegemann P 2014a Absorption and emission spectroscopic characterization of photodynamics of photoactivated adenylyl cyclase mutant bPAC-Y7F of *Beggiatoa* sp *J. Photochem. Photobiol. B: Biol.* **140** 182–93

Penzkofer A, Tanwar M, Veetil S K, Kateriya S, Stierl M and Hegemann P 2014b Photodynamics of BLUF domain containing adenylyl cyclase NgPAC3 from the amoeboflagellate *Naegleria gruberi* NEG-M strain *J. Photochem. Photobiol. A: Chem.* **287** 19–29

Penzkofer A, Tanwar M, Veetil S K and Kateriya S 2014c Photo-dynamics of photoactivated adenylyl cyclase LiPAC from the spirochete bacterium *Leptonema illini* strain 3055^T *Trends Appl. Spectr.* **11** 39–62

Penzkofer A, Luck M, Mathes T and Hegemann P 2014d Bistable retinal Schiff base photo-dynamics in histidine kinase rhodopsin HKR1 from *Chlamydomonas reinhardtii Photochem. Photobiol.* **90** 773–85

Penzkofer A, Tanwar M, Veetil S K and Kateriya S 2015 Photo-dynamics of photoactivated adenylyl cyclase TpPAC from the spirochete bacterium *Turneriella parva* strain H^T *J. Photochem. Photobiol. B: Biol.* **153** 90–102

Penzkofer A 2016 Absorption and emission spectroscopic investigation of alloxazine in aqueous solutions and comparison with lumichrome *J. Photochem. Photobiol. A: Chem.* **314** 114–24

Penzkofer A, Atri Y, Sharma K, Veetil S K and Kateriya S 2016a Photodynamics of photo–activated BLUF coupled endonuclease III mutant RmPAE from mesophilic, pigmented bacterium *Rubellimicrobium mesophilum* strain MSL-20^T *BAOJ Chem.* **2** 008

Penzkofer A, Kateriya S and Hegemann P 2016b Photodynamics of the optogenetic BLUF coupled photoactivated adenylyl cyclases (PACs) *Dyes and Pigments* **135** 102–12

Penzkofer A, Scheib U, Hegemann P and Stehfest K 2016c Absorption and emission spectro-scopic investigation of thermal dynamics and photo-dynamics of the rhodopsin domain of the rhodopsin–guanylyl cyclase from the aquatic fungus *Blastocladiella emersonii BAOJ Physics* **2/1** 006

Penzkofer A, Scheib U, Stehfest K and Hegemann P 2017 Absorption and emission spectroscopic investigation of thermal dynamics and photo-dynamics of the rhodopsin domain of the rhodopsin–guanylyl cyclase from the nematophagous fungus *Catenaria anguillulae Int. J. Mol. Sci.* **18** 2099

Penzkofer A, Silapetere A and Hegemann P 2019 Absorption and emission spectroscopic investigation of the thermal dynamics of the Archaerhodopsin 3 based fluorescent voltage sensor QuasAr1 *Int. J. Mol. Sci.* **20** 4086

Penzkofer A, Silapetere A and Hegemann P 2020a Absorption and emission spectroscopic investigation of the thermal dynamics of the Archaerhodopsin 3 based fluorescent voltage sensor Archon2 *Int. J. Mol. Sci.* **21** 6576

Penzkofer A, Silapetere A and Hegemann P 2020b Photocycle dynamics of the Archaerhodopsin 3 based fluorescent voltage sensor QuasAr1 *Int. J. Mol. Sci.* **21** 160

Penzkofer A, Silapetere A and Hegemann P 2021 Photocycle dynamics of the Archaerhodopsin 3 based fluorescent voltage sensor Archon2 *J. Photochem. Photobiol. B: Biol.* **225** 112331

Penzkofer A, Silapetere A and Hegemann P 2023 Theoretical investigation of the photocycle dynamics of the Archaerhodopsin 3 based fluorescent voltage sensor Archon2 *J. Photochem. Photobiol. A: Chem.* **437** 114366

Peron S, Chen T-W and Svoboda K 2015 Comprehensive imaging of cortical networks *Curr. Opin. Neurobiol.* **32** 115–23

Perriono A P, Miyagi A and Scheuring S 2021 Single molecule kinetics of bacteriorhodopsin by HS-AFM *Nat. Commun.* **12** 7225

Peter E, Dick B and Baeurle S A 2012 Regulatory mechanism of the light-activable allosteric switch LOV-TAP for the control of DNA binding: a computer simulation study *Proteins* **81** 394–405

Peter F, Herbst J, Tittor J, Oesterhelt D and Diller R 2006 Primary reaction dynamics of halorhodopsin observed by sub-picosecond IR – vibrational spectroscopy *Chem. Phys.* **323** 109–16

Peterson A, Baskett C, Ratcliff W C and Burnetti A 2023 Using light for energy: examining the evolution of phototrophic metabolism through synthetic construction *bioRxiv preprint*

Petryayeva E and Krull U J 2011 Localized surface plasmon resonance: nanostructures, bioassays and biosensing – a review *Anal. Chim. Acta* **706** 8–24

Pham E, Mills E and Truong K 2011 A synthetic photoactivated protein to generate local or global Ca^{2+} signals *Chem. Biol.* **18** 880–90

Pham V N, Kathare P K and Huq E 2018 Phytochromes and phytochrome interacting factors *Plant Physiol.* **176** 1025–38

Piatkevich K D, Subach F V and Verkhusha V V 2013 Engineering of bacterial phytochromes for near-infrared imaging, sensing, and light-control in mammals *Chem. Soc. Rev.* **42** 3441–52

Piatkevich K D *et al* 2018 A robotic multidimensional directed evolution approach applied to fluorescent voltage reporters *Nat. Chem. Biol.* **14** 352–60

Piatkevich K D, Murdock M H and Subach F 2019 Advances in engineering and application of optogenetic indicators for neuroscience *Appl. Sci.* **9** 562

Piccinini L, Iacopini S, Cazzaniga S, Ballottari M, Giuntoli B and Licausi F 2022 A synthetic switch based on orange carotenoid protein to control blue-green light responses in chloroplasts *Plant Physiol.* **189** 1153–68

Pigni N B, Clark K L, Beck W F and Gascón J A 2020 Spectral signatures of canthaxanthin translocation in the orange carotenoid protein *J. Phys. Chem.* B **124** 11387–95

Pilla V, Alves L P, de Santana J F, da Silva L G, Ruggiero R and Munin E 2012 Fluorescence quantum efficiency of CdSe/ZnS quantum dots in biofluids: pH dependence *J. Appl. Phys.* **112** 1104704

Pisanello F *et al* 2017 Dynamic illumination of spatially restricted or large brain volumes via a single tapered optical fiber *Nat. Neurosci.* **20** 1180–8

Platisa J and Pieribone V A 2018 Genetically encoded fluorescent voltage indicators: are we there yet? *Curr. Opin. Neurobiol.* **50** 146–53

Poddar H, Heyes D J, Zhang S, Hardman S J, Sakuma M and Scrutton N S 2022 An unusual light-sensing function for coenzyme B_{12} in bacterial transcription regulator CarH *Methods Enzymol.* **668** 349–72

Poddar H, Heyes D J, Schirò G, Weik M, Leys D and Scrutton N S 2022 A guide to time-resolved structural analysis of light-activated proteins *FEBS J.* **289** 576–95

Podolec R, Demarsy E and Ulm R 2021 Perception and signaling of ultraviolet-B radiation in plants *Annu. Rev. Plant Biol.* **72** 793–822

Podor B, Hu Y-l, Ohkura M, Nakai J, Croll R and Fine A 2015 Comparison of genetically encoded calcium indicators for monitoring action potentials in mammalian brain by two-photon excitation fluorescence microscopy *Neurophotonics* **2** 021014

Polívka T, Kerfeld C A, Pascher T and Sundström V 2005 Spectroscopic properties of the carotenoid 3'-hydroxyechinenone in the orange carotenoid protein from the cyanobacterium *Arthrospira maxima Biochem* **44** 3994–4003

Polívka T and Sundström V 2004 Ultrafast dynamics of carotenoid excited states – from solution to natural and artificial systems *Chem. Rev.* **104** 2021–71

Polland H-J, Franz M A, Zinth W, Kaiser W, Hegemann P and Oesterhelt D 1985 Picosecond events in the photochemical cycle of the light-driven chloride-pump halorhodopsin *Biophys. J.* **47** 55–9

Ponnu J and Hoecker U 2022 Signaling mechanisms by *Arabidopsis* cryptochromes *Front. Plant Sci.* **13** 844714

Pooam M, Arthaut L D, Burdick D, Link J, Martino C F and Ahmad M 2019 Magnetic sensitivity mediated by the *Arabidopsis* blue-light receptor cryptochrome occurs during flavin reoxidation in the dark *Planta* **249** 319–32

Pooam M, El-Esawi M, Aguida B and Ahmad M 2020 *Arabidopsis* cryptochrome and quantum biology: new insights for plant science and crop improvement *J. Plant Biochem. Biotechnol.* **29** 634–51

Porta-de-la-Riva M *et al* 2023 Neural engineering with photons as synaptic transmitters *Nat. Methods* **20** 761–9

Potter L R 2011 Guanylyl cyclase structure, function and regulation *Cell Signal.* **23** 1921–6

Prakash R *et al* 2012 Two-photon optogenetic toolbox for fast inhibition, excitation and bistable modulation *Nat. Meth.* **9** 1171–9

Prasher D C, Eckenrode V K, Ward W W, Prendergast F G and Cormier M J 1992 Primary structure of the *Aequorea victoria* green fluorescent protein *Gene* **111** 229–33

Prestori F, Montagna I, D'Angelo E and Mapelli L 2020 The optogenetic revolution in cerebellar investigations *Int. J. Mol. Sci.* **21** 2494

Procopio M, Link J, Engle D, Witczak J, Ritz T and Ahmad M 2016 Kinetic modeling of the *Arabidopsis* cryptochrome photocycle: $FADH^0$ accumulation correlates with biological activity *Front. Plant Sci.* **7** 888

Prosseda P P, Alvarado J A, Wang B, Kowal T J, Ning K, Stamer W D, Hu Y and Sun Y 2020 Optogenetic stimulation of phosphoinositides reveals a critical role of primary cilia in eye pressure regulation *Sci. Adv.* **6** eaay8699

Prosseda P P, Tran M, Kowal T, Wang B and Sun Y 2022 Advances in ophthalmic optogenetics: approaches and applications *Biomolecules* **12** 9269

Prukala D, Sikorska E, Koput J, Khmelinsii I, Karolczak J, Gierszewski M and Sikoski M 2012 Acid–base equilibriums of lumichrome and its 1-methyl, 3-methyl, and 1,3-dimethyl derivatives *J. Phys. Chem A.* **116** 7474–90

Pudasaini A, El-Arab K K and Zoltowski B D 2015 LOV-based optogenetic devices: Light-driven modules to impart photoregulated control of cellular signaling *Front. Mol. Biosci.* **2** 18

Pudasaini A, Shim J S, Song Y H, Shi H, Kiba T, Somers D E, Imaizumi T and Zoltowski B D 2017 Kinetics of the LOV domain of ZEITLUPE determine its circadian function in *Arabidopsis eLife* **6** e21646

Pudasaini A, Green R, Song Y H, Blumenfeld A, Karki N, Imaizumi T and Zoltowski B D 2020 Steric and electronic interactions at Gln154 in ZEITLUPE induce reorganization of the LOV domain dimer interface *Biochemistry* **60** 95–103

Pushkarev A *et al* 2018 A distinct abundant group of microbial rhodopsins discovered using functional metagenomics *Nature* **558** 595–9

Putschögl M, Zirak P and Penzkofer A 2008 Absorption and emission behaviour of *trans-p*-coumaric acid in aqueous solutions and some organic solvents *Chem. Phys.* **343** 107–20

Pyari G, Bansal H and Roy S 2023 Optogenetically mediated large volume suppression and synchronized excitation of human ventricular cardiomyocytes *Pflügers Archiv. Eur. J. Physiol.* **475** 1479–503

Qin S *et al* 2016 A magnetic protein biocompass *Nat. Mat.* **15** 217–26

Qing T, Zhang K, Qing Z, Wang X, Long C, Zhang P, Hu H and Feng B 2019 Recent progress in copper nanocluster-based fluorescent probing: a review *Microchim. Acta* **186** 670

Quail P H 2002 Phytochrome photosensory signalling networks *Nat. Rev. Mol. Cell Biol.* **3** 85–93

Quail P H 2010 Phytochromes *Curr. Biol.* **20** R504–7

Rabinowitz J C 1960 Folic acid *The Enzymes* vol 2 2nd edn ed P D Boyer, H Lardy and K Myrbäck (New York: Academic) pp 185–252

Raber H F *et al* 2020 Azulitox – a *Pseudomonas aeruginosa* P28-derived cancer-cell-specific protein photosensitizer *Biomacromolecules* **21** 5067–76

Rad M S, Choi Y, Cohen L B, Baker B J, Zhong S, Storace D A and Braubach O R 2017 Voltage and calcium imaging of brain activity *Biophys. J.* **113** 2160–7

Raffelberg S, Wang L, Gao S, Losi A, Gärtner W and Nagel G 2013 A LOV-domain-mediated blue-light-activated adenylate (adenylyl) cyclase from the cyanobacterium *Microcoleus chthonoplastes* PCC 7420 *Biochem. J.* **455** 359–65

Rai N, Morales L O and Aphalo P J 2021 Perception of solar UV radiation by plants: photoreceptors and mechanisms *Plant Physiol.* **186** 1382–96

Rajagopal S, Key M J, Purcell E B, Boerema D J and Moffat K 2004 Purification and initial characterization of a putative blue light-regulated phosphodiesterase from *Escherichia coli Photochem. Photobiol.* **80** 542–7

Rajamani R, Lin Y-L and Gao J 2010 The opsin shift and mechanism of spectral tuning in rhodopsin *J. Comput. Chem.* **32** 854–65

Ramírez-Gamboa D *et al* 2022 Photolyase production and current applications: a review *Molecules* **27** 5998

Ranzani A T, Wehrmann M, Kaiser J, Juraschitz M, Weber A M, Pietruschka G, Gerken U, Mayer G and Möglich A 2022 Light-dependent control of bacterial expression at the mRNA level *ACS Synth. Biol.* **11** 3482–92

Rasmussen E S, Takahashi J S and Green C B 2022 Time to target the circadian clock for drug discovery *Trends Biochem. Sci.* **47** 745–58

Ratner M 2021 Light-activated genetic therapy to treat blindness enters clinic *Nat. Biotechn.* **39** 126–7

Redhi A, Petersen J, Wagner V, Vuong T, Li W, Li W, Schrader L and Mittag M 2024 The UV-A receptor CRY-DASH1 up- and down regulates proteins in different plastidial pathways *J. Mol. Biol.* **436** 168271

Redmond R W and Gamlin J N 1999 A Compilation of singlet oxygen yields from biologically relevant molecules *Photochem. Photobiol.* **70** 391–475

Rehpenn A, Walter A and Storch G 2021 Molecular editing of flavins for catalysis *Synthesis* **53** 2583–93

Reiffers A, Ziegenbein C T, Engelhardt A, Kühnemuth R, Gilch P and Czekelius C 2018 Impact of mono-fluorination on the photophysics of the flavin chromophore *Photochem. Photobiol.* **94** 667–76

Reis J M, Burns D C and Woolley G A 2014 Optical control of protein–protein interactions via blue light-induced domain swapping *Biochemistry* **53** 5008–16

Reis J M and Woolley G A 2016 Photocontrol of protein function using photoactive yellow protein *Optogenetics: Methods and Protocols* ed A Kianianmomeni (New York: Humanity Press) pp 79–92

Reiter S, Bäuml L, Hauer J and de Vivie-Riedle R 2022 Q-band relaxation in chlorophyll: new insights from multireference quantum dynamics *Phys. Chem. Chem. Phys.* **24** 27212–23

Reja S I, Minoshima M, Hori Y and Kikuchi K 2021 Near-infrared fluorescent probes: next-generation tool for protein-labeling applications *Chem. Sci.* **12** 3437–47

Remeeva A *et al* 2020 Effects of proline substitutions on the thermostable LOV domain from *Chloroflexus aggregans* *Crystals* **10** 256

Remington S J, Wachter R M, Yarbrough D K, Branchaud B, Anderson D C, Kallio K and Lukyanov K A 2005 zFP538, a yellow-fluorescent protein from *Zoanthus*, contains a novel three-ring chromophore *Biochemistry* **44** 202–12

Remington S J 2011 Green fluorescent protein: a perspective *Prot. Sci.* **20** 1509–19

Repina N A, Rosenbloom A, Mukherjee A, Schaffer D V and Kane R S 2017 At light speed: advances in optogenetic systems for regulating cell signaling and behavior *Annu. Rev. Chem. Biomol. Eng.* **8** 13–39

Resch-Genger U, Grabolle M, Cavaliere-Jaricot S and Nitschke R 2008 Quantum dots versus organic dyes as fluorescent labels *Nat. Meth.* **5** 763–75

Rickgauer J P and Tank D W 2009 Two-photon excitation of channelrhodopsin-2 at saturation *PNAS* **106** 15025–30

Rinaldi J, Gallo M, Klinke S, Paris G, Bonomi H R, Bogomolni R A, Cicero D O and Goldbaum F A 2012 The β-scaffold of the LOV domain of the *Brucella* light-activated histidine kinase is a key element for signal transduction *J. Mol. Biol.* **420** 112–27

Ritter E, Piwowarski P, Hegemann P and Bartl F J 2013 Light-dark adaption of channelrhodopsin C128T mutant *J. Biol. Chem.* **288** 10451–8

Ritz T, Yoshii T, Helferich-Foerster C and Ahmad M 2010 Cryptochrome: a photoreceptor with the properties of a magnetoreceptor? *Commun. Integ. Biol.* **3** 24–7

Rivera-Cancel G, Ko W-H, Tomchick D R, Correa F and Gardner K H 2014 Full-length structure of a monomeric histidine kinase reveals basics for sensory regulation *PNAS* **111** 17839–44

Rizzini L *et al* 2011 Perception of UV-B by the *Arabidopsis* UVR8 protein *Science* **332** 103–6

Rizzo G and Laganà A S 2020 A review of vitamin B12 *Molecular Nutrition Vitamins* ed V B Patel (London: Academic Press) pp 105–29

Robbins R J, Fleming G R, Beddard G S, Robinson G W, Thistlethwaite P J and Woolfe G J 1980 Photophysics of aqueous tryptophan: pH and temperature effects *J. Am. Chem. Soc.* **102** 6271–9

Robeson C D, Blum W P, Dieterle J M, Cawley J D and Baxter J G 1955 Chemistry of Vitamin A. XXV. Geometrical isomers of vitamin A aldehyde and an isomer of its α-ionone analog *J. Am. Chem. Soc.* **77** 4120–5

Rocha-Rinza T, Christiansen O, Rahbek D B, Klaerke B, Andersen L H, Lincke K and Nielsen M B 2010 Spectroscopic implications of the electron donor-acceptor effect in the photoactive yellow protein chromophore *Chem. Eur. J.* **16** 11977–84

Rockwell N C, Su Y-S and Lagarias J C 2006 Phytochrome structure and signaling mechanisms *Annu. Rev. Plant Biol.* **57** 837–58

Rockwell N C and Lagarias J C 2010 A brief history of phytochromes *ChemPhysChem* **11** 1172–80

Rockwell N C, Martin S S, Gulevich A G and Lagarias J C 2012 Phycoviolobilin formation and spectral tuning in the DXCF cyanobacteriochrome subfamily *Biochemistry* **51** 1449–63

Rockwell N C, Duanmu D, Martin S S, Bachy C, Price D C, Bhattacharya D, Worden A Z and Lagarias J C 2014 Eukaryotic algal phytochromes span the visible spectrum *PNAS* **111** 3871–6

Rockwell N C, Martin S S and Lagarias J C 2015 Identification of DXCF cyanobacteriochrome lineages with predictable photocycles *Photochem. Photobiol. Sci.* **14** 929–41

Rockwell N C, Martin S S and Lagarias J C 2016 Identification of cyanobacteriochromes detecting far-red light *Biochem.* **55** 3907–19

Rockwell N C and Lagarias J C 2017 Phytochrome diversification in cyanobacteria and eukaryotic algae *Curr. Opin. Plant Biol.* **37** 87–93

Rockwell N C, Martin S S and Lagarias J C 2017 There and back again: loss and reacquisition of two-Cys photocycles in cyanobacteriochromes *Photochem. Photobiol.* **93** 741–54

Rockwell N C and Lagarias J C 2020 Phytochrome evolution in 3D: deletion, duplication, and diversification *New Physiologist* **225** 2283–300

Rodgers J *et al* 2021 Using a bistable animal opsin for switchable and scalable optogenetic inhibition of neurons *EMBO Rep.* **22** e51866

Rodriguez-Rozada S, Wietek J, Tenedini F, Sauter K, Dhiman N, Hegemann P, Soba P and Wiegert J S 2022 Aion is a bistable anion-conducting channelrhodopsin that provides temporarily extended and reversible neuronal silencing *Commun. Biol.* **5** 687

Rodriguez E A, Tran G N, Gross L A, Crisp J L, Shu X, Lin J Y and Tsien R Y 2016 A far-red fluorescent protein evolved from a cyanobacterial phycobiliprotein *Nat. Meth.* **13** 763–9

Romeo M V, López-Martinez E, Berganza-Granda J, Goñi-de-Cerio F and Cortajarena A L 2021 Biomarker sensing platforms based on fluorescent metal nanoclusters *Nanoscale Adv.* **3** 1331–41

Romling U, Gomelsky M and Galperin M Y 2005 C-di-GMP: the dawning of a novel bacterial signalling system *Mol. Microbiol.* **57** 629–39

Rose T, Goltstein P M, Portugues R and Griesbeck O 2014 Putting a finishing touch on GECIs *Front. Mol. Neurosci.* **7** 88

Ross W N 2012 Understanding calcium waves and sparks in central neurons *Nat. Rev.* **13** 157–68

Rost B R, Wietek J, Yizhar O and Schmitz D 2022 Optogenetics at the presynapse *Nat. Neurosci.* **25** 984–98

Rost B R, Schneider-Warme F, Schmitz D and Hegemann P 2017 Optogenetic tools for subcellular applications in neuroscience *Neuron* **96** 572–603

Rousseau B J G, Migliore A, Stanley R J and Beratan D N 2018 Determinants of photolyase's DNA repair mechanism in mesophiles and extremophiles *J. Am. Chem. Soc.* **140** 2853–61

Rousseau B J G, Shafei S, Migliore A, Stanley R J and Beratan D N 2023 Adenine fine-tunes DNA photolyase's repair mechanism *J. Phys. Chem.* B **217** 2941–54

Rowan R, Warshel A, Sykes B D and Karplus M 1974 Conformations of retinal isomers *Biochemistry* **13** 970–80

Rowland C E *et al* 2015 Electric field modulation of semiconductor quantum dot photoluminescence: insights into the design of robust voltage-sensitive cellular imaging probes *Nano Lett.* **15** 6848–54

Roy P P, Abe-Yoshizumi R, Kandori H and Buckup T 2019 Point mutation of *Anabaena* sensory rhodopsin enhances ground-state hydrogen out-of-plane wag Raman activity *J. Phys. Chem. Lett.* **10** 1012–7

Rozenberg A, Oppermann J, Wietek J, Lahore R G F, Sandaa R-A, Bratbak G, Hegemann P and Béjà O 2020 Lateral gene transfer of anion-conducting channelrhodopsins between green algae and giant viruses *Curr. Biol.* **30** 4910–20

Rozenberg A, Inoue K, Kandori H and Béjà O 2021 Microbial rhodopsins: the last two decades *Annu. Rev. Microbiol.* **75** 427–47

Rozenberg A *et al* 2022 Rhodopsin-bestrophin fusion proteins from unicellular algae form gigantic pentameric ion channels *Nat. Struct. Mol. Biol.* **29** 592–603

Rredhi A, Petersen J, Schubert M, Li W, Oldemeyer S, Li W, Westermann M, Wagner V, Kottke T and Mittag M 2021 DASH cryptochrome 1, a UV-A receptor, balances the photosynthetic machinery of *Chlamydomonas reinhardtii New Phytol.* **232** 610–24

Rupenyan A, van Stokkum I H M, Arents J C, van Grondelle R, Hellingwerf K and Groot M L 2008 Characterization of the primary photochemistry of proteorhodopsin with femtosecond spectroscopy *Biophys. J.* **94** 4020–30

Rury A S, Wiley T E and Sension J 2015 Energy cascades, excited state dynamics, and photochemistry in Cob(III)alamins and ferric porphyrins *Acc. Chem. Res.* **48** 860–7

Rusaczonek A, Czarnocka W, Willems P, Sujkowska-Rybkowska M, van Breusegem F and Karpiński S 2021 Phototropin 1 and 2 influence photosynthesis, UV-C induced photo-oxidative stress responses, and cell death *Cells* **10** 200

Ryu M-H, Moskvin O V, Siltberg-Liberles J and Gomelsky M 2010 Natural and engineered photoactivated nucleotidyl cyclases for optogenetic applications *J. Biol. Chem.* **8** 41501–8

Ryu M-H, Kang I-H, Nelson M D, Jensen T M, Lyuksyutova A I, Siltberg-Liberles J, Raizen D M and Gomelsky M 2014 Engineering adenylate cyclases regulated by near-infrared window light *PNAS* **111** 10167–72

Sabnis R W 2015 *Handbook of Fluorescent Dyes and Probes* (Hoboken, NJ: Wiley)

Sabzalipoor H, Karimi E, Nikkhah M, Abbasian S and Moshaii A 2022 Metal enhanced fluorescence of different metallic nanoclusters deposited on silver dendritic nanostructures *Micro Nano Lett.* **17** 114–23

Sahel J-A *et al* 2021 Partial recovery of visual function in a blind patient after optogenetic therapy *Nat. Med.* **27** 1223–9

Salcedo E, Zheng L, Phistry M, Bagg E E and Britt S G 2003 Molecular basis for ultraviolet vision in invertebrates *J. Neurosci.* **23** 10873

Saliminasab M, Yamazaki Y, Palmateer A, Harris A, Schubert L, Langner P, Heberle J, Bondar A-N and Brown L S 2023 A proteorhodopsin-related photosensor extends the repertoire of structural motifs employed by sensory rhodopsins *J. Phys. Chem.* B **127** 7872–86

Salinus F, Rojas V, Delgado V, López J, Agosin E and Larrondo L F 2018 Fungal light-oxygen-voltage domains for optogenetic control of gene expression and flocculation in yeast *mBio* **9** e00626–18

Salomon M, Christie J M, Knieb E, Lempert U and Briggs W R 2000 Photochemical and mutational analysis of the FMN-binding domains of the plant blue light receptor, photo-tropin *Biochemistry* **39** 9401–10

Salzmann S and Marian C M 2009 The photophysics of alloxazine: a quantum chemical investigation in vacuum and solution *Photochem. Photobiol. Sci.* **8** 1655–66

Samhadaneh D M, Mandl G A, Han Z, Mahjoob M, Weber S C, Tuznik M, Rudko D A, Capobianco J A and Stochaj U 2020 Evaluation of lanthanide-doped upconverting nano-particles for *in vitro* and *in vivo* applications *ACS Appl. Bio. Mater.* **3** 4358–69

Sample V, Newman R H and Zhang J 2009 The structure and function of fluorescent proteins *Chem. Soc. Rev.* **38** 2852–64

Sancar A 2003 Structure and function of DNA photolyase and cryptochrome blue-light photo-receptors *Chem. Rev.* **103** 2203–37

Sancar A 2004 Regulation of the mammalian circadian clock by cryptochrome *J. Biol. Chem.* **279** 34079–82

Sanyal S K, Sharma K, Bisht D, Sharma S, Sushmita K, Kateriya S and Pandey G K 2023 Role of calcium sensor protein module CBL-CIPK in abiotic stress and light signaling responses in green algae *Int. J. Biol. Macromol.* **237** 124163

Sato Y, Iwata T, Tokutomi S and Kandori H 2005 Reactive cysteine is protonated in the triplet excited state of the LOV2 domain in *Adiantum* phytochrome3 *J. Am. Chem. Soc.* **127** 1088–9

Sawa M, Nusinow D A, Kay S A and Imaizumi T 2007 FKF1 and GIGANTEA complex formation is required for day-length measurement in *Arabidopsis Science* **318** 261–5

Scheib U, Broser M, Constantin O M, Yang S, Gao S, Mukherjee S, Stehfest K, Nagel G, Gee C E and Hegemann P 2018 Rhodopsin-cyclases for photocontrol of cGMP/cAMP and 2.3 Å structure of the adenylyl cyclase domain *Nat. Commun.* **9** 2046

Scheer H 2006 An overview of chlorophylls and bacteriochlorophylls: biochemistry, biophysics, functions and applications *Chlorophylls and Bacteriochlorophylls: Biochemistry, Biophysics, Functions and Applications* ed B Grimm, R J Porra, W Rüdiger and H Scheer (Dordrecht: Springer) pp 1–26

Schleissner P, Szundi I, Chen E, Li H, Spudich J L and Kliger D S 2023 Isospectral intermediates in the photochemical reaction cycle of anion channelrhodopsin *GtACR1 Biophysical J.* **122** 4091–103

Schleicher E, Kowalczyk R M, Kay C W M, Hegemann P, Bacher A, Fischer M, Bittl R, Richter G and Weber S 2004 On the reaction mechanism of adduct formation in LOV domains of the plant blue-light receptor phototropin *J. Am. Chem. Soc.* **126** 11067–76

Schmidt A J, Ryjenkov D A and Gomelsky M 2005 The ubiquitous protein domain EAL is a cyclic diguanylate-specific phosphodiesterase: enzymatically active and inactive EAL domains *J. Bacteriol.* **187** 4774–81

Schmidt D and Cho Y K 2015 Natural photoreceptors and their application to synthetic biology *Trends Biotechnol.* **33** 80–91

Schnabel J, Hombach P, Waksman T, Giuriani G and Petersen J 2018 A chemical genetic approach to engineer phototropin kinases for substrate labeling *J. Biol. Chem.* **293** 5613–23

Schneidewind J, Krause F, Bocola M, Stadler A M, Davari M D, Schwaneberg U, Jaeger K-E and Krauss U 2019 Consensus model of a cyanobacterial light-dependent protochlorophyllide oxidoreductase in its pigment-free apo-form and photoactive ternary complex *Commun. Biol.* **2** 351

Schnermann M J and Lavis L D 2023 Rejuvenating old fluorophores with new chemistry *Curr. Opin. Chem. Biol.* **75** 102335

Schneider F, Grimm C and Hegemann P 2015 Biophysics of channelrhodopsin *Annu. Rev. Biophys.* **44** 167–86

Scholtz L, Eckert J G, Elahi T, Lübkemann F, Hübner O, Bigall N C and Resch-Genger U 2022 Luminescence encoding of polymer microbeads with organic dyes and semiconductor quantum dots during polymerization *Sci. Rep.* **12** 12061

Schröder-Lang S, Schwärzel M, Seifert R, Stünker T, Kateriya S, Looser J, Watanabe M, Kaupp U B, Hegemann P and Nagel G 2007 Fast manipulation of cellular cAMP level by light *in vivo Nat. Meth.* **4** 39–42

Schroeder C, Werner K, Otten H, Krätzig S, Schwalbe H and Essen L-O 2008 Influence of a joining helix on the BLUF domain of the YcgF photoreceptor from *Escherichia coli ChemBioChem* **9** 2463–73

Schweitzer C and Schmidt R 2003 Physical mechanisms of generation and deactivation of singlet oxygen *Chem. Rev.* **103** 1685–758

Schwerdtfeger C and Linden H 2003 VIVID is a flavoprotein and serves as a fungal blue light photoreceptor for photoadaption *EMBO J.* **22** 4846–55

Sen S, Kar R K, Borin V A and Schapiro I 2022 Insight into the isomerization mechanism of retinal proteins from hybrid quantum mechanics/molecular mechanics simulations *WIREs Comput. Mol. Sci.* **12** e1562

Senge M O, Sergeeva N N and Hale K J 2021 Classic highlights in porphyrin and porphyrinoid total synthesis and biosynthesis *Chem. Soc. Rev.* **50** 4730–89

Seong J and Lin M Z 2021 Optobiochemistry: genetically encoded control of protein activity by light *Annu. Rev. Biochem.* **90** 475–501

Sephus C D, Fer E, Garcia A K, Adam Z R, Schwieterman E W and Kacar B 2022 Earliest photic zone niches probed by ancestral microbial rhodopsins *Mol. Biol. Evol.* **39** msac100

Shah A M, Ishizaka S, Cheng M Y, Wang E H, Bautista A R, Levy S, Smerin D, Sun G and Steinberg G K 2017 Optogenetic neuronal stimulation of the lateral cerebellar nucleus promotes persistent functional recovery after stroke *Sci. Rep.* **7** 46612

Shah S, Liu J-J, Pasquale N, Lai J, McGowan H, Pang Z P and Lee K-B 2015 Hybrid upconversion nanomaterials for optogenetic neuronal control *Nanoscale* **7** 16571–7

Shamseldin H E *et al* 2020 The morbid genome of ciliopathies: an update *Genet. Med.* **22** 1051–60

Shaner N C, Campbell R E, Steinbach P A, Giepmans B N G, Palmer A E and Tsien R Y 2004 Improved monomeric red, orange and yellow fluorescent proteins derived from *Discosoma* sp. red fluorescent protein *Nat. Biotechnol.* **22** 1567–72

Shao J, Wang M, Yu G, Zhu S, Yu Y, Heng B C, Wu J and Ye H 2018 Synthetic far-red light-mediated CRISPR-dCas9 device for inducing functional neuronal differentiation *Proc. Natl. Acad. Sci.* **115** E6722

Shao K, Zhang X, Li X, Hao Y, Huang X, Ma M, Zhang M, Yu F, Liu H and Zhang P 2020 The oligomeric structure of plant cryptochromes *Nat. Struct. Mol. Biol.* **27** 480–8

Sharma A S, Ali S, Sabarinathan D, Murugavelu M, Li H and Chen Q 2021 Recent progress on graphene quantum dots-based fluorescence sensors for food safety and quality assessment applications *Compr. Rev. Food Sci. Food Saf.* **20** 5765–801

Sharma M (ed) 2023 *Fluorescent Proteins: Methods and Protocols* (New York: Humana)

Shcherbakova D M and Verkhusha V V 2014 Chromophore chemistry of fluorescent proteins controlled by light *Curr. Opin. Chem. Biol.* **20** 60–8

Shcherbakova D M, Shemetov A A, Kaberniuk A A and Verkhusha V V 2015 Natural photoreceptors as a source of fluorescent proteins, biosensors, and optogenetic tools *Annu. Rev. Biochem.* **84** 519–50

She M, Wang Z, Luo T, Yin B, Liu P, Liu J, Chen F, Zhang S and Li J 2018 Fluorescent probes guided by a new practical performance regulation strategy to monitor glutathione in living systems *Chem. Sci.* **9** 8065–70

Shemesh O A, Tanese D, Zampini V, Linghu C, Piatkevich K, Ronzitti E, Papagiakoumou E, Boyden E S and Emiliani V 2017 Temporally precise single-cell-resolution optogenetics *Nat. Neurosci.* **20** 1796–806

Shemetov A A *et al* 2021 A near-infrared genetically encoded calcium indicator for *in vivo* imagining *Nat. Biotechnol.* **39** 368–77

Shen Y, Campbell R E, Côte D C and Paquet M-E 2020 Challenges for therapeutic applications of opsin-based optogenetic tools in humans *Front. Neural Circuits* **14** 41

Shevchenko *et al* 2017 Inward H^+ pump xenorhodopsin: mechanism and alternative optogenetic approach *Sci. Adv.* **3** e1603187

Shi F, Kawano F, Park S-h E, Komazaki S, Hirabayashi Y, Polleux F and Yazawa M 2018 Optogenetic control of endoplasmic reticulum-mitochondria tethering *ACS Synth. Biol.* **7** 2–9

Shi Y, Lyu Z, Chen R, Nguyen Q N and Xia Y 2021 Noble-metal nanocrystals with controlled shapes for catalytic and electrocatalytic applications *Chem. Rev.* **121** 649–735

Shibata K, Nakasone Y and Terazima M 2018 Photoreaction of BlrP1: the role of a nonlinear photo-intensity sensor *Phys. Chem. Chem. Phys.* **20** 8133–42

Shibukawa A, Kojima K, Nakajima Y, Yoshizawa S and Sudo Y 2019 Photochemical character-ization of a new heliorhodopsin from the Gram-negative eubacterium *Bellilinea caldifistulae* (BcHeR) and comparison with heliorhodopsin-48C12 *Biochemistry* **58** 2934–43

Shihoya W *et al* 2019 Crystal structure of heliorhodopsin *Nature* **574** 132–6

Shimoda Y *et al* 2022 Optogenetic stimulus-triggered acquisition of seizure resistance *Neurobiol. Dis.* **163** 105602

Shimomura O, Johnson F H and Saiga Y 1962 Extraction, purification and properties of Aequrin, a bioluminescent protein from luminous hydromedusan, *Aequorea J. Cell. Comp. Physiol.* **59** 223–39

Shimomura O 1979 Structure of the chromophore of *Aequorea* green fluorescent protein *FEBS Lett.* **104** 220–2

Shin G *et al* 2017 Flexible near-field wireless optoelectronics as subdermal implants for broad applications in optogenetics *Neuron* **93** 509–21

Shinkai S, Kameoka K, Honda N, Ueda K and Manabe O 1985 Reactivity studies of roseoflavin analogues: a correlation between reactivity and absorption maxima *J. Chem. Soc. Chem. Commun.* **1985** 673–4

Shionoya T, Singh M, Mizuno M, Kandori H and Mizutani Y 2021 Strongly hydrogen-bonded Schiff base and adjoining polyene twisting in the retinal chromophore of schizorhodopsins *Biochemistry* **60** 3050–7

Shirai F and Hayashi-Takagi A 2017 Optogenetics: applications in psychiatric research *Psych. Clin. Neurosci.* **71** 363–72

Shirdel J, Zirak P, Penzkofer A, Breitkreuz H and Wolf E 2008 Absorption and fluorescence spectroscopic characterization of the ciradian blue-light photoreceptor cryptochrome from *Drosophila melanogaster* (dCry) *Chem. Phys.* **352** 35–47

Shu X, Lev-Ram V, Deerinck T J, Qi Y, Ramko E B, Davidson M W, Jin Y, Ellisman M H and Tsien R Y 2011 A genetically encoded tag for correlated light and electron microscopy of intact cells, tissues and organisms *PLOS Biol.* **9** e1001041

Shu X, Royant A, Lin M Z, Aguilera T A, Lev-Ram V, Steinbach P A and Tsien R Y 2009 Mammalian expression of infrared fluorescent proteins engineered from bacterial phytochrome *Science* **324** 804–7

Sidor M M, Davidson T J, Tye K M, Warden M R, Deisseroth K and McClung C A 2015 *In vivo* optogenetic stimulation of the rodent central nervous system *J. Vis. Exp.* **95** e 51483

Silva E and Edwards A M (ed) 2006 *Flavins: Photochemistry and Photobiology* (London: Royal Society of Chemistry)

Sikorski M, Khmelinsskii I and Sikorska E 2021 Spectral properties of flavins *Flavin-Based Catalysis: Principles and Applications* ed R Cibulka and M W Fraaije (Weinheim: Wiley-VCH) pp 67–96

Silapetere A *et al* 2022 QuasAr Odyssey: the origin of fluorescence and its voltage sensitivity in microbial rhodopsins *Nat. Commun.* **13** 5501

Silva P J and Cheng Q 2022 An alternative proposal for the reaction mechanism of light-dependent protochlorophyllide oxidoreductase *ACS Catal.* **12** 2589–605

Silverman A D, Karim A S and Jewett M C 2020 Cell-free gene expression: an expanded repertoire of applications *Nat. Rev. Genet.* **21** 151–70

Simkin A J, Kapoor L, Doss C G P, Hofmann T A, Lawson T and Ramamoorthy S 2022 The role of photosynthesis related pigments in light harvesting, photoprotection and enhancement of photosynthetic yield in planta *Photosynth. Res.* **152** 23–42

Simon C-J, Sahel J-A, Duebel J, Herlitze S and Dalkara D 2020 Opsins for vision restoration *Biochem. Biophys. Res. Commun.* **527** 325–30

Simon J D 1990 Solvation dynamics: new insights into chemical reaction and relaxation processes *Pure Appl. Chem.* **62** 2243–50

Sineshchekov V A 1995 Photophysics and photochemistry of the heterogeneous phytochrome system *Biochim. Biophys. Acta* **1228** 125–64

Sineshcheov O A, Govorunova E G, Wang Y and Spudich J L 2023 Sequential absorption of two photons creates a bistable form of RubyACR responsible for its strong desensitization *PNAS* **120** e2301521120

Singh S K, Mazumder S, Vincy A, Hiremath N, Kumar R, Banerjee I and Vankayala R 2023 Review of photoresponsive plasmonic nanoparticles that produce reactive chemical species for photodynamic therapy of cancer and bacterial infections *ACS Appl. Nano Mater.* **6** 1508–21

Sivaprakasam S, Mani V, Balasubramaniyan N and Abraham D R 2022 Cyanobacterial phytochromes in optogenetics *Epigenetics to Optogenetics – A new Paradigm in the Study of Biology* ed M Anwar, Z Farooq, R A Rather, M Tauseef and T Heinbockel (London: IntechOpen) ch 6

Skopintsev P *et al* 2020 Femtosecond-to-millisecond structural changes in a light-driven sodium pump *Nature* **583** 314–8

Slonimskiy Y B, Maksimov E G and Sluchannko N N 2020 Fluorescence recovery protein: a powerful yet underexplored regulator of photoprotection in cyanobacteria *Photochem. Photobiol. Sci.* **19** 763–75

Smart A D, Pache R A, Thomsen N D, Kortemme T and Davis G W 2017 Engineering a light-activated caspase-3 for precise ablation of neurons *in vivo PNAS* **114** E8174–83

Smetters D, Majewska A and Yuste R 1999 Detecting action potentials in neuronal populations with calcium imaging *Methods* **18** 215–21

Smirnova E Y and Zaitsev A V 2019 Use of optogenetic methods to study and suppress epileptic activity (review) *Neuroscience and Behavioral Physiology* **49** 1083–8

Soloviev V N, Eichhöfer A, Fenske D and Banin U 2001 Size-dependent optical spectrocopy of a homologous series of CdSe cluster molecules *J. Am. Chem. Soc.* **123** 2354–64

Song S-H, Dick B, Penzkofer A, Pokorny R, Batschauer A and Essen L-O 2006 Absorption and fluorescence spectroscopic characterization of cryptochrome 3 from *Arabidopsis thaliana J. Photochem. Photobiol. B: Biol.* **85** 1–16

Song S-H, Dick B and Penzkofer A 2007a Photoinduced reduction of flavin mononucleotide in aqueous solutions *Chem. Phys.* **332** 55–65

Song S-H, Dick B, Penzkofer A and Hegemann P 2007b Photo-reduction of flavin mononucleotide to semiquinone form in LOV domain mutants of blue-light receptor phot from *Chlamydomonas reinhardtii J. Photochem. Photobiol. B: Biol.* **87** 37–48

Song Y H, Estrada D A, Johnson R S, Kim S K, Lee S Y and MacCoss M J 2014 Distinct roles of FKF1, GIGANTEA, and ZEITLUPE proteins in the regulation of CONSTANS stability in *Arabidopsis* photoperiodic flowering *PNAS* **111** 17672–7

Soumya K, More N, Choppadandi M, Aishwarya D A, Singh G and Kappusetti G 2023 A comprehensive review on carbon quantum dots as an effective photosensitizer and drug delivery system for cancer treatment *Biomed. Technol.* **4** 11–20

Sovdat T, Bassolino G, Liebel M, Schnedermann C, Fletcher P and Kukura P 2012 Backbone modification of retinal induces protein-like excited state dynamics in solution *J. Am. Chem. Soc.* **134** 8318–20

Sowmya N, Mohanty D, Nirosha B, Kumar N U and Patra P K 2023 A systematic review of metallic nanoparticles: synthesis, biological activities & applications *J. Pharmaceut. Neg. Results* **14** 2525–33

Spagnuolo G, Genovese F, Fortunato L, Simeone M, Rengo C and Tatullo M 2020 The impact of optogenetics on regenerative medicine *Appl. Sci.* **10** 173

Spangler S M and Bruchas M R 2017 Optogenetic approaches for dissecting neuromodulation and GPCR signaling in neural circuits *Curr. Opin. Pharmacol.* **32** 56–70

Spiess E *et al* 2005 Two-photon excitation and emission spectra of green fluorescent protein variants ECFP, EGFP and EYFP *J. Microsc.* **217** 200–4

Sprenger W W, Hoff W D, Armitage J P and Hellingwerf K J 1993 The eubacterium *Ectothiorhodospira halophila* is negatively phototactic, with a wavelength dependence that fits the absorption spectrum of the photoactive yellow protein *J. Bacteriol.* **175** 3096–104

Squires A H, Wang Q, Dahlberg P D and Moerner W E 2022 A bottom-up perspective on photodynamics and photoprotection in light-harvesting complexes using anti-Brownian trapping *J. Chem. Phys.* **156** 070901

Stanley R J and MacFarlane A W IV 2000 Ultrafast excited state dynamics of oxidized flavins: direct observations of quenching by purines *J. Phys. Chem.* A **104** 6899–906

Stasiuk R, Krucorí T and Matlakowska R 2021 Biosynthesis of tetrapyrrole cofactors by bacterial community inhabiting porphyrin-containing shale rock (fore-sudetic monocline) *Molecules* **26** 6746

St-Pierre F, Chavarha M and Lin M Z 2015 Designs and sensing mechanisms of genetically encoded fluorescent voltage indicators *Curr. Opin. Chem. Biol.* **27** 31–8

Stern M A, Skelton H, Fernandez A M, Gutekunst C-A N, Berglund K and Gross R E 2022a Bioluminescence-optogenetics: a practical guide *Bioluminescence: Methods and Protocols* Vol 2 4th Ed. ed Sung-Bae Kim (New York: Humana Press) pp 333–46

Stern M A, Skelton H, Fernandez A M, Gutekunst C-A N, Gross R E and Berglund K 2022b Applications of bioluminescence-optogenetics in rodent models *Bioluminescence: Methods and Protocols* **Vol 2** 4th Ed. ed Sung-Bae Kim (New York: Humana Press) pp 347–63

Stevenson K, Massey V and Williams C (ed) 1997 *Flavins and Flavoproteins* (Calgary: University of Calgary)

Stevens F C 1983 Calmodulin: an introduction *Can. J. Biochem. Cell Biol.* **61** 906–10

Stewart S, Wei Q and Sun Y 2021 Surface chemistry of quantum-sized metal nanoparticles under light illumination *Chem. Sci.* **12** 1227–39

Stirbet A, Lazár D, Guo Y and Govindjee G 2020 Photosynthesis: basics, history and modelling *Ann. Bot.* **126** 511–37

Stich T A, Brooks A J, Buan N R and Brunold T C 2003 Spectroscopic and computational studies of Co^{3+}-corrinoids: spectroscopic and electronic properties of the B_{12} cofactors and biologically relevant precursors *J. Am. Chem. Soc.* **125** 5897–914

Stierl M *et al* 2011 Light modulation of cellular cAMP by a small bacterial photoactivated adenylyl cyclase, bPAC, of the soil bacterium *Beggiatoa J. Biol. Chem.* **286** 1181–8

Stierl M, Penzkofer A, Kennis J T M, Hegemann P and Mathes T 2014 Key residues for the light regulation of the blue light-activated adenylyl cyclase from *Beggiatoa* sp *Biochem.* **53** 5121–30

Storace D, Rad M S, Kang B, Cohen L B, Hughes T and Baker B J 2016 Toward better genetically encoded sensors of membrane potential *Trends Neurosci.* **39** 277–89

Streng C, Hartmann J, Leister K, Krauß N, Lamparter T, Krankenberg-Dinkel N, Weth F, Bastmeyer M, Yu Z and Fischer R 2021 Fungal phytochrome chromophore biosynthesis at mitochondria *EMBO J.* **40** e108083

Strickland D, Moffat K and Sosnick T R 2008 Light-activated DNA binding in a designed allosteric protein *PNAS* **105** 10709–14

Strickland D, Yao X, Gawlak G, Rosen M K, Gardner K H and Sosnick T R 2010 Rationally improving LOV domain-based photoswitches *Nat. Meth.* **7** 623–6

Strickland D, Lin Y, Wagner E, Hope C M, Zayner J, Antoniou C, Sosnick T R, Weiss E L and Glotzer M 2012 TULIPs: tunable, light-controlled interacting protein tags for cell biology *Nat. Meth.* **9** 379–84

Stroh A (ed) 2018 *Optogenetics: A Roadmap* (Berlin: Springer)

Subach O M *et al* 2022 cNTnC and fYTnC2, genetically encoded green calcium indicators based on troponin C from fast animals *Int. J. Mol. Sci.* **23** 14614

Subach O M, Vlaskina A V, Agapova Y K, Nikolaeva A Y, Anokhin K V, Piatkevich K D, Patrushev M V, Boyko K M and Subach F V 2023a Blue-to-red TagFT, mTagFT, mTsFT, and green-to-FarRed mNeptusFT2 proteins, genetically encoded true and tandem fluorescent timers *Int. J. Mol. Sci.* **24** 3279

Subach O M, Vlaskina A V, Agapova Y K, Nikolaeva A Y, Varizhuk A M, Podgorny O V, Piatkevich K D, Patrushev M V, Boyko K M and Subach F V 2023b YTnC2, an improved genetically-encoded calcium indicator based on toadfish troponin C *FEBS Open Bio* **13** 2047–60

Subach O M, Varfolomeeva L, Vlaskina A V, Agapova Y K, Nikolaeva A Y, Piatkevich K D, Patrushev M V, Boyko K M and Subach F V 2023c FNCaMP, ratiometric green calcium indicator based on mNeonGreen protein *Biochem. Biophys. Res. Commun.* **665** 169–77

Sudo Y, Mizuno M, Wei Z, Takeuchi S and Tahara T 2014 The early steps in the photocycle of photosensor protein sensory rhodopsin I from *Salinibacter ruber J. Phys. Chem.* B **118** 1510–8

Suetsugu N and Wada M 2020 Signalling mechanism of phototropin-mediated chloroplast movement in *Arabidopsis J. Plant Biochem. Biotechnol.* **29** 580–9

Sugiura M, Tsunoda S P, Hibi M and Kandori H 2020 Molecular properties of new enzyme rhodopsins with phosphodiesterase activity *ACS Omega* **5** 10602–9

Sung Y-L, Wang T-W, Lin T-T and Lin S-F 2022 Optogenetics in cardiology: methodology and future applications *Int. J. Arrhythmia* **23** 9

Sureda-Vives M and Sarkisyan K S 2020 Bioluminescence-driven optogenetics *Life* **10** 318

Suzuki S, Sasaki S, Sairi A S, Iwai R, Tang B Z and Konishi G-i 2020 Principles of aggregation-induced emission: design of deactivation pathways for advanced AIEgens and applications *Angew. Chem. Int. Ed.* **59** 9856–67

Swartz T E, Corchnoy S B, Christie J M, Lewis J W, Szundi I, Briggs W R and Bogomolni R A 2001 The photocycle of a flavin-binding domain of the blue light photoreceptor phototropin *J. Biol. Chem.* **276** 36493–500

Swartz T E *et al* 2007 Blue-light-activated histidine kinases: two-component sensors in bacteria *Science* **317** 1090–3

Syed A J and Anderson J C 2021 Applications of bioluminescence in biotechnology and beyond *Chem. Soc. Rev.* **50** 5668–705

Takala H, Edlund P, Ihalainen J A and Westenhoff S 2020 Tips and turns of bacteriophytochrome photoactivation *Photochem. Photobiol. Sci.* **19** 1488–510

Takakado A, Nakasone Y and Terazima M 2017 Photoinduced dimerization of a photosensory DNA-binding protein EL222 and its LOV domain *Phys. Chem. Chem. Phys.* **19** 24855–65

Takakado A, Nakasone Y and Terazima M 2018 Sequenial DNA binding and dimerization processes of the photosensory protein EL222 *Biochemistry* **57** 1603–10

Takeuchi S and Tahara T 1997 Ultrafast fluorescence study on the excited singlet-state dynamics of all-*trans* retinal *J. Phys. Chem.* A **101** 3052–60

Takano H *et al* 2011 Involvement of CarA/LitR and CRP/FNR family transcriptional regulators in light-induced carotenoid production in *Thermus thermophilus J. Bacteriol.* **193** 2451–9

Takano H, Mise K, Hagiwara K, Hirata N, Watanabe S, Toriyabe M, Shiratori-Takano H and Ueda K 2015 Role and function of LitR, an adenosyl B_{12}-bound light-sensitive regulator of *Bacillus megaterium* QM B1551, in regulation of carotenoid production *J. Bacteriol.* **197** 2301–15

Takao T, Yamada D and Takarada T 2022 Mouse model for optogenetic genome engineering *Acta Med. Okayama* **76** 1–5

Takeda J 2021 Molecular mechanisms of UVR8-mediated photomorphogenesis derived from revaluation of action spectra *Photochem. Photobiol.* **97** 903–10

Takeda N, Yamashita A and Maeda Y 2003 Structure of the core domain of human cardiac troponin in the Ca^{2+}-saturated form *Nature* **424** 35–41

Takemoto K 2021 Optical manipulation of molecular function by chromophore-assisted light inactivation *Proc. Jpn. Acad. Ser.* B **97** 197–209

Tamayo R, Tischler A D and Camilli A 2005 The EAL domain protein VieA is a cyclic diguanylate phosphodiesterase *J. Biol. Chem.* **280** 33324–30

Tamayo R, Pratt J T and Camilli A 2007 Roles of cyclic diguanylate in the regulation of bacterial pathogenesis *Annu. Rev. Microbiol.* **61** 131–48

Tamogami J and Kikukawa T 2021 Functional mechanism of proton pump-type rhodopsins found in various microorganisms as a potential effective tool in optogenetics *Epigenetics to Optogenetics – A New Paradigm in the Study of Biology* ed M Anwar, Z Farooq, R A Rather, M Tanseef and T H Heinbockel (London: IntechOpen) ch 7

Tan P, He L, Huang Y and Zhou Y 2022 Optophysiology: illuminating cell physiology with optogenetics *Physiol. Rev.* **102** 1263–325

Tanaka R and Tanaka A 2007 Tetrapyrrole biosynthesis in higher plants *Annu. Rev. Plant Biol.* **58** 321–46

Tang K, Beyer H M, Zurbriggen M D and Gärtner W 2021 The red edge: bilin-binding photoreceptors as optogenetic tools and fluorescence reporters *Chem. Rev.* **121** 14906–56

Taniguchi M and Lindsey J S 2021 Absorption and fluorescence spectral database of chlorophylls and analogues *Photochem. Photobiol.* **97** 136–65

Tantama M, Hung Y P and Yellen G 2012 Optogenetic reporters: fluorescent protein-based genetically encoded indicators of signaling and metabolism in the brain *Prog. Brain Res.* **196** 235–63

Tanwar M, Khera L, Haokip N, Kaul R, Naorem A and Kateriya S 2017 Modulation of cyclic nucleotide-mediated cellular signaling and gene expression using photoactivated adenylyl cyclase as an optogenetic tool *Sci. Rep.* **7** 12048

Tanwar M, Sharma K, Moar P and Kateriya S 2018 Biochemical characterization of the engineered photoactivated guanylate cyclases from microbes expands optogenetic tools *Appl. Biochem. Biotechnol.* **185** 1014–28

Taslimi A, Vrana J D, Chen D, Borinskaya S, Mayer B J, Kennedy M J and Tucker C L 2014 An optimized optogenetic clustering tool for probing protein interaction and function *Nat. Commun.* **5** 4925

Tavan P and Schulten K 1987 Electronic excitations in finite and infinite polyenes *Phys. Rev.* B **36** 4337–58

Tenboer J *et al* 2014 Time-resolved serial crystallography captures high resolution intermediates of photoactive yellow protein *Science* **346** 1242–6

Terai Y, Sato R, Matsumara R, Iwai S and Yamamoto J 2020 Enhanced DNA repair by DNA photolyase bearing an artificial light-harvesting chromophore *Nucleic Acids Res.* **48** 10076–86

Terazima M 2021 Time-resolved detection of association/dissociation reactions and conformation changes in photosensor proteins for application in optogenetics *Biophys. Rev.* **13** 1053–9

Terskikh A *et al* 2000 'Fluorescent timer': protein that changes color with time *Science* **290** 1585–8

Thestrup T *et al* 2014 Optimized ratiometric calcium sensors for functional *in vivo* imaging of neurons and T lymphocytes *Nat. Methods* **11** 175–82

Thomson A J 1969 The polarized fluorescence spectra of some naturally occurring corrins *J. Am. Chem. Soc.* **91** 2780–5

Tian Y, Gao S, von der Heyde E L, Hallmann A and Nagel G 2018 Two-component cyclase opsins of green algae are ATP-dependent and light-inhibited guanylyl cyclases *BMC Biol.* **16** 144

Tian Y, Gao S and Nagel G 2021 An engineered membrane-bound guanylyl cyclase with light-switchable activity *BMC Biol.* **19** 54

Tian Y, Yang S, Nagel G and Gao S 2022a Characterization and modification of light-sensitive phosphodiesterases from choanoflagellates *Biomolecules* **12** 88

Tian Y, Gao S and Nagel G 2022b *In vivo* and *in vitro* characterization of cyclase and phosphodiesterase rhodopsins *Rhodopsin: Methods and Protocols* ed V Gordeliy (New York: Humana Press) pp 325–38

Tilbrook K, Arongaus A B, Binkert M, Heijde M, Yin R and Ulm R 2013 The UVR8 UV-B photoreceptor: perception, signaling and response *The Arabidopsis Book* **11** e0164

Tischer D and Weiner O D 2014 Illuminating cell signalling with optogenetic tools *Nat. Rev. Mol. Cell Biol.* **15** 551–8

Tiwari P and Tolwinski N S 2023 Using optogenetics to model cellular effects of Alzheimer's disease *Int. J. Mol. Sci.* **24** 4300

Toda M J, Mamun A A, Lodowski P and Kozlowski P M 2020 Why is CarH photolytically active in comparison to other B_{12}-dependent enzymes? *J. Photochem. Photobiol. B: Biol.* **209** 111919

Toda M J, Lodowski P, Mamun A A and Kozlowski P M 2022 Photoproduct formation in coenzyme B_{12}-dependent CarH via a singlet pathway *Photochem. Photobiol. B: Biol.* **232** 112471

Toettcher J E, Voigt C A, Weiner O D and Lim W A 2011a The promise of optogenetics in cell biology: interrogating molecular circuits in space and time *Nat. Meth* **8** 35–8

Toettcher J E, Gong D, Lim W A and Weiner O D 2011b Light control of plasma membrane recruitment using the Phy-PIF system *Methods Enzymol.* **497** 409–23

Tokonami S, Onose M, Nakasone Y and Terazima M 2022 Slow conformational changes of blue light sensor BLUF proteins in milliseconds *J. Am. Chem. Soc.* **144** 4080–90

Tomida S, Kitagawa S, Kandori H and Furutani Y 2021 Inverse hydrogen-bonding change between the protonated retinal Schiff base and water molecules upon photoisomerization in heliorhodopsin 48C12 *J. Phys. Chem.* B **125** 8331–41

Tomita H and Sugano E 2021 Optogenetics-mediated gene therapy for retinal diseases *Optogenetics* (Berlin: Springer) pp 535–43

Tønnesen J and Kobala M 2017 Epilepsy and optogenetics: can seizures be controlled by light? *Clin. Sci.* **131** 1605–16

Too L K *et al* 2022 Optogenetic restoration of high sensitivity vision with bReaChES, a red-shifted channelrhodopsin *Sci. Rep.* **12** 19312

Toohey J I 1965 A vitamin B_{12} compound containing no cobalt *PNAS* **54** 934–42

Torra J, Lafaye C, Signor L, Aumonier S, Flors C, Shu X, Nonell S, Gotthard G and Royant A 2019 Tailoring miniSOG: structural bases of the complex photophysics of a flavin-binding singlet oxygen photosensitizing protein *Sci. Rep.* **9** 2428

Torres A C, Vannini V, Font G, Saavdra L and Taranto 2018 Novel pathway for corrinoid compounds production in *Lactobacillus Front. Microbiol.* **9** 2256

Tossi V E, Regalado J J, Iannicelli J, Laino L E, Burrieza H B, Escandón A S and Pitta-Álvarez S I 2019 Beyond Arabidopsis: differential UV-B response mediated by UVR8 in diverse species *Front. Plant Sci.* **10** 780

Touriño C, Eban-Rothschild A and de Lecea L 2013 Optogenetics in psychiatric diseases *Cur. Opin. Neurobiol.* **23** 430–5

Tretyakova Y A, Pakhomov A A and Martynov V I 2007 Chromophore structure of the kindling fluorescent protein asFP595 from *Anemonia sulcata J. Am. Chem. Soc.* **129** 7748–9

Trozzi F, Wang F, Verkhivker G, Zoltowski B D and Tao P 2021 Dimeric allostery mechanism of the plant circadian clock photoreceptor ZEITLUPE *PLoS Comput. Biol.* **17** e1009168

Tschowri N, Busse S and Hengge R 2009 The BLUF-EAL protein YcgF acts as a direct anti-repressor in a blue-light response of *Escherichia coli Genes Dev.* **23** 522–34

Tschowri N, Lindenberg S and Heengge R 2012 Molecular function and potential evolution of the biofilm-modulating blue light-signalling pathway of *Escherichia coli Mol. Microbiol.* **85** 893–906

Tsien R Y 1980 New calcium indicators and buffers with high selectivity against magnesium and protons: design, synthesis, and properties of prototype structures *Biochemistry* **19** 2396–404

Tsien R Y 1981 A non-disruptive technique for loading calcium buffers and indicators into cells *Nature* **290** 527–8

Tsien R Y 1989 Fluorescent probes for cell signaling *Ann. Rev. Neurosci.* **12** 227–53

Tsien R Y 1998 The green fluorescent protein *Annu. Rev. Biochem.* **67** 509–44

Tsuboi T, Penzkofer A, Slyusareva E and Sizykh A 2011 Photoluminescence properties of fluorone dyes in bio-related films at low temperatures *J. Photochem. Photobiol. A: Chem.* **222** 336–42

Tsujimura M, Kojima K, Kawanishi S, Sudo Y and Ishikita H 2021 Proton transfer pathway in anion channelrhodopsin-1 *eLife* **10** e72264

Tsumjimura M, Chiba Y, Saito K and Ishikita H 2022 Proton transfer and conformational changes along the hydrogen bond network in heliorhodopsin *Commun. Biol.* **5** 1336

Tsujimura M, Tamura H, Saito K and Ishikita H 2022 Absorption wavelength along chromophore low-barrier hydrogen bonds *iScience* **25** 104247

Tsunoda S P, Sugiura M and Kandori H 2021 Molecular properties and optogenetic applications of enzymerhodopsins *Optogenetics: Light-Sensing Proteins and Their Applications in Neuroscience and Beyond* 2nd edn ed H Yawo, H Kandori, A Koizumi and R Kageyama (Berlin: Springer) pp 153–65

Tsutsui K, Imai H and Shichida Y 2007 Photoisomerization efficiency in UV-absorbing visual pigments: protein-directed isomerization of an unprotonated retinal Schiff base *Biochemistry* **46** 6437–45

Tsvetkov P O, Protasevich I I, Gilli R, Lafitte D, Lobachov V M, Haiwech J, Briand C and Makarov A A 1999 Apocalmodulin binds to the myosin light chain kinase calmodulin target site *J. Biol. Chem.* **274** 18161–4

Tu S-L and Lagarias J C 2005 The phytochromes *Handbook of Photosensory Receptors* ed W R Briggs and J L Spudich (Weinheim: Wiley) pp 121–49

Tucker K, Sridharan S, Adesnik H and Brohawn S G 2022 Cryo-EM structures of the channelrhodopsin ChRmine in lipid nanodiscs *Nat. Commun.* **13** 4842

Turkowyd B, Balinovic A, Virant D, Gölz Carnero H G, Caldana F, Endesfelder M, Bourgeois D and Endesfelder U 2017 A general mechanism of photoconversion of green-to-red fluorescent proteins based on blue and infrared light reduces phototoxity in live-cell single-molecule imaging *Angew. Chem. Int. Ed.* **56** 11634–9

Turner G K 1985 Measurement of light from chemical or biochemical reactions *Bioluminescence and Chemiluminescence: Instruments and Applications* ed K Van Dyke (Boca Raton, FL: CRC Press) Vol. I pp 43–78

Turro N J, Ramamurthy V and Scaiano H C 2009 *Principles of Molecular Photochemistry. An Introduction* (Sausalito, CA: University Science Books)

Tyagi A, Penzkofer A, Griese J, Schlichting I, Kirienko N V and Gomelsky M 2008 Photodynamics of blue-light-regulated phosphodiesterase BlrP1 protein from *Klebsiella pneumoniae* and its photoreceptor BLUF domain *Chem. Phys.* **354** 130–41

Tyagi A, Penzkofer A, Batschauer A and Wolf E 2009a Thermal degradation of (6R,S)-5,10-methenyltetrahydrofolate in aqueous solution at pH 8 *Chem. Phys.* **358** 132–6

Tyagi A, Penzkofer A, Batschauer A and Wolf E 2009b Fluorescence behavior of 5,10-methenyltetrahydrofolate, 10-formyltetrahydrofolate, 10-formyldihydrofolate, and 10-formylfolate in aqueous solution at pH 8 *Chem. Phys.* **361** 75–82

Tyagi A, Zirak P, Penzkofer A, Mathes T, Hegemann P, Mack M and Ghisla S 2009c Absorption and emission spectroscopic characterization of 8-amino-riboflavin *Chem. Phys.* **364** 19–30

Tyagi A and Penzkofer A 2010a pH dependence of the absorption and emission behaviour of lumiflavin in aqueous solution *J. Photochem. Photobiol. A: Chem.* **215** 108–17

Tyagi A and Penzkofer A 2010b Fluorescence spectroscopic behavior of folic acid *Chem. Phys.* **367** 83–92

Tyagi A, Penzkofer A, Mathes T and Hegemann P 2010c Photo-degradation behaviour of roseoflavin in some aqueous solutions *Chem. Phys.* **369** 27–36

Tyagi A and Penzkofer A 2011 Absorption and emission spectroscopic characterization of lumichrome in aqueous solutions *Photochem. Photobiol.* **87** 524–33

Tye K M and Deisseroth K 2012 Optogenetic investigation of neural circuits underlying brain disease in animal models *Nat. Rev. Neurosci.* **13** 251–66

Ueda Y and Sato M 2018 Induction of signal transduction by using non-channelrhodopsin-type optogenetic tools *ChemBioChem* **19** 1217–31

Ugalde J A, Podell S, Narasingarao P and Allen E E 2011 Xenorhodopsins, an enigmatic new class of microbial rhodopsins horizontally transferred between archaea and bacteria *Biol. Dir.* **6** 52

Ujj L, Devanathan S, Meyer T E, Cusanovich M A, Tollin G and Atkinson G H 1998 New photocycle intermediates in the photoactive yellow protein from *Ectothiorhodospira halophila*: picosecond transient absorption spectroscopy *Biophys. J.* **75** 406–12

Ulijasz A T, Cornilescu G, von Stetten D, Cornilescu C, Escobar F V, Zhang J, Stankey R J, Rivera M, Hildebrandt P and Vierstra R D 2009 Cyanochromes are blue/green light photoreversible photoreceptors defined by a stable double cysteine linkage to a phycoviolobilin-type chromophore *J. Biol. Chem.* **284** 29757–72

Umezawa M, Ueya Y, Ichihashi K, Dung D T K and Soga K 2023 Controlling molecular dye encapsulation in the hydrophobic core of core-shell nanoparticles for *in vivo* imaging *Biomed. Mater. Devices* **1** 605–17

Ung K and Arenkiel B R 2012 Fiber-optic implementation for chronic optogenetic stimulation of brain tissue *J. Vis. Exp.* **(68)** e50004

Vaidya A T, Chen C H, Dunlap J C, Loros J J and Crane B R 2011 Structure of a light-activated LOV protein dimer that regulates transcription *Sci. Signal* **4** ra50

Valeur B 2002 *Molecular Fluorescence. Principles and Applications* (Weinheim: Wiley)

Valverde S, Vandecasteele M, Piette C, Derousseaux W, Gangarossa G, Arbelaiz A A, Touboul J, Degos B and Venance L 2020 Deep brain stimulation-guided optogenetic rescue of parkinsonian symptoms *Nat. Commun.* **11** 2388

Van Bergeijk P, Adrian M, Hoogenraad C C and Kapitein L C 2016 Optogenetic control of organelle transport and positioning *Nature* **518** 111–4

Van den Berg P A W, Feenstra K A, Mark A E, Berendsen H J C and Visser A J W G 2002 Dynamic conformations of flavin adenine dinucleotide: simulated molecular dynamics of the flavin cofactor related to the time-resolved fluorescence characteristics *J. Phys. Chem.* B **106** 8858–69

Van der Steen J B, Nakasone Y, Hendriks J and Hellingwerf K J 2013 Modeling the function of YtvA in the general stress response in *Bacillus subtilis* *Mol. Biosyst.* **9** 2331–43

Van der Horst M A, Hendriks J, Vreede J, Yeremenko S, Crielaard W and Hellingwerf K J 2005a Photoactive yellow protein, the xanthopsin *Handbook of Photosensory Receptors* ed W R Briggs and J L Spudich (Weinheim: Wiley) pp 391–415

Van der Horst M A, Laan W, Yeremenko S, Wende A, Palm P, Oesterhelt D and Hellingwerf K J 2005b From primary photochemistry to biological function in blue-light photoreceptors PYP and AppA *Photochem. Photobiol. Sci.* **4** 688–93

Van der Horst M A, Key J and Hellingwerf K J 2007 Photosensing in chemotrophic, non-phototrophic bacteria: let there be light sensing too *Trends Microbiol.* **15** 554–62

Van Gelder R N, Chiang M F, Dyer M A, Greenwell T N, Levin L A, Wong R O and Svendsen C N 2022 Regenerative and restorative medicine for eye disease *Nat. Med.* **28** 1149–56

Van Stokkum I H M, Hontani Y, Vierock J, Krause B S, Hegemann P and Kennis J T M 2023 Reaction dynamics in the Chrimson channelrhodopsin: observation of product-state evolution and slow diffusive protein motions *J. Phys. Chem. Lett.* **14** 1485–93

Vankayala R, Sagadevan A, Vijayaraghavan P, Kuo C-L and Hwang K C 2011 Metal nanoparticles sensitize the formation of singlet oxygen *Angew. Chem. Int. Ed.* **50** 10640–4

Vechtomova Y L, Telegina T A, Buglak A A and Kritsky M S 2021 UV radiation in DNA damage and repair involving DNA-photolyases and cryptochromes *Biomedicines* **9** 1564

Vedalankar P and Tripathy B C 2019 Evolution of light-independent protochlorophyllide oxidoreductase *Protoplasma* **256** 293–312

Verhoefen M-K, Bamann C, Blöcher R, Förster U, Bamberg E and Wachtveitl J 2010 The photocycle of channelrhodopsin-2: ultrafast reaction dynamics and subsequent reaction steps *ChemPhysChem* **11** 3113–22

Vermeulen A J and Bauer C E 2015 Members of the PpaA/AerR antirepressor family bind cobalamin *J. Bacteriol.* **197** 2694–703

Vickstrom C R, Snarrenberg S T, Friedman V and Liu Q-s 2022 Application of optogenetics and *in vivo* imaging approaches for elucidating the neurobiology of addiction *Mol. Psychiatry* **27** 640–51

Vide U, Kasapović D, Fuchs M, Heimböck M P, Totaro M G, Zenzmaier E and Winkler A 2023 Illuminating the inner workings of a natural protein switch: blue-light sensing in LOV-activated diguanylate cyclases *Sci. Adv.* **9** eadh4721

Vierock J *et al* 2021 BiPOLES is an optogenetic tool developed for bidirectional dual-color control of neurons *Nat. Commun.* **12** 4527

Vierock J *et al* 2022 WiChR, a highly potassium selective channelrhodopsin for low-light one- and two-photon inhibition of excitable cells *Sci. Adv.* **8** eadd77229

Vierstra R D and Karniol B 2005 Phytochromes in microorganisms *Handbook of Photosensory receptors* ed W R Briggs and J L Spudich (Weinheim: Wiley) pp 171–95

Vierstra R D and Zhang J 2011 Phytochrome signaling: solving the Gordian knot with microbial relatives *Trends Plant Sci.* **16** 417–26

Visaveliya N R and Köhler J M 2021 Softness meets with brightness: dye-doped multifunctional fluorescent polymer particles via microfluidics for labeling *Adv. Optical Mater.* **9** 2170050

Voet D and Voet J G 2004 *Biochemistry* (Hoboken, NJ: Wiley)

Wachter R M 2017 Photoconvertible fluorescent proteins and the role of dynamics in protein evolution *Int. J. Mol. Sci.* **18** 1792

Wachten D and Mick D U 2021 Signal transduction in primary cilia – analyzing and manipulating GPCR and second messenger signaling *Pharmacol. Therapeut.* **224** 107836

Walker L A, Jarrett J T, Anderson N A, Pullen S H, Mathews R G and Sension R J 1998 Time-resolved spectroscopic studies of B$_{12}$ coenzymes: the identification of a metastable Cob(III) alamin photoproduct in the photolysis of methylcobalamin *J. Am. Chem. Soc.* **120** 3597–603

Wan J, Zhang X, Zhang K and Su Z 2020 Biological nanoscale fluorescent probes: from structure and performance to bioimaging *Rev. Anal. Chem.* **39** 209–21

Wan M, Zhu Y and Zou J 2020 Novel near-infrared fluorescent probe for live cell imaging *Exp. Ther. Med.* **19** 1213–8

Wand A, Rozin R, Eliash T, Jung K-H, Sheves M and Ruhman S 2011 Asymmetric toggling of a natural photoswitch: ultrafast spectroscopy of *Anabena* sensory rhodopsin *J. Am. Chem. Soc.* **133** 20922–32

Wand A, Gdor I, Zhu J, Sheves M and Ruhman S 2013 Shedding new light on retinal protein photochemistry *Annu. Rev. Phys. Chem.* **64** 437–58

Wang F and Liu X G 2009 Recent advances in the chemistry of lanthanide-doped upconversion nanocrystals *Chem. Soc. Rev.* **38** 976–89

Wang F 2020 Optogenetics: the key to deciphering and curing neurological diseases *Sci. Insights* **35** 224–35

Wang H and Wang H 2015 Phytochrome signaling: time to tighten up the loose ends *Mol. Plant* **8** 540–51

Wang J 2019 Visualization of H atoms in the X-ray crystal structure of photoactive yellow protein: does it contain low-barrier hydrogen bonds? *Protein Sci.* **28** 1966–72

Wang J and Liu Y-J 2023a Vibrationally resolved absorption and fluorescence spectra of flavins: a theoretical simulation in the gas phase *J. Chin. Chem. Soc.* **70** 669–79

Wang J and Liu Y 2023b Systematic theoretical study on the pH-dependent absorption and fluorescence spectra of flavins *Molecules* **28** 3315

Wang M, Da Y and Tian Y 2023 Fluorescent proteins and genetically encoded biosensors *Chem. Soc. Rev.* **52** 1189–214

Wang Q, Liu Q, Wang X, Zuo Z, Oka Y and Lin C 2018 New insights into the mechanisms of phytochrome-cryptochrome coaction *New Phytol.* **217** 547–51

Wang Q and Lin C 2020 Mechanisms of cryptochrome-mediated photoresponses in plants *Annu. Rev. Plant Biol.* **71** 103–29

Wang W, Wildes C P, Pattarabanjird T, Sanchez M I, Glober G, Matthews G A, Tye K M and Ting A Y 2017a A light- and calcium-gated transcription factor for imaging and manipulating activated neurons *Nat. Biotechnol.* **35** 864–71

Wang X, Valiev R R, Ohulchanskyy T Y, Ågren H, Yang C and Chen G 2017b Dye-sensitized lanthanide doped upconversion nanoparticles *Chem. Soc. Rev.* **46** 4150–67

Wang X, Jing C, Selby C P, Chiou Y-Y, Yang Y, Wu W, Sancar A and Wang J 2018 Comparative properties and functions of type 2 and type 4 pigeon cryptochromes *Cell. Mol. Life Sci.* **75** 4629–41

Wang Y, Feng L and Wang S 2019 Conjugated polymer nanoparticles for imaging, cell activity regulation, and therapy *Adv. Funct. Mater.* **29** 1806818

Wang Y, Lin J, Zhang Q, Chen X, Luan H and Gu M 2022 Fluorescence nanoscopy in neuroscience *Engineering* **16** 29–38

Wang Z, Hu M, Ai Z, Zhang Z and Xing B 2019 Near-infrared manipulation of membrane ion channels via upconversion optogenetics *Adv. Biosys.* **3** 1800233

Weber M, von der Emde H, Leutenegger M, Gunkel P, Sambandan S, Khan T A, Keller-Findeisen J, Cordes V C and Hell S W 2023 MINISTED nanoscopy enters the Ångström localization range *Nat. Biotechnol.* **41** 569–76

Wehler P, Armbruster D, Gunter A, Schleicher E, Di Ventura B and Ozturk A 2022 Experimental characterization of *in silico* red-shift-predicted iLOV$^{L470T/Q489K}$ and iLOV$^{V392K/F410V/A426S}$ mutants *ACS Omega* **7** 19555–60

Wei W, Zhang X, Zhang S, Wei G and Su Z 2019 Biomedical and bioactive engineered nanomaterials for targeted tumor photothermal therapy: a review *Mater. Sci. Eng.* C **104** 109891

Weiss C 1978 Optical spectra of chlorophyll *The Porphyrins* ed D Dolphin (New York: Academic) pp 211–24

Weller J-P and Gossauer A 1980 Synthese und Photoisomerisierung des racem. Phytochromobilin-dimethylesters *Chem. Ber.* **113** 1603–11

Wen S, Zhou J, Zheng K, Bednarkiewicz A, Liu X and Jin D 2018 Advances in highly doped upconversion nanoparticles *Nat. Commun.* **9** 2415

Wend S, Wagner H J, Müller K, Zurbriggen M D, Weber W and Radziwill G 2014 Optogenetic control of protein kinase activity in mammalian cells *ACS Synth. Biol.* **3** 280–5

Westberg M, Etzerodt M and Ogilby P R 2019 Rational design of genetically encoded singlet oxygen photosensitizing proteins *Curr. Opin. Struct. Biol.* **57** 56–62

Wichert N, Witt M, Blume C and Scheper T 2021 Clinical applicability of optogenetic gene regulation *Biotechnol. Bioeng.* **118** 4168–85

Wickstrand C, Dods R, Royant A and Neutze R 2015 Bacteriorhodopsin: would the real structural intermediates please stand up? *Biochim. Biophys. Acta* **1850** 536–53

Wickstrand C, Nogly P, Nango E, Iwata S, Standfuss J and Neutze R 2019 Bacteriorhodopsin: structural insights revealed using X-ray lasers and synchrotron radiation *Annu. Rev. Biochem.* **88** 59–83

Wiegert J S, Mahn M, Prigge M, Printz Y and Yizhar O 2017 Silencing neurons: tools, applications, and experimental constraints *Neuron* **95** 504–29

Wietek J, Wiegert J S, Adeishvili N, Schneider F, Watanabe H, Tsunoda S P, Vogt A, Elstner M, Oertner T G and Hegemann P 2014 Conversion of channelrhodopsin into a light-gated chloride channel *Science* **344** 409–12

Wietek J and Prigge M 2016 Enhancing channelrhodopsins: an overview *Optogenetics* (Methods in Molecular Biology vol 1408) ed A Kianianmomeni (New York: Humana Press) pp 141–65

Wietek J, Beltramo R, Scanziani M, Hegemann P, Oertner T G and Wiegert J S 2015 An improved chloride-conducting channelrhodopsin for light-induced inhibition of neural activity *in vivo Sci. Rep.* **5** 14807

Wietek J, Rodriguez-Rozada S, Tutas J, Tenedini F, Grimm C, Oertner T G, Soba P, Hegemann P and Wiegert J S 2017 Anion-conducting channelrhodopsins with tuned spectra and modified kinetics engineered for optogenetic manipulation of behavior *Sci. Rep.* **7** 14957

Wilms C D and Häusser M 2014 Twitching towards the ideal calcium sensor *Nat. Methods* **11** 139–40

Wilson A *et al* 2008 A photoactive carotenoid protein acting as light intensity sensor *PNAS* **105** 12075–80

Wiltbank L B and Kehoe D M 2019 Diverse light responses of cyanobacteria mediated by phytochrome superfamily photoreceptors *Nat. Rev.* **17** 37–50

Wiltschko R and Wiltschko W 2014 Sensing magnetic directions in birds: radical pair processes involving cryptochrome *Biosensors* **4** 221–42

Wiltschko R and Wiltschko W 2019 Magnetoreception in birds *J. R. Soc. Interface* **16** 20190295

Wingen M, Potzkei J, Endres S, Casini G, Rupprecht C, Fahlke C, Krauss U, Jaeger K-E, Drepper T and Gensch T 2014 The photophysics of LOV-based fluorescent proteins – new tools for cell biology *Photochem. Photobiol. Sci.* **13** 875–83

Winkler A, Heintz U, Lindner R, Reinstein J, Shoeman R L and Schlichting I 2013 A ternary AppA-PpsR-DANN complex mediates light-regulation of photosynthesis-related gene expression *Nat. Struct. Mol. Biol.* **20** 859–67

Winkler A, Udvarhelyi A, Hartmann E, Reinstein J, Menzel A, Shoeman R L and Schlichting I 2014 Characterization of elements involved in allosteric light regulation of phosphodiesterase activity by comparison of different functional BlrP1 states *J. Mol. Biol.* **426** 853–68

Wolfbeis O S 2015 An overview of nanoparticles commonly used in fluorescent imaging *Chem. Soc. Rev.* **44** 4743–68

Wolff A R, Bygrave A M, Sanderson D J, Boyden E S, Bannerman D M, Kullmann D M and Kätzel D 2018 Optogenetic induction of the schizophrenia-related endophenotype of ventral hippocampal hyperactivity causes rodent correlates of positive and cognitive symptoms *Sci. Rep.* **8** 12871

Wolff J A, Scheiderer L, Engelhardt T, Engelhardt J, Matthias J and Hell S W 2023 MINFLUX dissects the unimpeded walking of kinesin-1 *Science* **379** 1004–10

Wong S, Mosabbir A A and Truong K 2015 An engineered split intein for photoactivated protein *trans*-splicing *PLoS One* **10** e0135965

Woodhouse J *et al* 2020 Photoswitching mechanism of a fluorescent protein revealed by time-resolved crystallography and transient absorption spectroscopy *Nat. Commun.* **11** 741

Wu D *et al* 2012 Structural basis of ultraviolet-B perception by UVR8 *Nature* **484** 214–20

Wu C and Chiu D T 2013 Highly fluorescent semiconducting polymer dots for biology and medicine *Angew. Chem. Int. Ed. Engl.* **52** 3086–109

Wu Q, Ko W-H and Gardner K H 2008 Structural requirements for key residues and auxiliary portions of a BLUF domain *Biochem.* **47** 10271–80

Wu Q and Gardner K H 2009 Structure and insight into blue light-induced changes in the BlrP1 BLUF domain *Biochemistry* **48** 2620–9

Wu S-Y, Shen Y, Shkolnikov I and Campbell R E 2022 Fluorescent indicators for biological imaging of monatomic ions *Front. Cell Dev. Biol.* **10** 885440

Wu X *et al* 2015 Tailoring dye-sensitized upconversion nanoparticle excitation bands towards excitation wavelength selective imaging *Nanoscale* **7** 18424–8

Wu X *et al* 2016 Dye-sensitized core/active shell upconversion nanoparticles for optogenetics and bioimaging applications *ACS Nano* **10** 1060–6

Wusigale H L, Cheng H, Gao Y and Liang L 2020 Mechanism for inhibition of folic acid photodecomposition by various antioxidants *J. Agric. Food Chem.* **68** 340–50

Wusigale and Liang L 2020 Folates: stability and interaction with biological molecules *J. Agric. Food Res.* **2** 100039

Wykes R C, Kullmann D M, Pavlov I and Magloire V 2016 Optogenetic approaches in epilepsy *J. Neurosci. Med.* **260** 215–20

Xia G, Shi H, Su Y, Han B, Shen C, Gao S, Chen Z and Xu C 2022 Photoactivated cyclases attenuate sepsis-induced cardiomyopathy by suppressing macrophage-mediated inflammation *Front. Immunol.* **13** 1008702

Xiao Y, Deng P, Zhao Y, Yang S and Li B 2023 Three-photon excited fluorescence imaging in neuroscience: from principles to applications *Front. Neurosci.* **17** 1085682

Xing J, Gumerov V M and Zhulin I B 2022 Photoactive yellow protein represents a distinct, evolutionarily novel family of PAS domains *J. Bacteriology* **204** e00300–22

Xu R, Cao H, Lin D, Yu B and Qu J 2022 Lanthanide-doped upconversion nanoparticles for biological super-resolution fluorescence imaging *Cell Rep. Phys. Sci.* **3** 100922

Xu X, Mee T and Jia X 2020 New era of optogenetics: from the central to peripheral nervous system *Crit. Rev. Biochem. Mol. Biol.* **55** 1–16

Xu X, Paik I, Zhu L and Huq E 2015 Illuminating progress in phytochrome-mediated light signaling pathways *Trends Plant Sci.* **20** 641–50

Xu Y *et al* 2018 Hybrid indicators for fast and sensitive voltage imaging *Angew. Chem. Int. Ed.* **57** 3949–53

Xu Y, Zou P and Cohen A E 2017 Voltage imaging with genetically encoded indicators *Curr. Opin. Chem. Biol.* **39** 1–10

Yadav A, Singh D, Lingwan M, Yadukrishnan P, Masakapalli S K and Datta S 2020 Light signaling and UV-B-medicated plant growth regulation *J. Integrative Plant Biol.* **62** 1270–92

Yadav A K and Chan J 2023 Activity-based bioluminescence probes for *in vivo* sensing applications *Curr. Opin. Chem. Biol.* **74** 102310

Yamada M, Nagasaki S C, Ozawa T and Imayoshi I 2020 Light-mediated control of gene expression in mammalian cells *Neurosci. Res.* **152** 66–77

Yamaguchi S and Hamaguchi H 1998 Femtosecond ultraviolet-visible absorption study of all-*trans* → 13-*cis* 9-*cis* photoisomerization of retinal *J. Chem. Phys.* **109** 1397–408

Yamamoto H, Fang M, Dragnea V and Bauer C E 2018 Differing isoforms of the cobalamin binding photoreceptor AerR oppositely regulate photosystem expression *eLife* **7** e39028

Yamauchi Y, Konno M, Ito S, Tsunoda S, Inoue K and Kandori H 2017 Molecular properties of a DTD channelrhodopsin from *Guillardia theta Biophys. Physicobiol.* **14** 57–66

Yan F, Sun Z, Zhang H, Sun X, Jiang Y and Bai Z 2019 The fluorescence mechanism of carbon dots, and methods for tuning their emission color: a review *Microchim. Acta* **186** 583

Yan L, Yang H, Zhang S, Zhou C and Lei C 2023a A critical review on organic small fluorescent probes for monitoring carbon monoxide in biology *Crit. Rev. Anal. Chem.* **53** 1792–806

Yan Y, Aierken A, Wang C, Jin W, Quan Z, Wang Z, Qing H, Ni J and Zhao J 2023b Neural circuits associated with fear memory: potential therapeutic targets for posttraumatic stress disorder *The Neuroscientist* **29** 332–51

Yang C, Kim S O, Kim Y, Yun S R, Choi J and Ihee H 2017a Photocycle of photoactive yellow protein in cell-mimetic environments: molecular volume changes and kinetics *J. Phys. Chem. B* **121** 769–79

Yang F, Moss L G and Phillips G N J 1996 The molecular structure of green fluorescent protein *Nat. Biotechnol.* **14** 1246–51

Yang X, Montano S and Ren Z 2015 How does photoreceptor UVR8 perceive a UV-B signal? *Photochem. Photobiol.* **91** 993–1003

Yang X, Qin X, Ji H, Du L and Li M 2022 Constructing firefly luciferin bioluminescence probes for *in vivo* imaging *Org. Biomol. Chem.* **20** 1360–72

Yang W, Carrillo-Reid L, Bando Y, Peterka D S and Yuste R 2018 Simultaneous two-photon imaging and two-photon optogenetics of cortical circuits in three dimensions *eLIFE* **7** e32671

Yang X, Jost A P-T, Weiner O D and Tang C 2013 A light-inducible organelle-targeting system for dynamically activating and inactivating signaling in budding yeast *Mol. Biol. Cell* **24** 2419–30

Yang X, Montano S and Ren Z 2014 How does photoreceptor UVR8 perceive a UV-B signal? *Photochem. Photobiol.* **91** 993–1003

Yang Z, Liu B, Su J, Liao J, Lin C and Oka Y 2017b Cryptochromes orchestrate transcription regulation of diverse blue light responses in plants *Photochem. Photobiol.* **93** 112–27

Yao Z, Zhang B S and Prescher J A 2018 Advances in bioluminescence imaging: new probes from old recipes *Curr. Opin. Chem. Biol.* **45** 148–56

Yao Z, Wang X, Liu J, Zhou S, Zhang Z, He S, Liu J, Wu C and Fang X 2023 Photoswitchable semiconducting polymer dots for pattern encoding and superresolution imaging *Chem. Commun.* **59** 2469–72

Yaroshevich I A *et al* 2021 Role of hydrogen bond alteration and charge transfer states in photoactivation of the orange carotenoid protein *Commun. Biol.* **4** 539

Yasukawa H, Sato A, Kita A, Kodaira K, Iseki M, Takahashi T, Shibusawa M, Watanabe M and Yagita K 2013 Identification of photoactivated adenylyl cyclases in *Naegleria australiensis* and BLUF-containing protein in *Naegleria fowleri J. Gen. Appl. Microbiol.* **59** 361–9

Yawo H, Kandori H, Koizumi A and Kageyama R (ed) 2021 *Optogenetics: Light-Sensing Proteins and Their Applications in Neuroscience and Beyond* 2nd edn (Singapore: Springer)

Yawo H, Kandori H and Koizumi A (ed) 2015 *Optogenetics: Light-sensing Proteins and their Applications* (Berlin: Springer)

Yazawa M, Sadaghiani A M, Hsueh B and Dolmetsch R 2009 Induction of protein-protein interactions in live cells using light *Nat. Biotechnol.* **27** 941–5

Ye H and Fussenegger M 2019 Optogenetic medicine: synthetic therapeutic solutions precision-guided by light *Cold Spring Harb. Perspect. Med.* **9** a034371

Yee E F, Chandrasekaran S, Lin C and Crane B R 2019 Physical methods for studying flavoprotein photoreceptors *Methods Enzymol.* **620** 509–44

Yi Z, All A H and Liu X 2021 Upconversion nanoparticle-mediated optogenetics *Optogenetics: Light-Sensing Proteins and Their Applications in Neuroscience and Beyond* 2nd edn ed H Yawo, H Kandori, A Koizumi and R Kageyama (Singapore: Springer) pp 641–63

Yin R, Arongaus A B, Binkert M and Ulm R 2015 Two distinct domains of the UVR8 photoreceptor interact with COP1 to initiate UV-B signaling in *Arabidopsis Plant Cell* **27** 202–13

Yizhar O, Fenno L E, Davidson T J, Mogri M and Deisseroth K 2011 Optogenetics in neural systems *Neuron* **71** 9–34

Yizhar *et al* 2011 Neocortical excitation/inhibition balance in information processing and social dysfunction *Nature* **477** 171–8

Yodo M, Houjou H, Inoue Y and Sakurai M 2001 Spectral tuning of photoactive yellow protein. Theoretical and experimental analysis of medium effects on the absorption spectrum of the chromophore *J. Phys. Chem.* B **105** 9887–95

Yoshida K, Tsunoda S P, Brown L S and Kandori H 2017 A unique choanoflagellate enzyme rhodopsin exhibits light dependent cyclic nucleotide phosphodiesterase activity *J. Biol. Chem.* **292** 7531–41

Yu H, Peng Y, Yang Y and Li Z-Y 2019 Plasmon-enhanced light-matter interactions and applications *npj Computational Materials* **5** 45

Yu J, Rong Y, Kuo C-T, Zhou X-H and Chiu D T 2017 Recent advances in the development of highly luminescent semiconducting polymer dots and nanoparticles for biological imaging and medicine *Anal. Chem.* **89** 42–56

Yu J K, Liang R, Liu F and Martínez T J 2019 First-principles characterization of the elusive I fluorescent state and the structural evolution of retinal protonated Schiff base in bacterio-rhodopsin *J. Am. Chem. Soc.* **141** 18193–203

Yu N, Huang L, Zhou Y, Xue T, Chen Z and Han G 2019 Near-infrared-light activatable nanoparticles for deep-tissue-penetrating wireless optogenetics *Adv. Healthcare Mater.* **8** 1801132

Yu P, Wen X, Toh Y-R, Ma X and Tang J 2015 Fluorescent metallic nanoclusters: electron dynamics, structure, and applications *Part. Part. Syst. Charact.* **2015** 142–63

Yu W W, Qu L, Guo W and Peng X 2003 Experimental determination of the extinction coefficient of CdTe, CdSe, and CdS nanocrystals *Chem. Mater.* **15** 2854–60

Yu Z and Fischer R 2019 Light sensing and responses in fungi *Nat. Rev. Microbiol.* **17** 25–36

Yuan H, Anderson S, Masuda S, Dragnea V, Moffat K and Bauer C 2006 Crystal structures of the *Synechocystis* photoreceptor Slr1694 reveal distinct structural states related to the signaling *Biochemistry* **45** 12687–94

Yudenko A, Smolentseva A, Maslov I, Semenov O, Goncharov I M, Nazarenko V V, Maliar N L, Gordeliy V and Remeeva A 2021 Rational design of a split flavin-based fluorescent reporter *ACS Synth. Biol.* **19** 72–83

Yun J-H *et al* 2020a Pumping mechanism of NM-R3, a new light-driven bacterial chloride importer in the rhodopsin family *Sci. Adv.* **6** eaay2042

Yun J-H, Park J-H, Jin Z, Ohki M, Wang Y, Lupala C S, Liu H, Park S-Y and Lee W 2020b Structure-based functional modification study of a cyanobacterial chloride pump for transporting multiple anions *J. Mol. Biol.* **432** 5273–86

Yun J-H *et al* 2021 Early-stage dynamics of chloride ion-pumping rhodopsin revealed by a femtosecond X-ray laser *PNAS* **118** e2020486118

Zeng Z, Wei J, Liu Y, Zhang W and Mabe T 2018 Magnetoreception of photoactivated cryptochrome 1 in electrochemistry and electron transfer *ACS Omega* **3** 4752–9

Zgrablić G, Voïtchovsky K, Kindermann M, Haacke S and Chergui M 2005 Ultrafast excited state dynamics of the protonated Schiff base of all-*trans* retinal in solvents *Biophys. J.* **88** 2779–88

Zgrablić G, Novello A M and Parmigiani F 2012 Population branching in the conical intersection of the retinal chromophore revealed by multipulse ultrafast optical spectroscopy *J. Am. Chem. Soc.* **134** 955–61

Zhang D, Ma X-l G Y, Huang H and Zhang G-w 2020 Green synthesis of metallic nanoparticles and their potential applications to treat cancer *Front. Chem.* **8** 799

Zhang D, Redington E and Gong Y 2021 Rational engineering of ratiometric sensors with bright green and red fluorescent proteins *Commun. Biol.* **4** 924

Zhang H, Chen Z-H, Liu X and Zhang F 2020 A mini-review on recent progress of new sensitizers for luminescence of lanthanide doped nanomaterials *Nano Res.* **13** 1795–809

Zhang H, Zhao M, Ábrahám I M and Zhang F 2021 Super-resolution with lanthanide luminescent nanocrystals: progress and prospect *Front. Bioeng. Biotechnol.* **9** 692075

Zhang H, Fang H, Liu D, Zhan Y, Adu-Amankwaah J, Yuan J, Tan R and Zhu J 2022 Applications and challenges of rhodopsin-based optogenetics in biomedicine *Front. Neurosci.* **16** 966772

Zhang H *et al* 2023 Quantitative assessment of near-infrared fluorescent proteins *Nat. Methods* **20** 1605–16

Zhang L, Yang X, Yin Z and Sun L 2022 A review on carbon quantum dots: synthesis, photoluminescence mechanisms and applications *Luminescence* **37** 1612–38

Zhang L, Patel N H, Lappe J W and Wachter R M 2006 Reaction progress of chromophore biogenesis in green fluorescent protein *J. Am. Chem. Soc.* **128** 4766–72

Zhang F *et al* 2007 Multimodal fast optical interrogation of neural circuitry *Nature* **446** 633–9

Zhang F *et al* 2011 The microbial opsin family of optogenetic tools *Cell* **147** 1446–57

Zhang H and Cohen A E 2017 Optogenetic approaches to drug discovery in neuroscience and beyond *Trends Biotechnol.* **35** 625–39

Zhang K and Cui B 2015 Optogenetic control of intracellular signaling pathways *Trends Biotechnol.* **33** 92–100

Zhang M, Wang L and Zhong D 2017 Photolyase: dynamics and mechanisms of repair of sun-induced DNA damage *Photochem. Photobiol.* **93** 78–92

Zhang N-n, Lu C-y, Chen M-j, Xu X-l, Shu G-f, Du Y-z and Ji J-s 2021 Recent advances in near-infrared II imaging technology for biological detection *J. Nanobiotechnol.* **19** 132

Zhang S *et al* 2019 Structural basis for enzymatic photocatalysis in chlorophyll biosynthesis *Nature* **574** 722–5

Zhang X M, Yokoyama T and Sakamoto M 2021 Imaging voltage with microbial rhodopsins *Front. Mol. Biosci.* **8** 738829

Zhang Y, Benz P, Stehle D, Yang S, Kurz H, Feil S, Nagel G, Feil R, Gao S and Bender M 2022 Optogenetic manipulation of cyclic guanosine monophosphate to probe phosphodiesterase activities in megakaryocytes *Open Biol.* **12** 220058

Zhang Y *et al* 2023a Fast and sensitive GCaMP calcium indicators for imaging neural populations *Nature* **615** 884–91

Zhang Y, Cai N and Chan V 2023b Recent advances in silicon quantum dot-based fluorescent biosensors *Biosensors* **13** 311

Zhang Z *et al* 2020 Semiconducting polymer dots with dual-enhanced NIR-IIa fluorescence for through-skull mouse-brain imaging *Angew. Chem. Int. Ed.* **59** 3691–8

Zhao Y *et al* 2011 An expanded palette of genetically encoded Ca^{2+} indicators *Science* **333** 1888–91

Zhao C *et al* 2023 Miniature three-photon microscopy maximized for scattered fluorescence collection *Nat. Methods* **20** 617–22

Zhao H, Ma B, Ji L, Wang H and Chen D 2017 Coexistence of light-driven Na^+ and H^+ transport in a microbial rhodopsin from *Nonlabens dokdonensis J. Photochem. Photobiol. B: Biol.* **172** 70–6

Zheng J, Nicovich P R and Dickson R M 2007 Highly fluorescent noble metal quantum dots *Annu. Rev. Phys. Chem.* **58** 409–31

Zhou X X and Lin M Z 2013 Photoswitchable fluorescent proteins: ten years of colorful chemistry and exciting applications *Curr. Opin. Chem. Biol.* **17** 682–90

Zhou B, Shi B, Jin D and Liu X 2015a Controlling upconversion nanocrystals of emerging applications *Nat. Nanotechnol.* **10** 924–36

Zhou H, Zoltowski B D and Tao P 2017 Revealing hidden conformational space of LOV protein VIVID through rigid residue scan simulations *Sci. Rep.* **7** 46626

Zhou X X, Pan M and Lin M Z 2015b Investigating neuronal function with optically controllable proteins *Front. Mol. Neurosci.* **8** 37

Zhou Y, Ding M, Duan X, Konrad K R, Nagel G and Gao S 2021 Extending the anion channelrhodopsin-based toolbox for plant optogenetics *Membranes* **11** 287

Zhou Y, Ding M, Nagel G, Konrad K Y and Gao S 2021 Advances and prospects of rhodopsin-based optogenetics in plant research *Plant Physiol.* **187** 572–89

Zhou Z *et al* 2022 The nature of proton-coupled electron transfer in a blue light using flavin domain *PNAS* **119** e2203996119

Zhu M H, Jang J, Milosevic M M and Antic S D 2021 Population imaging discrepancies between a genetically-encoded calcium indicator (GECI) versus a genetically-encoded voltage indicator (GEVI) *Sci. Rep.* **11** 5295

Zhuang B, Liebl U and Vos M H 2022 Flavoprotein photochemistry: fundamental processes and photocatalytic perspectives *J. Phys. Chem.* B **126** 3199–207

Ziegler T and Möglich A 2015 Photoreceptor engineering *Front. Mol. Biosci.* **2** 30

Zirak P, Penzkofer A, Schiereis T, Hegemann P, Jung A and Schlichting I 2005 Absorption and fluorescence spectroscopic characterization of BLUF domain AppA from *Rhodobacter sphaeroides Chem. Phys.* **315** 142–54

Zirak P, Penzkofer A, Schiereis T, Hegemann P, Jung A and Schlichting I 2006 Photodynamics of the small BLUF protein BlrB from *Rhodobacter sphaeroides J. Photochem. Photobiol. B: Biol.* **83** 180–94

Zirak P, Penzkofer A, Hegemann P and Mathes T 2007a Photodynamics of BLUF domain mutant H44R of AppA from *Rhodobacter sphaeroides Chem. Phys.* **335** 15–27

Zirak P, Penzkofer A, Lehmpfuhl C, Mathes T and Hegemann P 2007b Absorption and emission spectroscopic characterization of blue-light receptor Slr1694 from *Synechocystis* sp. PCC6803 *J. Photochem. Photobiol. B: Biol.* **86** 22–34

Zirak P, Penzkofer A, Mathes T and Hegemann P 2009 Photo-dynamics of roseoflavin and riboflavin in aqueous and organic solvents *Chem. Phys.* **358** 111–22

Zoltowski B D, Schwertfeger C, Widom J, Loros J J, Bilwes A M, Dunlap J C and Crane B R 2007 Conformational switching in the fungal light sensor Vivid *Science* **316** 1054–7

Zoltowski B D and Crane B R 2008 Light activation of the LOV protein Vivid generates a rapidly exchanging dimer *Biochemistry* **47** 7012–9

Zoltowski B D and Gardner K H 2011 Tripping the light fantastic: blue-light photoreceptors as examples of environmentally-modulated protein–protein interactions *Biochem.* **50** 4–16

Zoltowski B D, Motta-Mena L B and Gardner K H 2013 Blue-light induced dimerization of a bacterial LOV-HTH DNA-binding protein *Biochem.* **52** 6653–61

Zoltowski B D *et al* 2019 Chemical and structural analysis of a photoactive vertebrate cryptochrome from pigeon *PNAS* **116** 19449–57

Zou P, Zhao Y, Douglass A D, Hochbaum D R, Brinks D, Werley C A, Harrison D J, Campell R E and Cohen A E 2014 Bright and fast multicoloured voltage reporters via electrochromic FRET *Nat. Commun.* **5** 4625

Zulfiqar S, Sharif S, Saeed M and Tahir A 2021 Role of carotenoids in photosynthesis *Carotenoids: Structure and Function in the Human* ed M Zia-Ul-Haq, S Dewanjee and M Riaz (Springer Nature Switzerland AG) pp 147–87

IOP Publishing

Organic Lasers and Organic Photonics (Second Edition)

F J Duarte

Chapter 14

Tunable organic lasers, light sheet illumination, and organic molecules for biology and medicine

F J Duarte

Tunable organic lasers, and light sheet illumination, for biological and medical applications are described and discussed. Emphasis is given to tunable lasers emitting in the visible spectrum. The subject of multiple-prism instrumentation yielding extremely elongated Gaussian beams (up to 3000:1), also known as light sheet illumination, for nanoscopy and microscopy is discussed in detail. Organic molecules for photodymanic therapy are also surveyed and discussed.

14.1 Introduction

In this chapter, tunable organic lasers, and optical techniques, applicable to biology and medicine are considered while highlighting benign organic dye molecules applicable to photodynamic therapy (PDT). Furthermore, the fundamentals of light sheet microscopy are outlined with an emphasis on illumination techniques and some of the organic dyes utilized for fluorescence purposes.

Organic dye lasers have played a pioneering role in laser medicine and are applicable to numerous branches of medicine including:

1. Cancer diagnosis (Tsuchiya *et al* 1983, Duarte 1988, Andersson-Engels *et al* 1990)
2. Cardiology (Leon *et al* 1988, Goldman 1990, Aldag 1994, Aldag and Titterton 2005)
3. Dermatology (Morelli *et al* 2000, Garden *et al* 1988, Goldman 1990, Aldag 1994, Costela *et al* 2016)
4. Lithotripsy (Dretler *et al* 1987, Watson *et al* 1987, Floratos and de la Rosette 1999).
5. Photodynamic therapy (PDT) (Hayata *et al* 1982, Goldman 1990, Goldman 1991, Aldag 1994, Costela *et al* 2016)
6. Tatoo removal (Diettle *et al* 1985, Costela *et al* 2016)

7. Urology (Tsuchiya *et al* 1983, Hisazumi *et al* 1983, Goldman 1990, Aldag 1994, Floratos and de la Rosette 1999)
8. Vascular lesions (Greenwald *et al* 1981, Anderson and Parrish 1981, Clement *et al* 2002, Clement and Kiernan 2003)

One caveat: it should be noted that the perspective on medical applications offered in this chapter is from a physics-engineering view point rather than from a biological or medical vantage point.

References are cited mainly as examples and do not necessarily include the first work generated in a particular area of research.

14.2 Organic lasers for medicine

Organic dye lasers have been applied in biotechnology related fields since the early 1970s. One of the first publications in this area, if not the earliest, utilized a ruby laser-pumped infrared organic dye laser to study chromatophores (Chance *et al* 1970). Since that initiation other variations of the organic dye laser came into the scene. More specifically, flashlamp-pumped dye lasers to study methemoglobin derivatives (Asher *et al* 1977) and visible laser-pumped dye lasers to study acridine–DNA complexes (Andreoni *et al* 1978).

In this chapter it is assumed that the liquid organic dye laser so apt to yield very large pulse energies (Baltakov *et al* 1974) and very high average powers (Bass *et al* 1992) is replaced by compact solid-state organic dye lasers for laboratory and clinical applications. These solid-state organic dye lasers span with their emission in the electromagnetic spectrum throughout the visible spectrum (Costela *et al* 2016).

Furthermore, they are capable of high-performance emission in the 100 ns range (Duarte *et al* 1998) and in the sub 10 ns range (Duarte 1999).

Informative and recommended reviews on organic lasers for medical applications include the works of Aldag (1994), King (2004), Aldag and Titterton (2005), and Costela *et al* (2016).

14.2.1 High-performance solid-state organic lasers for medicine

Solid-state organic tunable lasers are introduced, described, and discussed in chapters 5–7 of this book. Here, attention is given to two classes of high-performance tunable solid-state organic lasers that albeit have not been explicitly applied to medicine (but homologous versions have) they have the potential to significantly enhance the effectiveness of the application.

High-performance solid-state lasers utilizing organic dye-doped polymer gain media were introduced by Duarte (1994). Using an optimized tunable narrow-linewidth multiple-prism grating (MPG) oscillator the narrow-linewidth emission is tunable in the $550 \leqslant \lambda \leqslant 603$ nm region (Duarte 1999). This optimized MPG, or dispersive, oscillator delivered TEM_{00} emission with a beam divergence of $\Delta\theta \approx 2.2$ mrad (1.5 times the diffraction limit) and a laser linewidth of $\Delta\nu \approx 350$ MHz, which is equivalent to $\Delta\lambda \approx 0.0004$ nm at $\lambda = 590$ nm. The temporal pulse length of this laser emission is $\Delta t \approx 3$ ns, which for a laser linewidth of $\Delta\nu \approx 350$ MHz, is near the limit allowed by the Heisenberg uncertainty principle (Duarte 1999). This optimized oscillator yields pulses with a peak power of ~33 kW which translates into a spectral power density of $\rho_{SP} \approx 8.33 \times 10^7$ W nm^{-1}.

An additional high-performance narrow-linewidth solid-state organic tunable laser configuration delivering pulses of a similar duration ($\Delta t \approx 3$ ns) is of the hybrid multiple-prism grazing-incidence (HMPGI) grating design (Duarte 1997).

The high-performance solid-state organic laser just described utilized rhodamine 6G dye-doped modified poly methyl methacrylate (MPMMA) as the gain medium. The next high-performance solid-state organic laser uses rhodamine 6G doped 2-hydroxyethyl methacrylate:methyl methacrylate (HEMA:MMA) as the gain medium (Duarte *et al* 1998).

Using a long-pulse laser as the excitation source, Duarte *et al* (1998) demonstrated a long-pulse high-performance solid-state laser yielding pulses in excess of 100 ns. Using a tunable narrow-linewidth four-prism grating oscillator the narrow-linewidth emission is tunable in the $564 \leqslant \lambda \leqslant 602$ nm region (Duarte *et al* 1998). This MPG, or dispersive, oscillator delivered TEM$_{00}$ emission with a beam divergence of $\Delta\theta \approx 3.5$ mrad and a laser linewidth of $\Delta\nu \approx 600$ MHz, which is equivalent to $\Delta\lambda \approx 0.0007$ nm at $\lambda = 590$ nm. The temporal pulse length of this laser emission is $\Delta t \approx 105$ ns. With a pulse energy of $E \approx 0.4$ mJ this MPG oscillator yields a spectral power density of $\rho_{SP} \approx 5.44 \times 10^6$ W nm^{-1}.

The basic characteristics of the tunable solid-state organic lasers just described are:

1. Tunability;
2. Low-divergence TEM$_{00}$ beam emission;
3. Narrow-linewidth emission;
4. Temporal pulse lengths in the $3 \leqslant \Delta t \leqslant 100$ ns range;
5. Spectral power densities in the $5.44 \times 10^6 \leqslant \rho_{SP} \leqslant 8.33 \times 10^7$ W nm^{-1}.

Although the tuning wavelength range quoted here is $550 \leqslant \lambda \leqslant 603$ nm the available tuning range, using additional organic dyes is $550 \leqslant \lambda \leqslant 750$ nm (Duarte 1999, Costela *et al* 2016). Inclusion of coumarin dye-doped polymer gain media should extend the lasing wavelength range down to the blue–green region (Duarte and James 2004).

Although the high-performance laser characteristics enlisted here are beyond the usual parameter availability from solid-state organic lasers used in medical applications, it is obvious that they can only improve the efficiency and effectiveness of the transfer of energy from the laser source to the biological site being irradiated. This is particularly relevant in laser medical systems utilizing optical fibers for delivery of the radiation to the biological tissue.

The engineering of solid-state high-performance tunable organic lasers excited by diode-laser arrays is a field of endeavor in waiting which could have a significant impact in compactness and cost reduction in the area of coherent sources for medical applications (see also chapter 8).

At a different realm of biomedical applications, miniature coherent electrically-excited organic semiconductor interferometric emitters offer attractive emission characteristics for applications in ophthalmology. This is particularly relevant given its spatially coherent smooth TEM_{00} beam and single-longitudinal-mode emission centered around $\lambda \approx 540$ nm (Duarte *et al* 2005, Duarte 2005, 2008). These integrated interferometric devices are described in detail in chapter 11.

14.2.2 Additional organic lasers for medicine

Additional organic lasers for medical applications and/or potential uses in bio-medicine include:

1. Organic optofluidic lasers as biosensors (Chen *et al* 2010, Choi *et al* 2013);
2. Organic semiconductor DFB laser for medical Raman spectroscopy (Liu *et al* 2013);
3. Organic semiconductor DFB laser as biosensors (Haughey *et al* 2016).

The main advantage of organic coherent sources such as the organic semi-conductor lasers is their enormous potential to be mass produced at low cost in a miniaturized configuration which could open a myriad of applications in the biomedical field.

Furthermore, miniaturized coherent electrically-excited organic semiconductor interferometric emitters (Duarte 2008), at this stage, represent a dark horse with enormous potential (see chapter 11).

14.3 Light sheet microscopy illumination

Light sheet illumination, using incoherent light, was introduced to microscopy by Siendentopf and Zsigmondy (1903). To form the light sheet they used traditional microscopic optics including lenses and microscope objectives.

In 1987, extremely elongated and ultra thin Gaussian laser beams were introduced for illumination in both imaging and microscopy applications (Duarte 1987, 1993a, 1993b) via a metrology instrument classified as an N-slit laser interferometer (NSLI).

This illumination technique uses telescopic beam expansion in conjunction with prismatic beam expansion and a focusing lens (Duarte 1987, 2015). This coherent illumination system is illustrated in figure 14.1. More specifically, this coherent extremely elongated Gaussian beam illumination (E^2GBI) can be 20 µm in the y-axis and 60 000 µm in the x-axis while propagation is in the z-axis (Duarte 1987). That means a width-to-height ratio of 3000:1. A schematics of this geometry is provided in figure 14.2.

An alternative method of light sheet illumination, using cylindrical lens optics, was introduced around 1993 under the description of orthogonal plane illumination (OPI) microscopy (Voie *et al* 1993).

At this stage it is it is necessary to clarify that E^2GBI is homologous to alternative terminology introduced in the literature such as OPI, light sheet illumination (LSI), and selective planar illumination (SPI). The main difference is that E^2GBI provides width-to-height ratios of 3000:1 while alternative illumination techniques offer lesser expansion ratios. For instance the illumination technique used in LSI is reported to

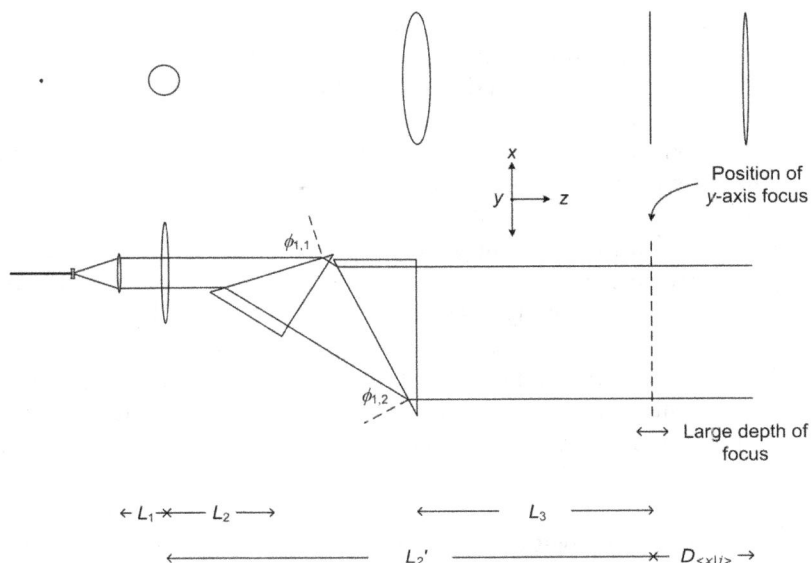

Figure 14.1. Schematics of the telescope-lens-multiple-prism beam expander optics used in generating extremely wide coherent light sheets or extremely-expanded Gaussian beam illumination (E^2GBI). Spatial beam profiles on top as seen along the z-axis. Telescope, lens, and multiple-prism beam expander as seen down the y-axis. The prismatic beam expansion occurs along the x-axis. The focal action occurs along the y-axis and the depth of focus, along the z-axis, is singularly extensive. In an example given by Duarte (2016) for a beam $32 \times 52\ 700$ μm the depth of focus, along the z-axis, is about 2000 μm. Detection takes place at the end of the distance $D_{\langle d|j\rangle}$ where a high resolution digital detector, such as a CMOS or CCD array, is located.

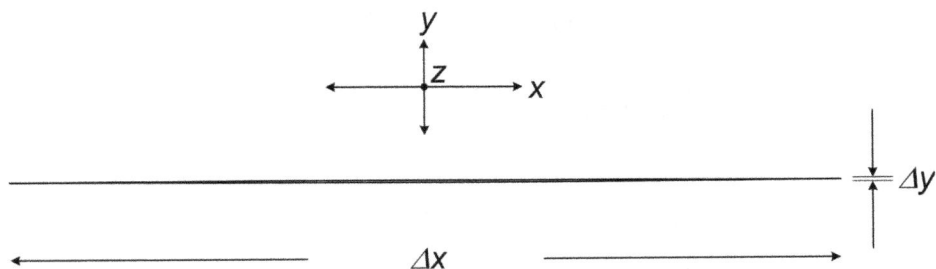

Figure 14.2. Schematics of the orientation of the E^2GBI coherent light sheet relative to the xyz-axes. In practice these extremely elongated coherent light sheets have been reported to be between 25 μm high × 25 000 μm wide and up to 20 μm high × 60 000 μm wide (Duarte 1987, 2016).

yield a 283:1 (Fuchs *et al* 2002) while the illumination technique applied in SPI offers a ratio of 110:1 (Huisken *et al* 2004). In this regard, even state-of-the-art LSI techniques used to study large specimens yield beams 180×8 μm which translate into an approximate ratio of 23:1 (Russell *et al* 2016). Nevertheless, since in the literature the designation associated with *light sheet* became pervasive, the title in this section also includes that two-word description. Further innovations in this field include the use of propagation-invariant light such as Airy beams and Bessel beams (Nylk *et al* 2013). It should be noted that with Bessel beams typical light sheets are approximately 100 × 1 μm (Dean *et al* 2022).

Shifting attention from the illumination techniques to the microscopy it should be mentioned that organic dye molecules are used on the samples to induce fluorescence under either one-photon (for excitation in the visible spectrum) or two-photon laser illumination (for excitation with longer wavelengths). In this regard, anthraquinone ($C_{24}H_{28}N_4O_4$) derivatives have been utilized in SPI microscopy (Verveer *et al*, 2007) in addition to organic green dyes (Holekamp *et al* 2008). In LSI microscopy, fluorescein ($C_{20}H_{12}O_5$) is also utilized (Fahrbach *et al* 2008). A new palette of fine-tuned rhodamine dyes to be used as fluorophores for live-cell and *in-vivo* biomedical imaging has been introduced by Grimm *et al* (2017). One possible caveat with the use of rhodamine dyes in medicine is that they include S, Cl, or F as part of their molecular structure.

14.3.1 Multiple-prism propagation matrix equations for extremely wide light sheets

The ray transfer matrix for a multiple-prism beam expander composed of r prisms is given by Duarte (1995, 2015, 2016)

$$\begin{pmatrix} A & B \\ C & D \end{pmatrix} = \begin{pmatrix} M_1 M_2 & B \\ 0 & (M_1 M_2)^{-1} \end{pmatrix} \tag{14.1}$$

where

$$B = M_1 M_2 \sum_{m=1}^{r-1} L_m \left(\prod_{j=1}^{m} k_{1,j} \prod_{j=1}^{m} k_{2,j} \right)^{-2}$$
$$+ \left(\frac{M_1}{M_2} \right) \sum_{m=1}^{r} (l_m/n_m) \left(\prod_{j=1}^{m} k_{1,j} \right)^{-2} \left(\prod_{j=m}^{r} k_{2,j} \right)^{2} \tag{14.2}$$

$$M_1 = \prod_{m=1}^{r} k_{1,m} \tag{14.3}$$

$$M_2 = \prod_{m=1}^{r} k_{2,m} \tag{14.4}$$

$$k_{1,m} = \frac{\cos \psi_{1,m}}{\cos \phi_{1,m}} \tag{14.5}$$

$$k_{2,m} = \frac{\cos \phi_{2,m}}{\cos \psi_{2,m}} \tag{14.6}$$

Here, $\phi_{1,m}$ and $\phi_{2,m}$ are the angles that are related to $\psi_{1,m}$ and $\psi_{2,m}$ via n_m and the law of refraction, or Snell's law. For a rigorous discussion on this topic, see Duarte (2006).

In equation (14.2), L_m is the prism-to-prism separation distance, while l_m is the optical path distance within each prism.

14.3.2 Propagation physics of the optics to generate extremely wide light sheets for microscopy

As illustrated in figure 14.1 the optics to generate E^2GBI laser beams consists of a two-dimensional telescope, a focusing lens, and a multiple-prism beam expander (Duarte 1987, 1993a, 1993b). The propagation is from left to right so that the ray matrices representing the corresponding optical elements are deployed in reverse order from right to left. The corresponding multiplication of the $ABCD$ matrices for the expanded beam on the horizontal, or plane of incidence, that is the plane lying on the x-axis is given by (Duarte 1993b, 1995)

$$\begin{pmatrix} A' & B' \\ C' & D' \end{pmatrix} = \begin{pmatrix} M_t\left[M - \left(\frac{\varsigma}{f}\right)\right] & B_t\left[M - \left(\frac{\varsigma}{f}\right)\right] + L_1\left(\frac{M}{M_t}\right) + \left(\frac{\varsigma}{M_t}\right)\left[1 - \left(\frac{L_1}{f}\right)\right] \\ -\left(\frac{M_t}{Mf}\right) & (MM_t)^{-1}\left[1 - \left(\frac{L_1}{f}\right) - \left(\frac{B_t}{Mf}\right)\right] \end{pmatrix} \tag{14.7}$$

Here, $M = M_1M_2$, while M_t and B_t are related to the A and B terms of the ray matrix for the Galilean telescope, and

$$\varsigma = ML_2 + B + \frac{L_3}{M} \tag{14.8}$$

For the vertical, or y-axis, component (Duarte 1995)

$$\begin{pmatrix} A' & B' \\ C' & D' \end{pmatrix} = \begin{pmatrix} M_t\left[1 - \left(\frac{L'_2}{f}\right)\right] & \left[1 - \left(\frac{L'_2}{f}\right)\right]\left[B_t + \left(\frac{L_1}{M_t}\right)\right] + \left(\frac{L'_2}{M_t}\right) \\ -\frac{M_t}{f} & M_t^{-1}\left[1 - \left(\frac{L_1}{f}\right)\right] - \left(\frac{B_t}{f}\right) \end{pmatrix} \tag{14.9}$$

where L'_2 is the distance between the convex lens and its focal point. The width of the extremely elongated Gaussian beam, on the x-axis, is given by (Duarte 1995)

$$w(B') = w_0\left[(A')^2 + \left(\frac{B'}{L_R}\right)^2 \right]^{1/2} \tag{14.10}$$

where A' and B' are given from equation (14.7) and the Rayleigh length L_R is defined as

$$L_R = \frac{\pi w_0^2}{\lambda} \tag{14.11}$$

Equation (14.10) also applies to the dimensions of the beam in the vertical y-axis in which case the A' and B' terms are given from equation (14.9)

In this optical configuration the telescope has multiple functions:
1. It minimizes beam divergence.
2. It decreases the requirements on beam expansion at the multiple beam expander thus augmenting its transmission efficiency.
3. It reduces the magnitude of the B' component in equation (14.9) for the y-axis component, as discussed in detail by Duarte (2016).

For a complete discussion of 2×2 and 4×4 propagation ray matrices for optics the reader should consult Duarte (2015).

The physics of the electro-optical microscope configuration described here has been found relevant by researchers developing instrumentation for structural analysis of organic compounds (Chrastil *et al* 1996), cytological research (Ortyn *et al* 2000), x-ray imaging (Sliski 1996, 1997, Kwonk and Lee 2005), and homodyne densitometry (Castile 2013).

14.4 Organic dye molecules for photodynamic therapy

Phototherapy endeavors were initiated by N R Finsen in the late 1800s. More specifically, photodynamic therapy (PDT) begun via the analysis of, and the utilization of, hematoporphyrin derivative (HpD) (Lipson and Baldes 1960, Lipson *et al* 1961).

The next milestone was the actual application of PDT, using the hematoporphyrin derivative (HpD), to treat cancerous tissue (Dougherty 1987, Dougherty *et al* 1978).

Recommended reviews on organic laser driven PDT include the writings of Goldman (1990) and Costela *et al* (2008, 2016).

The hematoporphyrin molecule ($C_{34}H_{38}N_4O_6$), depicted in figure 14.3, is manufactured from hemoglobin (Gomer and Dougherty 1979) and its molecular weight is 598.69 amu. HpD is synthesized from hematoporphyrin and has a symmetrical structure. Visually, the structure of this molecule appears as if conformed by the mirror images of two $C_{34}H_{38}N_4O_6$ molecules joined by a C–O–C chain linked to the HN–NH rings. Thus, the molecular weight of HpD is at least twice the molecular weight of $C_{34}H_{38}N_4O_6$, however, impurities can add to its weight. A detailed review on porphyrin-based photosensitizers for PDT is given by Stenberg *et al* (1998).

In addition to HpD, acridine orange, and rose bengal are also included as potential therapeutical agents in laser powered PDT (Goldman 1990). The molecular structure of these dyes is illustrated in figures 14.4 and 14.5. Like HpD, acridine orange does not contain potentially side-effect causing elements (such as Br, S, or F) while rose Bengal contains Cl which might limit its applicability with some class of allergic patients.

Briefly: following injection of HpD, or an alternative photosensitizer, the healthy cells in the targeted tissue, or organ, evacuate the photosensitizer while the malignant cells retain it. In other words, via a mechanism not yet fully elucidated,

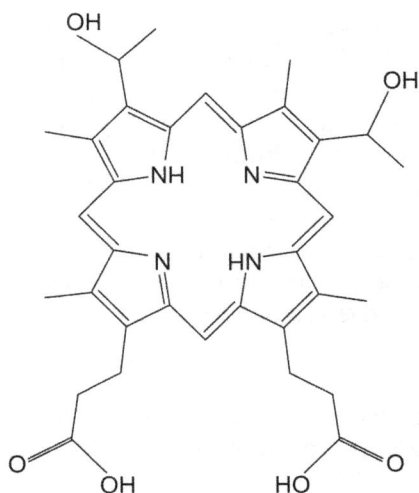

Figure 14.3. The Hematoporphyrin molecule ($C_{34}H_{38}N_4O_6$).

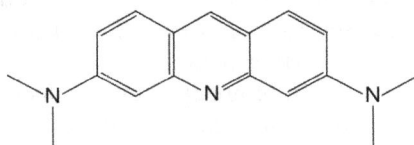

Figure 14.4. Acridine orange ($C_{17}H_{19}N_3$).

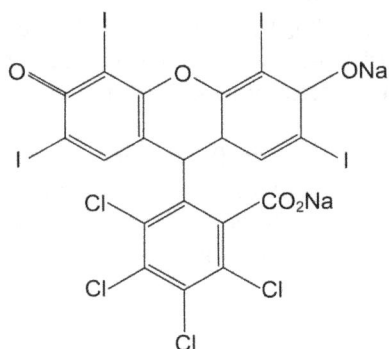

Figure 14.5. Rose bengal ($C_{20}H_4Cl_4I_4O_5$).

the photosensitizer congregates preferentially at the site of the tumor cells. This is a crucial feature of PDT.

The irradiation of HpD with laser emission, at $\lambda \approx 630$ nm, initiates a photo-chemical reaction that produces singlet oxygen species ($^1O_2$). The oxygen singlet interacts with the irradiated tissue generating a localized toxic reaction (Dougherty 1993). A revealing paper on the mechanism that photosensitizes singlet molecular oxygen from organic dyes such as rhodamine 6G and rose bengal was published by Stracke *et al* (1999). The killing ability of $^1O_2$, derived from photosensitized rhodamine 123, on carcinoma cells from mammalian ovaries has been documented by Richmond and O'Hara (1993).

Some observations about the generation of the oxygen singlet:

1. $^1O_2$ refers to the $^1\Delta_g$ state of the molecule O_2.
2. As explained by Jones (1990), the following reaction illustrates the formation of $^1O_2$ in an excited dye surrounded by molecular oxygen $^3O_2$

$$\uparrow D(T_1) + {}^3O_2 \rightarrow \downarrow D(S_0) + {}^1O_2 \tag{14.13}$$

   where $\uparrow D(T_1)$ is the first triplet state of the excited dye molecule sensitizer (see chapter 3) and $\downarrow D(S_0)$ represents the de-excited dye molecule back to the ground electronic state (S_0).
3. Relevant to reaction (14.13) is the rate of quenching $k_{S,T}$, in units of s^{-1}, this is the rate of inter-system crossing from the first excited electronic state S_1 to the triplet state T_1. On the other hand, $\tau_{T,S}$ in units of s is the time of decay from T_1 to the ground electronic state S_0. These quantities are measurable variables that can be used to monitor the concentration of $^1O_2$ (Stracke *et al* 1999). See chapter 3 for further details.
4. Increasing the concentration of ground state molecular oxygen $^3O_2$ increases the inter-system crossing $S_1 \rightarrow T_1$ and stimulates the creation of $^1O_2$
5. $^1O_2$ is said to be unstable and highly reactive (Dolphin 1993) which would explain its lethality toward malignant cells (Weishaupt *et al* 1976).
6. Use of the well known quencher of excited dye molecules 1,3,5,7-cyclooctate-traene (COT) prevents the formation of $^1O_2$ (Stracke *et al* 1999)
7. It is interesting to note that the presence of $^3O_2$ near the organic dye, known as aeration, is detrimental to laser action but $^3O_2$ is crucial to create the singlet $^1O_2$, in PDT, necessary to kill the malignant cells.

In the area of cancer diagnosis via fluorescence techniques: Lahoz *et al* (2013) report on amplified spontaneous emission from a complex formed by the anti-cancer drug tamoxifen and the nitro-2-1,3-benzoxadiazol-4-yl (NBD) dye while Yap and Neuhaus (2016) review the use of organic dyes as fluorophores for detection applications in oncology.

14.4.1 Additional organic molecules for photodynamic therapy

A second generation photosensitizer reported to show promise in PDT, when embedded into silica nanoparticles, is meta-tetra(hydroxyphenyl)-chlorin (*m*-THPC) (Yan and Kopelman 2003). This organic compound is also known as temoporfin and its molecular formula is $C_{44}H_{32}N_4O_4$. The motivation to use *m*-THPC embedded into silica nanoparticles is to overcome some of the shortcomings of HpD which include limitations in tumor-tissue normal-tissue selectivity, questionable stability in biological tissue, and preferred absorption at $\lambda \approx 630$ nm which results in poor transmission of laser radiation through tissue (Yan and Kopelman 2003).

Furthermore, the use of chromophores such as 4,4-difluoro-4-bora-3a,4a-diaza-s-indacene, also known as BODIPY, to generate $^1O_2$ has been discussed by Epelde-Elezcano et al (2016) and Palao et al (2016). The use of BODIPYs, in conjunction with core–shell silica nanoparticles (SiO_2), for the production of singlet oxygen is discussed by Epelde-Elezcano et al (2017). The photophysics of BODIPY dimers is discussed by Montero et al (2018).

14.5 Dual laser system for diagnosis and photodynamic therapy

As an application example, consider HpD in conjunction with a copper-laser-pumped dye laser (Duarte and Piper 1984). The main emission wavelength of the copper laser occurs via the transition $^2P_{3/2} - ^2D_{5/2}$ atomic copper transition at $\lambda = 510.554$ nm. The pulsed radiation at this wavelength can be used to excite a high-performance organic dye laser using rhodamine 640 as the gain medium capable of emitting in the $620 \leqslant \lambda \leqslant 650$ nm range (Costela et al 2016). This is perfectly compatible with HpD which strongly absorbs in the $500 \leqslant \lambda \leqslant 650$ nm region. In this regard, this laser and molecular system is applicable for a double therapeutical purpose: the principal emission from the pump laser can also be used for excitation of the HpD at $\lambda = 510.554$ nm, thus generating fluorescence for detection of the malignant area. Once the malignant area is located then emission from the organic dye laser at $\lambda \approx 630$ nm can be used for illumination of the HpD thus initiating the desired photochemical reaction associated with PDT.

A dual wavelength laser PDT apparatus for detection and treatment of cancer was disclosed by Duarte (1988) and is depicted in figure 14.6. As in the example given above, the first wavelength can be fixed and is available from the excitation laser. The second wavelength is variable and is provided by a high-performance tunable organic laser. This high-performance tunable laser belong of the class of laser yielding a low-divergence TEM_{00} beam, and narrow-linewidth emission, as previously described. The pump laser emits in the blue–green while the organic tunable laser emits in the red region of the spectrum.

In this dual detection-treatment laser system the radiation from the shorter wavelength pump laser passes through a polarization rotator (PR) (Duarte 1989) prior to encountering a variable beam-splitter (VBS) which can allow a totality of the radiation to continue to a beam shaping optics (BSO_2) telescope in its way to the optical fiber to illuminate the HpD for fluorescence detection purposes. For treatment, the VBS allows the laser pump radiation towards BSO_1 and on to excitation of the high-performance (HP) tunable organic laser. Subsequently, the HP tunable organic laser irradiates, via the optical fiber, or even directly, the biological tissue containing the HpD.

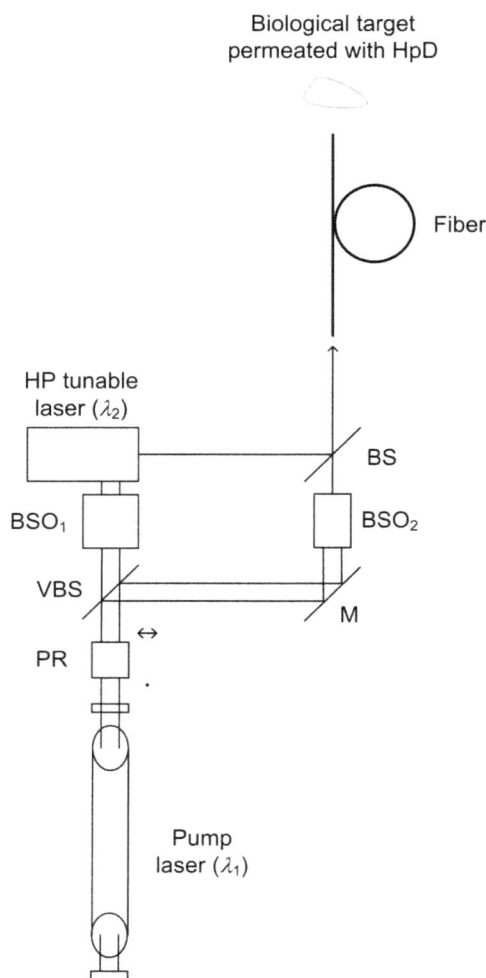

Figure 14.6. Double laser system for detection of HpD fluorescence and PDT applications. The polarization of the pump laser is rotated from perpendicular to the plane of incidence to parallel to the plane of incidence by a collinear prismatic polarization rotator (PR) (Duarte 1989). In the fluorescence mode the radiation from the excitation laser is reflected, as desired by the experimenter, by the variable beam-splitter (VBS) to propagate right toward the beam shaping optics (BSO$_2$). The collimated beam with the desired dimensions exits the BSO$_2$ to illuminate the optical fiber. For PDT treatment the VBS is set to excite the organic laser, via BSO$_1$, to produce wavelengths in the red compatible with efficient excitation of the photosensitizer. The corresponding laser beams are delivered directly to the biological target, 'doped' with the photosensitizer dye, via a broadband transmission optical fiber. Alternatively, wavelength specific fibers compatible with λ_1 and λ_2 can be used. BS: beam-splitter, PR: polarization rotator, M: mirror.

The principles being the same, alternative forms of this dual laser approach can be configured using a pulsed green diode-laser array excitation laser and a solid-state organic dye laser emitting in the red. This is depicted in figure 14.7. The semiconductor array can be from II–VI materials known to emit in the $450 \leqslant \lambda \leqslant 530$ nm

Biological target
permeated with HpD

Fiber

HP tunable
solid-state laser
(λ_2)

BS

BSO$_2$

VBS

M

BSO$_1$

Waveguide

Diode laser
array (λ_1)

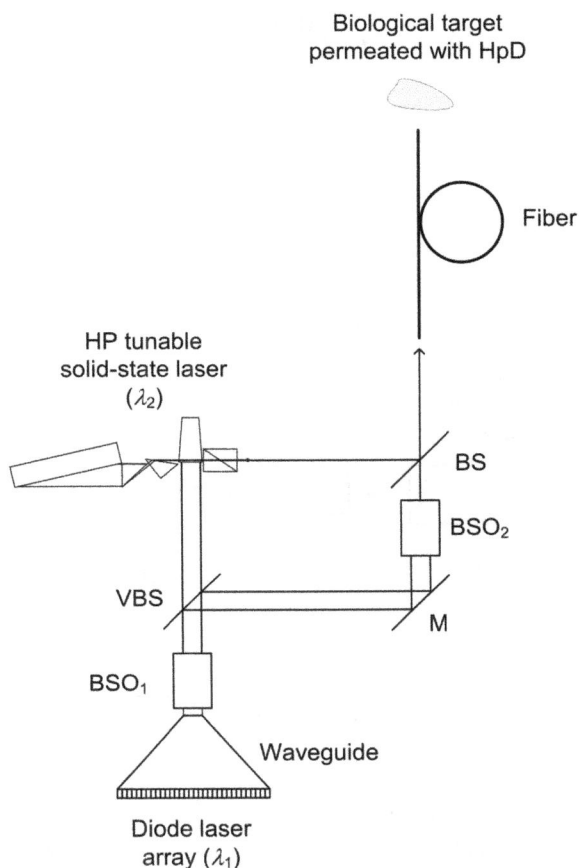

Figure 14.7. Hybrid dual wavelength laser system utilizing a pulsed green laser array as the pump laser and a solid-state organic dye laser as the therapeutic source. Here, λ_1 is provided by a diode laser array using semiconductors of the II–VI type known to emit in the $450 \leqslant \lambda \leqslant 530$ nm range. The tunable solid state organic laser is set to emit in the $620 \leqslant \lambda \leqslant 650$ nm region (see text).

region (Duarte 2015). The solid state organic laser is described in section 14.2, and chapter 5, with the polymer matrix being doped with rhodamine 640. Such a laser is expected to emit in the $620 \leqslant \lambda \leqslant 650$ nm region.

Yet another alternative is a CW green diode laser array and a red diode laser array, as illustrated in figure 14.8. Here the diode laser array providing λ_1 can be from II–VI semiconductors known to emit in the $450 \leqslant \lambda \leqslant 530$ nm region (Duarte 2015). The second wavelength λ_2 can be tunable in the $610 \leqslant \lambda \leqslant 690$ nm region using narrow-linewidth dispersive cavity designs as described by Zorabedian (1992, 1995) and AlGaInP/GaAs semiconductors.

If an alternative photosensitizer agent is used, such as m-THPC embedded into silica nanoparticles, λ_1 is required to be in the $500 \leqslant \lambda \leqslant 550$ nm region and $\lambda_2 \approx 650$ nm (Yan and Kopelman 2003).

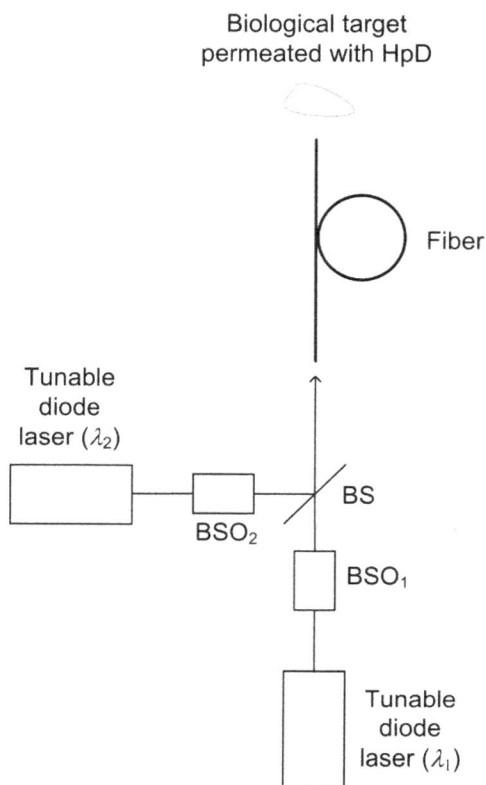

Figure 14.8. All semiconductor laser version of the dual wavelength laser system for diagnosis and PDT. Again, λ_l is designed to be provided by a diode laser array using semiconductors of the II–VI type which emit in the $450 \leqslant \lambda \leqslant 530$ nm range. The second beam, at λ_2, can be made tunable in the $610 \leqslant \lambda \leqslant 690$ nm region using narrow-linewidth dispersive cavity designs (see text).

14.6 Conclusion

High-performance tunable dispersive narrow-linewidth organic solid-state lasers, pumped with blue–green diode laser arrays, offer an unexplored field of laser development with enormous potential for biomedical and spectroscopy applications. In this regard, organic solid-state lasers utilizing dye-doped polymer, or nano-particle-polymer, gain media in either MPL grating, or HMPGI grating configurations, are a success waiting to be harvested.

In the area of light sheet microscopy laser beams with demonstrated high ratios of width to height (2000:1, or better) offer the fields of microscopy and nanoscopy unprecedented opportunities to map microscopic specimens with unusual width-to-height to ratios. Resolution extension to the nano spatial regime (nanoscopy), can be established theoretically using measurements in the microscopic regime, as was demonstrated with the disclosure of coherent interferometric imaging and the application of the quantum based generalized interferometric equation (Duarte 1993, 2016).

As far as the use of organic molecules as photosensitizers in PDT applications: it is important to encourage the discovery and synthesis of new compounds that ideally circumvent the use of halogens, with the possible exception of iodine ($_{53}$I), and other allergy causing elements such as sulfur ($_{16}$S). However, here it is useful to observe that although some patients might be allergic to $_{16}$S, let's say, they might well tolerate some of the halogens. Again, in general, the inclusion of known allergy-producing elements as part of the molecular structure of the therapeutical photosensitizer should preferably be avoided. The destruction of the malignant cells should be the sole result of the photochemical reaction, via the oxygen singlet, following the laser–photosensitizer interaction. In this regard, the interest in *m*-THPC embedded into silica nanoparticles appears to be a step in the right direction.

In addition to the outstanding chapter on organic dyes in optogenetics included in this book (Penzkofer *et al* 2024) the literature on this field has also been augmented by a discussion on the use of organic molecules in photonics, PDT, and neuro-photonics (Duarte 2019). Furthermore, recent developments in photodynamic therapy for skin cancer are reviewed by Farberg *et al* (2023) and Sun *et al* (2023) while photodynamic therapy to treat oral cancer is reviewed by Mosaddad (2023) and photodynamic therapy treatments for cervical cancer are considered by Alimu *et al* (2023).

14.7 Problems

- 14.1 Calculate the Rayleigh length L_R for $w_0 \approx 250$ μm and $\lambda \approx 632.82$ nm.
- 14.2 Refer to the optical schematics for the telescope-lens-multiple-prism beam expander depicted in figure 14.1: given that for a multiple-prism beam expander $M = 5.13$ and $B = 0.0167$ m, calculate A' as defined in the matrix equation (14.7) given that $M_t = 20$, $f = 0.3$ m, $L_2 = 0.0475$ m, and $L_3 = 0.2$ m.
- 14.3 Refer to the optical schematics for the telescope-lens-multiple-prism beam expander depicted in figure 14.1: using ζ calculated in the previous problem calculate B' as defined in the matrix equation (14.7) given that $M = 5.13$, $M_t = 20$, $B_t = 0.14$ m, $f = 0.3$ m, and $L_1 = 0.0475$ m.
- 14.4 Use the calculated A', B', and L_R to estimate $w(B')$ as given in equation (14.10) for the *horizontal* (or parallel to the plane of incidence) component of the Gaussian beam.
- 14.5 Using equation (14.9) and the same methodology as in problems (14.1)–(14.4) calculate A', B' for the *vertical* (or perpendicular to the plane of incidence) component of the Gaussian beam.
- 14.6 Using the A' and B' calculated in problem 14.5, and L_R, estimate $w(B')$ using equation (14.10), for the *vertical* component of the Gaussian beam.
- 14.7 The overall horizontal and vertical dimensions of the expanded Gaussian beam are given by $2w_x(B')$ and $2w_y(B')$. For the calculations performed above the ratio of these two components should be approximately 1300:1. Comment.

References

Aldag H R 1994 Solid-state dye laser for medical applications *SPIE Proc.* **2115** 184–9

Aldag H R and Titterton D H 2005 From flashlamp-pumped liquid dye lasers to diode-pumped solid-state dye lasers *SPIE Proc* **5707** 194–208

Anderson R R and Parrish J A 1981 Microvasculature can be selectively damaged using dye lasers: a basic theory and experimental evidence in human skin *Lasers Surg. Med.* **1** 263–76

Andersson-Engels S, Johansson J, Stenram U, Svanber K and Svanber S 1990 Malignant tumor and atherosclerotic plaque diagnosis using laser induced fluorescence *IEEE J. Quantum Electron.* **QE-26** 2207–17

Andreoni A, Sacchi C A, Svelto O, Bottiroli G and Prenna G 1978 Laser-induced fluorescence of acridine–DNA complexes *Sov. J. Quantum Electron.* **8** 1255–9

Asher S A, Vickery L E, Schuster T M and Sauer K 1977 Resonance Raman spectra of methemoglobin derivatives *Biochemistry* **16** 5849–56

Alimu G *et al* 2023 Liposomes with dual clinical photosensitizers for enhanced photodynamic therapy of cervical cancer *RSC Adv.* **13** 3459–67

Baltakov F N, Barikhin B A and Sukhanov L V 1974 400 J pulsed laser using a solution of rhodamine 6 G in ethyl alcohol *JETP Lett.* **19** 174–5

Bass I L, Bonano R E, Hackel R H and Hammond P R 1992 High-average-power dye laser at Lawrence Livermore National Laboratory *Appl. Opt.* **31** 6993–7006

Castile B 2013 Method and apparatus incorporating an optical homodyne into a self diffraction densitometer *US Patent* 8451455 B2

Chance B, McCray J A and Bunkenburg J 1970 Fast spectrophotometric measurement of H^+ in chromatium chromatophores activated by a liquid dye laser *Nature* **225** 705–8

Chen Y, Lei L, Zhang K, Shi J, Wang L, Li H, Zhang X M, Wang Y and Chan H L W 2010 Optofluidic microcavities: dye lasers and biosensors *Biomicrofluidics* **4** 043002

Choi E Y, Mager L, Cham T T, Dorkenoo K D, Fort A, Wu J W, Barsella A and Ribierre J-C 2013 Solven-free fluid organic dye lasers *Opt. Express* **21** 11368–75

Chrastil J 1996 Spectrophotometric method for structural analysis of organic compounds, polymers, nucleotides, and peptides *US Patent* 5550630

Clement R M, Kiernan M N and Donne K 2002 Treatment of vascular lesions *US Patent* 6398801 B1

Clement R M and Kiernan N M 2003 Reduction of vascular blemishes by selective thermolisis *US Patent* 6605083 B2

Costela A, García-Moreno I and Gómez C 2016 Medical applications of organic dye lasers *Tunable Laser Applications* ed F J Duarte (New York: CRC Press) 3rd edn ch 8

Costela A, García-Moreno I and Sastre R 2008 Medical applications of dye lasers *Tunable Laser Applications Tunable Laser Applications* ed F J Duarte (New York: CRC Press) 2nd edn ch 8

Dean K M *et al* 2022 Isotropic imaging across spatial scales with axially swept light-sheet microscopy *Nat. Protoc.* **17** 2025–53

Diette K M, Bronstein B R and Parrish J A 1985 Histologic comparison of argon and tunable dye lasers in the treatment of tattoos *J. Invest. Dermatol.* **85** 368–73

Dolphin D 1993 Photomedicine and photodynamic therapy *Can. J. Chem.* **72** 1005–13

Dougherty T J 1987 Photosensitizers: therapy and detection of malignant tumors *Photochem. Photobiol.* **45** 879–89

Dougherty T J 1993 Photodynamic therapy *Photochem. Photobiol.* **58** 895–900

Dougherty T J, Kaufman J E, Goldfarb A, Weishaupt K R, Boyle D and Mittleman A 1978 Photoradiation therapy for the treatment of malignant tumors *Cancer Res.* **38** 2628–35

Dretler S P, Watson G, Parrish J A and Murray S 1987 Pulsed dye laser fragmentation of ureteral calculi: initial clinical experience *J. Urol.* **137** 386–9

Duarte F J 1987 Beam shaping with telescopes and multiple-prism beam expanders *J. Opt. Soc. Am. A* **4** P30

Duarte F J 1988 Two-laser therapy and diagnosis device *European Patent* 0284330 A1

Duarte F J 1989 Optical device for rotating the polarization of a light beam *US Patent* 4822150

Duarte F J 1993a On a generalized interference equation and interferometric measurements *Opt. Commun.* **103** 8–14

Duarte F J 1993b Electro-optical interferometric microdensitometer system *US Patent* 5255069

Duarte F J 1994 Solid-state multiple-prism grating dye-laser oscillators *Appl. Opt.* **33** 3857–60

Duarte F J 1995 Interferometric imaging *Tunable Laser Applications* ed F J Duarte (New York: Marcel-Dekker) ch 5

Duarte F J 1997 Multiple-prism near-grazing-incidence grating solid-state dye-laser oscillator *Opt. Laser Tech.* **29** 513–6

Duarte F J 1999 Multiple-prism grating solid-state dye laser oscillator: optimized architecture *Appl. Opt.* **38** 663–5

Duarte F J 2006 Multiple-prism dispersion equations for positive and negative refraction *Appl. Phys. B* **82** 35–8

Duarte F J 2007 Coherent electrically-excited organic semiconductors: visibility of interferograms and emission linewidth *Opt. Lett.* **32** 412–4

Duarte F J 2008 Coherent electrically excited semiconductors: coherent or laser emission? *Appl. Phys. B* **90** 101–8

Duarte F J 2015 *Tunable Laser Optics* 2nd edn (New York: CRC Press)

Duarte F J 2016 Interferometric imaging *Tunable Laser Applications* ed F J Duarte (New York: CRC Press) 3rd edn ch 10

Duarte F J 2019 Organic molecules in photonics, cancer phototherapy, and neurophotonics *Neurophotonics and Biomedical Spectroscopy* ed R R Alfano and L Shi (New York: Elsevier) ch 16

Duarte F J and James R O 2004 Spatial structure of dye-doped polymer-nanoparticle laser media *Appl. Opt.* **43** 4088–90

Duarte F J, Liao L S and Vaeth K M 2005 Coherence characteristics of electrically excited tandem organic light-emitting diodes *Opt. Lett.* **30** 3072–4

Duarte F J and Piper J A 1984 Narrow-linewidth high-prf copper laser-pumped dye laser oscillators *Appl. Opt.* **23** 1391–4

Duarte F J, Taylor T S, Costela A, García-Moreno I and Sastre R 1998 Long-pulse narrow-linewidth dispersive solid-state dye-laser oscillator *App. Opt.* **37** 3987–9

Epelde-Elezcano N, Martínez-Martínez V, Peña-Cabrera E, Gómez-Durán C F A, López Arbeloa I and Lacombe S 2016 Modulation of singlet oxygen generation in halogenated BODIPY dyes by substitution at their meso position: towards a solvent-independent standard in the vis region *RSC Adv.* **6** 41991

Epelde-Elezcano N *et al* 2017 Adapting BODIPYs to singlet oxygen production on silica nanoparticles *Phys. Chem. Chem. Phys.* **19** 13746

Fahrbach F O, Gurchenkov V, Alessandri K, Nassoy P and Rohrbach A 2008 Light-sheet microscopy in thick media using scanned Bessel beams and two-photon fluorescence excitation *Opt. Express* **21** 13824–39

Farberg A S, Marson J W and Soleymani T 2023 Advances in photodynamic therapy for the treatment of acnitic keratosis and nonmelanoma skin cancer: a narrative review *Dermat. Ther.* **13** 689–716

Floratos D L and de la Rosette J J M C H 1999 Lasers in urology *BJU Int.* **84** 204–10

Fuchs E, Jaffe J S, Long R A and Azam F 2002 Thin laser light sheet microscope for microbial oceanography *Opt. Express* **10** 145–54

Garden J M, Polla L L and Tan O T 1988 The treatment of port-wine stains by the pulsed dye laser: analysis of pulse duration and long-term therapy *Arch. Dermatol.* **124** 889–96

Gomer C J and Dougherty T J 1979 Determination of [^{3}H]- and [^{14}C] hematoporphyrin derivative distribution in malignant and normal tissue *Cancer Res.* **39** 146–51

Greenwald J, Rosen S, Anderson R R, Harrist T, MacFarland M, Noe J and Parrish J A 1981 Comparative histological studies of the tunable dye (at 577 nm) laser and argon laser: the specific vascular effect of the dye laser *J. Invest. Dermatol.* **77** 305–10

Grimm J B *et al* 2017 A general method to fine-tune fluorophores for live-cell and *in-vivo* imaging *Nat. Methods* **14** 987–94

Goldman L 1990 Dye lasers in medicine *Dye Laser Principles* ed F J Duarte and L W Hillman (New York: Academic) ch 10

Goldman L (ed) 1991 *Laser Non-Surgical Medicine* (Lancaster, PA: Technomic)

Haughey A-M, McConnel G, Guilhabert B, Burley G A, Dawson M D and Laurand N 2016 Organic semiconductor laser biosensor *IEEE J. Sel. Top. Quantum Electron.* **22** 1300109

Hayata Y, Kato H, Konata C, Ono J and Takizawa N 1982 Hematoporphyrin derivative and laser photoradiation in the treatment of lung cancer *Chest* **81** 269–77

Hisazumi H, Misaki T and Miyoshi M 1983 Photoradiation therapy of bladder tumors *J. Urol.* **130** 685–7

Holekamp T F, Diwakar T and Holy T E 2008 Fast three dimensional fluorescence imaging of activity in neural populations by objective-coupled planar illumination microscopy *Neuron* **57** 661–72

Huisken J, Swoger J, Del Bene F, Wittbrodt J and Stelzer E H 2004 Optical sectioning deep inside live embryos by selective plane illumination microscopy *Science* **305** 1007–9

Jones II G 1990 Photochemistry of laser dyes *Dye Laser Principles* ed F J Duarte and L W Hillman (New York: Academic) ch 7

King T A 2004 Liquid state lasers *Lasers and Current Optical Techniques in Biology* ed G Palumbo and R Pratesi (Cambridge: Royal Society of Chemistry) ch 2

Kwok C S and Lee K Y 2005 Microdensitometer system with micrometer resolution for reading radiochromic films *US Patent* 6927859

Lahoz F, Oton C J, Lpez D, Marrero-Alonso J, Boto A and Daz M 2013 High efficiency amplified spontaneous emission from a fluorescent anticancer drug-dye complex *Org. Electron* **14** 1225–30

Leon M B, Lu D Y, Prevosti L G, Macy W W, Smith P D, Granovsky M, Bonner R F and Balaban R S 1988 Human arteria surface fluorescence: atherosclerotic plaque identification and effects of laser atheroma ablation *J. Am. Coll. Cardiol.* **12** 94–102

Lipson R L and Baldes E J 1960 The photodynamic properties of a particular hematoporphyrin derivative *Arch. Dermatol.* **82** 508–16

Lipson R L, Baldes E J and Olsen A M 1961 The use of a derivative of hematoporhyrin in tumor detection *J. Nat. Cancer Inst.* **26** 1–11

Liu X, Stefanau P, Wang B, Woggon T, Mappes T and Lemmer U 2013 Organic semiconductor distributed feedback (DFB) laser as excitation source in Raman spectroscopy *Opt. Express* **21** 28941–7

Montero R *et al* 2018 Singlet fission mediated photophysics of BODIPY dimmers *J. Phys. Chem. Lett.* **9** 641–6

Morelli J D, Tan O T, Margolis R, Seki Y, Boll J, Carney J M, Anderson R R, Ortyn W E, Piloco L R and Hayenga J W 2000 Cytological system illumination integrity checking apparatus and method *US Patent* 6011861

Mosaddad A S *et al* 2023 Photodynamic therapy in oral cancer *Photobiomod. Photomed. Laser Surg.* **41** 1–17

Nylk J *et al* 2018 Light sheet microscopy with attenuation compensated propagation-invariant beams *Sci. Adv.* **4** eaar4817

Palao E, Sola-Llano R, Tabero A, Manzano H, Agarrabeitia A R, Villanueva A, López Arbeloa I, Martínez-Martínez V and Ortiz M J 2016 Acetylacetonate BODIPY-biscyclometalated iridium(III) complexes effective strategy towards smarter fluorescent photosensitizer agents *Chem. Eur. J.* **23** 10139–47

Parrish J A, Furumoto H and Garden J 1986 Tunable dye laser (577 nm) treatment of port wine stains *Lasers Surg. Med.* **6** 94–9

Penzkofer A, Hegemann P and Katerlya S 2024 Organic dyes in optogenetics *Organic Lasers and Organic Photonics* ed F J Duarte (Bristol: Institute of Physics) 2nd edn ch 13

Richmond R C and O'Hara J A 1993 Effective photodynamic action by rhodamine 123 leading to photosensitized killing of Chinese hamster ovary cells in tissue culture and proposed mechanism *Photochem. Photobiol.* **57** 292–7

Russell C T, Rees E J and Kaminski C F 2016 Homographically generated light sheets for the microscopy of large specimens *Opt. Lett.* **43** 663–6

Siedentopf H and Zsigmondy R 1903 Über sichtbarmachung und größenbestimmung ultra-mikroskopischer teilchen, mit besonderer anwendung auf goldrubingläser *Ann. Phys.* **315** 1–39

Sliski A P 1996 X-ray phantom apparatus *US Patent* 5511107

Sliski A P 1997 CCD x-ray microdensitometer system *US Patent* 5623139

Sun J, Zhao H, Fu L, Cui J and Yang Y 2023 Global trends and research progress of photodynamic therapy in skin cancer: a bibliometric analysis and literature review *Clin. Cosmet. Invest. Dermatol.* **16** 479–98

Sternberg E D, Dolphin D and Brückner D 1998 Porphyrin-based photosensitizers for use in photodynamic therapy *Tetrahedron* **54** 4151–202

Stracke F, Heupel M and Thiel E 1999 Single molecular oxygen photosensitized by rhodamine dyes: correlation with photophysical properties of the sensitizers *J. Photocehem. Photobiol.* A **126** 51–8

Tsuchiya A, Obara N, Miwa M, Ohi T, Kato H and Hayata Y 1983 Hematoporphyrin derivative and laser photoradiation in the diagnosis and treatment of bladder cancer *J. Urol.* **130** 79–82

Verveer P J, Swoger J, Pampaloni F, Greger K, Marcello M and Stelzer E H K 2007 High-resolution three dimensional imaging of large specimens with light sheet–based microscopy *Nat. Methods* **4** 311–3

Voie A H, Burns D H and Spelman F A 1993 Orthogonal-plane fluorescence optical sectioning: three dimensional imaging of macroscopic biological specimens *J. Microsc.* **170** 229–36

Watson G, Murray S, Dretler S P and Parrish J A 1987 The pulsed dye laser for fragmenting urinary calculi *J. Urol.* **138** 195–8

Weishaupt K R, Gomer C J and Dougherty T J 1976 Identification of singlet oxygen as the cytotoxic agent in photo-inactivation of a murine tumor *Cancer Res.* **36** 2326–9

Yan F and Kopelman R 2003 The embedding of meta-tetra(hydroxyphenyl)-chlorin into silica nanoparticle platforms for photodynamic therapy and thei singlet oxygen production and pH-dependent optical properties *Photochem. Photobiol.* **78** 587–91

Yap K K and Neuhaus S J 2016 Making cancer visible – dyes in surgical oncology *Surg. Oncol.* **25** 30–6

Zorabedian P 1992 Characteristics of a grating-external-cavity semiconductor laser containing intracavity prism beam expanders *IEEE J. Lightwave Technol.* **10** 330–5

Zorabedian P 1995 Tunable external-cavity semiconductor lasers *Tunable Lasers Handbook* ed F J Duarte (New York: Academic) ch 8

IOP Publishing

Organic Lasers and Organic Photonics (Second Edition)

F J Duarte

Chapter 15

Organic lasers for N-channel quantum entanglement

F J Duarte

Sources of correlated indistinguishable ensembles of photons based on organic laser emission, for quantum entanglement experiments, are described and analyzed. In addition, the subject of quantum entanglement is also considered for $n = N = 2^2$ and the generalized case of $n = N = 2^1, 2^2, 2^3, 2^4, 2^5, ..., 2^r$.

15.1 Introduction

The first experimental configuration to measure the polarization quantum entanglement of two quanta propagating in opposite directions was provided by Pryce and Ward (1947) and is depicted in figure 15.1. This was the experimental arrangement adopted in the first quantum entanglement experiments (Hanna 1948, Wu and Shaknov 1950). This experimental configuration, albeit originally designed to use radiation generated in the annihilation of a positron–electron pair

$$e^+e^- \rightarrow \gamma_1\gamma_2 \qquad (15.1)$$

includes all the essentials of polarization quantum entanglement experiments for two quanta propagating in opposite directions.

Following Dirac (1930), Wheeler (1946) outlined the essentials of the experiment with a source emitting two quanta in opposite directions. The correlated quanta are polarized with their polarizations orthogonal to each other (Wheeler 1946, Pryce and Ward 1947).

Following the description of various relevant correlated quanta emission sources a review of the physics for pairs of quanta ($n = 2$), propagating in two different directions ($N = 2$), with entangled polarizations $|x\rangle$ and $|y\rangle$, is given. Furthermore, generalized probability amplitudes applicable to $n = N = 2^1, 2^2, 2^3, 2^4, 2^5, ..., 2^r$ are also given.

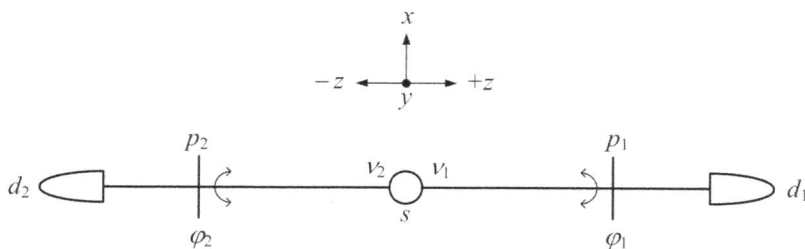

Figure 15.1. Optical equivalent of the Pryce–Ward γ-ray quantum entanglement experimental configuration.

Ideal optical configurations for $n = N = 2^2$ and $n = N = 2^3$ are described and analyzed.

15.2 Sources for quantum entanglement experiments

In this section a brief review on some of the sources used to generate quanta pairs with orthogonal polarizations is given. First, the current, and most widely used method of parametric down conversion is described, followed by the use of atomic transitions, and concluding with a description of an organic laser emitting highly coherent ensembles of quanta with orthogonal polarizations in opposite directions.

15.2.1 Parametric down conversion

According to the published literature the most obvious approach to produce several pairs of entangled photons is via type II matched spontaneous parametric down conversion (SPDC). The emission geometry for a multiple type II crystal configuration is shown in figure 15.2. For $n = N = 2^2$ (figure 15.3), two **BBO** crystals in series can be used (Bitton *et al* 2002). For $n = N = 2^3$ four crystals in cascade, and excited by a single laser, should be oriented to be in conjunction with the necessary optics, as illustrated in figure 15.4. Note that this is the source applicable to the optical configuration depicted in figure 15.5.

Disadvantages of SPDC as a generation technique of quanta pairs with orthogonal polarizations are its low conversion efficiency (Hübel *et al* 2010) and its emission geometry which does not yield quanta traveling in opposite directions, see figures 15.2 and 15.4. Furthermore, even though $\nu_p = \nu_i + \nu_s$ the two propagating quanta are not always indistinguishable since $\nu_i \neq \nu_s$ (Hübel *et al* 2010). Nevertheless, using a periodically poled KTiOPO$_4$ crystal, Yin *et al* (2017) report on the production of quanta pairs at $\lambda_i \approx \lambda_s \approx 810$ nm for excitation at $\lambda_p \approx 405$ nm at a linewidth of $\Delta\nu \approx 160$ MHz.

15.2.2 CW organic dye lasers

The realization of a visible optical version of the Pryce–Ward experiment was reported by Aspect *et al* (1981). These authors obtained visible polarization correlated quanta from the $4p^{2\,1}S_0 \rightarrow 4s4p^1P_1$ and the $4s4p^1P_1 \rightarrow 4s^{2\,1}S_0$ calcium transitions at $\lambda_1 \approx 551.3$ nm $\lambda_2 \approx 422.7$ nm, respectively. The two-photon

Four BBO crystals in series yielding 4 pairs
of entangled photons ($n = 8$)

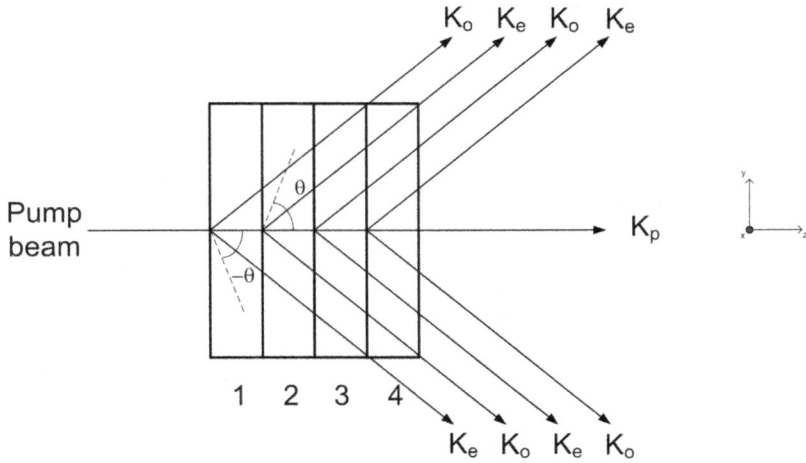

Figure 15.2. Emission geometry of multiple type II matched spontaneous parametric down conversion (SPDC) crystals.

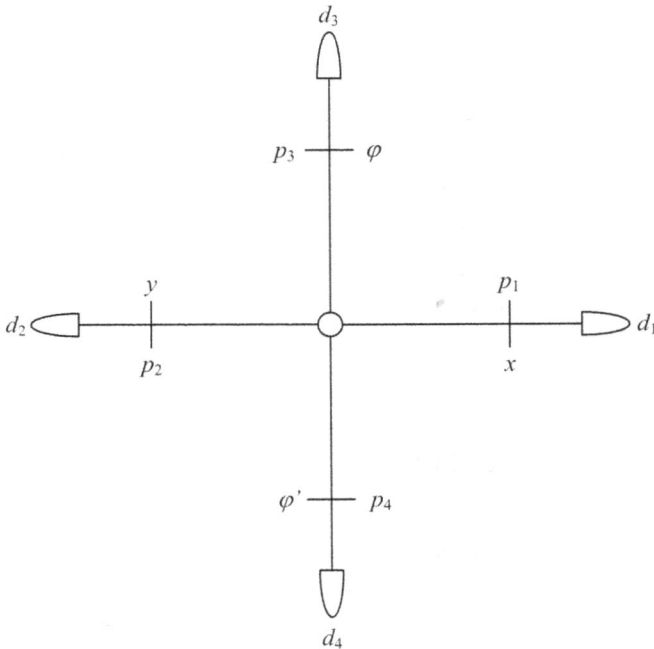

Figure 15.3. Optimized optical configuration for $n = N = 2^2$. Coordinates are (x, y) and (φ, φ').

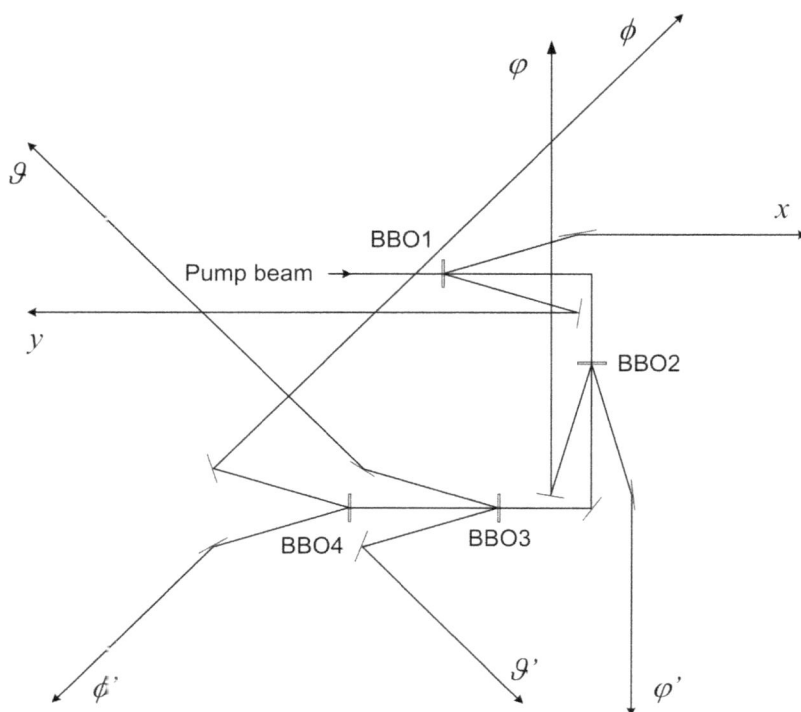

Figure 15.4. Simplified experimental diagram for the emission stage, applicable for the $n = N = 2^3$ configuration, utilizing type II phase matching spontaneous parametric down conversion (SPDC). At each change of trajectory of a beam a mirror is deployed. In total, nine mirrors are necessary. The first set of beams propagating in axes (x, y) and (φ, φ') are generated by BBO1 and BBO2. The second set of beams propagating in axes (ϕ, ϕ'), and (ϑ, ϑ') are generated by BBO3 and BBO4. This schematics corresponds to the source s in figure 15.5. For relative long propagation paths between the source and the detectors the geometry depicted in figure 15.5 is approached.

$4s^{12}S_0 \rightarrow 4p^{12}S_0$ excitation was performed using single-longitudinal-mode excitation from a Kr$^+$ laser, at $\lambda \approx 406.7$nm, and single-longitudinal-mode excitation from a rhodamine 6G organic dye laser at $\lambda \approx 581$ nm (Aspect *et al* 1981).

A notable caveat in this method is that the two quanta thus produced clearly are *not indistinguishable* given that $\lambda_1 \neq \lambda_2$.

An earlier use of calcium transitions to generate correlated quanta in polarization was reported by Freedman and Clauser (1972).

15.2.3 Organic lasers for *N*-channel quantum entanglement

It is well-known in the literature (Schäfer 1990, Duarte 1990) that laser dye molecules offer definite polarization emission characteristics in response to given polarization orientation of the excitation beam. For instance, rhodamine 6G emits nearly unpolarized radiation when excited with a beam polarized parallel to the plane of incidence at $\lambda = 510.554$ nm (Duarte 1990).

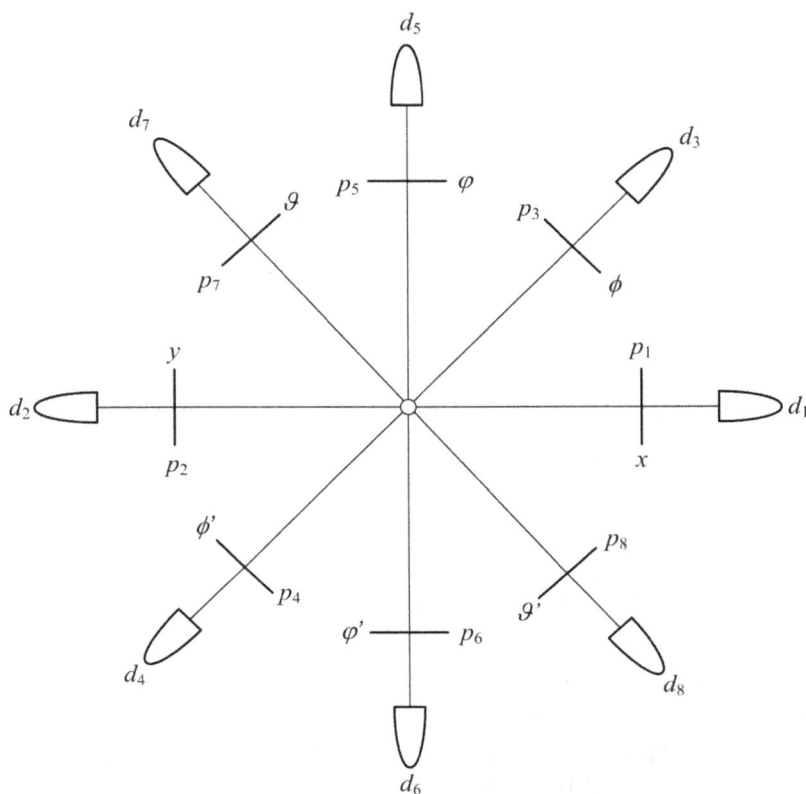

Figure 15.5. Optimized optical configuration for $n = N = 2^3$. Coordinates are (x, y), (φ, φ'), (ϕ, ϕ'), and (ϑ, ϑ').

Using narrow-linewidth excitation a definite vibrational-rotational level at the S_1 electronic state of the laser dye molecule can be populated. Furthermore, in reference to the cavity configuration illustrated in figure 15.6:

1. Using two partially reflective mirrors, M_1 and M_2, unpolarized narrow-linewidth emission can be established in both directions.
2. Using two or more intracavity etalons, E_1 and E_2 in this case, laser oscillation can be confined to very narrow-linewidth emission centered at a particular wavelength, let's say $\lambda = 590.00$ nm.
3. Because the laser emission is narrow-linewidth, it is highly coherent and it corresponds to *ensembles of indistinguishable photons*.
4. Inclusion of an intracavity linear polarizer p_r ensures that the polarization of the emission can be either in the $|x\rangle$ or $|y\rangle$ state, as desired.
5. To ensure that the indistinguishable ensembles of quanta, propagating in the $+z$ and $-z$ directions, are truly quantum randomized the intracavity polarizer p_r is rotated by $\pi/2$ at the mercy of a binary $(0, 1)$ quantum random number generator QRNG (Symul *et al* 2011, Haw *et al* 2015), so that $0 \rightarrow |y\rangle$, $1 \rightarrow |x\rangle$, lets say.

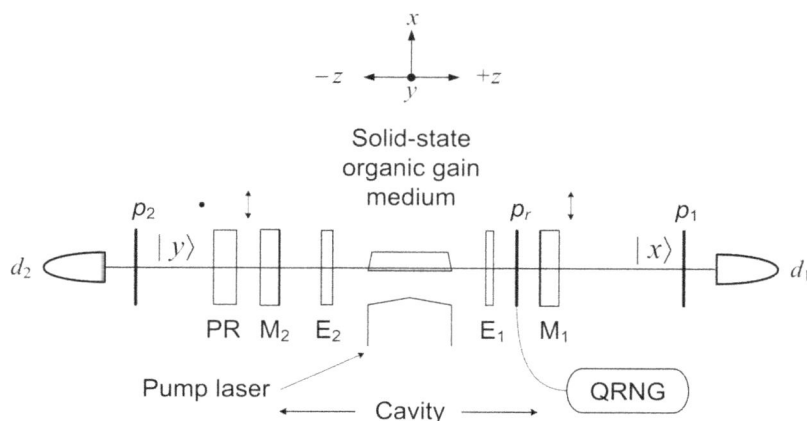

Figure 15.6. Organic laser configuration for $n = N = 2^1$ and coordinates (x, y). This device utilizes binary quantum random number generator (QRNG), where $0 \rightarrow |y\rangle$ and $1 \rightarrow |x\rangle$, and yields ensembles of indistinguishable quanta with orthogonal polarizations. Every time that d_1 detects a particular polarization, the receiver d_2 detects the corresponding orthogonal polarization. Here, M_1 and M_2 are the output coupler mirrors, E_1 and E_2 are intracavity etalons, p_r is the intracavity polarizer controlled by the QRNG, p_1 and p_2 are the polarizers corresponding to detectors d_1 and d_2, respectively.

6. Deployment of a *collinear polarization rotator* (PR) (Duarte 1992), on the optical axis just outside the cavity, insures that the polarization of the radiation propagating towards d_2 *will always be orthogonal* to the polarization propagating towards d_1.

The correlation of the ensembles of indistinguishable photons, propagating in the $+z$ and $-z$ directions, is guaranteed by the following physical facts:
1. Narrow-linewidth optical pumping ensures excitation from the ground level to a specific vibrational-rotational level at the S_1 electronic state.
2. Laser emission, in the $+z$ and $-z$ directions, occurs from the same vibrational-rotational level at the S_1 electronic state.
3. Narrow-linewidth emission, in the $+z$ and $-z$ directions, ensures that the quanta emitted are indistinguishable, highly coherent, and thus, quantum mechanically speaking, *the same quanta*.
4. The ensemble of indistinguishable quanta also have the identical polarization as determined by the QRNG controlled intracavity polarizer p_r.

The spectral characteristics of the organic laser emission from the laser described in figure 15.6 are as follows: *indistinguishable emission* ($\lambda_1 \pm \Delta\lambda_1 \approx \lambda_2 \pm \Delta\lambda_2$, where $\Delta\lambda_1 = \Delta\lambda_2$ is very small) with *directly opposite directions*. In other words, the emission originates from the very same $S_1 \rightarrow S_0$ transition in the $+z$ and $-z$ directions. Conversion efficiency of this emission is comparatively very high.

By altering the direction of the excitation beam by $\pi/2$, relative to the emission optical axis in reference to figure 15.6, it can immediately be seen that this organic

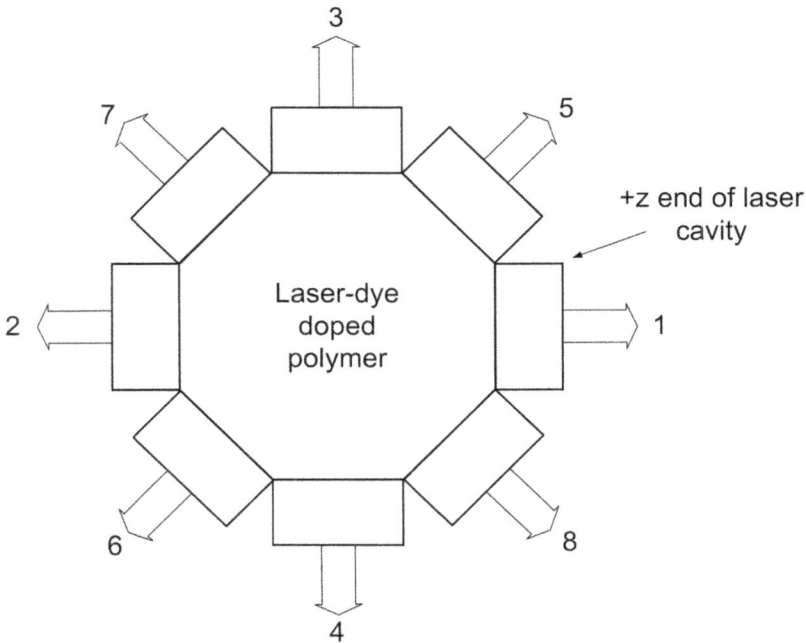

Figure 15.7. Organic laser configuration for $n = N = 2^3$. The narrow-linewidth laser excitation is rotated by $\pi/2$ relative to the excitation direction in figure 15.6.

laser approach can easily be adopted to generate $n = N = 2^2$ and $n = N = 2^3$, as outlined in figure 15.7.

Beyond the organic laser perspective, the conceptual vision of the laser configuration for the emission of orthogonally polarized ensembles of indistinguishable photons, described in figure 15.6, can be extended to other lasers in general, either electrically-pumped or optically-pumped. This is described in the laser optical configuration of figure 15.8. All that is required here is a laser medium that emits unpolarized amplified spontaneous emission that will then become narrow-linewidth polarized laser emission as controlled by the intracavity polarizer p_r which in turn is controlled by the binary inputs $0 \rightarrow |y\rangle$ and $1 \rightarrow |x\rangle$ generated by the QRNG. In the interest of compactness the laser medium of choice should be an electrically-excited semiconductor.

15.2.4 Entanglement of quantum indistinguishable ensembles

The beauty of the laser approach to quantum entanglement emission is that it removes the signal-to-noise problem associated with single-photon emission. Instead of single photons in the $|x\rangle$ or $|y\rangle$ states, we now have indistinguishable ensembles in the $|x\rangle$ or $|y\rangle$ states.

As discussed in chapter 16 and derived from the discussions of Duarte (2022) and Duarte and Taylor (2022) what we have now are highly coherent ensembles of indistinguishable quanta described by the ensemble states

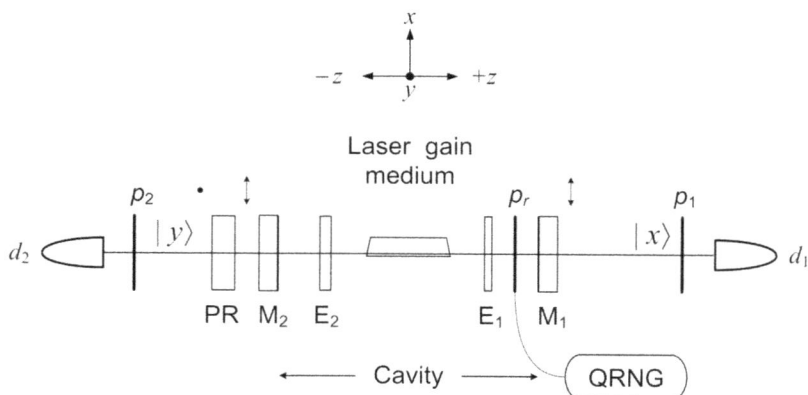

Figure 15.8. Generalized laser configuration for $n = N = 2^1$ and coordinates (x, y). This device utilizes binary quantum random number generator (QRNG), where $0 \rightarrow |y\rangle$ and $1 \rightarrow |x\rangle$, and yields ensembles of indistinguishable quanta with orthogonal polarizations. Every time that d_1 detects a particular polarization, the receiver d_2 detects the corresponding orthogonal polarization. Here, M_1 and M_2 are the output coupler mirrors, E_1 and E_2 are intracavity etalons, p_r is the intracavity polarizer controlled by the QRNG, p_1 and p_2 are the polarizers corresponding to detectors d_1 and d_2, respectively.

$$|x\rangle_I = |x\rangle_1 |x\rangle_2 |x\rangle_3 ... |x\rangle_g ... \tag{15.2}$$

$$|y\rangle_{II} = |y\rangle_1 |y\rangle_2 |y\rangle_3 ... |y\rangle_g ... \tag{15.3}$$

These states are directly applicable to laser quantum entanglement sources yielding ensembles of entangles states rather than single-photon pairs (Duarte 2022), that is

$$|\psi\rangle_- = 2^{-1/2} (|x\rangle_I |y\rangle_{II} - |y\rangle_I |x\rangle_{II}) \tag{15.4}$$

15.3 The physics of *N*-channel quantum entanglement

The probability amplitude equation $|\psi\rangle = (|x, y\rangle - |y, x\rangle)$ applies to two propagating quanta $(n = 2)$, moving in two different directions $(N = 2)$, with entangled polarizations x and y. For various important applications, including quantum communications, quantum information, and quantum computing, generalized probability amplitudes applicable to $n = N = 2^1, 2^2, 2^3, 2^4, 2^5, ...$ are of interest. In this review, a transparent, systematic, and rational statistical methodology, *à la Dirac*, is applied to derive the appropriate probability amplitudes. The interferometric approach is transparent and self-contained. A discussion on the implications of this derivational approach to interpretational approaches in quantum mechanics is also provided.

15.3.1 Background

One of the most ubiquitous and crucial equations of the quantum era, the mathematical expression stating the probability amplitude for two quanta propagating in different directions with entangled polarizations, became widely known via a philosophical route rather than via the pathway originating at its actual physical genesis (Duarte 2016). Briefly, although the relevant mathematical equation was introduced by Ward, using *bra ket* notation, in the 1947–49 period (Pryce and Ward 1947, Ward 1949) the discussion on quantum entanglement (Schrödinger 1935, 1936) followed a philosophical route (Bohm and Aharonov 1957) which initially incorporated the arguments of Einstein *et al* (1935), also known as EPR. Subsequently, the findings of Bell (1964) inspired a new generation of physicists to revisit the original experimental configuration (Pryce and Ward 1947, Wu and Shaknov 1950) using simpler polarization based optics (Aspect *et al* 1981). It was not until the early 1990s that the original *bra ket* quantum entanglement equation was *reintroduced* into the mainstream of the physics literature (Greenberger *et al* 1990).

In this review, the probability amplitude, for quanta (n) propagating in different directions (N) with entangled polarizations, is derived from first interferometric principles using Dirac's *bra ket* notation (Dirac 1930, 1958). Furthermore, a systematic generalized approach is described to derive probability amplitudes applicable to the binary sequence $n = N = 2^1, 2^2, 2^3, 2^4, 2^5, \ldots$. These cases are of interest for their applications to quantum communications, quantum information, and quantum computing.

15.3.2 Quantum entanglement probability amplitudes

The probability amplitude applicable to two quanta, 1 and 2, propagating in different directions with entangled polarizations, $|x\rangle$ and $|y\rangle$, was first discovered and applied in 1947 by Pryce and Ward (1947) and stated by Ward (1949), using *bra ket* notation, as

$$|\psi\rangle = (|x, y\rangle - |y, x\rangle) \tag{15.5}$$

which, using various Dirac identities (Dirac 1958), once normalized can be written as (Duarte 2013, 2014, 2015)

$$|\psi\rangle = 2^{-1/2}(|x\rangle_1|y\rangle_2 - |y\rangle_1|x\rangle_2) \tag{15.6}$$

The probability amplitude given in equation (15.6) can be derived using conservation arguments (Ward 1949), a Hamiltonian approach (Feynman *et al* 1965, Duarte 2014, 2015), or a straightforward interferometric approach (Duarte 2013, 2014). Historical aspects of this iconic equation have been exposed in detail elsewhere (Dalitz and Duarte 2000, Ward 2004, Duarte 2012, 2016). In this regard, it should be indicated that the probability amplitude discovered by Ward, using Dirac's *bra ket* notation, was independently confirmed by R H Dalitz and disclosed at a seminar in Cambridge (Ward 2004). Also independently, but subsequently, Snyder *et al* (1948) arrived at a probability amplitude of the form

$$|\psi\rangle = 2^{-1/2}(\psi_A(1)\psi_B(2) + \psi_C(1)\psi_D(2)) \tag{15.7}$$

using Schrödinger wave functions. This expression was adopted by Bohm and Aharonov (1957), to cast an equation of the form

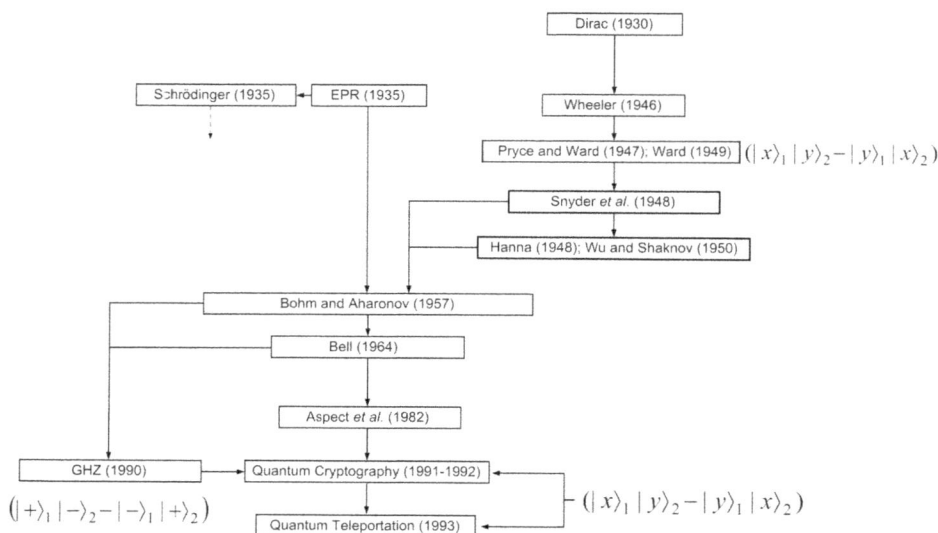

Figure 15.9. High-level literature pathway diagram depicting the rise of the field of quantum entanglement as we know it today. Solid vertical arrows in a downward direction indicate a direct citation pathway. The pathway initiated with the EPR paper also includes Bell's paper and an enormous number of publications focusing on philosophical and interpretational arguments. The pathway created by Dirac, Wheeler, and Ward focuses entirely on the physics in the absence of philosophical or interpretational arguments. The Schrödinger's papers which introduced the term of *entanglement* did not generate a unique pathway.

$$|\psi\rangle = 2^{-1/2}\big(\psi_+(1)\psi_-(2) - \psi_-(1)\psi_+(2)\big) \qquad (15.8)$$

in their widely cited paper. A visual representation of the literature history related to quantum entanglement and its probability amplitude equations is given in figure 15.9.

15.3.3 Interferometric approach to quantum entanglement

Feynman places interference at the *heart of quantum mechanics* (Feynman *et al* 1965). Ward discloses that Wheeler (1946), inspired by Dirac (1930) had attempted the quantum scattering calculation disclosed by Pryce and Ward (1947) but that ... 'through the neglect of interference terms he derived an incorrect ... value' (Ward 1949). Decades later Ward would reminisce ... '$|x, y\rangle - |y, x\rangle$ was my first lesson in quantum mechanics, and in a very real sense my last, since the rest is mere technique, which can be learnt from books' (Ward 2004).

The direct connection between the Dirac interferometric principle (Dirac 1939, 1958)

$$\langle d|s\rangle = \sum_{j=1}^{N}\langle d|j\rangle\langle j|s\rangle \qquad (15.9)$$

and the quantum entanglement probability amplitude equation

$$|\psi\rangle = 2^{-1/2}(|x\rangle_1|y\rangle_2 - |y\rangle_1|x\rangle_2)$$

was established and illustrated (Duarte 2013, 2014). Here, for background purposes and as an example, the case for $n = N = 4$, or $n = N = 2^2$, is re-exposed. In figure 15.3 a four-quanta source ($n = 4$) emits, in four distinct channels ($N = 4$), toward four detectors via four polarizers. Assuming four ideal and identical detectors ($d = d_1 = d_2 = d_3 = d_4$) (Duarte 2015)

$$\langle d|s\rangle = \langle d|p_4\rangle\langle p_4|s\rangle + \langle d|p_3\rangle\langle p_3|s\rangle + \langle d|p_2\rangle\langle p_2|s\rangle + \langle d|p_1\rangle\langle p_1|s\rangle \qquad (15.10)$$

In abstract form this can be expressed as

$$|s\rangle = |p_4\rangle\langle p_4|s\rangle + |p_3\rangle\langle p_3|s\rangle + |p_2\rangle\langle p_2|s\rangle + |p_1\rangle\langle p_1|s\rangle \qquad (15.11)$$

and defining

$$|C\rangle_1 = |p_1\rangle\langle p_1|s\rangle \qquad (15.12)$$

$$|C\rangle_2 = |p_2\rangle\langle p_2|s\rangle \qquad (15.13)$$

$$|C\rangle_3 = |p_3\rangle\langle p_3|s\rangle \qquad (15.14)$$

$$|C\rangle_4 = |p_4\rangle\langle p_4|s\rangle \qquad (15.15)$$

the overall probability amplitude, that is equation (15.11), becomes (Duarte 2015)

$$|s\rangle = |C\rangle_1 + |C\rangle_2 + |C\rangle_3 + |C\rangle_4 \qquad (15.16)$$

At this stage it is useful to introduce the normalized probability amplitude $|\psi\rangle_R$ where R is a Roman numeral. The relation between $|s\rangle$ and $|\psi\rangle_R$ is given by (Duarte 2015)

$$|\psi\rangle_R = N^{-1/2}|s\rangle \qquad (15.17)$$

where, for $N = 4$, the normalization condition

$$1 = ||\psi\rangle_I|^2 + ||\psi\rangle_{II}|^2 + ||\psi\rangle_{III}|^2 + ||\psi\rangle_{IV}|^2 \qquad (15.18)$$

has to be met. That normalization condition immediately means that

$$|\psi\rangle_I = 4^{-1/2}(|C\rangle_1 + |C\rangle_2 + |C\rangle_3 + |C\rangle_4) \qquad (15.19)$$

$$|\psi\rangle_{II} = 4^{-1/2}(|C\rangle_1 + |C\rangle_2 - |C\rangle_3 - |C\rangle_4) \qquad (15.20)$$

$$|\psi\rangle_{III} = 4^{-1/2}(|C\rangle_1 - |C\rangle_2 + |C\rangle_3 - |C\rangle_4) \qquad (15.21)$$

$$|\psi\rangle_{IV} = 4^{-1/2}(|C\rangle_1 - |C\rangle_2 - |C\rangle_3 + |C\rangle_4) \qquad (15.22)$$

Utilizing the Dirac identity (Dirac 1958) applicable to the state of several similar particles (Duarte 2013, 2014)

$$|X\rangle = |a\rangle_1|b\rangle_2|c\rangle_3...|g\rangle_n \qquad (15.23)$$

the probability amplitudes can be expressed explicitly as

$$
\begin{aligned}
|\psi\rangle_I = 4^{-1/2}(&|x\rangle_1|y\rangle_2|\varphi\rangle_3|\varphi'\rangle_4 + |y\rangle_1|x\rangle_2|\varphi'\rangle_3|\varphi\rangle_4 \\
&+ |\varphi\rangle_1|\varphi'\rangle_2|x\rangle_3|y\rangle_4 + |\varphi'\rangle_1|\varphi\rangle_2|y\rangle_3|x\rangle_4)
\end{aligned} \qquad (15.24)
$$

$$
\begin{aligned}
|\psi\rangle_{II} = 4^{-1/2}(&|x\rangle_1|y\rangle_2|\varphi\rangle_3|\varphi'\rangle_4 + |y\rangle_1|x\rangle_2|\varphi'\rangle_3|\varphi\rangle_4 \\
&- |\varphi\rangle_1|\varphi'\rangle_2|x\rangle_3|y\rangle_4 + |\varphi'\rangle_1|\varphi\rangle_2|y\rangle_3|x\rangle_4)
\end{aligned} \qquad (15.25)
$$

$$
\begin{aligned}
|\psi\rangle_{III} = 4^{-1/2}(&|x\rangle_1|y\rangle_2|\varphi\rangle_3|\varphi'\rangle_4 - |y\rangle_1|x\rangle_2|\varphi'\rangle_3|\varphi\rangle_4 \\
&+ |\varphi\rangle_1|\varphi'\rangle_2|x\rangle_3|y\rangle_4 - |\varphi'\rangle_1|\varphi\rangle_2|y\rangle_3|x\rangle_4)
\end{aligned} \qquad (15.26)
$$

$$
\begin{aligned}
|\psi\rangle_{IV} = 4^{-1/2}(&|x\rangle_1|y\rangle_2|\varphi\rangle_3|\varphi'\rangle_4 - |y\rangle_1|x\rangle_2|\varphi'\rangle_3|\varphi\rangle_4 \\
&- |\varphi\rangle_1|\varphi'\rangle_2|x\rangle_3|y\rangle_4 + |\varphi'\rangle_1|\varphi\rangle_2|y\rangle_3|x\rangle_4)
\end{aligned} \qquad (15.27)
$$

Observing these equations, a symmetry pattern emerges in the sign convention of the $|C\rangle_m$ states which expressed in a 4×4 arrangement it becomes

$$
\begin{aligned}
&+ |C\rangle_1 + |C\rangle_2 + |C\rangle_3 + |C\rangle_4 \\
&+ |C\rangle_1 + |C\rangle_2 - |C\rangle_3 - |C\rangle_4 \\
&+ |C\rangle_1 - |C\rangle_2 + |C\rangle_3 - |C\rangle_4 \\
&+ |C\rangle_1 - |C\rangle_2 - |C\rangle_3 + |C\rangle_4
\end{aligned} \qquad (15.28)
$$

For the basic $n = N = 2^1$ case applicable to equation (15.6) the corresponding 2×2 arrangement reduces to

$$
\begin{aligned}
&+ |C\rangle_1 + |C\rangle_2 \\
&+ |C\rangle_1 - |C\rangle_2
\end{aligned} \qquad (15.29)
$$

The mechanics to obtain explicit normalized probability amplitudes, for $n = N = 2^3$, is illustrated in the next section. The cases for $n = N = 3$ and $n = N = 6$ are examined in (Duarte 2014, 2015, 2022). For $n = N = 3$ the final explicit probability amplitudes include Hamilton's quaternions (Duarte 2015, 2016).

15.3.4 Generalized quantum entanglement probability amplitudes

For $n = N = 2^1, 2^2, 2^4, 2^5, ..., 2^r$, from an interferometric analysis, it can be established that the probability amplitude $|s\rangle$ can generally be expressed as

$$|s\rangle = \sum_{j=1}^{N} |C\rangle_{N+1-j} \qquad (15.30)$$

As an example now, the case of $n = N = 8$, or $n = N = 2^3$, as depicted in figure 15.5, is considered. For $N = 8$ equation (15.30) becomes

$$|s\rangle = |C\rangle_8 + |C\rangle_7 + |C\rangle_6 + |C\rangle_5 + |C\rangle_4 + |C\rangle_3 + |C\rangle_2 + |C\rangle_1 \quad (15.31)$$

Here, using one of Dirac's identities, the individual terms of the path probability amplitude can be written as

$$|C\rangle_{N+1-j} = \prod_{m=1,3,5\ldots}^{n} |a\rangle_m |b\rangle_{m+1} \quad (15.32)$$

These are a series of probability amplitudes starting at $|C\rangle_{N+1-j}$ with $j = 1$, and ending at $|C\rangle_1$ which is reached when $j = N$. Here, n is the total number of quanta which is an even number since quanta participate in pairs. For each pair, $(1, 2)$, $(3, 4)$, ..., $(m, m + 1)$, $|a\rangle_m$, $|b\rangle_{m+1}$ represent a set of orthogonal polarization alternatives such as (x, y), (φ, φ'), and so on. For instance, starting with $|C\rangle_1$ $(j = N - 0)$

$$|C\rangle_1 = |x\rangle_1|y\rangle_2|\varphi\rangle_3|\varphi'\rangle_4|\phi\rangle_5|\phi'\rangle_6|\vartheta\rangle_7|\vartheta'\rangle_8 \quad (15.33)$$

$$|C\rangle_2 \ (j = N - 1)$$

$$|C\rangle_2 = |y\rangle_1|x\rangle_2|\varphi'\rangle_3|\varphi\rangle_4|\phi'\rangle_5|\phi\rangle_6|\vartheta'\rangle_7|\vartheta\rangle_8 \quad (15.34)$$

$$|C\rangle_3 \ (j = N - 2)$$

$$|C\rangle_3 = |\varphi\rangle_1|\varphi'\rangle_2|\phi\rangle_3|\phi'\rangle_4|\vartheta\rangle_5|\vartheta'\rangle_6|x\rangle_7|y\rangle_8 \quad (15.35)$$

$$|C\rangle_4 \ (j = N - 3)$$

$$|C\rangle_4 = |\varphi'\rangle_1|\varphi\rangle_2|\phi'\rangle_3|\phi\rangle_4|\vartheta'\rangle_5|\vartheta\rangle_6|y\rangle_7|x\rangle_8 \quad (15.36)$$

$$|C\rangle_5 \ (j = N - 4)$$

$$|C\rangle_5 = |\phi\rangle_1|\phi'\rangle_2|\vartheta\rangle_3|\vartheta'\rangle_4|x\rangle_5|y\rangle_6|\varphi\rangle_7|\varphi'\rangle_8 \quad (15.37)$$

$$|C\rangle_6 \ (j = N - 5)$$

$$|C\rangle_6 = |\phi'\rangle_1|\phi\rangle_2|\vartheta'\rangle_3|\vartheta\rangle_4|y\rangle_5|x\rangle_6|\varphi'\rangle_7|\varphi\rangle_8 \quad (15.38)$$

$$|C\rangle_7 \ (j = N - 6)$$

$$|C\rangle_7 = |\vartheta\rangle_1|\vartheta'\rangle_2|x\rangle_3|y\rangle_4|\varphi\rangle_5|\varphi'\rangle_6|\phi\rangle_7|\phi'\rangle_8 \quad (15.39)$$

$$|C\rangle_8 \ (j = N - 7)$$

$$|C\rangle_8 = |\vartheta'\rangle_1|\vartheta\rangle_2|y\rangle_3|x\rangle_4|\varphi'\rangle_5|\varphi\rangle_6|\phi'\rangle_7|\phi\rangle_8 \quad (15.40)$$

These series indicate that the quanta pairs $(1, 2)$, $(3, 4)$, $(5, 6)$, $(7, 8)$ assume an original polarization order $(|C\rangle_1)$ and then, within pairs, the polarization is transposed $(|C\rangle_2)$. Next, the second pair assumes the position of the first pair and the first pair is displaced to the position of the last pair ($|C\rangle_3$). In this new order, transposition of the polarizations follows ($|C\rangle_4$), and so on.

For an even number of paths the *normalized probability amplitudes* are designated by $|\psi\rangle_R$, where R is a *Roman numeral*, and have the form of (Duarte 2015, 2016, Duarte and Taylor 2017)

$$|\psi\rangle_R = N^{-1/2} \sum_{m=1}^{N} \pm|C\rangle_m \tag{15.41}$$

and, as hinted previously, see equation (15.14), must satisfy the normalization condition (Duarte 2015, 2016)

$$1 = \sum_{R=I}^{N} ||\psi\rangle_R|^2 \tag{15.42}$$

Expanding equation (15.41) while observing the normalization condition yields (Duarte and Taylor 2017)

$$|\psi\rangle_I = 8^{-1/2}(|C\rangle_1 + |C\rangle_2 + |C\rangle_3 + |C\rangle_4 + |C\rangle_5 + |C\rangle_6 + |C\rangle_7 + |C\rangle_8) \tag{15.43}$$

$$|\psi\rangle_{II} = 8^{-1/2}(|C\rangle_1 + |C\rangle_2 + |C\rangle_3 + |C\rangle_4 - |C\rangle_5 - |C\rangle_6 - |C\rangle_7 - |C\rangle_8) \tag{15.44}$$

$$|\psi\rangle_{III} = 8^{-1/2}(|C\rangle_1 + |C\rangle_2 - |C\rangle_3 - |C\rangle_4 + |C\rangle_5 + |C\rangle_6 - |C\rangle_7 - |C\rangle_8) \tag{15.45}$$

$$|\psi\rangle_{IV} = 8^{-1/2}(|C\rangle_1 + |C\rangle_2 - |C\rangle_3 - |C\rangle_4 - |C\rangle_5 - |C\rangle_6 + |C\rangle_7 + |C\rangle_8) \tag{15.46}$$

$$|\psi\rangle_V = 8^{-1/2}(|C\rangle_1 - |C\rangle_2 + |C\rangle_3 - |C\rangle_4 + |C\rangle_5 - |C\rangle_6 + |C\rangle_7 - |C\rangle_8) \tag{15.47}$$

$$|\psi\rangle_{VI} = 8^{-1/2}(|C\rangle_1 - |C\rangle_2 + |C\rangle_3 - |C\rangle_4 - |C\rangle_5 + |C\rangle_6 - |C\rangle_7 + |C\rangle_8) \tag{15.48}$$

$$|\psi\rangle_{VII} = 8^{-1/2}(|C\rangle_1 - |C\rangle_2 - |C\rangle_3 + |C\rangle_4 + |C\rangle_5 - |C\rangle_6 - |C\rangle_7 + |C\rangle_8) \tag{15.49}$$

$$|\psi\rangle_{VIII} = 8^{-1/2}(|C\rangle_1 - |C\rangle_2 - |C\rangle_3 + |C\rangle_4 - |C\rangle_5 + |C\rangle_6 + |C\rangle_7 - |C\rangle_8) \tag{15.50}$$

As seen previously, for $n = N = 2^2$, observing these equations leads to a symmetry in the sign convention of the $|C\rangle_m$ states. This is more easily observed if the $|C\rangle_m$ amplitudes are expressed in an 8×8 format with the *bra ket* ($|\rangle$) symbols abstracted, so that

$$\begin{array}{l} + C_1 + C_2 + C_3 + C_4 + C_5 + C_6 + C_7 + C_8 \\ + C_1 + C_2 + C_3 + C_4 - C_5 - C_6 - C_7 - C_8 \\ + C_1 + C_2 - C_3 - C_4 + C_5 + C_6 - C_7 - C_8 \\ + C_1 + C_2 - C_3 - C_4 - C_5 - C_6 + C_7 + C_8 \\ + C_1 - C_2 + C_3 - C_4 + C_5 - C_6 + C_7 - C_8 \\ + C_1 - C_2 + C_3 - C_4 - C_5 + C_6 - C_7 + C_8 \\ + C_1 - C_2 - C_3 + C_4 + C_5 - C_6 - C_7 + C_8 \\ + C_1 - C_2 - C_3 + C_4 - C_5 + C_6 + C_7 - C_8 \end{array} \tag{15.51}$$

For instance, the coefficients in the first row and the first column are all positive. For the second row and second column the first four are positive and the second four are negative. For the third row and third column there are alternating positive and negative pairs. The same is true for the terms along the diagonal. For the fourth row and fourth column the middle four terms are negative. For the fifth row and fifth column the sign alternates positive and negative. The remaining three rows and columns complete the established sign pattern.

A general feature of these arrays is that the sign sequence order for rows and columns originating at a common term at the first row and column are identical. For example, for the row and column originating at $+C_1$ the sign sequences (right and down) are $+, +, +, +, +, +, +, +$ and $+, +, +, +, +, +, +, +$. On the other hand for the row and column originating at $-C_8$ the sign sequences (left and up) are $-, +, +, -, +, -, -, +$ and $-, +, +, -, +, -, -, +$.

Straightforward substitution of the corresponding $|C\rangle_{N+1-j}$ amplitudes, that is $|C\rangle_1, |C\rangle_2, |C\rangle_3 \ldots |C\rangle_8$ (from equations (15.43) to (15.50)), yield explicit expressions for $|\psi\rangle_I, |\psi\rangle_{II}, |\psi\rangle_{III} \ldots |\psi\rangle_{VIII}$ in terms of the polarization coordinates (x, y), (φ, φ'), (ϕ, ϕ'), and (ϑ, ϑ'), see figure 15.5. These expressions are

$$
\begin{aligned}
|\psi\rangle_I = 8^{-1/2}(&|x\rangle_1|y\rangle_2|\phi\rangle_3|\phi'\rangle_4|\varphi\rangle_5|\varphi'\rangle_6|\vartheta\rangle_7|\vartheta'\rangle_8 \\
&+ |y\rangle_1|x\rangle_2|\phi'\rangle_3|\phi\rangle_4|\varphi'\rangle_5|\varphi\rangle_6|\vartheta'\rangle_7|\vartheta\rangle_8 \\
&+ |\phi\rangle_1|\phi'\rangle_2|\varphi\rangle_3|\varphi'\rangle_4|\vartheta\rangle_5|\vartheta'\rangle_6|x\rangle_7|y\rangle_8 \\
&+ |\phi'\rangle_1|\phi\rangle_2|\varphi'\rangle_3|\varphi\rangle_4|\vartheta'\rangle_5|\vartheta\rangle_6|y\rangle_7|x\rangle_8 \\
&+ |\varphi\rangle_1|\varphi'\rangle_2|\vartheta\rangle_3|\vartheta'\rangle_4|x\rangle_5|y\rangle_6|\phi\rangle_7|\phi'\rangle_8 \\
&+ |\varphi'\rangle_1|\varphi\rangle_2|\vartheta'\rangle_3|\vartheta\rangle_4|y\rangle_5|x\rangle_6|\phi'\rangle_7|\phi\rangle_8 \\
&+ |\vartheta\rangle_1|\vartheta'\rangle_2|x\rangle_3|y\rangle_4|\phi\rangle_5|\phi'\rangle_6|\varphi\rangle_7|\varphi'\rangle_8 \\
&+ |\vartheta'\rangle_1|\vartheta\rangle_2|y\rangle_3|x\rangle_4|\phi'\rangle_5|\phi\rangle_6|\varphi'\rangle_7|\varphi\rangle_8)
\end{aligned}
\tag{15.52}
$$

$$
\begin{aligned}
|\psi\rangle_{II} = 8^{-1/2}(&|x\rangle_1|y\rangle_2|\phi\rangle_3|\phi'\rangle_4|\varphi\rangle_5|\varphi'\rangle_6|\vartheta\rangle_7|\vartheta'\rangle_8 \\
&+ |y\rangle_1|x\rangle_2|\phi'\rangle_3|\phi\rangle_4|\varphi'\rangle_5|\varphi\rangle_6|\vartheta'\rangle_7|\vartheta\rangle_8 \\
&+ |\phi\rangle_1|\phi'\rangle_2|\varphi\rangle_3|\varphi'\rangle_4|\vartheta\rangle_5|\vartheta'\rangle_6|x\rangle_7|y\rangle_8 \\
&+ |\phi'\rangle_1|\phi\rangle_2|\varphi'\rangle_3|\varphi\rangle_4|\vartheta'\rangle_5|\vartheta\rangle_6|y\rangle_7|x\rangle_8 \\
&- |\varphi\rangle_1|\varphi'\rangle_2|\vartheta\rangle_3|\vartheta'\rangle_4|x\rangle_5|y\rangle_6|\phi\rangle_7|\phi'\rangle_8 \\
&- |\varphi'\rangle_1|\varphi\rangle_2|\vartheta'\rangle_3|\vartheta\rangle_4|y\rangle_5|x\rangle_6|\phi'\rangle_7|\phi\rangle_8 \\
&- |\vartheta\rangle_1|\vartheta'\rangle_2|x\rangle_3|y\rangle_4|\phi\rangle_5|\phi'\rangle_6|\varphi\rangle_7|\varphi'\rangle_8 \\
&- |\vartheta'\rangle_1|\vartheta\rangle_2|y\rangle_3|x\rangle_4|\phi'\rangle_5|\phi\rangle_6|\varphi'\rangle_7|\varphi\rangle_8)
\end{aligned}
\tag{15.53}
$$

$$
\begin{aligned}
|\psi\rangle_{III} = 8^{-1/2}(&|x\rangle_1|y\rangle_2|\phi\rangle_3|\phi'\rangle_4|\varphi\rangle_5|\varphi'\rangle_6|\vartheta\rangle_7|\vartheta'\rangle_8 \\
&+ |y\rangle_1|x\rangle_2|\phi'\rangle_3|\phi\rangle_4|\varphi'\rangle_5|\varphi\rangle_6|\vartheta'\rangle_7|\vartheta\rangle_8 \\
&- |\phi\rangle_1|\phi'\rangle_2|\varphi\rangle_3|\varphi'\rangle_4|\vartheta\rangle_5|\vartheta'\rangle_6|x\rangle_7|y\rangle_8 \\
&- |\phi'\rangle_1|\phi\rangle_2|\varphi'\rangle_3|\varphi\rangle_4|\vartheta'\rangle_5|\vartheta\rangle_6|y\rangle_7|x\rangle_8 \\
&+ |\varphi\rangle_1|\varphi'\rangle_2|\vartheta\rangle_3|\vartheta'\rangle_4|x\rangle_5|y\rangle_6|\phi\rangle_7|\phi'\rangle_8 \\
&+ |\varphi'\rangle_1|\varphi\rangle_2|\vartheta'\rangle_3|\vartheta\rangle_4|y\rangle_5|x\rangle_6|\phi'\rangle_7|\phi\rangle_8 \\
&- |\vartheta\rangle_1|\vartheta'\rangle_2|x\rangle_3|y\rangle_4|\phi\rangle_5|\phi'\rangle_6|\varphi\rangle_7|\varphi'\rangle_8 \\
&- |\vartheta'\rangle_1|\vartheta\rangle_2|y\rangle_3|x\rangle_4|\phi'\rangle_5|\phi\rangle_6|\varphi'\rangle_7|\varphi\rangle_8)
\end{aligned}
\tag{15.54}
$$

$$\begin{aligned}
|\psi\rangle_{IV} = 8^{-1/2}(&|x\rangle_1|y\rangle_2|\phi\rangle_3|\phi'\rangle_4|\varphi\rangle_5|\varphi'\rangle_6|\vartheta\rangle_7|\vartheta'\rangle_8 \\
&+ |y\rangle_1|x\rangle_2|\phi'\rangle_3|\phi\rangle_4|\varphi'\rangle_5|\varphi\rangle_6|\vartheta'\rangle_7|\vartheta\rangle_8 \\
&- |\phi\rangle_1|\phi'\rangle_2|\varphi\rangle_3|\varphi'\rangle_4|\vartheta\rangle_5|\vartheta'\rangle_6|x\rangle_7|y\rangle_8 \\
&- |\phi'\rangle_1|\phi\rangle_2|\varphi'\rangle_3|\varphi\rangle_4|\vartheta'\rangle_5|\vartheta\rangle_6|y\rangle_7|x\rangle_8 \\
&- |\varphi\rangle_1|\varphi'\rangle_2|\vartheta\rangle_3|\vartheta'\rangle_4|x\rangle_5|y\rangle_6|\phi\rangle_7|\phi'\rangle_8 \\
&- |\varphi'\rangle_1|\varphi\rangle_2|\vartheta'\rangle_3|\vartheta\rangle_4|y\rangle_5|x\rangle_6|\phi'\rangle_7|\phi\rangle_8 \\
&+ |\vartheta\rangle_1|\vartheta'\rangle_2|x\rangle_3|y\rangle_4|\phi\rangle_5|\phi'\rangle_6|\varphi\rangle_7|\varphi'\rangle_8 \\
&+ |\vartheta'\rangle_1|\vartheta\rangle_2|y\rangle_3|x\rangle_4|\phi'\rangle_5|\phi\rangle_6|\varphi'\rangle_7|\varphi\rangle_8)
\end{aligned}$$

(15.55)

$$\begin{aligned}
|\psi\rangle_{V} = 8^{-1/2}(&|x\rangle_1|y\rangle_2|\phi\rangle_3|\phi'\rangle_4|\varphi\rangle_5|\varphi'\rangle_6|\vartheta\rangle_7|\vartheta'\rangle_8 \\
&- |y\rangle_1|x\rangle_2|\phi'\rangle_3|\phi\rangle_4|\varphi'\rangle_5|\varphi\rangle_6|\vartheta'\rangle_7|\vartheta\rangle_8 \\
&+ |\phi\rangle_1|\phi'\rangle_2|\varphi\rangle_3|\varphi'\rangle_4|\vartheta\rangle_5|\vartheta'\rangle_6|x\rangle_7|y\rangle_8 \\
&- |\phi'\rangle_1|\phi\rangle_2|\varphi'\rangle_3|\varphi\rangle_4|\vartheta'\rangle_5|\vartheta\rangle_6|y\rangle_7|x\rangle_8 \\
&+ |\varphi\rangle_1|\varphi'\rangle_2|\vartheta\rangle_3|\vartheta'\rangle_4|x\rangle_5|y\rangle_6|\phi\rangle_7|\phi'\rangle_8 \\
&- |\varphi'\rangle_1|\varphi\rangle_2|\vartheta'\rangle_3|\vartheta\rangle_4|y\rangle_5|x\rangle_6|\phi'\rangle_7|\phi\rangle_8 \\
&+ |\vartheta\rangle_1|\vartheta'\rangle_2|x\rangle_3|y\rangle_4|\phi\rangle_5|\phi'\rangle_6|\varphi\rangle_7|\varphi'\rangle_8 \\
&- |\vartheta'\rangle_1|\vartheta\rangle_2|y\rangle_3|x\rangle_4|\phi'\rangle_5|\phi\rangle_6|\varphi'\rangle_7|\varphi\rangle_8)
\end{aligned}$$

(15.56)

$$\begin{aligned}
|\psi\rangle_{VI} = 8^{-1/2}(&|x\rangle_1|y\rangle_2|\phi\rangle_3|\phi'\rangle_4|\varphi\rangle_5|\varphi'\rangle_6|\vartheta\rangle_7|\vartheta'\rangle_8 \\
&- |y\rangle_1|x\rangle_2|\phi'\rangle_3|\phi\rangle_4|\varphi'\rangle_5|\varphi\rangle_6|\vartheta'\rangle_7|\vartheta\rangle_8 \\
&+ |\phi\rangle_1|\phi'\rangle_2|\varphi\rangle_3|\varphi'\rangle_4|\vartheta\rangle_5|\vartheta'\rangle_6|x\rangle_7|y\rangle_8 \\
&- |\phi'\rangle_1|\phi\rangle_2|\varphi'\rangle_3|\varphi\rangle_4|\vartheta'\rangle_5|\vartheta\rangle_6|y\rangle_7|x\rangle_8 \\
&- |\varphi\rangle_1|\varphi'\rangle_2|\vartheta\rangle_3|\vartheta'\rangle_4|x\rangle_5|y\rangle_6|\phi\rangle_7|\phi'\rangle_8 \\
&+ |\varphi'\rangle_1|\varphi\rangle_2|\vartheta'\rangle_3|\vartheta\rangle_4|y\rangle_5|x\rangle_6|\phi'\rangle_7|\phi\rangle_8 \\
&- |\vartheta\rangle_1|\vartheta'\rangle_2|x\rangle_3|y\rangle_4|\phi\rangle_5|\phi'\rangle_6|\varphi\rangle_7|\varphi'\rangle_8 \\
&+ |\vartheta'\rangle_1|\vartheta\rangle_2|y\rangle_3|x\rangle_4|\phi'\rangle_5|\phi\rangle_6|\varphi'\rangle_7|\varphi\rangle_8)
\end{aligned}$$

(15.57)

$$\begin{aligned}
|\psi\rangle_{VII} = 8^{-1/2}(&|x\rangle_1|y\rangle_2|\phi\rangle_3|\phi'\rangle_4|\varphi\rangle_5|\varphi'\rangle_6|\vartheta\rangle_7|\vartheta'\rangle_8 \\
&- |y\rangle_1|x\rangle_2|\phi'\rangle_3|\phi\rangle_4|\varphi'\rangle_5|\varphi\rangle_6|\vartheta'\rangle_7|\vartheta\rangle_8 \\
&- |\phi\rangle_1|\phi'\rangle_2|\varphi\rangle_3|\varphi'\rangle_4|\vartheta\rangle_5|\vartheta'\rangle_6|x\rangle_7|y\rangle_8 \\
&+ |\phi'\rangle_1|\phi\rangle_2|\varphi'\rangle_3|\varphi\rangle_4|\vartheta'\rangle_5|\vartheta\rangle_6|y\rangle_7|x\rangle_8 \\
&+ |\varphi\rangle_1|\varphi'\rangle_2|\vartheta\rangle_3|\vartheta'\rangle_4|x\rangle_5|y\rangle_6|\phi\rangle_7|\phi'\rangle_8 \\
&- |\varphi'\rangle_1|\varphi\rangle_2|\vartheta'\rangle_3|\vartheta\rangle_4|y\rangle_5|x\rangle_6|\phi'\rangle_7|\phi\rangle_8 \\
&- |\vartheta\rangle_1|\vartheta'\rangle_2|x\rangle_3|y\rangle_4|\phi\rangle_5|\phi'\rangle_6|\varphi\rangle_7|\varphi'\rangle_8 \\
&+ |\vartheta'\rangle_1|\vartheta\rangle_2|y\rangle_3|x\rangle_4|\phi'\rangle_5|\phi\rangle_6|\varphi'\rangle_7|\varphi\rangle_8)
\end{aligned}$$

(15.58)

$$
\begin{aligned}
|\psi\rangle_{VIII} = 8^{-1/2}(&|x\rangle_1|y\rangle_2|\phi\rangle_3|\phi'\rangle_4|\varphi\rangle_5|\varphi'\rangle_6|\vartheta\rangle_7|\vartheta'\rangle_8 \\
&- |y\rangle_1|x\rangle_2|\phi'\rangle_3|\phi\rangle_4|\varphi'\rangle_5|\varphi\rangle_6|\vartheta'\rangle_7|\vartheta\rangle_8 \\
&- |\phi\rangle_1|\phi'\rangle_2|\varphi\rangle_3|\varphi'\rangle_4|\vartheta\rangle_5|\vartheta'\rangle_6|x\rangle_7|y\rangle_8 \\
&+ |\phi'\rangle_1|\phi\rangle_2|\varphi'\rangle_3|\varphi\rangle_4|\vartheta'\rangle_5|\vartheta\rangle_6|y\rangle_7|x\rangle_8 \\
&- |\varphi\rangle_1|\varphi'\rangle_2|\vartheta\rangle_3|\vartheta'\rangle_4|x\rangle_5|y\rangle_6|\phi\rangle_7|\phi'\rangle_8 \\
&+ |\varphi'\rangle_1|\varphi\rangle_2|\vartheta'\rangle_3|\vartheta\rangle_4|y\rangle_5|x\rangle_6|\phi'\rangle_7|\phi\rangle_8 \\
&+ |\vartheta\rangle_1|\vartheta'\rangle_2|x\rangle_3|y\rangle_4|\phi\rangle_5|\phi'\rangle_6|\varphi\rangle_7|\varphi'\rangle_8 \\
&- |\vartheta'\rangle_1|\vartheta\rangle_2|y\rangle_3|x\rangle_4|\phi'\rangle_5|\phi\rangle_6|\varphi'\rangle_7|\varphi\rangle_8)
\end{aligned}
\tag{15.59}
$$

An experimental schematics for the emission stage, applicable for $n = N = 2^3$, using type II phase matching spontaneous parametric down conversion (SPDC) is outlined in figure 15.4. In this figure, the source s is depicting as consisting of four BBO crystals in series. The first two crystals give origin to the emission in the (x, y) and (φ, φ') directions while the third and fourth crystals give origin to the emission in the (ϕ, ϕ') and (ϑ, ϑ') axes which are oriented at a $\pi/4$ angle relative to the first set of axes. For sufficiently long propagation distances the geometry depicted in figure 15.5 is approached. Certainly, an equivalent experimental configuration, for $n = N = 2^3$, is also attainable via the use of an $N = 2^3$ channel narrow-linewidth laser organic laser source, as described in figure 15.7.

For $n = N = 2^4$, the 16×16 array for the corresponding $|C\rangle_m$ states, with the *bra ket* ($|\rangle$) symbols abstracted, is given by

$$
\begin{aligned}
&+ c_1 + c_2 + c_3 + c_4 + c_5 + c_6 + c_7 + c_8 + c_9 + c_{10} + c_{11} + c_{12} + c_{13} + c_{14} + c_{15} + c_{16} \\
&+ c_1 + c_2 + c_3 + c_4 + c_5 + c_6 + c_7 + c_8 - c_9 - c_{10} - c_{11} - c_{12} - c_{13} - c_{14} - c_{15} - c_{16} \\
&+ c_1 + c_2 + c_3 + c_4 - c_5 - c_6 - c_7 - c_8 + c_9 + c_{10} + c_{11} + c_{12} - c_{13} - c_{14} - c_{15} - c_{16} \\
&+ c_1 + c_2 + c_3 + c_4 - c_5 - c_6 - c_7 - c_8 - c_9 - c_{10} - c_{11} - c_{12} + c_{13} - c_{14} + c_{15} + c_{16} \\
&+ c_1 + c_2 - c_3 - c_4 + c_5 + c_6 - c_7 - c_8 + c_9 + c_{10} - c_{11} - c_{12} + c_{13} - c_{14} - c_{15} - c_{16} \\
&+ c_1 + c_2 - c_3 - c_4 + c_5 + c_6 - c_7 - c_8 - c_9 - c_{10} + c_{11} + c_{12} - c_{13} - c_{14} + c_{15} + c_{16} \\
&+ c_1 + c_2 - c_3 - c_4 - c_5 - c_6 + c_7 + c_8 + c_9 + c_{10} - c_{11} - c_{12} - c_{13} - c_{14} + c_{15} + c_{16} \\
&+ c_1 + c_2 - c_3 - c_4 - c_5 - c_6 + c_7 + c_8 - c_9 - c_{10} + c_{11} + c_{12} + c_{13} + c_{14} - c_{15} - c_{16} \\
&+ c_1 - c_2 + c_3 - c_4 + c_5 - c_6 + c_7 - c_8 + c_9 - c_{10} + c_{11} - c_{12} + c_{13} - c_{14} + c_{15} - c_{16} \\
&+ c_1 - c_2 + c_3 - c_4 + c_5 - c_6 + c_7 - c_8 - c_9 + c_{10} - c_{11} + c_{12} - c_{13} + c_{14} - c_{15} + c_{16} \\
&+ c_1 - c_2 + c_3 - c_4 - c_5 + c_6 - c_7 + c_8 + c_9 - c_{10} + c_{11} - c_{12} - c_{13} + c_{14} - c_{15} + c_{16} \\
&+ c_1 - c_2 + c_3 - c_4 - c_5 + c_6 - c_7 + c_8 - c_9 + c_{10} - c_{11} + c_{12} + c_{13} - c_{14} + c_{15} - c_{16} \\
&+ c_1 - c_2 - c_3 + c_4 + c_5 - c_6 - c_7 + c_8 + c_9 - c_{10} - c_{11} + c_{12} + c_{13} - c_{14} - c_{15} + c_{16} \\
&+ c_1 - c_2 - c_3 + c_4 + c_5 - c_6 - c_7 + c_8 - c_9 + c_{10} + c_{11} - c_{12} - c_{13} + c_{14} + c_{15} - c_{16} \\
&+ c_1 - c_2 - c_3 + c_4 - c_5 + c_6 + c_7 - c_8 + c_9 - c_{10} - c_{11} + c_{12} - c_{13} + c_{14} + c_{15} - c_{16} \\
&+ c_1 - c_2 - c_3 + c_4 - c_5 + c_6 + c_7 - c_8 - c_9 + c_{10} + c_{11} - c_{12} + c_{13} - c_{14} - c_{15} + c_{16}
\end{aligned}
$$

Finally, observation of the $|C\rangle_m$ arrays indicate that for $n = N = 2^1$, 2^3 the diagonals are a combination of $+$ and $-$ signs, while for $n = N = 2^2$, 2^4 they are all $+$.

15.3.5 Alternative probability amplitude methodologies

A survey of the literature containing expressions for the probability amplitudes related to $n = N = 2^1$, 2^2, 2^4 indicates two main avenues of approach to the subject. Some researchers provide expressions with a straightforward absence of derivation

(Gaetner *et al* 2007) while other authors refer to the product of so-called GHZ states (Pan *et al* 2001, Zhang *et al* 2016). In both cases no explicit details on the mechanics of the derivaticn are given.

15.4 The interferometric equation and quantum entropy

The Dirac interferometric principle (Dirac 1939, 1958, Feynman *et al* 1965)

$$\langle d|s \rangle = \sum_{j=1}^{N} \langle d|j \rangle \langle j|s \rangle$$

also gives origin to the generalized probability interferometric equation (Duarte 1991, 1993)

$$\langle d|s \rangle \langle d|s \rangle^* = \sum_{j=1}^{N} \Psi(r_j)^2 + 2\sum_{j=1}^{N} \Psi(r_j) \left(\sum_{m=j+1}^{N} \Psi(r_m)\cos(\Omega_m - \Omega_j) \right) \qquad (15.60)$$

This equation, which applies to single-photon propagation or the propagation of an ensemble of indistinguishable photons, in conjunction with interferometric measurements for any intra-interferometric distances $D_{\langle d|j \rangle}$, provides an alternative approach to long-distance space-to-space quantum communications (Duarte 2002, 2016, Duarte and Taylor 2015).

The generalized probability interferometric equation can also be used to study the entropy characteristics of a propagating interferograms from very short intra-interferometric distances to extremely long intra-interferometric distances. In this regard, the overall intra-interferometric range of interest is

$$10^{-3} \leqslant D_{\langle d|j \rangle} \leqslant 10^7 \text{ m}$$

which corresponds to an approximate intra-interferometric propagation time $t_{\langle d|j \rangle}$ range of

$$3.33 \times 10^{-12} \leqslant t_{\langle d|j \rangle} \leqslant 3.33 \times 10^{-2} \text{ s}$$

Interferometrically speaking, for visible quanta, a post emission intra-interferometric propagation time in the picosecond domain yields highly structured and distinctive zeroth order interferograms, which are associated with a low entropy regime. On the other hand, intra-interferometric propagation times in the microseconds, and beyond, produce zeroth order interferograms that have structureless and uniform near-Gaussian propagation profiles. These structureless profiles are representative of a higher entropy regime.

Hence, N-slit quantum interferometry can be used to characterize *the arrow of time*, forth and back[1], at a fundamental level. This topic is discussed further by Duarte (2022).

---

[1] Measurement of a quantum interferogram at a time $t_{\langle d|j \rangle}$ corresponding to an intra-interferometric distance $D_{\langle d|j \rangle} \gg D_{\langle d|j \rangle_i}$ where $D_{\langle d|j \rangle_i}$ is an infinitesimal initial distance from j allows the measurer use to equation (15.60), in conjunction with information about $(D_{\langle d|j \rangle} - D_{\langle d|j \rangle_i})$, N, λ, slit width, and slit separation, to 'go back in time' and recover the interferogram corresponding to $t_{\langle d|j \rangle_i}$ where $t_{\langle d|j \rangle_i} \ll t_{\langle d|j \rangle}$.

15.5 Implications for the interpretations of quantum mechanics

Previously it has been noted and documented that the original equation for the probability amplitude of entanglement in quantum mechanics, that is,

$$(|x\rangle_1|y\rangle_2 - |y\rangle_1|x\rangle_2)$$

was discovered by Pryce and Ward (Pryce and Ward 1947, Ward 1949) in a *complete vacuum of philosophical arguments* while following foundational physics guidelines outlined first by Dirac (1930) and then by Wheeler (1946), as illustrated in figure 15.9.

Likewise, the various probability amplitudes for quantum entanglement applicable to the propagation of n quanta in N pathways for $n = N = 2^1, 2^2, 2^3, 2^4, 2^5, \ldots$ can be shown to originate from interferometric arguments first articulated by Dirac (1958) that were then adapted and extended to single-photon, and indistinguishable-photon, N-slit interference (Duarte 1991, 1993) via narrow-linewidth laser illumination. The relatively straightforward theoretical structure (Duarte 2013, 2014, Duarte and Taylor 2017) necessary to derive these equations is provided by the various principles outlined in this review, all of which are based on the ideas exposed by Dirac in his book classic (Dirac 1958). These principles provide a direct, effortless, and transparent route to a most crucial probability amplitude in all of quantum optics. A probability amplitude with enormous impact in physics and application areas of high practical interest such as quantum communications and quantum computing. These principles are consistent with the framework of quantum mechanics as laid down by Dirac and beautifully articulated by Feynman.

It is also relevant to point out that Bell's theorem (Bell 1964) *plays no role* in the derivation of the probability amplitude of quantum entanglement. Bell's theorem only becomes relevant once so-called 'hidden variable theories' are brought into the discussion. But the essential physics is unaffected. This is discussed in detail by Duarte (2022).

The transparency and ease of derivation of these quantum entanglement probability amplitudes ... *free of paradoxes* ... using the indeterministic principles of quantum mechanics, *à la Dirac*, raise considerably the bar for proposed 'alternative interpretations' of quantum mechanics relying on deterministic concepts such as those suggested by Bohm and Bub (1966) and 't Hooft (2016).

15.6 Perspective

Besides the fascinating aspects of the physics of photon pair-based quantum entanglement, its practical realization in the communications domain continues to demand a fair degree of complexity even in its latest more streamlined versions. A specific challenge persists in the detection of single photons where signal-to-noise ratio are unfavorable.

Here, we have described swiftly and in detail the optical configuration of a visible source of entangled ensembles of indistinguishable photons such as $|x\rangle_I = |x\rangle_1|x\rangle_2|x\rangle_3\ldots|x\rangle_g\ldots$ and $|y\rangle_{II} = |y\rangle_1|y\rangle_2|y\rangle_3\ldots|y\rangle_g\ldots$ that should significantly improve the signal-to-noise ratio for quantum entanglement experiments. The description given here is an improvement on the original disclosure done in the first edition of this book in 2018. Although the description was done here from an organic laser vantage point, the optical configuration is also applicable to semi-conductor lasers.

From a pragmatic perspective, for space-to-space quantum communications, we remind the readership that the opportunity persists for the quantum interferometric approach which is also based on the emission of ensembles of indistinguishable quanta (Duarte 2002, 2016, Duarte and Taylor 2015).

15.7 Problems

- 15.1. Use the Dirac–Feynman interferometric principle stated in equation (15.9) to derive equation (15.6) for $N = 4$.
- 15.2. Use the generalized equation for quantum entanglement, that is, equation (15.41), and the condition for normalization given in equation (15.42), to arrive at equations (15.19)–(15.22) for $n = N = 2^2$.
- 15.3. Show that using the Dirac identity given in equation (15.23), equations (15.19)–(15.22) can be re-expressed as equations (15.24)–(15.27)
- 15.4. Use the generalized equation for quantum entanglement, that is equation (15.41), and the condition for normalization given in equation (15.42), to arrive at equations (15.43)–(15.50) for $n = N = 2^3$.
- 15.5. Use the generalized equation for quantum entanglement, that is, equation (15.41), and the condition for normalization given in equation (15.42), to derive the quantum entanglement probability amplitudes for $n = N = 3$. Note: this time Hamilton quaternions have to be used (Duarte 2022).

References

Aspect A, Grangier P and Roger G 1981 Experimental tests of realistic local theories via Bell's theorem *Phys. Rev. Lett.* **47** 460–3

Bell J S 1964 On the Einstein–Podolsky–Rosen paradox *Physics (N.Y.)* **1** 195–200

Bitton G, Grice W P, Moreau J and Zhang L 2002 Cascaded ultrabright source polarization-entangled photons *Phys. Rev.* A **65** 063805

Bohm D and Aharonov Y 1957 Discussion of experimental proof for the paradox of Einstein, Rosen, and Podolsky *Phys. Rev.* **108** 1070–6

Bohm D and Bub J 1966 A proposed solution to the measurement problem in quantum mechanics by hidden variable theory *Rev. Mod. Phys.* **38** 453–69

Dalitz R H and Duarte F J 2000 John Clive Ward *Phys. Today* **53** 99–100

Dirac P A M 1930 On the annihilation of electrons and protons *Math. Proc. Camb. Phil. Soc.* **2** 361–75

Dirac P A M 1939 A new notation for quantum mechanics *Math. Proc. Camb. Phil. Soc.* **35** 416–8

Dirac P A M 1958 *The Principles of Quantum Mechanics* 4th edn (Oxford: Oxford University)

Duarte F J 1990 Technology of pulsed dye lasers *Dye Laser Principles* ed F J Duarte and L W Hillman (New York: Academic) ch 6

Duarte F J 1991 Dispersive dye lasers *High Power Dye Lasers* ed F J Duarte (Berlin: Springer) ch 2

Duarte F J 1992 Beam transmission characteristics of a collinear polarization rotator *Appl. Opt.* **31** 3377–8

Duarte F J 1993 On a generalized interference equation and interferometric measurements *Opt. Commun.* **103** 8–14

Duarte F J 2002 Secure interferometric communications in free space *Opt. Commun.* **205** 313–9

Duarte F J 2012 The origin of quantum entanglement experiments based on polarization measurements *Euro. Phys. J.* H **37** 311–3

Duarte F J 2013 The probability amplitude for entangled polarizations: an interferometric approach *J. Mod. Opt.* **60** 1585–7

Duarte F J 2014 *Quantum Optics for Engineers* (New York: CRC Press)

Duarte F J 2015 *Tunable Laser Optics* 2nd edn (New York: CRC Press)

Duarte F J 2016 Secure space-to-space interferometric communications and its nexus to the physics of quantum entanglement *Appl. Phys. Rev.* **3** 041301

Duarte F J 2022 *Fundamentals of Quantum Entanglement* 2nd edn (Bristol: Institute of Physics Publishing))

Duarte F J and Taylor T S 2015 Quantum entanglement physics secures space-to-space interferometric communications *Laser Focus World* **51** 54–8

Duarte F J and Taylor T S 2017 Quantum entanglement probability amplitudes in multiple propagation channels: an interferometric approach *Optik* **139** 222–30

Duarte F J and Taylor T S 2022 Quantum coherence in electrically-pumped organic interferometric emitters *Appl. Phys.* B **128** 11

Einstein A, Podolsky B and Rosen N 1935 Can quantum mechanical description of physical reality be considered complete? *Phys. Rev.* **47** 777–80

Feynman R P, Leighton R B and Sands M 1965 *The Feynman Lectures on Physics* **vol III** (Reading, MA: Addison-Wesley)

Freedman S J and Clauser J F 1972 Experimental test of local hidden-variable theories *Phys. Rev. Lett.* **28** 938–41

Gaertner S, Kurtsiefer C, Bourennane M and Weinfurter H 2007 Experimental demonstration of four-party quantum secret sharing *Phys. Rev. Lett.* **98** 020503

Greenberger D N, Horne M A, Shimony A and Zeilinger A 1990 Bell's theorem without inequalities *Am. J. Phys.* **58** 1131–43

Hanna R C 1948 Polarization of annihilation radiation *Nature* **162** 332

Haw J Y, Assad S M, Lance A M, Ng N H Y, Sharma V, Lam P K and Symul T 2015 Maximization of extractable randomness in a quantum random number generator *Phys. Rev. Appl.* **3** 054004

Hübel H, Hamel D R, Fedrizzi A, Ramelow S, Resch K J and Jennewein T 2010 Direct generation of photon triplets using cascade photon-pair sources *Nature* **466** 601–3

Pan J-W, Daniell M, Gasparoni S, Weihs G and Zeilinger A 2001 Experimental demonstration of four-photon entanglement and high-fidelity teleportation *Phys. Rev. Lett.* **86** 4435–8

Pryce M L H and Ward J C 1947 Angular correlation effects with annihilation radiation *Nature* **160** 435

Schäfer F P (ed) 1990 *Dye Lasers* 3rd edn (Berlin: Springer)

Schrödinger E 1935 Discussion of probability relations between separated systems *Math. Proc. Camb. Phil. Soc.* **31** 555–63

Schrödinger E 1936 Probability relations between separated systems *Math. Proc. Camb. Phil. Soc.* **32** 446–52

Snyder H S, Pasternack S and Hornbostel J 1948 Angular correlation of scattered annihilation radiation *Phys. Rev.* **73** 440–8

't Hooft G 2016 *The Cellular Automaton Interpretation of Quantum Mechanics* (Berlin: Springer)

Symul T, Assad S M and Lam P K 2011 Real time demonstration of a high bitrate quantum random number generation with coherent laser light *Appl. Phys. Lett.* **98** 231103

Ward J C 1949 *Some Properties of the Elementary Particles* (Oxford: Oxford University Press)

Ward J C 2004 *Memoirs of a Theoretical Physicist* (Rochester: Optics Journal)

Wheeler J A 1946 Polyelectrons *Ann. N. Y. Acad. Sci.* **48** 219–38

Wu C S and Shaknov I 1950 The angular correlation of scattered annihilation radiation *Phys. Rev.* **77** 136

Yin J *et al* 2017 Satellite based entanglement distribution over 1200 kilometers *Science* **356** 1140–4

Zhang C, Huang Y-F, Zhang C-J, Wang J, Liu B-H, Li C-F and Guo G-C 2016 Generation and applications of an ultrahigh-fidelity four-photon Greenberger–Horne–Zeilinger state *Opt. Express* **24** 27059–69

IOP Publishing

Organic Lasers and Organic Photonics (Second Edition)

F J Duarte

Chapter 16

Intrinsic quantum coherence in electrically-pumped organic interferometric emitters: Diracian emission

F J Duarte and K M Vaeth

Coherent emission from electrically-pumped organic interferometric emitters is explained in terms of intrinsic quantum coherence via Dirac's identities.

16.1 Introduction

In chapter 11 it was advocated and demonstrated, via experimental data, that the coherent emission from electrically-pumped interferometric emitters (Duarte *et al* 2005) is indistinguishable from laser emission. This indistinguishability is in terms of interferometric coherence and spatial coherence. This coherence is characterized via the following directly measured parameters:

$$\Delta\theta_R \approx (1.09) \times \left(\frac{\lambda}{\pi w}\right) \tag{16.1}$$

$$\mathcal{V} \approx 0.9 \tag{16.2}$$

In this regard, equation (16.1) describes spatial coherence, and equation (16.2) describes interferometric coherence as determined via visibility measurements (Michelson 1927)

$$\mathcal{V} = \frac{I_1 - I_2}{I_1 + I_2} \tag{16.3}$$

which characterize spectral visibility.

In this chapter it is explained that it is not necessary to utilize traditional laser concepts to describe the quantum nature of this emission. This coherence can also be interpreted via Dirac's identities which are central to fundamental phenomena such as quantum entanglement (Duarte 2022).

16.2 Quantum interferometric probabilities

Quantum interferometric probabilities can be described starting from the fundamental Dirac–Feynman interferometric principle (Dirac 1958, Feynman *et al* 1965) which is in itself a superposition probability amplitude where s is the photon source and d is the detector or interferometric plane. The index j refers to the jth slit in the N-slit array ($j = 1, 2, 3...N$). Multiplication of equation (16.4) with its complex conjugate, via Born's rule (Born 1926), yields the interferometric probability (Duarte 1993, 2003)

$$\langle d|s \rangle = \sum_{j=1}^{N} \langle d|j \rangle \langle j|s \rangle \tag{16.4}$$

$$|\langle d|s \rangle|^2 = \langle d|s \rangle \langle d|s \rangle^* \tag{16.5}$$

$$\langle d|s \rangle \langle d|s \rangle^* = \left(\sum_{j=1}^{N} \langle d|j \rangle \langle j|s \rangle \right)\left(\sum_{j=1}^{N} \langle d|j \rangle \langle j|s \rangle \right)^* \tag{16.6}$$

$$\langle d|s \rangle \langle d|s \rangle^* = \sum_{j=1}^{N} \Psi(r_j) \sum_{m=1}^{N} \Psi(r_m) e^{i(\Omega_m - \Omega_j)} \tag{16.7}$$

$$|\langle d|s \rangle|^2 = \sum_{j=1}^{N} \Psi(r_j)^2 + 2\sum_{j=1}^{N} \Psi(r_j)\left(\sum_{m=j+1}^{N} \Psi(r_m)\cos(\Omega_m - \Omega_j) \right) \tag{16.8}$$

Equations (16.5)–(16.8) are equivalent quantum probabilities that apply either to single photons or *ensembles of indistinguishable photons* (Duarte 1993). These quantum probabilities are directly related to measured intensities via (Duarte 2014, 2022)

$$I(\nu)_r = rKh\nu\langle d|s \rangle \langle d|s \rangle^* \tag{16.9}$$

where $r = 1$ for a single photon or quantum. The intensity$I(\nu)_r$ has units of J s^{-1} m^{-2}, or W m^{-2}, and K is a quantity that has the cross-section in the denominator and has units of s^{-1} m^{-2} (Duarte 2014, 2022). The main message here is that the spatial profile of the measured intensity depends entirely on the dimensionless quantum probability $\langle d|s \rangle \langle d|s \rangle^*$.

The measured intensity profile for the emission from the electrically-pumped organic interferometric emitter is shown in figure 16.1 while the measured intensity profile from the $3s_2 - 2p_{10}$ transition of a He–Ne laser emission within an identical interferometric configuration is shown in figure 16.2. The corresponding interferometric calculated profile, using equation (16.8), is shown in figure 16.3. In this theoretical interferogram the separation of the two secondary peaks is $\Delta d \approx 490$ μm

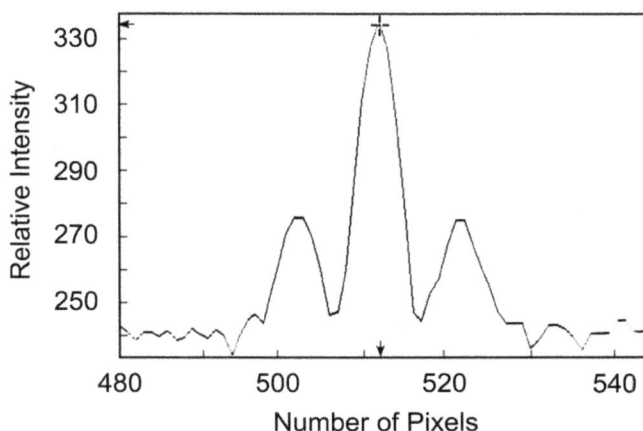

Figure 16.1. Digital image of the double-slit interferogram of the highly directional on-axis emission from the interferometric electrically excited organic semiconductor device. Each pixel is 25 μm. (Reproduced from Duarte F J *et al* (2005), with permission from Optica.)

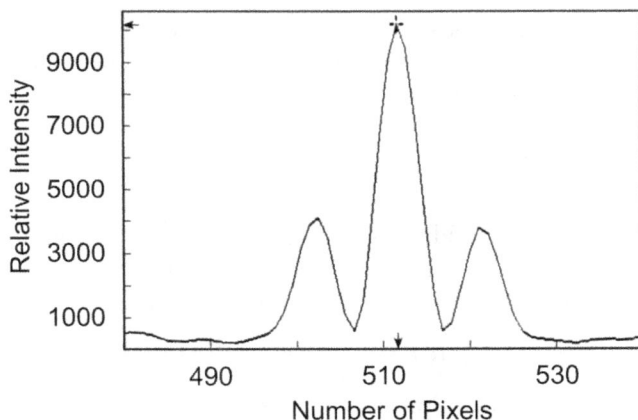

Figure 16.2. Digital image of the double-slit interferogram of the highly directional on-axis emission from the $3s_2 - 2p_{10}$ transition of the He–Ne laser using an identical interferometric configuration. Each pixel is 25 μm. (Reproduced from Duarte F J *et al* (2005), with permission from Optica.)

which should be compared to $\Delta d \approx 484$ μm from the measured interferogram depicted in figure 16.1 and $\Delta d \approx 485$ μm from the measured interferogram depicted in figure 16.3 (Duarte and Taylor 2021).

As already mentioned in chapter 11, the measured visibility from the electrically-pumped organic semiconductor interferometric emitter is $\mathcal{V} \approx 0.901 \pm 0.088$ while the visibility for the He–Ne laser, at $\lambda = 543.30$ nm it is $\mathcal{V} \approx 0.952 \pm 0.031$ (Duarte *et al* 2005). The published literature considers visibilities in the $0.85 \leqslant \mathcal{V} \leqslant 1.0$ range as corresponding to laser emission.

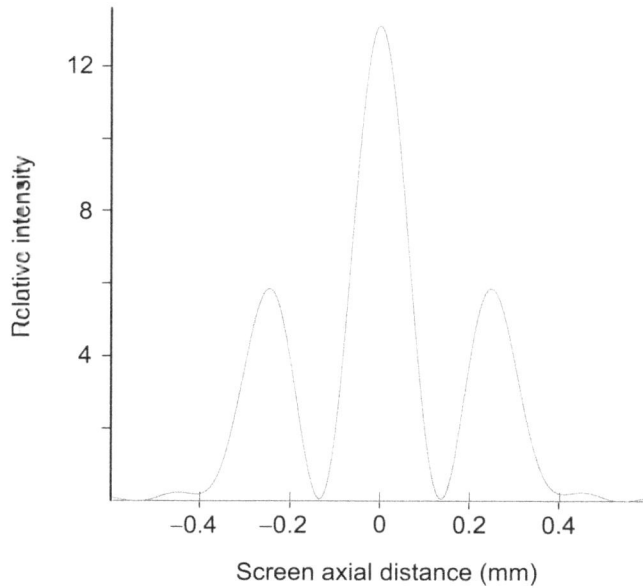

Figure 16.3. Calculated profile of the double-slit interferogram. Here, $N = 2$, slit width and slit separation is 50 μm, $\lambda = 540$ nm, and the intra interferometric distance is $D_{\langle d|j \rangle} = 50$ mm. (Reproduced from Duarte and Taylor (2022), with permission from Springer Nature.)

16.3 Dirac's identities

Here, a brief review of Dirac's identities (Dirac 1958) in *bra ket* notation is given to provide a background to the main subject matter. Transitioning from ψ to ϕ is registered as $\langle \phi | \psi \rangle$. Transitioning from ψ to ϕ via j is written as

$$\langle \phi | \psi \rangle = \langle \phi | j \rangle \langle j | \psi \rangle \tag{16.10}$$

where each of the *bra ket* expressions is a probability amplitude, or state, in complex notation (Duarte 2022). Next, is the crucial identity

$$\langle \phi | \psi \rangle = \langle \psi | \phi \rangle^* \tag{16.11}$$

where $\langle \psi | \phi \rangle^*$ is the complex conjugate of the probability amplitude, or state, $\langle \phi | \psi \rangle$.
Some abstraction identities (Dirac 1958):

$$|\psi\rangle = |j\rangle \langle j | \psi \rangle \tag{16.12}$$

$$\langle \chi | A | \phi \rangle = \langle \chi | i \rangle \langle i | A | j \rangle \langle j | \phi \rangle \tag{16.13}$$

$$A = |i\rangle \langle i | A | j \rangle \langle j | \tag{16.14}$$

$$A | \phi \rangle = |i\rangle \langle i | A | j \rangle \langle j | \phi \rangle \tag{16.15}$$

The following identities (Dirac 1958) are central to the issue at hand

$$|a\rangle|b\rangle = |b\rangle|a\rangle = |ab\rangle \qquad (16.16)$$

$$|a\rangle|b\rangle|c\rangle... = |abc...\rangle \qquad (16.17)$$

and the commutative property

$$|a\rangle|b\rangle = |b\rangle|a\rangle = |ab\rangle = |ba\rangle \qquad (16.18)$$

Also, crucial to quantum coherence are (Dirac 1958)

$$|a_1\rangle|b_2\rangle|c_3\rangle...|g_n\rangle = |a_1 b_2 c_3...g_n\rangle \qquad (16.19)$$

$$|b_1\rangle|a_2\rangle|c_3\rangle...|g_n\rangle = |b_1 a_2 c_3...g_n\rangle \qquad (16.20)$$

Identity (16.19) is equivalent to

$$|a\rangle_1|b\rangle_2|c\rangle_3...|g\rangle_n = |a_1 b_2 c_3...g_n\rangle \qquad (16.21)$$

For the combined state $|X\rangle$, defined as (Duarte 2022)

$$|X\rangle = |a_1 b_2 c_3...g_n\rangle \qquad (16.22)$$

we can write

$$|X\rangle = |a\rangle_1|b\rangle_2|c\rangle_3...|g\rangle_n \qquad (16.23)$$

This identity is central to the description of fundamental quantum coherence.

16.4 Quantum coherence in electrically-pumped interferometric emitters

This section is based on previous discussions on indistinguishability (Duarte and Taylor 2021, Duarte 2022). In this regard, the central result from the previous section is the identity representing the assembly state $|X\rangle$

$$|X\rangle = |a\rangle_1|b\rangle_2|c\rangle_3...|g\rangle_n$$

This means that indistinguishable quanta 1, 2, 3, ..., n can be in different states $|a\rangle$, $|b\rangle$, $|c\rangle$, ...$|g\rangle$.

Dirac describes this situation as a 'curious phenomena... having no analogue in classical theory'. He also emphasizes that he is writing about quanta of the '*same kind*' and '*absolutely indistinguishable from one another*' (Dirac 1958).

Although Dirac highlights indistinguishability in his description, quanta are labeled with numbers 1, 2, 3, ..., n. Dirac arrived at identity (16.19), which we express as (16.23), by first considering a single photon in different states. More explicitly: $|a\rangle_1$, $|b\rangle_1$, $|c\rangle_1...|g\rangle_1$ for photon 1, $|a\rangle_2$, $|b\rangle_2$, $|c\rangle_2...|g\rangle_2$ for photon 2, $|a\rangle_3$, $|b\rangle_3$, $|c\rangle_3...|g\rangle_3$ for photon 3, and so on.

For all available combinations of quanta, or bosons, in different states Dirac tabulates the various possible series of states for quanta 1, 2, 3, ..., n (Dirac 1958) as

$$\begin{vmatrix} |a\rangle_1 & |a\rangle_2 & |a\rangle_3 & \cdots & |a\rangle_n \\ |b\rangle_1 & |b\rangle_2 & |b\rangle_3 & \cdots & |b\rangle_n \\ |c\rangle_1 & |c\rangle_3 & |c\rangle_3 & \cdots & |c\rangle_n \\ \vdots & \vdots & \vdots & \vdots & \vdots \\ |g\rangle_1 & |g\rangle_2 & |g\rangle_3 & \cdots & |g\rangle_n \end{vmatrix} \quad (16.24)$$

Extending Dirac's identities, Duarte (2022) describes ensembles of indistinguishable quanta in an *identical* state of polarization: for example, the ensemble for indistinguishable quanta in the $|x\rangle$ polarization state can be expressed as

$$|\alpha\rangle = |x\rangle_1 |x\rangle_2 |x\rangle_3 \ldots |x\rangle_n \quad (16.25)$$

and for indistinguishable quanta in the $|y\rangle$ polarization state

$$|\beta\rangle = |y\rangle_1 |y\rangle_2 |y\rangle_3 \ldots |y\rangle_n \quad (16.26)$$

Hence, Dirac-type identities can be applied to describe single-transverse-mode single-longitudinal-mode coherent emission of indistinguishable quanta including both, $|x\rangle$ and $|y\rangle$, states of polarization. Indeed, Dirac-type identities can be applied to describe narrow-linewidth laser emission.

The identities in (16.25) and (16.26) describe lucidly and elegantly what *coherent emission is at a very fundamental level*: ensembles of indistinguishable quanta in a given state of polarization (Duarte and Taylor 2022, Duarte 2022). This way of thinking is in phase with Dirac's observation that '*with Bose statistics the probability of two particles being in the same state is greater than with classical statistics*' (Dirac 1958).

When considering optical quantum entanglement the word 'particles' should be replaced by quanta. Where quanta is the plural of photon. In turn, the photon is essentially non-local (Lamb 1995) and should be considered as a non-local form of energy $E = h\nu$ with frequency ν (Duarte 2022), or as a field of energy $E = h\nu$.

Dirac did not disclose the origin of the determinant (16.24). An array of $n \times n$ states can be arrived at via assembly states of the form (Duarte and Taylor 2021)

$$|Z\rangle = |a\rangle_1 |b\rangle_1 |c\rangle_1 \ldots |g\rangle_1 \quad (16.27)$$

$$|Y\rangle = |a\rangle_2 |b\rangle_2 |c\rangle_2 \ldots |g\rangle_2 \quad (16.28)$$

$$|\xi\rangle = |a\rangle_3 |b\rangle_3 |c\rangle_3 \ldots |g\rangle_3 \quad (16.29)$$

$$|\zeta\rangle = |a\rangle_n |b\rangle_n |c\rangle_n \ldots |g\rangle_n \quad (16.30)$$

and the transpose of this array can be written as

$$|a\rangle_1 |a\rangle_2 |a\rangle_3 \ldots |a\rangle_n$$
$$|b\rangle_1 |b\rangle_2 |b\rangle_3 \ldots |b\rangle_n$$
$$|c\rangle_1 |c\rangle_3 |c\rangle_3 \ldots |c\rangle_n \qquad (16.31)$$
$$\vdots \quad \vdots \quad \vdots \quad \vdots \quad \vdots$$
$$|g\rangle_1 |g\rangle_2 |g\rangle_3 \ldots |g\rangle_n$$

which has the same form as Dirac's determinant (Duarte and Taylor 2021). The rows describe quanta in the same state and the diagonal is the crucial Dirac identity

$$|X\rangle = |a\rangle_1 |b\rangle_2 |c\rangle_3 \ldots |g\rangle_n$$

In reference to quanta indistinguishability and the $n \times n$ state array given in (16.24) or (16.31) Dirac observed: 'one cannot say which particle is in which state, each quantum being equally likely to be in any state' and he adds that for bosons 'two or more particles can be in the same state' (Dirac 1958).

Identities (16.25) and (16.26) can also be expressed as either of which neatly describe the coherent emission from electrically-pumped organic semiconductors at a very fundamental level (Duarte and Taylor 2022).

$$|x\rangle_I = |x\rangle_1 |x\rangle_2 |x\rangle_3 \ldots |x\rangle_g \ldots \qquad (16.32)$$

$$|y\rangle_{II} = |y\rangle_1 |y\rangle_2 |y\rangle_3 \ldots |y\rangle_g \ldots \qquad (16.33)$$

These states are directly applicable to laser quantum entanglement sources yielding ensembles of entangles states rather than single photon pairs (Duarte 2022)

$$|\psi\rangle_+ = 2^{-1/2}(|x\rangle_I |y\rangle_{II} + |y\rangle_I |x\rangle_{II}) \qquad (16.34)$$

$$|\psi\rangle_- = 2^{-1/2}(|x\rangle_I |y\rangle_{II} - |y\rangle_I |x\rangle_{II}) \qquad (16.35)$$

$$|\psi\rangle^+ = 2^{-1/2}(|x\rangle_I |x\rangle_{II} + |y\rangle_I |y\rangle_{II}) \qquad (16.36)$$

$$|\psi\rangle^- = 2^{-1/2}(|x\rangle_I |x\rangle_{II} - |y\rangle_I |y\rangle_{II}) \qquad (16.37)$$

The emission described in equations (16.32) to (16.37) are directly applicable to the laser devices emitting entangled ensembles of indistinguishable quanta, as discussed in chapter 15.

16.5 Born's rule

As seen previously, Born's rule is crucial to the foundations of quantum mechanics. However, when Born first introduced what eventually became known as his rule he did so without derivation (Born 1926). Even much later when he exposed and discussed *indeterminism* as crucial to the foundations of quantum mechanics he again passed on providing a derivation (Born 1949) .

Here, Born's rule is derived from N-slit quantum interferometric principles applicable to straight-forward optical propagation from plane to plane. A basic

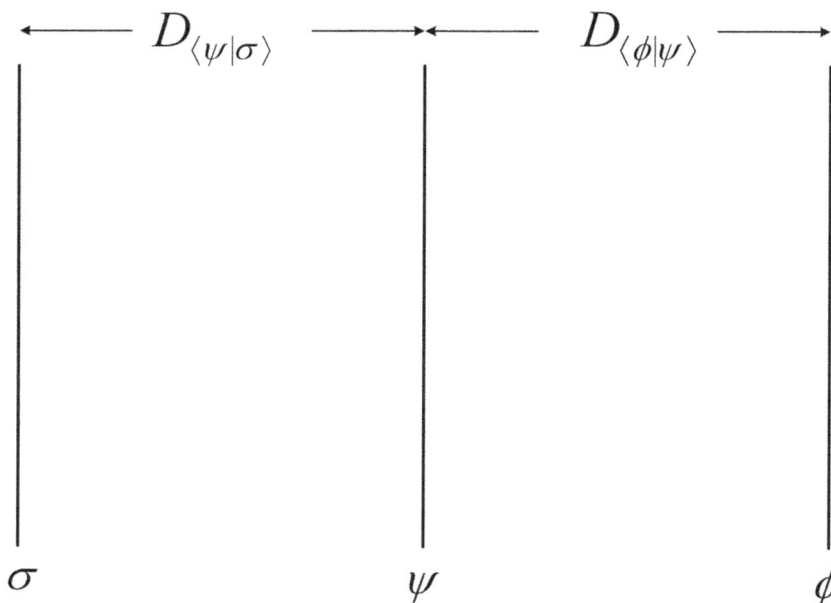

Figure 16.4. Propagation from plane σ to plane ψ to plane ϕ. The intra propagation distances are identical.

alternative geometrical version of the physics outlined by equation (16.4) is depicted in figure 16.4 and is described by the probability amplitude

$$\langle\phi|\sigma\rangle = \langle\phi|\psi\rangle\langle\psi|\sigma\rangle \tag{16.38}$$

Now, let us consider one further arrangement where propagation proceeds from plane ϕ to an identical plane ϕ, via ψ, as described in figure 16.5. That is, $\phi \equiv \phi$ and $D_{\langle\psi|\phi\rangle} \equiv D_{\langle\phi|\psi\rangle}$, so that the propagation physics is described by

$$\langle\phi|\phi\rangle = \langle\phi|\psi\rangle\langle\psi|\phi\rangle \tag{16.39}$$

Dirac's notation provides us with several useful identities including

$$\langle\psi|\phi\rangle = \langle\phi|\psi\rangle^* \tag{16.40}$$

so that equation (16.39) can be re expressed as

$$\langle\phi|\psi\rangle\langle\psi|\phi\rangle = \langle\phi|\psi\rangle\langle\phi|\psi\rangle^* \tag{16.41}$$

and ultimately as

$$\langle\phi|\psi\rangle\langle\phi|\psi\rangle^* = |\langle\phi|\psi\rangle|^2 \tag{16.42}$$

Abstracting the ϕ we can write

$$|\psi\rangle|\psi\rangle^* = \||\psi\rangle|^2 \tag{16.43}$$

which is Born's rule.

16-8

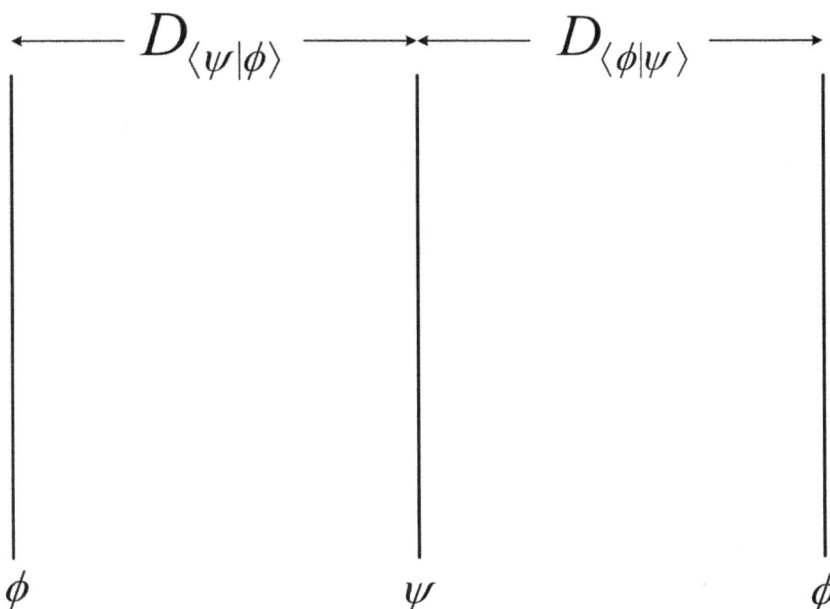

Figure 16.5. Propagation from plane ϕ to plane ψ to plane ϕ. Here, $\phi \equiv \phi$. The intra propagation distances are also identical $D_{\langle \psi | \phi \rangle} \equiv D_{\langle \phi | \psi \rangle}$.

16.6 Discussion

Duarte and Vaeth (2018) concluded that 'there are two possible explanations for the observed spatial and spectral coherence of the pulsed electrically excited organic semiconductor integrated interferometric emitter: it is either laser emission or it is a new type of yet to be discovered spatially coherent and spectrally coherent radiation.' And they went on to add: 'the coherent emission is intrinsically present and the interferometric configuration's only function is to *reveal it*.'

Both laser emission and intrinsic coherent quantum emission can be described via the assembly states $|x\rangle_I = |x\rangle_1 |x\rangle_2 |x\rangle_3 \ldots |x\rangle_g \ldots$ and $|y\rangle_{II} = |y\rangle_1 |y\rangle_2 |y\rangle_3 \ldots |y\rangle_g \ldots$.

Emission described by assembly states such as $|x\rangle_I$ and $|x\rangle_{II}$ we call *Diracian emission* given their original implicit description given in Dirac's determinant for bosons (Dirac 1958).

In the case of traditional lasing the origin of these states is population inversion. On the other hand, in the case of intrinsic coherent quantum emission the origin of these states is purely statistical. Given that the coherent emission observed in electrically-pumped organic semiconductor interferometric emitters is in the nW level... the odds appear to favor the concept of intrinsic coherent quantum emission, or Diracian emission.

16.7 Problems

- 16.1 Define the probability amplitudes $\langle j|s \rangle$ and $\langle d|j \rangle$ as wave functions $\langle j|s \rangle = \Psi(r_{j,s})e^{-i\theta_j}$ and $\langle d|j \rangle = \Psi(r_{x,j})e^{-i\phi_j}$ to write an explicit version of equation (16.4).

- 16.2 Using the explicit version of equation (16.4) apply Born's rule $|\langle d|s\rangle|^2 = \langle d|s\rangle\langle d|s\rangle^*$ to arrive at an explicit version of equation (16.6).
- 16.3 Expand the explicit version of equation (16.6) to arrive at the interferometric probability given in equation (16.7).
- 16.4 Use the identity $2\cos(\Omega_m - \Omega_j) = e^{-i(\Omega_m - \Omega_j)} + e^{i(\Omega_m - \Omega_j)}$ to express equation (16.7) in the explicit form of equation (16.8).
- 16.5 Use the complex wave functions $|a\rangle = \psi_1 e^{-i(\phi_1 + \theta_1)}$ and $|b\rangle = \psi_2 e^{-i(\phi_2 + \theta_2)}$ to verify the Dirac identity $|a\rangle|b\rangle = |b\rangle|a\rangle$.
- 16.6 Does the identity $|a\rangle|b\rangle = |b\rangle|a\rangle$ apply when using state vectors rather than straightforward complex wave functions?
- 16.7 Suggest an alternative approach to arrive at the matrix expressed in (16.31).

References

Born M 1926 Zur quantenmechanik der stoßvorgänge *Z. Phys.* **37** 863–27

Born M 1949 *Natural Philosophy of Cause and Chance* (Oxford: Clarendon Press)

Dirac P A M 1958 *The Principles of Quantum Mechanics* 4th edn (Oxford: Oxford University)

Duarte F J 1993 On a generalized interference equation and interferometric measurements *Opt. Comm.* **103** 8–14

Duarte F J 2003 *Tunable Laser Optics* 1st edn (New York: Elsevier Academic)

Duarte F J 2014 *Quantum Optics for Engineers* (New York: CRC Press)

Duarte F J 2022 *Fundamentals of Quantum Entanglement* 2nd edn (Bristol: Institute of Physics Publishing)

Duarte F J, Liao L S and Vaeth K M 2005 Coherence characteristics of electrically excited tandem organic light-emitting diodes *Opt. Lett.* **30** 3072–4

Duarte F J and Taylor T S 2021 *Quantum Entanglement Engineering and Applications* (Bristol: Institute of Physics Publishing)

Duarte F J and Taylor T S 2022 Quantum coherence in electrically-pumped organic interferometric emitters *Appl. Phys.* B **128** 11

Duarte F J and Vaeth K M 2018 Electrically-pumped organic semiconductor laser emission *Organic Lasers and Organic Photonics* ed F J Duarte (Bristol: Institute of Physics) ch 11

Feynman R P, Leighton R B and Sands M 1965 *The Feynman Lectures on Physics* **vol III** (Reading: Addison-Wesley)

Lamb W E 1995 Anti-photon *Appl. Phys.* B **60** 77–84

Michelson A A 1927 *Studies in Optics* (Chicago, IL: University of Chicago Press)

Index